Semigroups Underlying First-Order Logic

MEMOIRS
of the
American Mathematical Society

Number 866

Semigroups Underlying First-Order Logic

William Craig

November 2006 • Volume 184 • Number 866 (second of 4 numbers) • ISSN 0065-9266

American Mathematical Society
Providence, Rhode Island

2000 *Mathematics Subject Classification.*
Primary 03–XX, 03Gxx, 03G25; Secondary 06F05, 20Mxx, 20M05, 20M17.

Library of Congress Cataloging-in-Publication Data

Craig, William, 1918–
Semigroups underlying first-order logic / William Craig.
p. cm. — (Memoirs of the American Mathematical Society, ISSN 0065-9266 ; no. 866)
"Volume 184, number 866 (second of 4 numbers)."
Includes bibliographical references and indexes.
ISBN-13: 978-0-8218-4149-5 (alk. paper)
1. Algebraic logic. 2. First-order logic. 3. Semigroups. I. Title.

QA10.C74 2006
511.3′24—dc22 2006043004

Memoirs of the American Mathematical Society

This journal is devoted entirely to research in pure and applied mathematics.

Subscription information. The 2006 subscription begins with volume 179 and consists of six mailings, each containing one or more numbers. Subscription prices for 2006 are US$624 list, US$499 institutional member. A late charge of 10% of the subscription price will be imposed on orders received from nonmembers after January 1 of the subscription year. Subscribers outside the United States and India must pay a postage surcharge of US$31; subscribers in India must pay a postage surcharge of US$43. Expedited delivery to destinations in North America US$35; elsewhere US$130. Each number may be ordered separately; *please specify number* when ordering an individual number. For prices and titles of recently released numbers, see the New Publications sections of the *Notices of the American Mathematical Society.*

Back number information. For back issues see the *AMS Catalog of Publications.*

Subscriptions and orders should be addressed to the American Mathematical Society, P. O. Box 845904, Boston, MA 02284-5904, USA. *All orders must be accompanied by payment.* Other correspondence should be addressed to 201 Charles Street, Providence, RI 02904-2294, USA.

Memoirs of the American Mathematical Society is published bimonthly (each volume consisting usually of more than one number) by the American Mathematical Society at 201 Charles Street, Providence, RI 02904-2294, USA. Periodicals postage paid at Providence, RI. Postmaster: Send address changes to Memoirs, American Mathematical Society, 201 Charles Street, Providence, RI 02904-2294, USA.

This publication is indexed in *Science Citation Index®*, *SciSearch®*, *Research Alert®*, *CompuMath Citation Index®*, *Current Contents®/Physical, Chemical & Earth Sciences.*
Printed in the United States of America.

∞ The paper used in this book is acid-free and falls within the guidelines
established to ensure permanence and durability.
Visit the AMS home page at http://www.ams.org/

10 9 8 7 6 5 4 3 2 1 11 10 09 08 07 06

To my four children:
Ruth, Walter, Sarah, and Deborah Craig

Contents

Abstract ix

Acknowledgements xi

Overview xiii

Chapter I. Boolean, Relation-Induced, and Other Operations for Dealing with First-Order Definability 1

Chapter II. Uniform Relations Between Sequences 25

Chapter III. Diagonal Relations 61

Chapter IV. Uniform Diagonal Relations and Some Kinds of Bisections or Bisectable Relations 111

Chapter V. Presentation of $\mathbf{S}_q$, $\mathbf{S}_p$, and Related Structures 135

Chapter VI. Presentation of $\mathbf{S}_{pq}$, $\mathbf{S}_{pe}$ and Related Structures 165

Chapter VII. Presentation of $\mathbf{S}_{pqe}$ and Related Structures 209

Appendix. Presentation of ${}_{\delta}\mathbf{S}_q$ and Related Structures 233

Bibliography 243

Index of Symbols 245

Index of Phrases and Subjects 256

List of Relations Involved in Presentations 260

Synopsis of Presentations 262

Abstract

Let U be any nonempty set and let ${}^{[0,\omega)}U = \bigcup_{n<\omega} {}^nU$ be the set of sequences $\langle x_0, \ldots, x_{n-1}\rangle$ such that $0 \leq n < \omega$ and each x_m is in U. For $0 \leq i < \omega$, $0 \leq j < \omega$ and $\frown$ concatenation of sequences, let $\dot{q}_i$, p_i, and e_{ij} be the partial function on ${}^{[0,\omega)}U$ that when the associated if-clause holds satisfies the equality given below and that otherwise is undefined.

$$\begin{aligned}
\dot{q}_i(\langle x_0, \ldots, x_{n-1}\rangle) &= \langle x_0, \ldots, x_{i-1}\rangle^\frown\langle x_{i+1}, \ldots, x_{n-1}\rangle, \quad \text{if } n > i. \\
p_i(\langle x_0, \ldots, x_{n-1}\rangle) &= \langle x_0, \ldots, x_{i-1}\rangle^\frown\langle x_{i+1},\ x_i\rangle^\frown\langle x_{i+2} \ldots, x_{n-1}\rangle, \\
&\quad \text{if } n > i+1. \\
e_{ij}(\langle x_0, \ldots, x_{n-1}\rangle) &= \langle x_0, \ldots, x_{n-1}\rangle, \text{ if } n > i,\ n > j, \text{ and } x_i = x_j.
\end{aligned}$$

If $\overline{\overline{U}} \geq 2$, then the converse $\dot{q}_i^{\smile}$ of $\dot{q}_i$ is not single-valued, whereas $p_i^{\smile} = p_i$ and $e_{ij}^{\smile} = e_{ij}$. With $\circ$ being relative product let S_q, S_p, S_e be the closure under $\circ$ of $\{\dot{q}_i : i < \omega\} \cup \{\dot{q}_i^{\smile} : i < \omega\}$, $\{p_i : i < \omega\}$, or $\{e_{ij} : i, j < \omega, \{i, j\} \neq \{0\}\}$, respectively. Also, let S_{pq}, S_{pe}, S_{pqe} be the closure under $\circ$ of $S_p \cup S_q$, $S_p \cup S_e$, or $S_p \cup S_q \cup S_e$, respectively. For several of these six inductively defined sets S of binary relations on ${}^{[0,\omega)}U$ we will give an explicit characterization. For any of these six sets S the structure $\langle S, \circ, {}^{\smile}, \subseteq\rangle$, where $\subseteq$ is set inclusion, is an ordered semigroup with involution which, up to isomorphism, is the same for any U such that $\overline{\overline{U}} \geq 2$. For each of these structures except when S is S_e we will give a presentation. We will also give a presentation of those structures which result from these when one changes from sequences that are finite to those of length ω.

Let O_{pqe} be the set of those unary operations F on the set $\{W : W \subseteq {}^{[0,\omega)}U\}$ such that, for some R in S_{pqe}, given any W in the set, $F(W)$ is the direct image $\{y : \langle w, y\rangle \in R \text{ for some } w \text{ in } W\}$ of W under R. Consider any $\langle U,\ X_\delta\rangle_{\delta<\eta}$ such that each X_δ is an n_δ-ary relation on U for some n_δ, $0 \leq n_\delta < \omega$, and such that at least one X_δ is nonempty. Also consider any m-ary relation Y on U, $0 \leq m < \omega$. Then Y is definable in $\langle U,\ X_\delta\rangle_{\delta<\eta}$ by a first-order formula (with equality but without individual parameters) if and only if Y is in the closure of $\{X_\delta : \delta < \eta\}$ under relative complement and the operations F in O_{pqe}. (See Theorem I.2(b).)

Subject classification: (2000) 03G25, 06F05, 20M05, 20M17

Keywords: First-order definability, algebraic first-order logic, relation-induced operations on sets of sequences, uniform relations between sequences, diagonal relations, ordered semigroups with involution, normal forms, presentations.

Manuscript received by editor S. Lempp: 11/13/2003; revised ms. received 12/22/05.

This work conducted with support of the Committee on Research, University of California, Berkeley.

Acknowledgements

For various useful remarks I am indebted to the following: H.Andréka; W. Hanf; L. Henkin; D. Haskell; S. W. Margolis; I. Németi; I. Sain; B. M. Schein; R. J. Thompson.

Among meetings concerned with algebraic logic, the following have been the most helpful: Budapest, summer 1988; Mills College, Oakland, summer 1990; Banach Center, Warsaw, fall 1991. A series of talks I was invited to give in Budapest during fall 1995 has contributed toward clarifying my ideas and tightening their organization.

I am grateful to Johannes Hafner for his careful substantial help in preparing the indices. As a pleasant Christmas surprise, Deirdre Haskell jump-started the typing of the revisions of the manuscript. About one half of the revisions were typed with great care by Kenny Easwaran. For her careful, skillful, and sensitive typing of the original manuscript and of about one half of the revisions I am very grateful to Deborah Craig (who was formerly with the Mathematics Department at Berkeley and is a namesake but, alas, not a relative or my youngest daughter).

The editor, Steffen Lempp, has been very accommodating in putting up with a luddite with whom e-mail correspondence had to be carried on through intermediaries. Readers, and I, owe our greatest debt to the referee. His or her report of eleven closely-spaced typewritten pages pointed out various mathematical errors, some of which must have required very close reading to detect. It also raised some interesting general questions. Most of all, it made many thoughtful and very helpful suggestions concerning exposition and organization. In particular, owing to them, there were added to the revised manuscript an *Overview*, a *List of Relations*, and a *Synopsis of Presentations*. Also, the indices were expanded substantially. Readers and I are very lucky to have had such thoughtful and conscientious refereeing.

Overview

Tarski showed that first-order logic, as well as other sufficiently expressive languages, involves operations on sets of sequences (see [Ta], VI, VIII). He also showed that not all operations involved are Boolean. Jónsson and Tarski, in their joint paper [JT] on Boolean algebras with operators called attention to operations that arise from a certain correspondence. To be more precise, let V be any set and let $0 \leq n < \omega$. Then to any $n+1$-ary relation R on V there corresponds the n-ary operation R^* on the power set $\{X : X \subseteq V\}$ of V that, when $n = 1$ satisfies the following condition and, when $n \neq 1$, satisfies a similar condition:

For any set $X \subseteq V$, $R^*(X)$ is the direct image
$\{y : xRy$ for some $x \in X\}$ of X under R.

When this condition holds, then R^* shall be ***induced by*** R. Also, any operation that is induced by some R shall be ***relation-induced***. While these notions apply to any set V, in connection with first-order logic V is, for some set $U \neq \emptyset$ serving as universe or base set (or "alphabet"), the set V_U of sequences ("words") that are formed from elements of U.

Relation-induced operations play a major role in classical as well as non-classical first-order logic. For studying these logics it has been useful to study, by themselves, the Boolean operations involved or, respectively, their non-classical counterparts. This suggests that it may be good strategy to also study, by themselves, the pertinent relation-induced operations.

One of the relation-induced operations used in the theory of relation algebras, which deals with a fragment of first-order logic, is 2-ary, namely the operation of forming the relative product of two relations. In contrast, the survey given in Chapter 5 of [HMT] suggests that in most, and perhaps in all, algebraic theories of full first-order logic, every relation-induced operation serving as primitive is either 0-ary or 1-ary and hence every inducing relation is either 1-ary or 2-ary. For example, in the theory of cylindric set algebras as described in Section 1.1 of [HMT], every relation-induced operation serving as primitive is either a cylindrification and hence 1-ary or a diagonal set d_{ij} which, regarded as operation, is 0-ary.

For many purposes, a 1-ary relation R and the 0-ary operation R^* it induces can be replaced, respectively, by a 2-ary relation and the 1-ary operation it induces. For example, suppose that R is the diagonal set d_{ij} of those sequences in V_U whose ith and jth term are the same element of U. Let e_{ij} be the partial identity relation on V_U whose domain is d_{ij} and let e^*_{ij} be the 1-ary operation on $\{X : X \subseteq V_U\}$ that e_{ij} induces. Then $d_{ij} = e^*_{ij}(V_U)$ and, for any $X \subseteq V_U$, $e^*_{ij}(X) = d_{ij} \cap X$. Thus, when V_U and $\cap$ are available, there is no change in expressive power when d_{ij} is replaced as primitive by e^*_{ij}. This illustrates that, for many algebraic theories of first-order logic, the relation-induced operations serving as primitives can be so chosen that every inducing relation is 2-ary.

In [JT], Jónsson and Tarski called 1-ary operations f and g on the power set $\{X : X \subseteq V\}$ of a set V ***conjugates*** of each other if and only if they satisfy the following condition for any $X \subseteq V$ and $Y \subseteq V$:

$$f(X) \cap Y = \emptyset \text{ if and only if } X \cap g(Y) = \emptyset \ .$$

They showed that f and g are conjugate if and only if, for some 2-ary relation R on V and its converse $R^{\smile}$, both necessarily unique, $f = R^*$ and $g = R^{\smile *}$. Now, for any 2-ary relations R and S on V, let $R \circ S$ be their relative product $\{\langle x, z\rangle : \text{ for some } y, xRy \text{ and } ySz\}$. Also, for any 1-ary operations f and g on $\{X : X \subseteq V\}$, let $f \circ g$ be the 1-ary operation that, for any $X \subseteq V$, satisfies $(f \circ g)(X) = g(f(X))$. Further, for any 1-ary f such that $f = R^*$ for some 2-ary R, let $f^\dagger$ be the conjugate $R^{\smile *}$ of f. There follows that, for any 2-ary R and S, $(R \circ S)^* = R^* \circ S^*$. There also follows that * is an isomorphism of $\langle\{R : R \subseteq V \times V\}, \circ, ^{\smile}\rangle$ onto $\langle\{R^* : R \subseteq V \times V\}, \circ, ^\dagger\rangle$. Moreover, if one lets $\Subset$ be the partial ordering of the 1-ary operations on $\{X : X \subseteq V\}$ such that $f \Subset g$ if and only if, for every $X \subseteq V$, there holds $f(X) \subseteq g(X)$, then * is also an isomorphism of $\langle\{R : R \subseteq V \times V\}, \subseteq\rangle$ onto $\langle\{R^* : R \subseteq V \times V\}, \Subset\rangle$.

Suppose that all the relation-induced operations that have been chosen as primitives are 1-ary and hence all the inducing relations are 2-ary. Let G be the set of these inducing relations. Let $S = S_G$ be the closure of G under $\circ$ and $^{\smile}$. Let $O = O_G = \{R^* : R \in S_G\}$. There follows that $\mathbf{S}_G = \langle S_G, \circ, ^{\smile}\rangle$ is a semigroup with involution that is ordered by $\subseteq$ and that * is an isomorphism of $\langle \mathbf{S}_G, \subseteq\rangle = \langle S_G, \circ, ^{\smile}, \subseteq\rangle$ onto $\langle O_G, \circ, ^\dagger, \Subset\rangle$. Evidently, a study of the operations in O_G should include a study of the structures $\langle O_G, \circ, ^\dagger\rangle$ and $\langle O_G, \circ, ^\dagger, \Subset\rangle$. Often, it is more advantageous to study their isomorphs $\langle S_G, \circ, ^{\smile}\rangle$ and $\langle S_G, \circ, ^{\smile}, \subseteq\rangle$, since the elements R of S_G are more elementary and, in general, better understood than the elements R^* of O_G. For this reason, our emphasis throughout will be on structures whose elements are 2-ary relations between sequences rather than operations on sets of these sequences. From now on, unless we say otherwise, a relation will be 2-ary.

Four structures are the principal objects of our study: The semigroups with involution $\mathbf{S}_{pqe} = \langle S_{pqe}, \circ, ^{\smile}\rangle$, ${}_{[\omega]}\mathbf{S}_{pqe} = \langle {}_{[\omega]}S_{pqe}, \circ, ^{\smile}\rangle$ and their respective ordered expansions $\langle \mathbf{S}_{pqe}, \subseteq\rangle = \langle S_{pqe}, \circ, ^{\smile}, \subseteq\rangle$ and $\langle {}_{[\omega]}\mathbf{S}_{pqe}, \subseteq\rangle = \langle {}_{[\omega]}S_{pqe}, \circ, ^{\smile}, \subseteq\rangle$. To give a fuller description, consider any nonempty set U that is to serve as universe or base set (or "alphabet") from whose elements the sequences ("words") concerned are being formed. Also, for any ordinal γ, $0 \leq \gamma \leq \omega$, let ${}^{\gamma}U$ be the set of sequences x (formed from elements of U), whose length $|x|$ is γ. Further, let ${}^{[0,\omega)}U = \bigcup\{{}^{n}U : 0 \leq n < \omega\}$ so that ${}^{[0,\omega)}U$ is the set of those sequences x (formed from elements of U) such that $|x| = n$ for some $n < \omega$. For any $i < \omega$ and $j < \omega$, let $p_i = p_{(i,i+1)}$, $\dot{q}_i$, and $e_{ij} = e_{(i,j)}$ be the partial functions described early in Chapter I and in the Abstract. Then the universe S_{pqe} of the structures $\mathbf{S}_{pqe}$ and $\langle \mathbf{S}_{pqe}, \subseteq\rangle$ is the closure under $\circ$ and $^{\smile}$ of the union of the sets $\{p_i : i < \omega\}$, $\{\dot{q}_i : i < \omega\}$, $\{e_{ij} : i, j < \omega\}$. Also, the universe ${}_{[\omega]}S_{pqe}$ of the structures ${}_{[\omega]}\mathbf{S}_{pqe}$ and $\langle {}_{[\omega]}\mathbf{S}_{pqe}, \subseteq\rangle$ is the closure under $\circ$ and $^{\smile}$ of the union of the sets $\{\overset{\omega}{\rightarrow} p_i : i < \omega\}$, $\{\overset{\omega}{\rightarrow} \dot{q}_i : i < \omega\}$, $\{\overset{\omega}{\rightarrow} e_{ij} : i, j < \omega\}$, where, for any relation R on ${}^{[0,\omega)}U$, $\overset{\omega}{\rightarrow} R$ is that relation on ${}^{\omega}U$ that results from R by the process of ω-prolongation that is described early in Chapter II, and also briefly described a little later on in this Overview.

Each of the four structures is, up to isomorphism, the same for any set U serving as universe, provided that the cardinality of U is at least 2. A proof, not very difficult, of this independence from the size of U is given as part of the proof of Theorem III.11(a). Probably a similar proof of independence can be given for many of the structures $\langle S, \circ, ^{\smile}\rangle$ and $\langle S, \circ, ^{\smile}, \subseteq\rangle$ whose elements induce operations R^* of a first-order logic. This independence from the size of U is a major reason for studying relation-induced operations, or their inducing relations, by themselves.

The structures $\mathbf{S}_{pqe}$ and ${}_{[\omega]}\mathbf{S}_{pqe}$ are closely related, and so are the structures $\langle \mathbf{S}_{pqe}, \subseteq\rangle$ and $\langle {}_{[\omega]}\mathbf{S}_{pqe}, \subseteq\rangle$. In fact, for any n, $0 \leq n < \omega$, there is a process $\overset{n}{\rightarrow}$ of prolongation by sequences of length n that is analogous to the above-mentioned process $\overset{\omega}{\rightarrow}$ of prolongation by sequences of length ω. The relation that holds between members R and R' of S_{pqe} if and only if, for some $n < \omega$, either $R = \overset{n}{\rightarrow} R'$ or $R' = \overset{n}{\rightarrow} R$ turns out to be a congruence relation on $\langle \mathbf{S}_{pqe}, \circ, ^{\smile}, \subseteq\rangle$. According to Theorems II.11 and VII.17, it gives rise to a homomorphism of $\langle \mathbf{S}_{pqe}, \subseteq \rangle$ onto $\langle {}_{[\omega]}\mathbf{S}_{pqe}, \subseteq \rangle$.

Let $O_{pqe} = \{R^* : R \in S_{pqe}\}$ and ${}_{[\omega]}O_{pqe} = \{R^* : R \in {}_{[\omega]}S_{pqe}\}$. According to Theorem I.2(b), which does not go far beyond folklore, O_{pqe} is adequate for classical first-order logic (with equality but without function symbols or individual constants) in the following sense: Let $\langle U, X_\delta \rangle_{\delta<\eta}$ be any relational structure such that at least one X_δ is nonempty and let Y be any m-ary relation on U, $0 \leq m < \omega$. Then Y is first-order definable in $\langle U, X_\delta \rangle_{\delta<\eta}$ if and only if Y is in the closure of $\{X_\delta : \delta < \eta\}$ under $O_{pqe} \cup \{\overline{2}\}$, where $\overline{2}$ is relative complementation.

Since ${}_{[\omega]}O_{pqe}$ consists of operations on sets of sequences of length ω, its relationship to first-order definablility of m-ary relations Y, $0 \leq m < \omega$, is less direct. This is why the structures $\langle O_{pqe}, \circ, {}^\dagger, \Subset \rangle$ and $\langle S_{pqe}, \circ, {}^{\smile}, \subseteq \rangle$ may turn our to be more important than $\langle {}_{[\omega]}O_{pqe}, \circ, {}^\dagger, \Subset \rangle$ and $\langle {}_{[\omega]}S_{pqe}, \circ, {}^{\smile}, \subseteq \rangle$. They may turn out to be useful for constructing algebraic systems of first-order logic that are more closely related to mathematical practice. (See also the discussion in Chapter I of [Cr 74a].)

On the other hand, since S_{pqe} and $\langle S_{pqe} \subseteq \rangle$ have to take different lengths into account, they are more complex than ${}_{[\omega]}S_{pqe}$ and $\langle {}_{[\omega]}S_{pqe}, \subseteq \rangle$, although perhaps less so than might have been expected. This is why in our study of S_{pqe}, $\langle S_{pqe} \subseteq \rangle$ and various substructures, we generally first tried to develop intuitions by investigating ${}_{[\omega]}S_{pqe}$, $\langle {}_{[\omega]}S_{pqe}, \subseteq \rangle$ and their corresponding substructures.

In investigating semigroups with involution $\langle S, \circ, {}^{\smile} \rangle$ that are fairly complex, it is natural to also investigate subalgebras that are simpler. Moreover, generally a choice of subalgebras is advantageous such that S is the closure under $\circ$ of the union of the universes of the chosen subalgebras. In the case of $\langle {}_{[\omega]}S_{pqe}, \circ, {}^{\smile} \rangle$ this strategy led to investigating $\langle {}_{[\omega]}S_{pq}, \circ, {}^{\smile} \rangle$ and $\langle {}_{[\omega]}S_{pe}, \circ, {}^{\smile} \rangle$ and also, in turn, to investigating $\langle {}_{[\omega]}S_p, \circ, {}^{\smile} \rangle$ and $\langle {}_{[\omega]}S_q, \circ, {}^{\smile} \rangle$. This is done in Chapters VI and V respectively.

For any $\gamma = \{i : i < \gamma\}$ and any $\delta = \{j : j < \delta\}$, a mapping ${}_\gamma[\ ]_\delta$ is defined after Theorem III.4 which, to every subset M of $\gamma \times \delta$ assigns the following subset of ${}^\gamma U \times {}^\delta U$:

$${}_\gamma[M]_\delta = \{\langle x, y \rangle : x \in {}^\gamma U, y \in {}^\delta U, x_i = y_j \text{ for any } \langle i, j \rangle \text{ in } M\}.$$

Early in Chapter IV, a relation M is defined to be ***bisectable*** if and only if there are $s = \{0, \ldots, s-1\}$ and $t = \{0, \ldots, t-1\}$ such that

$$M = (M \cap (s \times t)) \cup \{\langle s+n, t+n \rangle : n < \omega\}.$$

From Theorems III.12(a), IV.2(b), IV.3(b) and IV.6(b) there follows that, if the cardinality of U is at least 2, then $\langle {}_{[\omega]}S_{pq}, \circ, {}^{\smile} \rangle$ is the isomorphic image under ${}_{\omega}[\ ]_{\omega}$ of $\langle \{M\colon M \text{ is bisectable and one-one}\}, \circ, {}^{\smile} \rangle$. Also, from Theorems III.14(e), IV.2(d), IV.3(d), and IV.5(c) there follows that, if $\overline{\overline{U}} \geq 2$, then $\langle {}_{[\omega]}S_{q}, \circ, {}^{\smile} \rangle$ is the isomorphic image under ${}_{\omega}[\ ]_{\omega}$ of $\langle \{M : M \text{ is bisectable, } M \text{ and } M^{\smile} \text{ are order-preserving}\}, \circ, {}^{\smile} \rangle$. Thus $\langle {}_{[\omega]}S_{pq}, \circ, {}^{\smile} \rangle$ and $\langle {}_{[\omega]}S_{q}, \circ, {}^{\smile} \rangle$ are isomorphic images under ${}_{\omega}[\ ]_{\omega}$ of structures that are more elementary and hence more easily understood. This was a major reason for investigating them before investigating $\langle {}_{[\omega]}S_{pqe}, \circ, {}^{\smile} \rangle$, for which there exists no similar isomorphism.

For similar reasons, an investigation of the structure $\langle {}_{[\omega]}S_{pe}, \circ, {}^{\smile} \rangle$ seemed advisable. As can be seen from Theorems III.12(b), III.16, IV.2(d)(e), IV.3(f) and IV.6(a)(c), $\langle {}_{[\omega]}S_{pe}, \circ, {}^{\smile} \rangle$ is the isomorphic image under ${}_{\omega}[\ ]_{\omega}$ of the following structure $\langle S, \bullet, {}^{\smile} \rangle$: S is the closure under $\bullet$ of those relations M such that M is bisectable and either an equivalence relation on ω or a permutation of ω, and for any M and M', $M \bullet M'$ is the least relation N that includes the relative product $M \circ M'$ and satisfies $N \circ N^{\smile} \circ N \subseteq N$ (so that, when M and M' are permutations, then $M \bullet M' = M \circ M'$, and when M and M' are equivalences, then $M \bullet M'$ is the least equivalence relation that includes $M \circ M'$).

As is essentially well-known, $\langle {}_{[\omega]}S_{p}, \circ, {}^{\smile} \rangle$ is the isomorphic image under ${}_{\omega}[\ ]_{\omega}$ of the group of those permutations of ω that are bisectable or, equivalently, have finite support. A presentation of this group is easily obtained from presentations, for $2 \leq n < \omega$, of the symmetric group on $n = \{0, \ldots, n-1\}$. As remarked before Lemma V.17, this presentation has to be adjusted if different finite lengths are to be taken into account. By Theorem V.18(a), to obtain a presentation of $\langle S_{p}, \circ, {}^{\smile} \rangle$, a fairly small change is sufficient, namely replacement of P_4 by $P_4^{\#}$.

Under the assumption that the cardinality of U is at least 2, presentations of $\langle S_{q}, \circ, {}^{\smile} \rangle$ and $\langle S_{p}, \circ, {}^{\smile} \rangle$ are given in Chapter V, along with presentations of the related structures $\langle S_{q}, \circ, {}^{\smile}, \subseteq \rangle$, $\langle {}_{[\omega]}S_{q}, \circ, {}^{\smile} \rangle$, $\langle {}_{[\omega]}S_{q}, \circ, {}^{\smile}, \subseteq \rangle$, $\langle S_{p}, \circ, {}^{\smile}, \subseteq \rangle$, and $\langle {}_{[\omega]}S_{p}, \circ, {}^{\smile} \rangle$. Presentations of $\langle S_{pq}, \circ, {}^{\smile} \rangle$, of $\langle S_{pe}, \circ, {}^{\smile} \rangle$, and of similarly related structures are given in Chapter VI. Making use of these presentations, and only 3 additional ordered pairs, Chapter VII gives a presentation of $\langle S_{pqe}, \circ, {}^{\smile} \rangle$ and of similarly related structures. The set of generators used in the presentation of $\langle S_{pqe}, \circ, {}^{\smile} \rangle$ is $\{p_i : i < \omega\} \cup \{\dot{q}_0{}^{\smile}\} \cup \{e_{0,1}\}$. As is often the case, these presentations provide only limited insight into the structure being presented. There are many natural questions concerning these structures that will not only be left unanswered but will not even be raised. The main purpose of giving presentations of $\langle S_{pqe}, \circ, {}^{\smile}, \subseteq \rangle$

and of $\langle {}_{[\omega]}S_{pqe}, \circ, {}^{\smile}, \subseteq \rangle$, and thus of their respective isomorphs $\langle O_{pqe}, \circ, \dagger, \Subset \rangle$ and $\langle {}_{[\omega]}O_{pqe}, \circ, \dagger, \Subset \rangle$, is to make them available to serve, perhaps after modifications, as a coherent part of an algebraic axiomatization of first-order logic.

(Hindsight has made it clear that by presenting S_q, S_{pq}, S_{pqe} and ${}_{[\omega]}S_q$, ${}_{[\omega]}S_{pq}$, ${}_{[\omega]}S_{pqe}$ as monoids, rather than as semigroups, with involution one obtains presentations that are more transparent. Also, for many applications, it may often be better to replace $q_0 = \dot{q_0}^{\smile}$ as generator by $\dot{q_0}$. The reader can easily make the needed changes.

A change similar to the first of these in the presentation of S_p and of S_{pe} would also yield a somewhat simpler presentation. However, since the identity element of S_p and that of S_{pe} differs from that of S_{pqe}, this simpler presentation could not be made part of a presentation of S_{pqe}. For similar reasons, adjoining a partial ordering to $\mathbf{S}_{pq}$ or to $\mathbf{S}_{pe}$ by definition, as is commonly done for inverse semigroups, would not be helpful for presenting $\langle \mathbf{S}_{pqe}, \subseteq \rangle$, since, as remarked after the proof of Theorem III.11, $\mathbf{S}_{pqe}$ is not an inverse semigroup.)

As the referee noted, different readers may be interested in different ones among the various presentations mentioned. Also, for their purpose it may be sufficient to have access to an accurate description of the presentation concerned without the need of knowing the proofs leading up to it or of becoming familiar with other material. Also, of course, it should be clear in each case where this desired information can be found. I hope that the **List of Relations** together with the **Synopsis of Presentations** (of the structures of central interest), both near the end of the book, supplemented by the remarks on $\equiv$ and $\leq$ at the beginning of Chapter V, will serve this purpose.

We now turn to the set-theoretic aspects of membership in S_{pqe} or in ${}_{[\omega]}S_{pqe}$, or in various subsets or supersets of these, that are discussed in Chapters II, III, IV. Two conditions, each interesting in its own right, will be seen to be necessary for membership in S_{pqe} or in ${}_{[\omega]}S_{pqe}$ respectively. One of these, about to be described, will be discussed more fully in Chapter II. Consider any $\gamma \leq \omega$ and any relation X on ${}^{[0,\omega)}U$. For any $x \in {}^{[0,\omega)}U$ and any $z \in {}^{[0,\omega)}U \cup {}^{\omega}U$, let $|z|$ be the length of z and let $xz = x^\frown z$, where $\frown$ is concatenation of sequences. Then ***the*** γ ***prolongation*** $\xrightarrow{\gamma} X$ ***of*** X shall be the following set of pairs of sequences, which is a relation on ${}^{[0,\omega)}U$ when $\gamma < \omega$ and a relation on ${}^{\omega}U$ when $\gamma = \omega$:

$$\xrightarrow{\gamma} X = \{\langle xz, yz \rangle : \langle x, y \rangle \in X, |z| = \gamma\}.$$

Given any $s < \omega$ and $t < \omega$, consider any $W \subseteq {}^sU \times {}^tU$ and let X be the following disjoint union:

$$X = \bigcup\{\overset{n}{\rightarrow} W : 0 \leq n < \omega\} = \bigcup\{\overset{n}{\rightarrow} (X \cap ({}^sU \times {}^tU)) : 0 \leq n < \omega\}.$$

Then X shall have ***grade*** $\langle s,t\rangle$ and W shall be the ***initial section*** of X. Thus, if X has grade $\langle s,t\rangle$ then, for any $s' < \omega$ and $t' < \omega$, $X \cap ({}^{s'}U \times {}^{t'}U)$ is related to the initial section $X \cap ({}^sU \times {}^tU)$ of X in a simple way. More explicitly, if $\langle s',t'\rangle$ differs from every $\langle s+n, t+n\rangle$, then $X \cap ({}^{s'}U \times {}^{t'}U) = \emptyset$ and if, for some $n < \omega$, $\langle s',t'\rangle = \langle s+n, t+n\rangle$, then the behaviour of $X \cap ({}^{s'}U \times {}^{t'}U)$ is rather similar to that of $X \cap ({}^sU \times {}^tU)$ and can be predicted from it.

A relation X shall be ***uniform on*** ${}^{[0,\omega)}U$ if and only if either $X = \emptyset$ or, for some $s < \omega$ and $t < \omega$, necessarily unique, X has grade $\langle s,t\rangle$. The set of uniform relations on ${}^{[0,\omega)}U$ will be denoted by ${}_{[0,\omega)}Un$. By Theorem II.7(a), ${}_{[0,\omega)}Un$ is closed under $\circ$ and $\smile$. As one can verify, each of the functions p_i, $\dot{q}_i$, e_{ij} has a grade (cf. Lemma II.2). From the definition of S_{pqe}, there now follows that $S_{pqe} \subseteq {}_{[0,\omega)}Un$.

A relation Y on ${}^\omega U$ shall be ***uniform on*** ${}^\omega U$ if and only if $Y = \overset{\omega}{\rightarrow} X$ for some $X \in {}_{[0,\omega)}Un$. The set of uniform relations on ${}^\omega U$ will be denoted by ${}_{[\omega]}Un$. From Theorem II.10(a) and the definition of ${}_{[\omega]}S_{pqe}$ there now follows that ${}_{[\omega]}S_{pqe} \subseteq {}_{[\omega]}Un$.

A second condition that will be seen to be necessary for membership in S_{pqe} or in ${}_{[\omega]}S_{pqe}$ is described in Chapter III. Recall that a little earlier there were defined, for any $\gamma \leq \omega$, $\delta \leq \omega$ and $M \subseteq \gamma \times \delta$, the subset ${}_\gamma[M]_\delta$ of ${}^\gamma U \times {}^\delta U$. It turns out, as is shown before Lemma III.6, that for the relations ${}_\omega[M]_\omega = \overset{\omega}{\rightarrow} e_{01}$ and ${}_\omega[M']_\omega = \overset{\omega}{\rightarrow} (\dot{q}_0^2)$ on ${}^\omega U$ there is no $N \subseteq \omega \times \omega$ such that ${}_\omega[M]_\omega \circ {}_\omega[M']_\omega = {}_\omega[N]_\omega$. One is therefore led to consider the closure of the set $\{{}_\omega[M]_\omega : M \subseteq \omega \times \omega\}$ under $\circ$ and $\smile$. Similar reasons lead one to consider the closure under $\circ$ and $\smile$ of the set of those uniform relations $X \neq \emptyset$ on ${}^{[0,\omega)}U$ whose initial section W is a set ${}_s[M]_t$.

Let $\gamma = \{m : m < \gamma\} \leq \omega$ and $\delta = \{n : n < \delta\} \leq \omega$. Near the beginning of Chapter III, a $\langle\gamma,\delta\rangle$ ***triple*** is defined to be any triple $\langle M_0, M_1, M_2\rangle$ such that M_0 is an equivalence relation on γ, $M_1 \subseteq \gamma \times \delta$, and M_2 is an equivalence relation on δ. Also, for any $\gamma \leq \omega$, $\delta \leq \omega$ and $\langle\gamma,\delta\rangle$ triple $\langle M_0, M_1, M_2\rangle$ we define there the subset $[M_0, M_1, M_2]$ of ${}^\gamma U \times {}^\delta U$ that is shown below. Further, for $\gamma = s < \omega$, $\delta = t < \omega$ and $\langle M_0, M_1, M_2\rangle$ an $\langle s,t\rangle$ triple, we define in addition the uniform relation $[\![M_0, M_1, M_2]\!]$ on ${}^{[0,\omega)}U$ that also is shown below.

$$\begin{aligned}
[M_0, M_1, M_2] &= \{\langle x, y\rangle : |x| = \gamma, |y| = \delta, x_i = x_j \text{ if } \langle i, j\rangle \in M_0, \\
&\qquad x_i = y_j \text{ if } \langle i, j\rangle \in M_1, y_i = y_j \text{ if } \langle i, j\rangle \in M_2\}. \\
[\![M_0, M_1, M_2]\!] &= \bigcup\{\overset{n}{\rightarrow} [M_0, M_1, M_2] : n < \omega\} \\
&= \{\langle xz, yz\rangle : |x| = s, |y| = t, |z| < \omega, x_i = x_j \text{ if } \langle i, j\rangle \in M_0, \\
&\qquad x_i = y_j \text{ if } \langle i, j\rangle \in M_1, y_i = y_j \text{ if } \langle i, j\rangle \in M_2\}.
\end{aligned}$$

We shall say of a $\langle\gamma, \delta\rangle$ triple $\langle M_0, M_1, M_2\rangle$ that it ***determines*** $[M_0, M_1, M_2]$ and, if $\gamma = s < \omega$ and $\delta = t < \omega$, that it ***mediately determines*** $[\![M_0, M_1, M_2]\!]$. A ***diagonal relation*** will be any relation that is either determined or mediately determined by some $\langle\gamma, \delta\rangle$ triple. This notion is a 2-ary analogue of the notion of a generalized diagonal set, used in the theory of cylindric set algebras.

Let ${}_{[0,\omega)}D$ be the set of diagonal relations that are uniform on ${}^{[0,\omega)}U$. Note that ${}_{[0,\omega)}D \subseteq {}_{[0,\omega)}Un$. Let ${}_{[\omega]}D$ be the set of diagonal relations on ${}^{\omega}U$. As is easily seen, ${}_{[\omega]}D \not\subseteq {}_{[\omega]}Un$. By Theorem III.11(a), ${}_{[0,\omega)}D$ and ${}_{[\omega]}D$ are closed under $\circ$ and $\smile$. As one can verify, every $\dot{q}_i$, p_i, e_{ij} is in ${}_{[0,\omega)}D$ and every $\overset{\omega}{\rightarrow} \dot{q}_i$, $\overset{\omega}{\rightarrow} p_i$, $\overset{\omega}{\rightarrow} e_{ij}$ is in ${}_{[\omega]}D$. There follows that $S_{pqe} \subseteq {}_{[0,\omega)}D$ and that ${}_{[\omega]}S_{pqe} \subseteq {}_{[\omega]}D$. Let ${}_{[\omega]}UD = {}_{[\omega]}Un \cap {}_{[\omega]}D$. Since ${}_{[\omega]}S_{pqe} \subseteq {}_{[\omega]}Un$, as we saw earlier, there follows that ${}_{[\omega]}S_{pqe} \subseteq {}_{[\omega]}UD$.

To prove that, conversely, ${}_{[0,\omega)}D \subseteq {}_{[\omega]}S_{pqe}$ and ${}_{[\omega]}UD \subseteq {}_{[\omega]}S_{pqe}$, it is convenient to let ${}_s[\![M]\!]_t = \bigcup\{\overset{n}{\rightarrow} {}_s[M]_t : n < \omega\}$, for any $s < \omega$, $t < \omega$. As one can verify, if $\langle M_0, M_1, M_2\rangle$ is any $\langle\gamma, \delta\rangle$ triple, where $\gamma \le \omega$ and $\delta \le \omega$, then $[M_0, M_1, M_2] = {}_\gamma[M_0]_\gamma \circ {}_\gamma[M_1]_\delta \circ {}_\delta[M_2]_\delta$ and if, in addition, $\gamma < \omega$ and $\delta < \omega$ then $[\![M_0, M_1, M_2]\!] = {}_\gamma[\![M_0]\!]_\gamma \circ {}_\gamma[\![M_1]\!]_\delta \circ {}_\delta[\![M_2]\!]_\delta$ (cf. Lemma III.6). Also, by Theorem III.2, for any $\gamma \le \omega$, $\delta \le \omega$ and $\langle\gamma, \delta\rangle$ triple $\langle M_0, M_1, M_2\rangle$, there is a $\langle\gamma, \delta\rangle$ triple $\langle N_0, N_1, N_2\rangle$ such that N_1 is one-one and such that $[M_0, M_1, M_2] = [N_0, N_1, N_2]$ and, when $\gamma < \omega$ and $\delta < \omega$, also $[\![M_0, M_1, M_2]\!] = [\![N_0, N_1, N_2]\!]$. First, suppose that $\gamma < \omega$ and $\delta < \omega$. Then ${}_\gamma[\![N_0]\!]_\gamma$ and ${}_\delta[\![N_2]\!]_\delta$ are equivalence relations on $\{0, \ldots, \gamma-1\}$ or $\{0, \ldots, \delta-1\}$ respectively, and hence in the closure under $\circ$ of $\{e_{ij} : i < \omega, j < \omega\}$. Also, since N_1 is one-one, ${}_\gamma[\![N_1]\!]_\delta$ is in the closure S_{pq} under $\circ$ and $\smile$ of $\{p_i : i < \omega\} \cup \{\dot{q}_i : i < \omega\}$. There follows that $[\![N_0, N_1, N_2]\!]$, and hence $[\![M_0, M_1, M_2]\!]$, is in S_{pqe}. This shows that ${}_{[0,\omega)}D \subseteq S_{pqe}$. Now suppose that $\gamma = \delta = \omega$ and that $[M_0, M_1, M_2]$ is in ${}_{[\omega]}Un$ and hence in ${}_{[\omega]}UD$. Then, by Theorem IV.2, the relations M_0, M_1, M_2, N_0, N_1, N_2 are bisectable. There follows that ${}_\omega[N_0]_\omega$ and ${}_\omega[N_2]_\omega$ are in the closure under $\circ$ of $\{\overset{\omega}{\rightarrow} e_{ij} : i, j < \omega\}$ and that ${}_\omega[N_1]_\omega$ is in the closure ${}_{[\omega]}S_{pq}$ under $\circ$ and $\smile$ of $\{\overset{\omega}{\rightarrow} p_i : i < \omega\} \cup \{\overset{\omega}{\rightarrow} \dot{q}_i : i < \omega\}$. Therefore $[N_0, N_1, N_2]$, and hence $[M_0, M_1, M_2]$, is in ${}_{[\omega]}S_{pqe}$. There now follows that ${}_{[\omega]}UD \subseteq {}_{[\omega]}S_{pqe}$. This concludes an outline of our proof that $S_{pqe} = {}_{[0,\omega)}D$ and ${}_{[\omega]}S_{pqe} = {}_{[\omega]}UD$.

Evidently, if R has grade $\langle s,t\rangle$, R' has grade $\langle s',t'\rangle$, and $s-t \neq s'-t'$, then $R \cap R' = \emptyset$. Since $\emptyset \notin {}_{[0,\omega)}D$, there follows that ${}_{[0,\omega)}D$ is not closed under $\cap$. In contrast, ${}_{[0,\omega)}D \cup \{\emptyset\}$ is closed under $\circ$, $\smile$, and $\cap$. This is briefly discussed at the end of Chapter III. A further study of $\langle {}_{[0,\omega)}D \cup \{\emptyset\}, \circ, \smile, \cap\rangle$ is highly desirable.

As also indicated at the end of Chapter III, the relationship to ${}_{[\omega]}UD$ of its closure $cl({}_{[\omega]}UD)$ under $\circ$, $\smile$, and $\cap$ is more complex. The relationship of $\langle cl({}_{[\omega]}UD), \circ, \smile, \cap\rangle$ to $\langle {}_{[0,\omega)}D \cup \{\emptyset\}\circ, \smile, \cap\rangle$ also seems to be complex.

For the proper subsets of S_{pqe} and ${}_{[\omega]}S_{pqe}$ that have been mentioned, their inductive definitions by a closure condition can likewise be replaced by an explicit set-theoretic characterization. For example, let a relation X on ${}^{\omega}U$ be ***full*** if and only if its domain $Do\ X$ and its range $Rg\ X$ both coincide with ${}^{\omega}U$, and let

$${}_{[\omega]}UDF = {}_{[\omega]}UD \cap \{X : X \subseteq {}^{\omega}U \times {}^{\omega}U, X \text{ is full }\}.$$

Then by Theorem IV.6(b), ${}_{[\omega]}S_{pq} = {}_{[\omega]}UDF$.

For the other proper subsets of S_{pqe} and ${}_{[\omega]}S_{pqe}$ that have been mentioned, a set-theoretic characterization can be extracted from different parts of Chapters III and IV, but is not always given explicitly. Each of these proper proper subsets S of S_{pqe} is a set of relations $[\![M]\!]$ and each of these proper subsets S' of ${}_{[\omega]}S_{pqe}$ is a set of relations ${}_{\omega}[M]_{\omega}$. Necessary and sufficient conditions on $[\![M]\!]$ to be a member of S or, respectively, of ${}_{\omega}[M]_{\omega}$ to be a member of S', are in a certain sense dual to conditions on M. (Cf. Theorem III.7 and the discussion preceding Theorem III.13.)

In obtaining a presentation of the structures $\mathbf{S} = \langle S, \circ, \smile\rangle$, and $\langle \mathbf{S}, \subseteq\rangle$ under consideration, set-theoretic conditions of the kind described played a major role. They were of some help in finding a set of generators that is fairly economical. More importantly, they provided guidance in finding a manageable normal form for products formed from these generators. As usual, showing that a set of equalities or of inclusion conditions is adequate for presenting $\mathbf{S}$ or $\langle \mathbf{S}, \subseteq\rangle$ amounted to showing that the set is adequate for conversion to a suitable normal form. In the case of ${}_{[\omega]}\mathbf{S}_{pqe}$, $\langle {}_{[\omega]}\mathbf{S}_{pqe}, \subseteq\rangle$ and their substructures this tended to be somewhat easier than in the case of the structures involving sequences of finite length, since all the sequences involved are of the same length ω. Also, in the case of the proper substructures of ${}_{[\omega]}\mathbf{S}_{pqe}$ or of $\langle {}_{[\omega]}\mathbf{S}_{pqe}, \subseteq\rangle$, it was of great heuristic value, as shown in Theorem IV.3, that they are isomorphic under ${}_{\omega}[\ ]_{\omega}$ to a structure of a kind with which one is more familiar. This is why presentations of ${}_{[\omega]}\mathbf{S}_{pqe}$, $\langle {}_{[\omega]}\mathbf{S}_{pqe}, \subseteq\rangle$, and their substructures were found first and then adapted to find presentations of $\mathbf{S}_{pqe}$, $\langle \mathbf{S}_{pqe}, \subseteq \rangle$, and their substructures. To save space, our exposition will largely invert this order of discovery.

(A presentation of ${}_{[\omega]}\mathbf{S}_{pqe} = \langle {}_{[\omega]}UD, \circ, {}^{\smile} \rangle$ should be mentioned here that makes use of different substructures and of a somewhat different normal form. Call a relation M on ω an ***almost constant shift*** if and only if M is single-valued, $Do\ M = \omega$, and M is bisectable. Let Acs be the set of almost constant shifts. By Lemma 7.5 and Theorem 7.6 of [Cr74a], or as one can verify directly, ${}_{[\omega]}S_{pqe}$ is the closure under $\circ$ of the set $\{{}_{\omega}[M]_{\omega} \colon M \in Acs\} \cup \{{}_{\omega}[M^{\smile}]_{\omega} \colon M \in Acs\}$. In Chapter 10 of [Cr74a] a presentation of ${}_{[\omega]}\mathbf{S}_{pqe}$ is given that uses this set as set of generators and that uses the equalities T1,…,T7 listed there on p.151. This presentation seems less economical than the one given in Chapter VII below and less likely to be useful. Also, it is not clear how to adapt it to obtain a presentation of $\mathbf{S}_{pqe}$. On the other hand, as is indicated there, a similar presentation can be given of any semigroup with involution $\mathbf{S} = \langle S, \circ, {}^{\smile} \rangle$ in a certain large class that contains ${}_{[\omega]}\mathbf{S}_{pqe}$ as a member. For any $\mathbf{S}$ in this class there exists a set I serving as index set and a set H of functions mapping I into I that satisfies certain conditions (listed on p.129 of [Cr74a]) such that the union of $\{f : f \in H\}$ and $\{f^{\smile} : f \in H\}$ generates $\mathbf{S}$.

The main purpose of Chapter I below is to bring out the relationship to algebraic first-order logic of the set O_{pqe} of those operations R^* on $\{X \colon X \subseteq {}^{[0,\omega)}U\}$ such that R is in S_{pqe}. According to Theorem I.2(b), O_{pqe} lends itself to an algebraic theory of first-order logic where each object X is a subset of some ${}^{n}U$, $0 \le n < \omega$, and where the only Boolean operations are relative complementation $\overline{\ \ }_2$ and intersection $\cap$ (which is definable in terms of $\overline{\ \ }_2$). The class of finite unions $X_0 \cup \cdots \cup X_{s-1}$ such that $1 \le s < \omega$ and, for every $r < s$, X_r is a subset of some ${}^{n}U$ is evidently closed under $\cap$, $\overline{\ \ }_2$, and $\cup$, but also, as will be seen, under every operation in O_{pqe}. Thus, O_{pqe} lends itself to a second kind of algebraic theory of first-order logic, namely one where, for every object X there is some n, $0 < n < \omega$, such that $X \subseteq {}^{[0,n)}U = \bigcup\{{}^{m}U \colon 0 \le m < n\}$. Since evidently, the power set $\{X \colon X \subseteq {}^{[0,\omega)}U\}$ of ${}^{[0,\omega)}U = \bigcup\{{}^{n}U \colon 0 \le n < \omega\}$ is closed under every operation in O_{pqe} and under every Boolean operation (including 1-ary complementation), therefore O_{pqe} lends itself to a third kind of algebraic theory of first-order logic, namely one where every object X is a set of sequences x, each of which is of some length $n < \omega$.

Whether a theory of the first or the second kind can be constructed that is fruitful remains to be seen. As will be seen from Theorem I.6, a theory that is very closely related to a theory of the third kind but where, in place of O_{pqe}, one uses a different set O of operations is described in Chapter 4 of [Cr74a], and its completeness is proved there in Chapter 5. The set O is obtained by adjoining, to the cylindrifications and diagonal sets of a cylindric algebra of subsets of ${}^{[0,\omega)}U$, the operation q_0^* and its conjugate $\dot{q}_0^* = (q_0{}^{\smile})^*$. Thus, although there may be few passages where this is directly apparent, the present work owes its origin to the theory of cylindric algebras. In particular, adjoining q_0^* and its conjugate $\dot{q}_0^*$ to the set of operations available in the theory of cylindric algebras was suggested by results of that theory concerning neat embedding.

An important aspect of $\langle S_{pqe}, \subseteq\rangle$ and of $\langle {}_{[\omega]}\mathbf{S}_{pqe}, \subseteq\rangle$ that, up to now, has been given little emphasis is the following: each of the relations $p_i, \dot{q}_i, e_{ij}$ is a partial function on ${}^{[0,\omega)}U$ and each of the relations $\overset{\omega}{\to} p_i, \overset{\omega}{\to} \dot{q}_i, \overset{\omega}{\to} e_{ij}$ is a partial function on ${}^{\omega}U$, with p_i and $\dot{q}_i$, in fact, being total functions. Let ${}_{[0,\omega)}\overset{\circ}{1} = \dot{q}_0{}^{\smile} \circ \dot{q}_0$ and ${}_{[\omega]}\overset{\circ}{1} = \overset{\omega}{\to} (\dot{q}_0{}^{\smile} \circ \dot{q}_0)$. Then ${}_{[0,\omega)}\overset{\circ}{1}$ is the identity relation on ${}^{[0,\omega)}U$ and the identity element of $\langle \mathbf{S}_{pqe}, \subseteq\rangle$, while ${}_{[\omega]}\overset{\circ}{1}$ is the identity relation on ${}^{\omega}U$ and the identity element of $\langle {}_{[\omega]}\mathbf{S}_{pqe}, \subseteq\rangle$. Now, for any ordered semigroup with involution $\langle S, \circ, {}^{\smile}, \leq\rangle = \langle \mathbf{S}, \leq\rangle$, let an element s of $\langle \mathbf{S}, \leq\rangle$ be a ***functional element*** if and only if $\mathbf{S}$ has an identity element ${}_{\mathbf{S}}\overset{\circ}{1}$ and, moreover, $s^{\smile} \circ s \leq {}_{\mathbf{S}}\overset{\circ}{1}$. Then each of the elements p_i, q_i, e_{ij} of S_{pqe} is a functional element of $\langle \mathbf{S}_{pqe}, \subseteq\rangle$. Likewise, every $\overset{\omega}{\to} p_i, \overset{\omega}{\to} \dot{q}_i, \overset{\omega}{\to} e_{ij}$ is a functional element of $\langle {}_{[\omega]}\mathbf{S}_{pqe}, \subseteq\rangle$. Thus, each of the structures $\langle \mathbf{S}_{pqe}, \subseteq\rangle$ and $\langle {}_{[\omega]}\mathbf{S}_{pqe}, \subseteq\rangle$ has a set of generators that consists entirely of functional elements.

Consider any ordered semigroup with involution $\langle \mathbf{S}, \leq\rangle = \langle S, \circ, {}^{\smile}, \leq \rangle$ having a set G of generators that consists entirely of functional elements. Let A be any nonempty set, let $B = B_A$ be the set of all 2-ary relations on A, let $\mathbf{B} = \langle B, \circ, {}^{\smile}\rangle$, and let $\overset{\circ}{1}_A = \{\langle a, a\rangle : a \in A\}$, so that $\overset{\circ}{1}_A$ is the identity element of $\mathbf{B}_A$. Suppose that h is a homomorphism of $\langle \mathbf{S}, \leq\rangle$ into $\langle \mathbf{B}, \subseteq\rangle$ such that $h({}_{\mathbf{S}}\overset{\circ}{1}) = \overset{\circ}{1}_A$. Then, for every $s \in G$, $h(s)$ is single-valued, since $(h(s))^{\smile} \circ h(s) \subseteq \overset{\circ}{1}_A$. Hence $\mathbf{A} = \langle A, h(s)\rangle_{s\in G}$ is a unary partial algebra. It shall be a ***unary representation of*** $\langle \mathbf{S}, \leq\rangle$ or, more explicitly, ***the unary representation by*** h ***of*** $\langle \mathbf{S}, \leq\rangle$ ***when*** G ***serves as set of generators***.

Assume that $\mathbf{A} = \langle A, h(s)\rangle_{s\in G}$ is thus related to $\langle \mathbf{S}, \leq\rangle$, G, and h. Then to any equality or inclusion condition concerning the elements of G that holds for $\langle \mathbf{S}, \leq\rangle$ there corresponds a condition concerning the partial functions $h(s), s \in G$, that holds for $\mathbf{A}$. For example, as one can verify, or as will be seen in Chapter VI, $\langle \mathbf{S}_{pq}, \subseteq\rangle$ and hence also $\langle \mathbf{S}_{pqe}, \subseteq\rangle$, satisfies the equality $\dot{q}_0 p_i = p_{i+1}\dot{q}_i$ and the inclusion condition $\dot{q}_0{}^{\smile} p_0^2 \dot{q}_0 \subseteq \dot{q}_0 \dot{q}_0{}^{\smile}$. There follows that if $\mathbf{A} = \langle A, h_s \rangle_{s\in G} = \langle A, h(p_i), h(\dot{q}_i), h(e_{ij})\rangle_{i,j<\omega}$ is a unary representation of $\langle \mathbf{S}_{pqe}, \subseteq\rangle$, then the partial functions $h(p_i)$ and $h(\dot{q}_0)$ on A satisfy the following two conditions:

$$h(\dot{q}_o) \circ h(p_i) = h(p_{i+1}) \circ h(\dot{q}_0)$$

$$(h(\dot{q}_0))^{\smile} \circ (h(p_0))^2 \circ h(\dot{q}_0) \subseteq h(\dot{q}_0) \circ (h(\dot{q}_0))^{\smile}.$$

Since $\circ$ and $^{\smile}$ are set-theoretic operations that can be expressed rather simply in first-order language, these two conditions are fairly simple first-order conditions on $\mathbf{A}$. Note however, that the second of these conditions involves the relations $(h(\dot{q}_o))^{\smile}$ and $h(\dot{q}_0) \circ (h(\dot{q}_0))^{\smile}$, which, in general, are not single-valued.

Let $\mathcal{K}_{p\dot{q}e}$ be the class of those unary partial algebras $\mathbf{A}$ such that when $G = \{p_i, \dot{q}_i, e_{ij}\}_{i,j<\omega}$ serves as set of generators, then $\mathbf{A}$ is a unary representation of $\langle \mathbf{S}_{pqe}, \subseteq\rangle$ by some h. Thus $\mathcal{K}_{p\dot{q}e}$ is a class of unary partial algebras that can be characterized by fairly simple first-order conditions. Since the particular h being used often does not matter and since there is little chance of confusion, it is often convenient to denote a member $\langle A, h(s)\rangle_{s\in G}$ of $\mathcal{K}_{p\dot{q}e}$ by $\langle A, p_i, \dot{q}_i, e_{ij}\rangle_{i,j<\omega}$. Then $\mathcal{K}_{p\dot{q}e}$ can be characterized as the class of those unary partial algebras $\mathbf{A} = \langle A, p_i, \dot{q}_i, e_{ij}\rangle_{i,j<\omega}$ whose primitive functions $p_i, \dot{q}_i, e_{ij}$ satisfy those first-order conditions that correspond to the equalities and inclusion conditions that are used in the presentation of $\langle \mathbf{S}_{pqe}, \subseteq\rangle$. Note that the class of unary representations of $\langle {}_{[\omega]}\mathbf{S}_{pqe}, \subseteq\rangle$ is a subclass of $\mathcal{K}_{p\dot{q}e}$, since $\langle {}_{[\omega]}S_{pqe}, \subseteq\rangle$ is a homomorphic image of $\langle \mathbf{S}_{pqe}, \subseteq\rangle$. This subclass consists of those $\mathbf{A} = \langle A, p_i, \dot{q}_i, e_{ij}\rangle_{i,j<\omega}$ in $\mathcal{K}_{p\dot{q}e}$ such that every p_i and $\dot{q}_i$ is a total function on A.

The presentation of $\langle \mathbf{S}_{pqe}, \subseteq\rangle$ given in Chapter VII is a study of the functions $p_i, \dot{q}_i, e_{ij}$ on ${}^{[0,\omega)}U$ that is quite abstract, considering them to be elements of a structure. It is highly desirable that there be a follow-up study of these functions that is less abstract. A study of $\mathcal{K}_{p\dot{q}e}$ would be a study of this kind. It would treat $p_i, \dot{q}_i, e_{ij}$ as functions on a set rather than as elements of a structure. Nevertheless, one would abstract to this extent: membership of $\mathbf{A} = \langle A, p_i, \dot{q}_i, e_{ij}\rangle_{i,j<\omega}$ in $\mathcal{K}_{p\dot{q}e}$ is characterized by conditions that are meaningful for any unary partial algebra of this similarity type and do not involve the notion of sequence.(Whereas the differences between $\mathbf{S}_{pqe}$ and ${}_{[\omega]}\mathbf{S}_{pqe}$ are fairly small, the differences between those structures A in $\mathcal{K}_{p\dot{q}e}$ that are most closely related to them may be much more significant.)

Given any $\mathbf{A}$ in $\mathcal{K}_{p\dot{q}e}$, let $\mathbf{C} = \mathbf{C_A}$ be the complex algebra of $\mathbf{A}$ as defined by 3.8 in [JT]. Since the characterization of $\mathcal{K}_{p\dot{q}e}$ involves only equalities and inclusion conditions between terms formed from $p_i, \dot{q}_i, e_{ij}$ by $\circ$ and $^{\smile}$ it is probable that, by using methods developed by [JT], one can give a fairly transparent characterization of the class of those algebras that are isomorphic to a subalgebra of some $\mathbf{C}_\mathbf{A}$ such that $\mathbf{A}$ is in $\mathcal{K}_{p\dot{q}e}$. In view of Theorem I.6 below, the equational theory of this class may turn out to be polynomially equivalent to the theory given in Chapter 4 of [Cr74a].

The class $\mathcal{K}_{p\dot{q}e}$ fails to capture some crucial properties possessed by the concrete relations $p_i, \dot{q}_i, \dot{q}_i{}^{\smile}, e_{ij}$ on ${}^{[0,\omega)}U$. Hence the class obtained from algebras $\mathbf{C}_\mathbf{A}$ in the way just sketched fails to capture the corresponding properties possessed by the induced operations $p_i^*, \dot{q}_i^*, (\dot{q}_i^*)^\dagger, e_{ij}^*$ on $\{X\colon X \subseteq^{[0,\omega)} U\}$. Two such properties of $p_i, \dot{q}_i, \dot{q}_i{}^{\smile}, e_{ij}$ are the following, the first of which was suggested by 2.7.45 and 3.2.16 of [HMT], while the second equates $e_{i,i+1}$ to the restriction of p_i to its set of fixed points.

$$(\dot{q}_i \circ \dot{q}_i{}^{\smile}) \cap (\dot{q}_j \circ \dot{q}_j{}^{\smile}) \subseteq \mathring{1}_A, \text{ if } i \neq j.$$

$$p_i \cap \mathring{1}_A = p_i \cap p_i^2 = e_{i,i+1}.$$

Both of these properties involve $\cap$. Much further work may be needed for giving a satisfactory account of the properties of $p_i^*, \dot{q}_i^*, (\dot{q}_i^*)^\dagger, e_{ij}^*$ to which they give rise.

CHAPTER I

Boolean, Relation-Induced, and Other Operations for Dealing with First-Order Definability

The notion of first-order definability in a relational structure deals with relations that often are of different finite rank and whose members therefore are sequences of corresponding different finite length. When investigating the use of algebraic methods for dealing with this notion, it therefore seems a natural and mild restriction to consider only operations F which, when applied to sets $X_0, \ldots, X_{n-1}$ of finite sequences yield as value a set $F(X_0, \ldots, X_{n-1})$ that, like each X_m, also consists of finite sequences.

Several sets of operations of this kind will be considered below. They agree with respect to those operations F in them for which the following holds: For some binary relation R between finite sequences, F is the unary operation R^* that maps each set X of finite sequences into the direct image $R^*(X) = \{y : \langle x, y\rangle \in R \text{ for some } x \in X\}$ of X under R. We shall say that R ***induces*** F. Any F that is induced by some R shall be ***relation-induced***.

Besides these relation-induced operations, each of the sets to be considered will contain at least one Boolean operation. Some of the sets also will contain an operation of a third kind, namely, the 0-ary operation $H_{\{\emptyset\}}$ whose constant value is the set $\{\emptyset\}$, i.e., the set whose only element is the null-sequence, i.e. is the sequence of length 0. This 0-ary operation is not relation-induced. Nor is the closely related unary operation F relation-induced that satisfies $F(X) = \{\emptyset\}$ for every X, since every operation G that is relation-induced satisfies $G(\emptyset) = \emptyset$.

(Let H' be the unary operation such that $H'(\emptyset) = \emptyset$ and $H'(X) = \{\emptyset\}$ for any $X \neq \emptyset$. Then, as the referee noted, H' is closely related to $H_{\{\emptyset\}}$ and, moreover, is induced by the binary relation $S_\emptyset$ such that $S_\emptyset$ is single-valued and for any finite sequence x as argument yields as value the empty sequence $\emptyset$. In contrast to the relations considered in Chapter II, functions such as $S_\emptyset$ that are constant or almost constant are not uniform.)

If every sequence in X is of the same length s then X shall be ***of rank*** s. Also, if X is of some rank s then X shall be ***ranked***. Thus, every nonempty set X that is ranked is of a unique rank s. Now every set that is first-order definable in a relational structure is ranked. In choosing operations F for dealing with these sets it therefore seems desirable, whenever it does not conflict with important other considerations, to allow for economies and to choose operations F under which the class of ranked sets is closed, i.e., which satisfy the following condition:

$$\text{If each of } X_0, \ldots, X_{n-1} \text{ is ranked, then } \ F(X_0, \ldots, X_{n-1}) \ \text{ is ranked.}$$

As we shall see, the class of ranked sets is closed under each of the relation-induced operations that we shall choose. It is evidently also closed under the 0-ary operation whose constant value is $\{\emptyset\}$ and under the two binary Boolean operations of intersection and relative complementation.

In contrast, there is no universe that includes all ranked sets such that the class of ranked sets is closed under the unary operation of complementation with respect to that universe. Nor is the class closed under union. Evidently, whenever X and Y are nonempty and of different rank, then $X \cup Y$ is not of any rank.

In dealing with first-order definability it may therefore turn out to be impractical, and perhaps even wrong-headed, to admit only those Boolean operations under which the class of ranked sets is closed. We shall therefore also consider a set of operations where this condition on the Boolean operations in the set has been weakened and a further set where the condition has been dropped. (The question of which Boolean operations to admit was first called to my attention by the late Bill Hanf.)

For most of the chapter, the notion of first-order definability discussed will be that where one does not use individual parameters. First-order definability with the use of individual parameters will be taken up near the end.

Some conventions will be useful. Let R and R' be any binary relations and let X be any set. Below we repeat the definition of R^* and introduce some other notation that is fairly standard. Henceforth, when convenient, we shall often omit $\circ$.

$$\begin{aligned} R^{\smile} &= \{\langle x, y\rangle : \langle y, x\rangle \in R\} \\ R \circ R' &= \{\langle x, z\rangle : \langle x, y\rangle \in R \text{ and } \langle y, z\rangle \in R', \text{ for some } y\} \\ \mathit{Do}\ R &= \{x : \langle x, y\rangle \in R \ \text{ for some } y\} \end{aligned}$$

$$\begin{aligned}
Rg\ R &= \{y : \langle x, y\rangle \in R \ \text{ for some } x\} \\
R^0 &= \{\langle x, x\rangle : x \in Do\ R \cup Rg\ R\} \\
R^{n+1} &= R \circ R^n \\
R^*(X) &= \{y : \langle x, y\rangle \in R \ \text{ for some } x \in X\} \\
R_x &= R(x) = y, \quad \text{if } \langle x, y\rangle \in R \text{ and } R \text{ is single-valued.}
\end{aligned}$$

An ordinal δ will be the set $\{\gamma : \gamma < \delta\}$ of its predecessors. We let $[\beta, \delta] = \{\gamma : \beta \leq \gamma \leq \delta\}$ and $[\beta, \delta) = \{\gamma : \beta \leq \gamma < \delta\}$. As usual, we let ω be the first infinite ordinal. In contrast to $\beta, \gamma, \ldots$, the following will always range over the finite ordinals: $i, j, \ldots, m, \ldots$. Sometimes, for emphasis, we may state that $m \geq 0$ or that $m < \omega$. A cardinal will be a finite ordinal or an infinite initial ordinal.

Henceforth we will assume that a nonempty finite or infinite set U has been given. It will serve as our universe or base set. It will sometimes be suggestive to think of U as an alphabet and of its elements as letters. In the last three chapters, but not now, it will be assumed that $\overline{\overline{U}} \geq 2$, where $\overline{\overline{X}}$ is the cardinality of X.

Let γ be any ordinal. A ***sequence of length*** γ will be any function, i.e. single-valued relation, whose domain is γ. Unless there are indications to the contrary, γ will be an ordinal satisfying $0 \leq \gamma \leq \omega$. At times, the range of a sequence will be a subset of ω. This will be made apparent by the context, or by the use of $f, g, \ldots$ for such sequences, or by applicable qualifications such as the following: strictly increasing, nondecreasing, numerical, ... Most of the time, however, the range of a sequence will be a subset of U, so that, for some γ such that $0 \leq \gamma \leq \omega$, the sequence belongs to the following set:

$$^{\gamma}U = \{x : Do\ x = \gamma,\ Rg\ x \subseteq U,\ x \ \text{ is single-valued}\}\ .$$

Henceforth, unless there are indications to the contrary, we let $v, \ldots, z \ldots$ range over all sequences of this kind. For x we will sometimes write $\langle x_0, x_1, \ldots\rangle$. If $x \in {}^nU$, we will sometimes write $\langle x_0, \ldots, x_{n-1}\rangle$. If $x \in {}^{\omega}U$, we will sometimes write $\langle x_0, \ldots, x_n, \ldots\rangle$. Thus, if $x_0, x_1, \ldots$ are regarded as letters, then x may be regarded as the word they form.

Concatenation will be indicated by $\frown$ or, perhaps more often, simply by juxtaposition. Thus, if $x = \langle x_0, \ldots, x_{m-1}\rangle$, $y = \langle y_0, \ldots, y_{n-1}\rangle$, and $z = \langle z_0, \ldots, z_n, \ldots\rangle$, then $xy = \langle x_0, \ldots, x_{m-1}, y_0, \ldots, y_{n-1}\rangle$ and $xz = \langle x_0, \ldots, x_{m-1}, z_0, \ldots, z_n, \ldots\rangle$.

For brevity, the length $Do\ x$ of a sequence x will also be denoted by $|x|$. Further, x shall be ***finite*** or ***infinite*** according to whether $|x| < \omega$ or $|x| = \omega$ and hence according to whether x, considered as a function, is a finite or an infinite set of ordered pairs. We will also use the notation that is shown below. Thus, in particular, ${}^{[0,\omega)}U$ is the set $\bigcup_{n<\omega} {}^nU = \{x : |x| < \omega\}$ of all sequences that are finite.

$$\begin{aligned} {}^{[m,\gamma)}U &= \bigcup\{{}^nU : m \le n < \gamma\} = \{x : m \le |x| < \gamma\} \\ {}^{[m,\gamma]}U &= \bigcup\{{}^\beta U : m \le \beta \le \gamma\} = \{x : m \le |x| \le \gamma\} \end{aligned}$$

The identity relation on ${}^{[m,\gamma)}U$, ${}^{[m,\gamma]}U$, or ${}^\gamma U$ will henceforth be denoted, respectively, by ${}_{[m,\gamma)}\overset{\circ}{1}$, ${}_{[m,\gamma]}\overset{\circ}{1}$, or ${}_\gamma\overset{\circ}{1}$.

Among the relations that induce one of our operations F, those of the following four kinds, where $0 \le i < \omega$ and $0 \le j < \omega$, will play a central role. Those of the first three kinds are single-valued.

$$\begin{aligned} p_{(i,j)} &= \{\langle x, y\rangle : \max(i,j) < |x| = |y| < \omega, \\ &\qquad x_i = y_j,\ x_j = y_i,\ x_k = y_k \text{ if } k \in |x| \cap -\{i,j\}\} \\ e_{(i,j)} &= \{\langle x, x\rangle : \max(i,j) < |x| < \omega, x_i = x_j\} \\ \dot{q}_i &= \{\langle x, y\rangle : i < |x| = |y| + 1 < \omega,\ y_n = x_n \text{ if } n < i, \\ &\qquad y_n = x_{n+1} \text{ if } i \le n < |y|\} \\ &= \{\langle vwz, vz\rangle : |v| = i,\ |w| = 1,\ |z| < \omega\} \\ q_i &= \{\langle x, y\rangle : i \le |x| = |y| - 1 < \omega,\ y_n = x_n \text{ if } n < i, \\ &\qquad y_{n+1} = x_n \text{ if } i \le n < |x|\} \\ &= \{\langle vz, vwz\rangle : |v| = i,\ |w| = 1,\ |z| < \omega\}\ . \end{aligned}$$

Let $k = \max(i,j)$. Then $p_{(i,j)}$ is a permutation of ${}^{[k+1,\omega)}U$ and hence is a partial permutation on ${}^{[0,\omega)}U$. If $i \ne j$, it is of order 2, and if $i = j$, it is the identity relation on ${}^{[i+1,\omega)}U$. It will be called ***(i,j)-interchange (for finite sequences)***.

Instead of $e_{(i,j)}$ we will often write, more briefly, $e_{i,j}$ or e_{ij}.

The domain $d_{ij} = \{x : \max(i,j) < |x| < \omega, x_i = x_j\}$ of e_{ij} shall be the $\langle i,j\rangle$ ***diagonal set (of finite sequences)*** and e_{ij} itself shall be the ***identity (relation) on d_{ij}***. Note that for $k = 1 + \max(i,j)$, $e_{ij} = e_{(i,j)}$ is the restriction $p_{(i,j)} \cap {}_{[k,\omega)}\overset{\circ}{1}$ of $p_{(i,j)}$ to the set d_{ij} of its fixed points.

Let $i < \omega$. Then $Do\ \dot{q}_i = {}^{[i+1,\omega)}U$ and, for any $x = vwz$ such that $|v| = i, |w| = 1$, and $0 \le |z| < \omega$, $\dot{q}_i$ yields the value $\dot{q}_i(x) = \dot{q}_i(vwz) = vz$. Hence $\dot{q}_i$ shall be ***excision at place*** i or, more briefly, ***i-excision***. More specifically, $\dot{q}_i$ shall be ***the i-excision from ${}^{[i+1,\omega)}U$ onto ${}^{[i,\omega)}U$***. For mnemonic and other purposes one may wish to think of $\dot{q}_i$ as the function that on sequences x satisfying $|x| > i$ carries out the following instruction: Puncture at place i, remove what is there, and, if $|x| > i + 1$, close up. (The term used on p.6 of [Lo] is ***deletion.***)

If $\overset{=}{U} \geq 2$, then q_i is not single-valued. For suggestiveness, q_i will be called ***multivalued*** i***-insertion***. More specifically, q_i shall be ***the multivalued*** i***-insertion from*** ${}^{[i,\omega)}U$ ***onto*** ${}^{[i+1,\omega)}U$. Note that $\emptyset$ is in $Do\ q_0 = Rg\ \dot{q}_0$, but not in $Do\ \dot{q}_0$, $Do\ e_{(0,0)}$, or $Do\ p_{(0,0)}$.

Evidently $p_{(j,i)} = p_{(i,j)}$ and $e_{j,i} = e_{i,j}$. Also, $p_{(i,j)}^{\smile} = p_{(i,j)}$, $e_{ij}^{\smile} = e_{ij}$, $\dot{q}_i^{\smile} = q_i$, and $q_i^{\smile} = \dot{q}_i$. We will often make tacit use of these equalities. Mostly, (but probably not consistently) we will use $\dot{q}_i$ when we think of it as i-excision, or more generally as in some sense fundamental, and also at times for typographical reasons, while we tend to use $q_i^{\smile}$ when we wish to stress the relation to q_i and to regard the latter as primary.

Relations of the following four kinds can now be introduced by definition. Those of the first kind, already mentioned earlier, will play a major role.

$$\begin{aligned}
{}_{[i,\omega)}\overset{\circ}{1} &= q_i \circ q_i^{\smile} = \{\langle x, x\rangle : i \leq |x| < \omega\} = \{\langle x, x\rangle : x \in {}^{[i,\omega)}U\} \\
c_i &= q_i^{\smile} \circ q_i \\
&= \{\langle x, y\rangle : i < |x| = |y| < \omega,\ x_n = y_n \text{ if } n \in |x| \cap -\{i\}\} \\
r_i &= q_i \circ e_{i,i+1} \\
&= \{\langle vwz, vwwz\rangle : |v| = i,\ |w| = 1,\ |z| < \omega\} \\
\dot{r}_i &= r_i^{\smile} = e_{i,i+1} \circ q_i^{\smile} \\
&= \{\langle vwwz, vwz\rangle : |v| = i,\ |w| = 1,\ |z| < \omega\}
\end{aligned}$$

Let $i < \omega$. Then ${}_{[i,\omega)}\overset{\circ}{1}$ is the identity relation on ${}^{[i,\omega)}U$ and hence is a subidentity on ${}^{[0,\omega)}U$. (For typographical reasons we will use a left subscript of $\overset{\circ}{1}$, rather than a left superscript, since, instead of i, we sometimes need to use rather lengthy expressions.) Algebraically, ${}_{[i,\omega)}\overset{\circ}{1}$ is the identity element of that semigroup whose universe is the set of binary relations on ${}^{[i,\omega)}U$ and whose primitive operation is $\circ$. (Our notation is intended to serve as a reminder of this.) The relation c_i is an equivalence relation on ${}^{[i+1,\omega)}U$ and hence is a partial equivalence on ${}^{[0,\omega)}U$. Both r_i and $r_i^{\smile}$ are one-one. We call r_i ***replication at place*** i or, more briefly, i***-replication***. More specifically, r_i shall be ***the*** i***-replication from*** ${}^{[i+1,\omega)}U$ ***into*** ${}^{[i+2,\omega)}U$. If x is a sequence such that $|x| > i$ and hence such that at place i of x there is the element x_i, then $r_i(x)$ results from x by making a copy of x_i and inserting it at place i, first having made room for this insertion by a one-place shift to the right of the end segment of x that begins at place i. We call $\dot{r}_i = r_i^{\smile}$ ***the inverse of*** i***-replication*** and also i***-excision restricted to*** *Do* $e_{i,i+1}$ or, more briefly, ***restricted*** i***-excision.*** Note that $e_{i,i+1}$ can be defined in terms of $r_i^{\smile}$ and r_i as follows:

$$e_{i,i+1} = r_i^{\smile} \circ r_i \ .$$

In accord with a practice now fairly well established, a relation R on ${}^{[0,\omega)}U$ shall be ***logical*** or ***invariant*** if and only if it satisfies the following condition:

> If f is any permutation of the universe U and $f^{\#}$ is the permutation of ${}^{[0,\omega)}U$ that changes each $x = \langle x_0, \ldots, x_{n-1}\rangle$ to $f^{\#}(\langle x_0, \ldots, x_{n-1}\rangle) = \langle f(x_0), \ldots, f(x_{n-1})\rangle$, then $\langle f^{\#}(x), f^{\#}(y)\rangle \in R$ if and only if $\langle x, y\rangle \in R$.

Evidently, each relation R of one of the kinds we have introduced is invariant in this sense. Thus each of the functions $p_{(i,j)}$, e_{ij}, $\dot{q}_i$, ${}_{[i,\omega)}\overset{\circ}{1}$, r_i, $\dot{r}_i$ amounts to a procedure that is applicable to words of any alphabet U. In contrast, most procedures that typesetters and others use for changing words are alphabet-dependent. For example, if the alphabet contains no Greek letters then the procedure of replacing the letter "α" by "β" is worthless.

For R being one of the relations e_{ij}, $p_{(i,j)}$, q_i, $\dot{q}_i$, ${}_{[i,\omega)}\overset{\circ}{1}$, c_i, r_i, or $\dot{r}_i$, where $i \leq j < \omega$, we now describe the unary operation R^* on the power set $\{X : X \subseteq {}^{[0,\omega)}U\}$ of ${}^{[0,\omega)}U$ that is induced by R. Recall that $e_{ji} = e_{ij}$ and $p_{(j,i)} = p_{(i,j)}$, so that $e^*_{ji} = e^*_{ij}$ and $p^*_{(j,i)} = p^*_{(i,j)}$.

$$\begin{aligned}
e^*_{ij}(X) &= X \cap \{vwxwz : |v|=i,\ |w|=1,\ |x|=j-i-1, |z|<\omega\} \\
&= X \cap d_{ij},\ \text{ if } i<j. \\
p^*_{(i,j)}(X) &= \{vwxyz : |v|=i,\ |w|=|y|=1,\ |x|=j-i-1, \\
&\qquad vyxwz \in X\},\ \text{if } i<j. \\
e^*_{ii}(X) &= p^*_{(i,i)}(X) = X \cap \{vwz : |v|=i, |w|=1, |z|<\omega\} \\
&= X \cap {}^{[i+1,\omega)}U \\
q^*_i(X) &= \{v^\frown\langle u\rangle^\frown z : |v|=i,\ u\in U,\ vz \in X\} \\
\dot{q}^*_i(X) &= \{vz : |v|=i,\ \text{and, for some } u \in U,\ v^\frown\langle u\rangle^\frown z \in X\}
\end{aligned}$$

$$
\begin{aligned}
{}_{[i,\omega)}\overset{\circ *}{1}(X) &= (q_i\dot{q}_i)^*(X) = \dot{q}_i^*(q_i^*(X)) = X \cap {}^{[i,\omega)}U \\
c_i^*(X) &= \{v^\frown\langle u\rangle^\frown z : |v| = i,\ u \in U, \text{ and, for some } u' \in U, \\
&\qquad v^\frown\langle u'\rangle^\frown z \in X\} \\
r_i^*(X) &= \{vwwz : |v| = i,\ |w| = 1,\ vwz \in X\} \\
\dot{r}_i^*(X) &= \{vwz : |v| = i,\ |w| = 1,\ vwwz \in X\}
\end{aligned}
$$

The operations e_{ij}^* and ${}_{[i,\omega)}\overset{\circ *}{1}$ shall be, respectively, ***intersection with*** d_{ij} or ***intersection with*** ${}^{[i,\omega)}U$. Also for $p_{(i,j)}^*$, q_i^*, $\dot{q}_i^*$, c_i^*, r_i^*, and $\dot{r}_i^*$ we shall use respectively: (i,j) ***interchange***, i-***insertion***, i-***excision***, ***partial*** i-***cylindrification***, i-***replication, restricted*** i-***excision (applied to sets of finite sequences).***

In a digression we want to call attention, in this and the next paragraph, to two interesting features that most of the above operations possess, even though these do not seem to have a direct bearing on our investigation. As to one of these features, assume that R^* is any of the above operations other than $\dot{q}_i^*$ or c_i^*. Then R^* in the following sense ***transmits computability***:

> Given any subset X of ${}^{[0,\omega)}U$ and any method whereby, given any sequence x in ${}^{[0,\omega)}U$, one can compute whether or not x belongs to X, by combining this method and the description of R^* one obtains a method whereby, given any sequence y in ${}^{[0,\omega)}U$, one can compute whether or not y belongs to $R^*(X)$.

As to the other feature, assume that R^* is any of the above operations other than q_i^* or c_i^*. Let V and W be any sets serving as universe and let ${}_VR^*$ and ${}_WR^*$ be related to V or W respectively as, up to now, we have been relating R^* to U. Then R^* is ***partly independent of the universe*** in the following sense:

$$
\text{For any } V \text{ and } W, \text{ if } X \subseteq {}^{[0,\omega)}V \cap {}^{[0,\omega)}W, \\
\text{then } {}_VR^*(X) = {}_WR^*(X).
$$

In contrast, as can be seen from their definition, q_i^* and c_i^* are not thus independent of the universe. (Of course, each of the operations R^* listed depends in the following sense on the universe: if $V \neq W$, then $Do\ {}_VR^* \neq Do\ {}_WR^*$.)

Let X be of rank s and let $j \leq i$. If $s \leq i$, then the sets $p_{(i,j)}^*(X)$, $e_{ij}^*(X)$, $\dot{q}_i^*(X)$, and $q_{i+1}^*(X)$ are empty. Also, if $s > i$ then $p_{(i,j)}^*(X)$ and $e_{ij}^*(X)$ are of rank s and $q_i^{\smile}{}^*(X)$ is of rank $s-1$. Further, if $s \geq i$ then $q_i^*(X)$ is of rank $s+1$. There follows that whenever X is ranked, then $p_{(i,j)}^*(X)$, $e_{ij}^*(X)$, $\dot{q}_i^*(X)$, and $q_i^*(X)$ also are ranked. In other words, the class of ranked sets is closed under each of the operations $p^*_{(i,j)}$, e^*_{ij}, $\dot{q}^*_i$, q^*_i. It is therefore also closed under ${}_{[i,\omega)}\overset{\circ}{1}{}^*$, c^*_i, r^*_i, and $\dot{r}^*_i$.

In contrast, d_{ij} is not ranked and, in fact, is not included in any ${}^{[0,s]}U$. This is one important reason for choosing e^*_{ij} rather than d_{ij} for dealing with first-order definability, even though, in some respects, d_{ij} is simpler. A second reason is the following: a change from d_{ij} to e^*_{ij} often allows one to reformulate a problem of how d_{ij} interacts with p^*_{ij}, q^*_i, and $\dot{q}^*_i$ as a problem of how the relations e_{ij}, $p_{(ij)}$, q_i, and $\dot{q}_i$ interact under $\circ$. Thus reformulated, the problem concerns objects that are simpler and an operation on them that has been widely studied. (Later on, in connection with individual parameters, we shall make a similar change from a set to the identity relation that has this set as domain and thereby codes it.)

We let S_q be the closure under $\circ$ and $\smile$ of $\{q_i : i < \omega\}$. Since $\dot{q}_i = q_i^{\smile}$, therefore S_q is also the closure under $\circ$ of $\{q_i : i < \omega\} \cup \{\dot{q}_i : i < \omega\}$. Note that S_q contains every ${}_{[i,\omega)}\overset{\circ}{1} = q_i \circ q_i^{\smile}$.

Henceforth we let $p_i = p_{(i,i+1)}$. Also, we let S_p be the closure under $\circ$ of $\{p_i : i < \omega\}$. Since $p_i = p_i^{\smile}$, therefore S_p is closed under $\smile$. Note that S_p contains ${}_{[i+2,\omega)}\overset{\circ}{1} = p_i \circ p_i = p_i^2$ for every $i \geq 0$, but does not contain ${}_{[1,\omega)}\overset{\circ}{1}$ or ${}_{[0,\omega)}\overset{\circ}{1}$. Thus, S_p differs from the closure under $\circ$ of $\{p_{(i,j)} : i < \omega,\ j < \omega\}$, which contains ${}_{[1,\omega)}\overset{\circ}{1} = p_{(0,0)}$ (but also does not contain ${}_{[0,\omega)}\overset{\circ}{1}$).

We let S_e be the closure under $\circ$ of $\{e_{ij} : \{i,j\} \neq \{0\}\}$. Since $e_{ij} = e_{ij}^{\smile}$, therefore S_e is closed under $\smile$. Note that, like S_p, S_e contains ${}_{[i+2,\omega)}\overset{\circ}{1} = e_{i+1,i+1}$ for every $i \geq 0$, but does not contain ${}_{[0,\omega)}\overset{\circ}{1}$ or ${}_{[1,\omega)}\overset{\circ}{1} = e_{0,0}$.

We let $S_{pq}, S_{pe}, S_{qe}, S_{pqe}$ be the closure under $\circ$ of $S_p \cup S_q$, $S_p \cup S_e$, $S_q \cup S_e$, $S_p \cup S_q \cup S_e$ respectively. Thus, they are also the closure under $\circ$ and $\smile$ of the following respectively:

$$\{p_i : i < \omega\} \cup \{q_i : i < \omega\}\ ,$$
$$\{p_i : i < \omega\} \cup \{e_{i,j} : \{i,j\} \neq \{0\}\}\ ,$$
$$\{q_i : i < \omega\} \cup \{e_{i,j} : \{i,j\} \neq \{0\}\}\ ,$$
$$\{p_i : i < \omega\} \cup \{q_i : i < \omega\} \cup \{e_{i,j} : \{i,j\} \neq \{0\}\}\ .$$

The definitions of S_p, S_e, and S_{pe} have been so chosen that each of the semigroups $\langle S_p, \circ\rangle$, $\langle S_e, \circ\rangle$, $\langle S_{pe}, \circ\rangle$ has $p_0^2 = e_{1,1}$ as identity element and so that, as can be seen from the proof of Theorem 1(b) below, $\langle S_{pe}, \circ\rangle$ is generated by $\{e_{0,1}\} \cup \{p_i : i < \omega\}$. Thus, as will be seen in Chapter VI, $\langle S_{pe}, \circ, \smile\rangle$ lends itself to a presentation that is probably simpler than any that one can give of $\langle S_{pe} \cup \{e_{0,0}\}, \circ, \smile\rangle$. Failure of S_p, S_e, S_{pe} to contain ${}_{[0,\omega)}\overset{\circ}{1} = q_0 \circ \dot{q}_0$ or ${}_{[1,\omega)}\overset{\circ}{1} = e_{0,0} = q_1 \circ \dot{q}_1$ will cause no harm, since these two relations do belong to S_q, S_{pq}, S_{qe}, S_{pqe}.

Some further sets will play an occasional role. We let S_r be the closure under $\circ$ and $\smile$ of $\{r_i : i < \omega\} \cup \{\dot{r}_i : i < \omega\}$. Whereas S_q contains every ${}_{[i,\omega)}\overset{\circ}{1}$, S_r contains ${}_{[i+1,\omega)}\overset{\circ}{1} = r_i \circ \dot{r}_i$ for every $i < \omega$, but does not contain ${}_{[0,\omega)}\overset{\circ}{1}$. Also, S_c shall be the closure under $\circ$ of $\{c_i : i < \omega\}$. Since $c_i = c_i^\smile$, therefore S_c is also closed under $\smile$. Since $c_i = \dot{q}_i \circ q_i$, therefore $S_c \subseteq S_q$. Since $r_i = q_i \circ e_{i,i+1}$, therefore $S_r \subseteq S_{qe}$. Also, we let, for example, S_{pr} and S_{prc} be the closure under $\circ$ of $S_p \cup S_r$ or $S_p \cup S_r \cup S_c$, respectively. Both of these sets are also closed under $\smile$.

At times we will be interested in the set that results from $\{q_i : i < \omega\}$, $\{r_i : i < \omega\}$, $\{p_i : i < \omega\} \cup \{q_i : i < \omega\}, \ldots$ by forming its closure under $\circ$ only. This set will be denoted by $T_q, T_r, T_{pq}, \ldots$, repectively.

We let O_q be the closure under $\circ$ of $\{q_i^* : i < \omega\} \cup \{\dot{q}_i^* : i < \omega\}$. Also, O_p and O_e shall be the closure under $\circ$ of $\{p_i^* : i < \omega\}$ or $\{e_{i,j}^* : \{i,j\} \neq \{0\}\}$, respectively. Further, $O_{pq}, O_{pe}, O_{qe}, O_{pqe}$ shall be the closure under $\circ$ of the following, respectively: $O_p \cup O_q$, $O_p \cup O_e$, $O_q \cup O_e$, $O_p \cup O_q \cup O_e$. Each of these sets is a set of 1-ary operations on the power set $\{X : X \subseteq {}^{[0,\omega)}U\}$ of ${}^{[0,\omega)}U$ such that, for every operation in the set, its conjugate also is in the set.

We let O_r be the closure under $\circ$ of $\{r_i^* : i < \omega\} \cup \{\dot{r}_i^* : i < \omega\}$ and O_c the closure under $\circ$ of $\{c_i^* : i < \omega\}$. Also, for example, O_{pr} and O_{prc} shall be the closure under $\circ$ of $O_p \cup O_r$ or of $O_p \cup O_r \cup O_c$, respectively.

Since $R_1^* \circ R_2^* = (R_1 \circ R_2)^*$, there follows that every operation in one of the sets O just defined is relation-induced. Also, since the class of ranked sets is closed under each of the operations $q_i^*, \dot{q}_i^*, p_i^*, e_{ij}^*, r_i^*, \dot{r}_i^*, c_i^*$, therefore it is also closed under each operation that belongs to one of these sets O. Further, since the relations $q_i, \dot{q}_i, \ldots, c_i$ are logical or invariant, so are, in a sense that should be obvious, the operations $q_i^*, \dot{q}_i^*, \ldots, c_i^*$, and hence so are the operations in any of these sets O.

Consider any R_1 and R_2 in S_{pqe}. Evidently, $R_1^* = R_2^*$ implies that $R_1 = R_2$. Also, as we already noted, $(R_1 \circ R_2)^* = R_1^* \circ R_2^*$. There follows that * is an isomorphism between $\langle S_{pqe}, \circ\rangle$ and $\langle O_{pqe}, \circ\rangle$. Similar remarks apply to $S_q, S_{pq}, S_{pe}, \ldots$ and $O_q, O_{pq}, O_{pe}, \ldots$, respectively.

Henceforth, for brevity, we often let $q = q_0$, $\dot{q} = \dot{q}_0$, $r = r_0$, $\dot{r} = \dot{r}_0$, and $e = e_{0,1}$.

THEOREM 1. (a) $S_{pqe} = S_{qe}$ *and* $O_{pqe} = O_{qe}$.

(b) S_{pqe} *coincides with the closure under* $\circ$ *of* $\{p_i : i < \omega\} \cup \{q, q^\smile, e\}$ *and* O_{pqe} *coincides with the closure under* $\circ$ *of* $\{p_i^* : i < \omega\} \cup \{q^*, q^{\smile *}, e^*\}$.

Proof. Since * is an isomorphism, it suffices to prove the assertions about S_{pqe}.

Since $Do(q_i e_{i(i+2)} \breve{q}_{i+2}) = {}^{[i+2,\omega)}U$, therefore, $Do(q_i e_{i(i+2)} \breve{q}_{i+2}) = Do\ p_i$. Now let $x = vww'z$, where $|v| = i$, $|w| = |w'| = 1$, and $0 \le |z| < \omega$. Then $\langle x, y\rangle$ is in p_i if and only if $y = vw'wz$. Also, $\langle x, y\rangle$ is in q_i if and only if there is some w'' such that $|w''| = 1$ and $y = vw''ww'z$. Hence $\langle x, y\rangle$ is in $q_i e_{i(i+2)}$ if and only if $y = vw'ww'z$. Therefore $\langle x, y\rangle$ is in $q_i e_{i(i+2)} \breve{q}_{i+2}$ if and only if $y = \breve{q}_{i+2}(vw'ww'z) = vw'wz$. There follows that $p_i = q_i e_{i(i+2)} \breve{q}_{i+2}$.

This shows that $S_p \subseteq S_{qe}$. There follows that $S_{pqe} = S_{qe}$.

In order to prove part (b), we first consider the closure under $\circ$ of $\{p_i : i < \omega\} \cup \{q, q^{\smile}\}$. It includes S_q since $q_{i+1} = q_i p_i$, as one can verify, and since therefore also $\breve{q}_{i+1} = p_i \breve{q}_i$.

Next, we consider the closure under $\circ$ of $\{p_i : i < \omega\} \cup \{e\}$. As one can verify, $p_{(i,i+2)} = p_i \circ p_{i+1} \circ p_i$. One can also verify that $e_{i+1,i+2} = p_{(i,i+2)} e_{i,i+1} p_{(i,i+2)}$ and that $i < j$ implies that $e_{i,j+1} = p_j e_{i,j} p_j$. There follows that S_e is included in this closure and hence that S_{pe} coincides with it.

Part (b) now follows from the last two paragraphs. □

Theorem 1(b) will play a role in almost all of our subsequent development, whereas the role of Theorem 1(a) will be fairly minor. To emphasize that relations from each of the sets S_q, S_q, and S_e play a role, we will ususally use S_{pqe}, rather than S_{qe}, to refer to the set $S_{pqe} = S_{qe}$. Similarly, we will generally use O_{pqe} rather than O_{qe}.

Note that $q_{i+1} = r_i \circ c_{i+1}$ and ${}_{[1,\omega)}\overset{\circ}{1} \circ q_0 = r_0 \circ c_0$. Hence $\{q_{i+1} : i < \omega\} \subseteq S_{rc}$ and q_0 comes close to being in S_{rc}. However, the domain and the range of every relation in $\{p_i : i < \omega\}$, $\{r_i : i < \omega\}$, or $\{c_i : i < \omega\}$, and hence that of every relation in S_{prc}, is included in ${}^{[1,\omega)}U$. Hence there are relations in $S_{qe} = S_{pqe}$, such as the relations q_0, $\breve{q}_0$, and ${}_{[0,\omega)}\overset{\circ}{1}$, that are not in S_{prc} and hence not in S_{rc}.

The Boolean operations on the power set $\{X : X \subseteq {}^{[0,\omega)}U\}$ of ${}^{[0,\omega)}U = \{x : |x| < \omega\}$ that will mainly concern us are the following three: relative complementation $\underset{2}{-}$ which, for any pair $\langle X, Y\rangle$ of subsets of ${}^{[0,\omega)}U$ as argument, yields the value $X \underset{2}{-} Y = \{w : w \in X,\ w \notin Y\}$; binary union $\cup$; the set ${}^{[0,\omega)}U$ of all finite sequences, which will also be regarded as the 0-ary operation whose constant value is ${}^{[0,\omega)}U$. In terms of $\underset{2}{-}$, binary intersection can be defined as follows: $X \cap Y = X \underset{2}{-} (X \underset{2}{-} Y)$. Also, in terms of ${}^{[0,\omega)}U$ and $\underset{2}{-}$, unary complementation with respect to ${}^{[0,\omega)}U$ can be defined as follows: $\underset{1}{-}X = {}^{[0,\omega)}U \underset{2}{-} X$. When it is clear from the context whether the complementation concerned is unary or binary, we may sometimes omit the numeral that we use elsewhere to distinguish them.

Since $\emptyset$ is of any rank, therefore, if X is of rank s, then $X \,\bar{_2}\, Y$ is of rank s for any $Y \subseteq {}^{[0,\omega)}U$. Hence, as was mentioned earlier, the class of ranked sets is closed under $\bar{_2}$. It is evidently also closed under $\cap$. It does not contain ${}^{[0,\omega)}U$ and hence is not closed under it and, as we saw earlier, it is also not closed under $\bar{_1}$ or $\cup$.

We now turn to first-order definability that does not involve the use of individual parameters. Until further notice it will be taken as understood that this is the notion of first-order definability under discussion.

A ***relational structure*** shall be any $\langle U, X_\delta \rangle_{\delta<\eta}$ such that U is our given universe, η is an ordinal, and for each $\delta < \eta$ there is some $\rho(\delta) < \omega$ such that X_δ is of rank $\rho(\delta)$. It shall be ***nonvacuous*** if and only if there is at least one $\delta < \eta$ such that $X_\delta \neq \emptyset$.

Our first-order language pertaining to $\langle U, X_\delta \rangle_{\delta<\eta}$ shall be as follows: Its predicate symbols are a symbol $\approx$ for equality and, for each $\delta < \eta$, a symbol $\mathbf{P}_\delta$ that takes $\rho(\delta)$ arguments; there are no function symbols, individual constants, or variables other than those in the set $\{\mathbf{v}_i : i < \omega\}$ of individual variables; also, for convenience, the only logical connectives are $\wedge$, $\neg$, and $\exists$.

The formulas of the language shall be formed in the usual way. For any formula ψ, the ***rank associated with*** ψ shall be the least $s = \{0, \ldots, s-1\}$ such that $i \in s$ for each $\mathbf{v}_i$ that has a free occurrence in ψ.

Satisfaction shall be defined in the usual way. When $\langle U, X_\delta \rangle_{\delta<\eta}$ and the language pertaining to it are given, it is a binary relation that obtains between formulas of the language and the infinite sequences $w = \langle w_0, \ldots, w_n, \ldots \rangle$ that are in ${}^\omega U$. Now let ψ be any formula of the language and let s be the rank associated with ψ. Then, often omitting reference to $\langle U, X_\delta \rangle_{\delta<\eta}$ when understood, we shall say that ψ ***defines*** (***in*** $\langle U, X_\delta \rangle_{\delta<\eta}$) the following set Y of rank s:

$$\begin{aligned} Y &= \{y \in {}^sU : \ yz \text{ satisfies } \ \psi \text{ in } \langle U, X_\delta \rangle_{\delta<\eta} \text{ for some } z \text{ in } {}^\omega U\} \\ &= \{y \in {}^sU : \ yz \text{ satisfies } \ \psi \text{ in } \langle U, X_\delta \rangle_{\delta<\eta} \text{ for each } z \text{ in } {}^\omega U\} \end{aligned}$$

Given $\langle U, X_\delta \rangle_{\delta<\eta}$ and our language pertaining to it, first-order definability in $\langle U, X_\delta \rangle_{\delta<\eta}$ can then be defined as follows:

> A set Y is ***first-order definable*** in $\langle U, X_\delta \rangle_{\delta<\eta}$ if and only if some formula ψ defines Y in $\langle U, X_\delta \rangle_{\delta<\eta}$

THEOREM 2. *(a) Y is first-order definable in $\langle U, X_\delta\rangle_{\delta<\eta}$ if and only if Y is in the closure of $\{X_\delta : \delta < \eta\}$ under $O_{pqe} \cup \{\{\emptyset\}, \bar{2}\}$.*

(b) Assume that $\langle U, X_\delta\rangle_{\delta<\eta}$ is nonvacuous. Then $\{\emptyset\}$ is in the closure of $\{X_\delta : \delta < \eta\}$ under O_q and hence Y is first-order definable in $\langle U, X_\delta\rangle_{\delta<\eta}$ if and only if Y is in the closure of $\{X_\delta : \delta < \eta\}$ under $O_{pqe} \cup \{\bar{2}\}$.

Proof. (a) To prove that each Y in the closure concerned is first-order definable one uses induction. If $Y = \{\emptyset\}$, then Y is defined by the sentence $\exists \mathbf{v}_0(\mathbf{v}_0 \approx \mathbf{v}_0)$. If $Y = X_\delta$ where $\delta < \eta$, then Y is defined by the atomic formula $\mathbf{P}_\delta \mathbf{v}_0 \cdots \mathbf{v}_{\rho(\delta)-1}$. Now let Y and Y' be in this closure and assume as inductive hypothesis that Y is defined by ψ and Y' by ψ'. Then, for some s and s', Y is of rank s and Y' is of rank s'. Without loss of generality we may also assume that $k > s$ for each $\mathbf{v}_k$ that has a bound occurrence in ψ. If Y and Y' are of the same rank, then $Y \bar{2} Y'$ is defined by $\psi \wedge \neg\psi'$. If Y and Y' are not of the same rank then $Y \bar{2} Y' = Y$ and hence is defined by ψ. If $i < s$, then $p_i^*(Y)$ is defined by the formula that results from ψ by interchanging everywhere $\mathbf{v}_i$ and $\mathbf{v}_{i+1}$. If $i \geq s$, then $p_i^*(Y) = \emptyset$ and hence is defined by $\neg\mathbf{v}_0 \approx \mathbf{v}_0$. If $i \leq s$, then $q_i^*(Y)$ is defined by $\mathbf{v}_i \approx \mathbf{v}_i \wedge \psi''$, where ψ'' results from ψ by simultaneously substituting $\mathbf{v}_{k+1}$ for $\mathbf{v}_k$ for each k, if any, such that $i \leq k < s$. If $i > s$, then $q_i^*(Y)$ is defined by $\neg\mathbf{v}_0 \approx \mathbf{v}_0$. If $i < s$, then $\dot{q}_i^*(Y)$ is defined by $\exists \mathbf{v}_s \psi'''$, where ψ''' results from ψ by simultaneously substituting $\mathbf{v}_s$ for $\mathbf{v}_i$ and $\mathbf{v}_{k-1}$ for $\mathbf{v}_k$ for each k, if any, such that $i < k < s$. If $i \geq s$, then $\dot{q}_0^*(Y) = \emptyset$ and hence is defined by $\neg\mathbf{v}_0 \approx \mathbf{v}_0$. Finally, if $\max(i,j) < s$, then $e_{ij}^*(Y)$ is defined by $\mathbf{v}_i \approx \mathbf{v}_j \wedge \psi$ and if $\max(i,j) \geq s$, then $e_{ij}^*(Y)$ is defined by $\neg\mathbf{v}_0 \approx \mathbf{v}_0$.

Now let A be the closure of $\{X_\delta : \delta < \eta\}$ under $O_{pqe} \cup \{\{\emptyset\}, \bar{2}\}$. Then, for every $s < \omega$, ${}^sU = q_0^{s*}\{\emptyset\}$ is in A. This fact will be used repeatedly in the proof, about to be given, that every Y that is first-order definable in $\langle U, S_\delta\rangle_{\delta<\eta}$ is in A. The proof uses induction on the length of formulas.

Let ψ be the formula $\mathbf{v}_i \approx \mathbf{v}_j$. Then ψ defines in $\langle U, X_\delta\rangle_{\delta<\eta}$ the set $Y = e_{ij}^*({}^sU)$, where $s = 1 + \max(i,j)$. Since ${}^sU$ is in A and A is closed under e_{ij}^*, therefore Y is in A.

Let $\gamma < \eta$ and let ψ be an atomic formula formed from $\mathbf{P}_\gamma$. Then there is a function λ with $Do\ \lambda = \rho(\gamma) = \{0, \ldots, \rho(\gamma) - 1\}$ and $Rg\ \lambda \subseteq \omega$ such that ψ is the formula $\mathbf{P}_\gamma \mathbf{v}_{\lambda(0)} \cdots \mathbf{v}_{\lambda(\rho(\gamma)-1)}$. In other words, ψ is formed from $\mathbf{P}_\gamma$ by appending that string of length $\rho(\gamma)$ of variables which, for $0 \leq m < \rho(\gamma)$, contains in place m the variable $\mathbf{v}_{\lambda(m)}$. Let s be the rank associated with ψ, so that $s = \{0, \ldots, s-1\}$ is the least number that includes $Rg\ \lambda$. Let

$$Z_\lambda = \{\lambda \circ y : y \in {}^sU\} = \{z : z \in {}^{\rho(\gamma)}U,\ z_m = z_n \text{ if } \lambda(m) = \lambda(n)\}\ .$$

Then ψ defines in $\langle U, X_\delta\rangle_{\delta<\eta}$ the following set Y of rank s:

$$\begin{aligned} Y &= \{\langle y_0, \ldots, y_{s-1}\rangle : \langle y_{\lambda(0)}, \ldots, y_{\lambda(\rho(\gamma)-1)}\rangle \in X_\gamma\} \\ &= \{y \in {}^sU : \lambda \circ y \in X_\gamma \cap Z_\lambda\} \ . \end{aligned}$$

We let W be the set of rank $s + \rho(\gamma)$ that is related to $X_\gamma \cap Z_\lambda$ as follows:

$$W = \{yz : y \in {}^sU, \ z = \lambda \circ y, \ z \in X_\gamma \cap Z_\lambda\}$$

It follows that

$$Y = \{y \in {}^sU, \ yz \in W \text{ for some } z\} = ((\dot{q}_s^*)^{\rho(\gamma)}(W) \ .$$

Since A is closed under $\dot{q}_s^*$, therefore to show that Y is in A it suffices to show that W is in A.

Since X_γ is in A and since A is closed under each e_{mn}^*, therefore $X_\gamma \cap Z_\lambda$ is in A. To show that W is in A, it therefore suffices to show that W can be obtained from $X_\gamma \cap Z_\lambda$ by operations belonging to O_{pqe}.

We shall only give an illustration, hoping that it will make apparent how in each particular case one may proceed. This will also give a concrete example of how ρ, λ, s, and Z_λ are related to ψ and of the fairly complex relationship of Y and of W to ψ and X_γ.

Let ψ be the formula $\mathbf{P}_\gamma \mathbf{v}_4 \mathbf{v}_2 \mathbf{v}_4$. Then $\rho(\gamma) = \{0, 1, 2\} = 3 = Do\ \lambda$. Also $\lambda(0) = 4$, $\lambda(1) = 2$, $\lambda(2) = 4$ and, furthermore, $s = 5$ and $Z_\lambda = \{z \in {}^3U : z_0 = z_2\}$. Assume now that x, x' and X_γ satisfy the conditions given below. Then $X_\gamma \cap Z_\lambda$, Y, and W satisfy the equalities

$$\begin{aligned} x &= \langle x_0, x_1, x_2\rangle \quad \text{and} \quad x_0 = x_2 \\ x' &= \langle x_0', x_1', x_2'\rangle \quad \text{and} \quad x_0' \neq x_2' \\ X_\gamma &= \{x, x'\} \\ X_\gamma \cap Z_\lambda &= \{x\} = \{\langle x_0, x_1, x_0\rangle\} \\ Y &= \{y \in {}^5U : \langle y_4, y_2, y_4\rangle \in X_\gamma \cap Z_\lambda\} \\ &= \{y \in {}^5U : \langle y_4, y_2, y_4\rangle = \langle x_0, x_1, x_0\rangle\} \\ &= \{\langle u, u', x_1, u'', x_0\rangle : u, u', u'' \in U\} \\ W &= \{yz : y \in {}^5U, \ z = \langle y_4, y_2, y_4\rangle = \langle x_0, x_1, x_0\rangle\} \\ &= \{\langle u, u', x_1, u'', x_0\rangle^\frown \langle x_0, x_1, x_0\rangle : u, u', u'' \in U\} \ . \end{aligned}$$

Letting $W_0 = X_\gamma \cap Z_\lambda = \{\langle x_0, x_1, x_0\rangle\}$ and $q = q_0$, we now form in succession the following four sets (where we could have used $(qe_{03})^*$, instead of $(qe_{01})^*$, in the first step):

$$\begin{aligned} W_1 &= (qe_{01})^*(W_0) = \{\langle x_0\rangle^\frown\langle x_0, x_1, x_0\rangle\} \ , \\ W_2 &= q^*(W_1) = \{\langle u'', x_0\rangle^\frown\langle x_0, x_1, x_0\rangle : u'' \in U\} \ , \\ W_3 &= (qe_{04})^*(W_2) = \{\langle x_1, u'', x_0\rangle^\frown\langle x_0, x_1, x_0\rangle : u'' \in U\} \ , \\ W_4 &= q^{*2}(W_3) = \{\langle u, u', x_1, u'', x_0\rangle^\frown\langle x_0, x_1, x_0\rangle : u, u', u'' \in U\} \end{aligned}$$

Since A is closed under q^* and under each e^*_{ij}, since W_0 is in A, and since $W = W_4$, therefore, for $X_\gamma = \{x, x'\}$, W is in A.

Now suppose that X_γ is such that $X_\gamma \cap Z_\lambda$ is not a singleton. Then for each x in $X_\gamma \cap Z_\lambda$ one forms $W_{0,x}, \ldots, W_{4,x}$ in the way $W_0, \ldots, W_4$ were formed above and now lets $W_i = \bigcup\{W_{i,x} : x \in X_\gamma \cap Z_\lambda\}$ for $i \leq 4$. Then each W_i is again in A and hence, in particular, W is in A.

Next, letting ψ and ψ' be any formulas of our first-order language for $\langle U, X_\delta\rangle_{\delta<\eta}$, we assume as inductive hypothesis that each of these defines in $\langle U, X_\delta\rangle_{\delta<\eta}$ a set that belongs to A. Let Y and Y', respectively, be the sets defined and let their respective ranks be s and s'. First, consider $\neg\psi$. The set it defines is ${}^sU \underset{2}{-} Y$. Since A contains ${}^sU$ and is closed under $\underset{2}{-}$ therefore ${}^sU \underset{2}{-} Y$ is in A. Next, consider $\psi \wedge \psi'$. Suppose that $s \geq s'$. Then the set defined by $\psi \wedge \psi'$ is $Y \cap (q^*_{s'})^{s-s'}(Y')$. Since A is closed under $q^*_{s'}$ and under $\cap$, therefore this set is in A. If $s \leq s'$ then, by a similar argument, the set defined by $\psi \wedge \psi'$ is again in A. Finally, given any i, consider $\exists \mathbf{v}_i\psi$. There are three cases to distinguish. If $s > i+1$, then $\exists \mathbf{v}_i\psi$ has s as its associated rank and defines in $\langle U, X_\delta\rangle_{\delta<\eta}$ the set $(q_i^{\smile} q_i)^*(Y) = q_i^*(q_i^{\smile *}(Y))$ of rank s, which is in A. If $s = i+1$ and r is the rank associated with $\exists \mathbf{v}_i\psi$, then $r \leq s$ and $\exists \mathbf{v}_i\psi$ defines in $\langle U, X_\delta\rangle_{\delta<\eta}$ the set $(q_r^{\smile *})^{s-r}(Y)$, which is in A. Finally, if $s \leq i$, then the initial quantifier in $\exists \mathbf{v}_i\psi$ is vacuous and $\exists \mathbf{v}_i\psi$ defines in $\langle U, X_\delta\rangle_{\delta<\eta}$ the set Y, which is in A. It now follows by induction that every formula of our first-order language defines in $\langle U, X_\delta\rangle_{\delta<\eta}$ a set that belongs to A.

(b) Assume that $\langle U, X_\delta\rangle_{\delta<\eta}$ is nonvacuous. Then $X_\delta \neq \emptyset$ for some $\delta < \eta$. Since X_δ is of rank $\rho(\delta)$, it follows that for each $m \leq \rho(\delta)$, $(\dot{q}_0^*)^m(X_\delta)$ is of rank $\rho(\delta) - m$ and is nonempty. In particular, $(q_0^{\smile *})^{\rho(\delta)}(X_\delta)$ is of rank 0 and is nonempty. Hence $(\dot{q}_0^*)^{\rho(\delta)}(X_\delta) = \{\emptyset\}$. It follows that $\{\emptyset\}$ is in the closure of $\{X_\delta\}_{\delta<\eta}$under O_q. The rest of Theorem 2(b) then follows from Theorem 2(a). □

A part of the proof of Theorem 2(a) shows that in the case of atomic formulas there is no need for use of Boolean operations. More precisely, it shows the following: If Y is first-order definable in $\langle U, X_\gamma\rangle$ by an atomic formula having $\mathbf{P}_\gamma$ as predicate symbol, then Y is in the closure of $\{X_\gamma\}$ under O_{qe}. This part of the proof was partly suggested by the semantical argument in [HT], p.86, to the effect that Y can be defined in $\langle U, X_\gamma\rangle$ by a formula in which the only atomic formula involving $\mathbf{P}_\gamma$ is $\mathbf{P}_\gamma \mathbf{v}_0 \mathbf{v}_1 \ldots \mathbf{v}_{\rho(\gamma)-1}$.

Neither for atomic formulas nor for formulas in general does our proof of Theorem 2(a) make use of the operations p_i^*. As the referee suggests, for atomic formulas a proof can be given that seems simpler than ours, makes use of Theorem 1 and thus allows the use of any operation in O_{pqe}, and proceeds along the following lines: One starts with the set X_γ denoted by $\mathbf{P}_\gamma$ and then, to obtain Y, applies, if needed, appropriate operations of the following kind, where $i < \omega$ and $j < \omega$: $q_i^*, e_{ij}^*, p_{(i,j)}^*$, and $\dot{q}_i^*$. The order in which one uses these is to some extent flexible.

If $\langle U, X_\delta\rangle_{\delta<\eta}$ is vacuous, then neither one of the conclusions of Theorem 2(b) holds. If $\langle U, X_\delta\rangle_{\delta<\eta}$ is vacuous, then $\{X_\delta : \delta < \eta\}$, the closure of $\{X_\delta : \delta < \eta\}$ under $O_{pqe} \cup \{\bar{2}\}$, and the closure of $\{X_\delta : \delta < \eta\}$ under O_q all coincide with $\{\emptyset\}$ and hence do not contain $\{\emptyset\}$ as an element. Now let ψ be a formula in which $\approx$ is the only predicate symbol. Then the set Y defined by ψ in $\langle U, X_\delta\rangle_{\delta<\eta}$ does not depend on $\{X_\delta : \delta < \eta\}$ and is the same, whether $\langle U, X_\delta\rangle_{\delta<\eta}$ is vacuous or not. For example, in every $\langle U, X_\delta\rangle_{\delta<\eta}$, $\exists\mathbf{v}_0(\mathbf{v}_0 \approx \mathbf{v}_0)$ defines $\{\emptyset\}$, $\mathbf{v}_i \approx \mathbf{v}_i$ defines ${}^{i+1}U$, and if $i \leq j$ then $\neg v_i \approx v_j$ defines $\{x \in {}^{j+1}U : x_i \neq x_j\}$. None of these sets is in $\{\emptyset\}$.

As we just saw, for each s there are nonempty sets of rank s that are first-order definable in $\langle U, X_\delta\rangle_{\delta<\eta}$. Now the set $\{\rho(\delta) : \delta < \eta\}$ may be any subset of ω and hence $\omega \cap -\{\rho(\delta) : \delta < \eta\}$ also may be any subset of ω. It follows that any set O of operations that can take the place of $O_{pqe} \cup \{\{\emptyset\}, \bar{2}\}$ in Theorem 2 must contain operations that raise rank and also operations that lower rank.

We now turn to the case where $\{\bar{2}, \cup\}$ is the set of Boolean operations that one may use. This leads from the class of ranked sets to the larger class of finite unions of ranked sets. To compare the two classes, let X be any finite union $X_0 \cup \cdots \cup X_n$ of ranked sets. For $m \leq n$ let X_m be of rank r_m. Let $s \geq r_m$ for each $m \leq n$. Then $X \subseteq \bigcup\{{}^{r_m}U : m \leq n\} \subseteq {}^{[0,s]}U$. Now suppose that $X \subseteq {}^{[0,s]}U$. For $r \leq s$ let $X_r = {}^rU \cap X$. Then X is the finite union $\bigcup\{X_r : r \leq s\}$ of pairwise disjoint ranked sets.

Let X be a finite union of ranked sets. Then, as we just saw, there is some s such that $X \subseteq {}^{[0,s]}U$ and hence such that any sequence x in X is of length $|x| \leq s$. We shall therefore say that X is ***length-bounded by*** s. If X is length-bounded by some $s < \omega$, then X shall be ***length-bounded***. Thus, the class of finite unions of ranked sets is the class of length-bounded sets.

The class of length-bounded sets is closed under each of the operations in $O_{pqe} \cup \{\{\emptyset\}, \bar{_2}, \cup\}$. That it is closed under $\{\emptyset\}$, $\bar{_2}$, and $\cup$ is evident. To see that it is closed under each of the operations in O_{pqe}, we consider any F in O_{pqe}, any $s < \omega$, and any $X \subseteq {}^{[0,s]}U$. For $r \leq s$, let $X_r = {}^rU \cap X$. Since the class of ranked sets is closed under F, therefore for each $r \leq s$ there is some t_r such that $F(X_r) \subseteq {}^{t_r}U$. Since F is relation-induced, therefore $F(Y \cup Y') = F(Y) \cup F(Y')$ for every Y and Y'. Hence $F(X) = \bigcup\{F(X_r) : r \leq s\} \subseteq \bigcup\{{}^{t_r}U : r \leq s\}$. Hence, if $t_r \leq t$ for each $r \leq s$, then $F(X) \subseteq {}^{[0,t]}U$.

Between those classes A of ranked sets that are closed under $O_{pqe} \cup \{\bar{_2}\}$ and those classes B of length-bounded sets that are closed under $O_{pqe} \cup \{\bar{_2}, \cup\}$, there holds the following simple one-one correspondence. Since $O_{pqe} = O_{qe}$, therefore similar remarks apply to O_{qe}.

THEOREM 3. (*a*) *Assume that A is a class of ranked sets that is closed under $O_{pqe} \cup \{\bar{_2}\}$. Let B be the closure of A under $\cup$. Then B is also closed under $O_{pqe} \cup \{\bar{_2}\}$ and the class of ranked sets in B is A. Furthermore,*

$$B = \bigcup_{s<\omega} \{X : X \subseteq {}^{[0,s]}U,\ {}^rU \cap X \in A \text{ for each } r \leq s\}\ .$$

(*b*) *Assume that B is a class of length-bounded sets that is closed under $O_{pqe} \cup \{\bar{_2}, \cup\}$. Let A be the class of ranked sets in B. Then A is closed under $O_{pqe} \cup \{\bar{_2}\}$ and the closure of A under $\cup$ is B.*

Proof. (a) The theorem holds trivially if $A = \emptyset$ or $A = \{\emptyset\}$. Assume therefore that $A \neq \emptyset$ and $A \neq \{\emptyset\}$. Then, as we saw in the proof of Theorem 2, A contains every ${}^rU$. Since A is closed under $\bar{_2}$ and hence under $\cap$, it follows that if X and Y are in A and are of the same rank, say r, then $X \cup Y = {}^rU \bar{_2}(({}^rU \bar{_2} X) \cap ({}^rU \bar{_2} Y))$ also is in A. Since every set X in B is a finite union $X_0 \cup \cdots \cup X_n$ of sets in A and since, if X is of rank r then each of the sets $X_0, \ldots, X_n$ is of rank r, it follows that the class of ranked sets in B is A. Now consider any X in B, whether ranked or not. There is some $s \geq 1$ such that $X = \bigcup_{r \leq s}({}^rU \cap X)$ and such that each ${}^rU \cap X$ is a finite union of members of rank r in A and therefore is itself in A. Conversely, if X satisfies these conditions, then it is in the closure B of A under $\cup$.

Consider any X and Y in B. There is some s such that $X = \bigcup_{r \leq s}({}^rU \cap X)$ and $Y = \bigcup_{r \leq s}({}^rU \cap Y)$. Then each of the sets ${}^rU \cap X$ and ${}^rU \cap Y$ is in A. Since A is closed under $\bar{_2}$, therefore each of the sets $({}^rU \cap X) \bar{_2} ({}^rU \cap Y)$ is in A and hence their union $X \bar{_2} Y$ is in B.

Consider any F in O_{pqe} and let X be as above. Since F is relation-induced, therefore $F(X) = F({}^0U \cap X) \cup \cdots \cup F({}^sU \cap X)$. Since A is closed under F, therefore each of the sets $F({}^0U \cap X), \ldots, F({}^sU \cap X)$ is in A. Hence their union $F(X)$ is in B.

(b) The class $\bigcup_{r<\omega}\{X : X \subseteq {}^rU\}$ of all sets that are ranked is closed under $O_{pqe} \cup \{\bar{_2}\}$. By assumption, so is the class B. Hence so is their intersection, namely A. It then follows from part (a) that the closure under $\cup$ of A is B. □

From $\{\bar{_2}, \cup\}$ as set of Boolean operations to be regarded as available we now turn to $\{{}^{[0,\omega)}U, \bar{_2}, \cup\}$. (Use of $\{\bar{_1}, \cup\}$ or of $\{{}^{[0,\omega)}U, \bar{_2}\}$ as Boolean primitives would make comparison with $\{\bar{_2}, \cup\}$ harder, although use of $\{{}^{[0,\omega)}U, \bar{_2}\}$ would make comparison with $\{\bar{_2}\}$ easier.) Changing to $O_{pqe} \cup \{{}^{[0,\omega)}U, \bar{_2}, \cup\}$ has the important advantage that Boolean and relation-induced operations suffice as primitives. In terms of these one can define $\{\emptyset\}$ as follows:

$$\{\emptyset\} \;=\; {}^{[0,\omega)}U \;\bar{_2}\; q_0^*({}^{[0,\omega)}U)\ .$$

As was remarked after the proof of Theorem 2, Theorem 2(b) relates classes closed under $O_{pqe} \cup \{\bar{_2}\}$ to first-order definability in relational structures that are nonvacuous but not to first-order definability in arbitrary relational structures. Similar remarks apply to the joint use of Theorem 2(b) and Theorem 3 for relating classes closed under $O_{pqe} \cup \{\bar{_2}, \cup\}$ to first-order definability. A change to classes C that are closed under $O_{pqe} \cup \{{}^{[0,\omega)}U,\ \bar{_2}, \cup\}$ and hence contain $\{\emptyset\}$ overcomes this limitation. Having $\{\emptyset\}$ available is also useful in another way which will be brought out in Theorem 6 below and the discussion that follows it.

A change of this kind also has disadvantages which may turn out to be important. As we shall see, in contrast to Theorem 3, each B gives rise to many C. Moreover, C cannot be obtained from B as simply as B is obtained from A in Theorem 3.

Let B be a class of length-bounded sets and let C be a class of subsets of ${}^{[0,\omega)}U$. Guided by analogy to the relationship of B to A in Theorem 3, we consider the following two conditions:

(1) The class of length-bounded sets in C is B.
(2) C is closed under $O_{pqe} \cup \{{}^{[0,\omega)}U,\ \bar{_2}, \cup\}$, and hence is the universe of an algebra whose operations are those in $O_{pqe} \cup \{{}^{[0,\omega)}U,\ \bar{_2}, \cup\}$.

An ***eligible extension of*** B shall be any C that satisfies (1) and (2).

THEOREM 4. *Assume that B is a class of length-bounded sets such that $B \neq \emptyset$, $B \neq \{\emptyset\}$, and B is closed under $O_{pqe} \cup \{\underset{2}{-}, \cup\}$.*

(a) *If $\{C_k : k \in K\}$ is a non-empty family of eligible extensions of B, then $\bigcap\{C_k : k \in K\}$ is an eligible extension of B.*

(b) *The least eligible extension of B is the closure C_0 of B under $O_{pqe} \cup \{{}^{[0,\omega)}U, \underset{2}{-}, \cup\}$ and the greatest eligible extension of B is the following set:*

$$C_1 = \{X : X \subseteq {}^{[0,\omega)}U, {}^{[0,s)}U \cap X \in B \textit{ for each } s < \omega\}\ .$$

Proof. (b) Let C be an eligible extension of B. Since $B \neq \emptyset$ and $B \neq \{\emptyset\}$ therefore, as we saw in the proof of Theorem 3, B contains every ${}^rU$ and hence every ${}^{[0,s]}U$. Since $B \subseteq C$, therefore C contains every ${}^{[0,s]}U$. Now consider any X in C. Since C is closed under $\underset{2}{-}$ and hence under $\cap$, therefore every ${}^{[0,s]}U \cap X$ is in C. Since every length-bounded set in C is also in B, therefore every ${}^{[0,s]}U \cap X$ is in B. It follows that X is in C_1. Hence $C \subseteq C_1$.

Let X be length-bounded. By the definition of C_1, if X is in C_1 then X is in B. Now let X be in B. Since B contains every ${}^{[0,s]}U$ and is closed under $\cap$, therefore ${}^{[0,s]}U \cap X$ is in B for every $s < \omega$. Hence X is in C_1. Thus, the class of length-bounded sets in C_1 is B.

Since ${}^{[0,s]}U \cap {}^{[0,\omega)}U$ is in B for every $s < \omega$, therefore ${}^{[0,\omega)}U$ is in C_1. Now consider any X and Y in C_1. Let $s < \omega$. Then ${}^{[0,s]}U \cap X$ and ${}^{[0,s]}U \cap Y$ are in B. Since B is closed under $\underset{2}{-}$ and $\cup$, therefore the sets ${}^{[0,s]}U \cap (X \underset{2}{-} Y) = ({}^{[0,s]}U \cap X) \underset{2}{-} ({}^{[0,s]}U \cap Y)$ and ${}^{[0,s]}U \cap (X \cup Y) = ({}^{[0,s]}U \cap X) \cup ({}^{[0,s]}U \cap Y)$ are in B. Since this holds for every $s < \omega$, therefore $X \underset{2}{-} Y$ and $X \cup Y$ are in C_1.

Let $i < \omega$ and $X \subseteq {}^{[0,\omega)}U$. As we saw earlier, from the definition of $\breve{q}_i{}^*$ there follows: If $r \leq i$ then $\breve{q}_i{}^*({}^rU \cap X) = \emptyset$ and if $r > i$ then $\breve{q}_i{}^*({}^rU \cap X) = {}^{r-1}U \cap \breve{q}_i{}^*X$. Now let $s < \omega$. Since $\breve{q}_i{}^*(Y \cup Y') = \breve{q}_i{}^*(Y) \cup \breve{q}_i{}^*(Y')$ it follows that ${}^{[0,s]}U \cap \breve{q}_i{}^*(X) = \breve{q}_i{}^*({}^{[0,s+1)}U \cap X)$. Now assume that X is in C_1. Then ${}^{[0,s+1)}U \cap X$ is in B. Since B is closed under $\breve{q}_i{}^*$, there follows that ${}^{[0,s]}U \cap \breve{q}_i{}^*(X)$ is in B. Since this holds for every $s < \omega$, therefore $\breve{q}_i{}^*(X)$ is in C_1.

Arguments that are fairly similar show that, for every i and j, if X is in C_1, then $q_i^*(X)$, $p_i^*(X)$, and $e_{ij}^*(X)$ are in C_1. It now follows that C_1 is closed under $O_{pqe} \cup \{{}^{[0,\omega)}U, \underset{2}{-}, \cup\}$ and that it is the greatest eligible extension of B.

Since $B \subseteq C_1$ and C_1 is closed under $O_{pqe} \cup \{{}^{[0,\omega)}U, \underset{2}{-}, \cup\}$, therefore $C_0 \subseteq C_1$. Since every length-bounded set in C_1 is in B and since $B \subseteq C_0$, therefore the class of length-bounded sets in C_0 is B. It now follows that C_0 is an eligible extension of B and hence that C_0 is the least eligible extension of B. □

Let B and C_1 be as in Theorem 4 and let J be any subset of ω. Then C_1 contains, for example, the set $\bigcup_{s\in J} e^*_{01}({}^sU) \cup \bigcup_{s\notin J}({}^sU \;\bar{_2}\; e^*_{01}({}^sU))$. It follows that if $\overline{\overline{U}} \geq 2$ then C_1 is non-denumerable.

Let B and C_0 be as in Theorem 4. In contrast to Theorem 3, where B is the closure of A under $\cup$ only, forming C_0 from B involves operations from both O_{pqe} and $\{{}^{[0,\omega)}U,\ \bar{_2}, \cup\}$. For example, operations from both sets are involved in forming $(q_0^{i+3})^*({}^{[0,\omega)}U) \;\bar{_2}\; e^*_{i(i+1)}({}^{[0,\omega)}U)$.

Thanks to gentle reminders by Philip Scowcroft and Elisabeth Bouscaren we now turn briefly to first-order definability with the aid of individual parameters. Thus, in the language of [MMT], pp. 142-145, from a notion of definability that involves a clone of term operations we now move to one that involves a clone of polynomial operations.

Henceforth, to have convenient names for the members of our universe U, we let κ be the cardinality of U and let $U = \bigcup\{u_\lambda \colon \lambda < \kappa\}$.

For any $\lambda < \kappa$, we let ι_λ be the subidentity on ${}^{[0,\omega)}U$ that is given below. It follows that $\{\iota_\lambda : \lambda < \kappa\}$ is a partition of ${}_{[1,\omega)}\overset{\circ}{1}$ in which every block ι_λ is of the same cardinality;

$$\iota_\lambda = \{\langle x, x\rangle : 1 \leq |x| < \omega,\ x_0 = u_\lambda\} = \{\langle\langle u_\lambda\rangle^\frown z, \langle u_\lambda\rangle^\frown z\rangle : |z| < \omega\}\ .$$

In contrast to the relations e_{ij}, $p_{(i,j)}$, $\dot{q}_i$, and q_i, if $\kappa \geq 2$ then ι_λ is not invariant. Let f be any permutation of U and, for every $x = \langle x_0, \ldots, x_{n-1}\}$ in ${}^{[0,\omega)}U$, let $f^{\#}(\langle x_0, \ldots, x_{n-1}\rangle) = \langle f(x_0), \ldots, f(x_{n-1})\rangle$. Then $f(u_\lambda) \neq u_\lambda$ implies that ι_λ and $\{\langle f^{\#}(x), f^{\#}(x)\rangle : \langle x, x\rangle \in \iota_\lambda\}$ are, in fact, disjoint.

For any $\lambda < \kappa$ and any $j < \omega$ we let $\iota_{(j/\lambda)} = p_{(0,j)} \circ \iota_\lambda \circ p_{(0,j)}$. Hence $\iota_{(j/\lambda)}$ is the following subidentity on ${}^{[0,\omega)}U$ and $\iota^*_{(j/\lambda)}$ is the operation on $\{X : X \subseteq {}^{[0,\omega)}U\}$ that is given underneath:

$$\begin{aligned} \iota_{(j/\lambda)} &= \{\langle x, x\rangle : j < |x| < \omega,\ x_j = u_\lambda\}\ , \\ \iota^*_{(j/\lambda)}(X) &= X \cap \{x : j < |x| < \omega,\ x_j = u_\lambda\}\ . \end{aligned}$$

We let O_i be the closure under $\circ$ of $\{\iota^*_{(j/\lambda)} : \lambda < \kappa,\ j < \omega\}$. Also, $O_{ei}, \ldots, O_{pqei}$ shall be the closure under $\circ$ of $O_e \cup O_i, \ldots, O_{pqe} \cup O_i$, respectively. Similarly, $S_i, S_{ei}, \ldots, S_{pqei}$ shall be the closure under $\circ$ of $\{\iota_{(j/\lambda)} : \lambda < \kappa,\ j < \omega\}$, of $S_e \cup S_i, \ldots,$ of $S_{pqe} \cup S_i$, respectively.

For any $\lambda < \kappa$ and any $j < \omega$, we let $s_{\lambda,j}$ be the following one-one mapping of ${}^{[j,\omega)}U$ into ${}^{[j+1,\omega)}U$, and call it ***insertion of*** u_λ ***at place*** j:

$$s_{\lambda,j} = \{\langle xz, x^\frown\langle u_\lambda\rangle^\frown z\rangle : |x| = j,\ |z| < \omega\}\ .$$

Its converse $\dot{s}_{\lambda,j} = s_{\lambda,j}{}^{\smile}$ shall be ***excision of*** u_λ ***at place*** j. It is the function that results from $\dot{q}_j = q_j{}^{\smile}$ by restricting it to $Rg\ \iota_{(j/\lambda)}$. Note that the following hold for any $\lambda < \kappa$ and any $j < \omega$:

$$\begin{aligned} s_{\lambda,j} &= q_j \circ \iota_{(j/\lambda)} \\ \iota_{(j/\lambda)} &= s_{\lambda,j}^{\smile} \circ s_{\lambda,j} \ . \end{aligned}$$

We let S_s be the closure under $\circ$ and $^{\smile}$ of $\{s_{\lambda,j} : \lambda < \kappa,\ j < \omega\}$. Also, $S_{qs}, S_{qes}, \ldots$ shall be the closure under $\circ$ and $^{\smile}$ of $S_q \cup S_s$, $S_q \cup S_e \cup S_s, \ldots$. Further, we let O_s be the closure under $\circ$ of $\{s^*_{\lambda,j} : \lambda < \kappa,\ j < \omega\} \cup \{(s_{\lambda,j}^{\smile})^* : \lambda < \kappa,\ j < i\}$, and we let $O_{qs}, O_{qes}, \ldots$ be the closure under $\circ$ of $O_q \cup O_s$, $O_q \cup O_e \cup O_s, \ldots$. There follows that $S_i \subseteq S_s$ and $S_s \subseteq S_{qi}$. Hence $S_{qi} = S_{qs}$. From Theorem 1(a) there then follows that the following sets coincide: S_{pqei}, S_{qei}, S_{pqes}, S_{qes}. The corresponding sets $O_{pqei}, \ldots, O_{qes}$ therefore also coincide.

There will only be some occasions later on where we shall deal with O_{pqei} or S_{pqei}. Specifically, we shall see in Chapter II that every relation in S_{pqei} is uniform. Also, while not every relation in S_{pqei} belongs to the class of diagonal relations as defined in Chapter III, a fairly natural generalization yields a class that does include S_{pqei}. Thus it is likely that notions and methods related to those being used for dealing with S_{pqe} will be useful for dealing with S_{pqei}.

First-order definability with the aid of individual parameters can be defined by proceeding as follows. Given any relational structure $\langle U, X_\delta\rangle_{\delta<\eta}$ whose universe is U, one forms the structure $\langle U, X_\delta, u_\lambda\rangle_{\delta<\eta,\lambda<\kappa}$ with distinguished elements and augments the first-order language accordingly, adding distinct individual constants $\mathbf{c}_\lambda$ as names for the elements u_λ of U. For this augmented language one defines in the usual way the notion of satisfaction in $\langle U, X_\delta, u_\lambda\rangle_{\delta<\eta,\lambda<\kappa}$. If ψ is a formula of the augmented language and s is the rank associated with ψ then ψ shall ***define in*** $\langle U, X_\delta, u_\lambda\rangle_{\delta<\eta,\lambda<\kappa}$ the following set Y of rank s:

$$Y = \{y \in {}^sU : yz \text{ satisfies } \psi \text{ in } \langle U, X_\delta, u_\lambda\rangle_{\delta<\eta,\lambda<\kappa} \text{ for some } z \text{ in } {}^{[\omega]}U\}\,.$$

A set Y shall be ***first-order definable in*** $\langle U, X_\delta\rangle_{\delta<\eta}$ ***with the aid of individual parameters*** if and only if some formula ψ of the augmented language defines Y in $\langle U, X_\delta, u_\lambda\rangle_{\delta<\eta,\lambda<\kappa}$.

By Theorem 1(a) and since $S_{qi} = S_{qs}$, therefore the following theorem also holds when O_{pqei} is replaced by one of the following: O_{pqes}, O_{qei}, O_{qes}.

THEOREM 5. *Y is first-order definable in $\langle U, X_\delta\rangle_{\delta<\eta}$ with the aid of individual parameters if and only if Y is in the closure of $\{X_\delta : \delta < \eta\}$ under $O_{pqei} \cup \{\{\emptyset\}, \bar{}_2\}$.*

If $\langle U, X_\delta \rangle_{\delta<\eta}$ *is non-vacuous, then this closure is also the closure of* $\{X_\delta : \delta < \eta\}$ *under* $O_{pqei} \cup \{\bar{\underset{2}{}}\}$.

Proof. Let A be the closure of $\{X_\delta\rangle_{\delta<\eta}\}$ under $O_{pqei} \cup \{\{\emptyset\}, \bar{\underset{2}{}}\}$. The proof that for every Y in A there is some formula ψ of the augmented language that defines Y in $\langle U, X_\delta, u_\lambda \rangle_{\delta<\eta,\lambda<\kappa}$ is similar to the proof given for Theorem 2, except that for the inductive step one must now also consider the case where $Y = \iota^*_{(j/\lambda)}(Y')$ and Y' is defined in $\langle U, X_\delta, u_\lambda \rangle_{\delta<\eta,\lambda<\kappa}$ by a formula ψ'. In that case Y is defined in $\langle U, X_\delta, u_\lambda \rangle_{\delta<\eta,\lambda<\kappa}$ by the formula $\mathbf{v}_j \approx \mathbf{c}_\lambda \wedge \psi'$.

Let ψ be any formula of the augmented language and let Y be the set defined by ψ in $\langle U, X_\delta, u_\lambda \rangle_{\delta<\eta,\lambda<\kappa}$. If there are no individual constants in ψ then, by Theorem 2, Y is in A. Now assume that the individual constants that have at least one occurrence in ψ are $\mathbf{c}_{\lambda_0}, \ldots, \mathbf{c}_{\lambda_{n-1}}$, where $n \geq 1$. Let s be the rank associated with ψ. Without loss of generality we may assume that none of the variables $\mathbf{v}_s, \ldots, \mathbf{v}_{s+n-1}$ occurs in ψ. Let χ result from ψ by replacing every occurrence of $\mathbf{c}_{\lambda_0}, \ldots, \mathbf{c}_{\lambda_{n-1}}$ by an occurrence of $\mathbf{v}_s, \ldots, \mathbf{v}_{s+n-1}$, respectively, and let W be the set defined in $\langle U, X_\delta, u_\lambda \rangle_{\delta<\eta,\lambda<\kappa}$ and hence also in $\langle U, X_\delta \rangle_{\delta<\eta}$, by χ. Then $y \in Y$ if and only if $y^\frown \langle u_{\lambda_0}, \ldots, u_{\lambda_{n-1}} \rangle \in W$.

Let $W' = (\iota^*_{(s/\lambda_0)} \circ \cdots \circ \iota^*_{(s+n-1/\lambda_{n-1})})(W)$. Then W' is the set of those sequences $y^\frown y'$ in W such that $|y| = s$ and $y' = \langle u_{\lambda_0}, \ldots, u_{\lambda_{n-1}} \rangle$. Hence W' is also the set of those sequences $y^\frown \langle u_{\lambda_0}, \ldots, \lambda_{n-1} \rangle$ such that $y \in Y$. There follows that $Y = ((q_s^\smile)^n)^*(W')$.

Since χ contains no individual constant, therefore, by Theorem 2, W is in A. Since A is closed under each $\iota^*_{(j/\lambda)}$, therefore W' also is in A. Since $Y = ((q_s^\smile)^n)^*(W')$, there follows that Y is in A. □

The proof just given, together with a part of the proof of Theorem 2(a), shows that in the atomic case there is no need for Boolean operations and that $O_{qei} = O_{pqei}$ is sufficient. More precisely, if Y is defined in $\langle U, X_\delta, u_\lambda \rangle_{\delta<\eta,\lambda<\kappa}$ by a formula ψ of the augmented language such that ψ is atomic and has $\mathbf{P}_\gamma$ as predicate symbol, then Y is in the closure of $\{X_\gamma\}$ under O_{qei}.

For O_{pqei}, and for the sets that coincide with it, there hold analogues of Theorem 3 and of Theorem 4. Their proof is similar to the proof of these theorems. We omit details.

In the language of [MMT], pp. 142-144, O_{pqe} is a ***clone*** of unary operations on $\{X : X \subseteq {}^{[0,\omega)}U\}$ and different, increasingly more comprehensive, clones of n-ary operations, $0 \leq n < \omega$, on $\{X : X \subseteq {}^{[0,\omega)}U\}$ are involved in Theorems 2,3,4 respectively. For the clone involved in Theorem 4, the next theorem gives three sets of generators.

THEOREM 6. *The following three sets generate the same clone of operations on the power set* $\{X\colon X \subseteq {}^{[0,\omega)}U\}$.

$$\begin{aligned}
G_1 &= \{{}^{[0,\omega)}U, {}_{\overline{2}}, \cup\} \cup O_{pqe}.\\
G_2 &= \{{}^{[0,\omega)}U, {}_{\overline{2}}, \cup\} \cup \{p_i^*\colon i<\omega\} \cup \{q_0^*, \dot{q}_0^*\} \cup \{e_{0,1}^*\}.\\
G_3 &= \{{}^{[0,\omega)}U, {}_{\overline{2}}, \cup\} \cup \{b_i^*\colon i<\omega\} \cup \{q_0^*, \dot{q}_0^*\} \cup \{d_{0,1}\},
\end{aligned}$$

where $d_{0,1}$ *is the diagonal set* $\{x\colon 2 \le |x| < \omega,\ x_0 = x_1\}$, b_i *is the equivalence relation* ${}_{[0,i)}\overset{\circ}{1} \cup (q_i{}^{\smile} \circ q_i)$ *on* ${}^{[0,\omega)}U$, *and* $b_i^*(X) = \{y\colon$ *for some* $x \in X, xb_iy\} = {}_{[0,i)}\overset{\circ}{1}{}^*(X) \cup q_i^*(\dot{q}_i^*(X))$.

Proof. For $1 \le k \le 3$, let O_k be the clone of operations that is generated by G_k. From Theorem 1(b) there follows that $O_1 = O_2$. Since $d_{01} = e_{01}^*({}^{[0,\omega)}U)$, therefore d_{01} is in O_2. Let $0 \le i < \omega$ and let $(q_0^*)^i$ be the i-th power of q_0^*. Then $(q_o^i)^*$, $(q_0^i)^*{}^{[0,\omega)}U$, and ${}^{[0,\omega)}U \;{}_{\overline{2}}\; (q_0^i)^*({}^{[0,\omega)}U) = {}^{[0,i]}U$ are in O_2. Hence so is the operation ${}_{[0,i]}\overset{\circ}{1}{}^*$, since ${}_{[0,i]}\overset{\circ}{1}{}^*(X) = {}^{[0,i]}U \cap X$. Also, since for every $i < \omega$, $q_{i+1} = q_i p_i$ and hence, for every $X \subseteq {}^{[0,\omega)}U$, $q_{i+1}^*(X) = p_i^*(q_i^*(X))$ and $\dot{q}_{i+1}^*(X) = \dot{q}_i^*(p_i^*(X))$, therefore, by induction, every q_i^* and every $\dot{q}_i^*$ is in O_2. There now follows that every b_i^* is in O_2. Thus $O_3 \subseteq O_2$.

Since d_{01} and $(q_0^i)^*$ are in O_3, therefore so is $d_{i,i+1} = (q_0^i)^* d_{0,1}$, $0 \le i < \omega$. Since $\cap$ is in O_3 and since $e_{i,i+1}^*(X) = d_{i,i+1} \cap X$, therefore $e_{i,i+1}^*$ is in O_3, $0 \le i < \omega$.

Now consider any $i < \omega$. One can verify that the following hold:

$$\begin{aligned}
q_i \circ e_{i,i+1} &= \{\langle vwz, vwwz\rangle\colon |v| = i,\ |w| = 1,\ 0 \le |z| < \omega\}.\\
q_i \circ e_{i,i+1} \circ b_{i+1} &= \{\langle vwz, vwyz\rangle\colon |v| = i,\ |w|{=}|y|{=}1,\ 0 \le |z| < \omega\}\\
&= q_{i+1}.
\end{aligned}$$

There follows that $q_{i+1}^*(X) = b_{i+1}^*(e_{i,i+1}^*(q_i^*(X)))$. Since q_0^* and every b_i^* and $e_{i,i+1}^*$ are in O_3, there follows by induction that every q_i^* is in O_3. Since $q_i{}^{\smile} = \dot{q}_i$, $e_{i,i+1}{}^{\smile} = e_{i,i+1}$, and $b_{i+1}{}^{\smile} = b_{i+1}$, therefore $\dot{q}_{i+1} = q_{i+1}{}^{\smile} = (q_i \circ e_{i,i+1} \circ b_{i,i+1})^{\smile} = b_{i,i+1} \circ e_{i,i+1} \circ \dot{q}_i$, and hence $\dot{q}_{i+1}^*(X) = \dot{q}_i^*(e_{i,i+1}^*(b_{i+1}^*(X)))$. Since $\dot{q}_0^*$ is in O_3, there follows by induction that every q_i^* is in O_3.

If $0 \le i < j < \omega$, then $d_{i,j+1} = b_j^*(d_{i,j} \cap d_{j,j+1})$. As we saw earlier, every $d_{j,j+1}$ is in O_3. There follows by induction that every $d_{i,j}$ such that $0 \le i < j < \omega$ is in O_3. Hence, so is every $e_{i,j}^*$, $0 \le i < j < \omega$. As we saw in the proof of Theorem 1(a),

$p_i = q_i e_{i(i+2)} q_{i+1}{}^{\smile}$. There follows that $p_i^*(X) = q_{i+1}^*(e_{i,i+2}^*(q_i^*(X)))$, $0 \le i < \omega$. Thus, every p_i^* is in O_3. There now follows that $O_2 \subseteq O_3$. $\square$

The operations $b_i^*, q_0^*, \dot{q}_0^*, d_{01}$ are the same as the operations C_i, Q, P, D respectively, on p. 27 of [Cr74a]. Hence by Chapter 4 of [Cr74a], the set E_{pq} of equalities given there on pp. 46,47, with its symbols $\mathbf{c}_i, \mathbf{q}, \mathbf{p}, \mathbf{d}$ interpreted as denoting $b_i^*, q_0^*, q_0{}^{\smile *}, d_{01}$, respectively, is a sound and complete axiomatization of those equalities involving the operations in O_3 that are valid. This axiomatization, perhaps together with concepts and results from the theory of cylindric algebras, should be helpful for finding a sound and complete axiomatization in which the operations in G_2 serve as primitives. An axiomatization based on G_2 may provide a different perspective and perhaps new insights.

Each of Theorems 2(b), 3 and 4 contains suggestions for an algebraic treatment of first-order logic with equality (and with no individual constants). By the remarks just made, those contained in Theorem 4 are perhaps the first one should try to carry out. Relevant to those contained in Theorem 3 and perhaps also to those contained in Theorem 2(b) is recent work in the theory of Boolean or distributive lattices with operators. Some references are given in [DGP], [GJ], and [Go], but apparently there is much further activity in this area. Theorems 2(b), 3, and 4 differ only with respect to the Boolean operations they involve. For each, the remaining set of operations is $O_{pqe} = O_{qe}$. Thus for the three suggested algebraic treatments of first-order logic there is a common area of investigation. It is the subject of the chapters that follow.

CHAPTER II

Uniform Relations Between Sequences

Ultimately, one should like to understand how the Boolean operations and the other operations discussed in Chapter I interact. As a step towards it, we will now begin to investigate how the operations in $O_{pqe} = O_{qe}$ interact with each other. Because the structures involved are isomorphic, we will concentrate on the corresponding questions concerning the relations in $S_{pqe} = S_{qe}$. This has the advantage that the objects we will consider are more elementary.

Each relation R in S_{pqe} or in S_{pqei} is a relation between sequences of finite length. With each R we will associate a relation $\overset{\omega}{\rightarrow} R$ between sequences of infinite length whose properties are in many respects similar to those of R. An investigation of either of the two kinds of relations throws light on the other kind. In the following chapters we will therefore investigate them both.

In the present chapter we will deal with one condition on relations between sequences and also with algebraic or relational structures that result from this condition. We will show that the structures that result when the sequences are finite and those that result when the sequences are infinite are quite closely related. In Chapter III, we will deal with another condition on relations between sequences. In Chapter IV, we will see that the relations in S_{pqe} are exactly those relations between finite sequences that satisfy both of these conditions. Each of these two conditions gives rise to a set that is much more comprehensive than their intersection. A discussion of these two sets that does them some justice requires more than what is strictly needed for discussing their intersection. Moreover, only parts of Chapter IV play a direct role in the presentations given in the chapters that follow it. A reader mainly interested in these presentations may therefore wish to read Chapters II, III and IV impressionistically, mainly paying attention to their bearing on normal forms, without worrying too much about details.

The condition to be discussed in the present chapter is, in the case of infinite sequences, an analogue that applies to binary relations between sequences of a condition that applies to unary sets of these. This condition on sets of sequences arises naturally in connection with the notion of satisfaction and has played a role in algebraic treatments of first-order logic. One way of arriving at it is through the operations $\xrightarrow[(1)]{\gamma}$ on sets of sequences that will now be introduced.

Let $0 \leq \gamma \leq \omega$ and let $Y \subseteq {}^{[0,\omega)}U = \bigcup_{n<\omega} {}^nU$ be any set of finite sequences. Then the set $\xrightarrow[(1)]{\gamma} Y$ given below shall be the γ-***prolongation of*** Y. Evidently, for every $m < \omega$, $\xrightarrow[(1)]{m} Y$ is a subset of ${}^{[0,\omega)}U$, while $\xrightarrow[(1)]{\omega} Y$ is a subset of ${}^\omega U$.

$$\xrightarrow[(1)]{\gamma} Y \ = \ \{yz : y \in Y,\ |z| = \gamma\}\ .$$

Assume that Y is of rank s, so that $Y \subseteq {}^sU = \{y : |y| = s\}$. Let $X = \xrightarrow[(1)]{\gamma} Y$. If $\gamma = m < \omega$, then X is a set of rank $s + m$ of a certain kind. If $\gamma = \omega$, then X is a subset of ${}^\omega U$ of a certain kind. In either case, the condition that $X = \xrightarrow[(1)]{\gamma} Y$ for some Y of rank s is equivalent to the condition that $X \subseteq {}^{s+m}U$ or $X \subseteq {}^\omega U$, respectively, and that the following holds:

For any y, z and z', if $|y| = s$ and $|z| = |z'| = \gamma$,
then $yz \in X$ if and only if $yz' \in X$.

If $\gamma = \omega$, then Halmos in [Ha] expresses the above condition by saying that $s = \{0, \ldots, s-1\}$ ***supports*** X. Related conditions play a role in [HMT I]. By letting s vary, one is thus led to the class or kind of subsets X of ${}^\omega U$ that satisfy either one of the following two equivalent conditions:

For some $s < \omega$, X is supported by s.
For some $s < \omega$, $X = \xrightarrow[(1)]{\omega} Y$ for some Y of rank s.

In working with the notion of satisfaction, one often encounters ω-prolongation of sets Y or related operations. For example, in Chapter I, in dealing with first-order definability in a relational structure $\langle U, X_\delta \rangle_{\delta<\eta}$, we, so to speak, inverted ω-prolongation. Given a formula ψ with associated rank s, we first considered the set X given below and then formed the set Y given underneath. This set X is supported by s.

$$\begin{aligned} X &= \{yz : |y| = s,\ |z| = \omega,\ yz \text{ satisfies } \psi \text{ in } \langle U, X_\delta \rangle_{\delta<\eta}\}\ . \\ Y &= \{y : |y| = s,\ yz \in X \text{ for each } z \text{ in } {}^\omega U\}\ . \end{aligned}$$

From the operation $\xrightarrow[(1)]{\gamma}$ on sets of finite sequences we now turn to an analogous operation on relations between finite sequences. Let $0 \leq \gamma \leq \omega$ and let X be any relation on ${}^{[0,\omega)}U$, so that X is a subset of $\{\langle x,y\rangle : |x|<\omega,\ |y|<\omega\}$. Then the set $\overset{\gamma}{\to} X$ of pairs of sequences that is given below shall be the ***γ-prolongation of*** X. Evidently, for every $m<\omega$, $\overset{m}{\to} X$ is a relation on ${}^{[0,\omega)}U$, while $\overset{\omega}{\to} X$ is a relation on ${}^{\omega}U$.

$$\overset{\gamma}{\to} X \ = \ \underset{(2)}{\overset{\gamma}{\longrightarrow}} X \ = \ \{\langle xz,yz\rangle : \langle x,y\rangle \in X,\ |z|=\gamma\}\ .$$

From this definition of $\overset{m}{\to}$ there follows that $\overset{0}{\to} X = X$ and $\overset{m+1}{\longrightarrow} X = \overset{1}{\to}(\overset{m}{\to} X)$ for every relation X on ${}^{[0,\omega)}U$ and every $m<\omega$, and hence that $\overset{m}{\longrightarrow}$ can be defined recursively in terms of $\overset{1}{\to}$. This fact, and some simple consequences of it, will be used tacitly.

It is important not to confuse $\overset{\gamma}{\to} = \underset{(2)}{\overset{\gamma}{\longrightarrow}}$ with the one-many prolongation operation $\underset{((2))}{\overset{\gamma}{\longrightarrow}}$ which is given below and which also is a natural 2-dimensional analogue of $\underset{(1)}{\overset{\gamma}{\longrightarrow}}$, since, in a certain sense, $\underset{(2)}{\overset{\gamma}{\longrightarrow}}$, also preserves support. Unlike $\overset{\gamma}{\to}$, $\underset{((2))}{\overset{\gamma}{\longrightarrow}}$ will be of no use to us. Among reasons for this is the fact that whenever $\overline{\overline{U}} \geq 2$, $\gamma \neq 0$ and $X \neq \emptyset$ then, even when X is single-valued, $\underset{((2))}{\overset{\gamma}{\longrightarrow}} X$ is not.

$$\underset{((2))}{\overset{\gamma}{\longrightarrow}} X \ = \{\langle xz,yz'\rangle : \langle x,y\rangle \in X,\ |z|=|z'|=\gamma\}\ .$$

Given any relation X on ${}^{[0,\omega)}U$, since every $\overset{m}{\to} X$ is also a relation on ${}^{[0,\omega)}U$, therefore so is the union $\bigcup\{\overset{m}{\to} X : m<\omega\}$. We let $\overset{[0,\omega)}{\longrightarrow}$ be the operation on $\{X : X \subseteq {}^{[0,\omega)}U \times {}^{[0,\omega)}U\}$ which to every argument X assigns this union. Thus

$$\overset{[0,\omega)}{\longrightarrow} X = \bigcup\{\overset{m}{\to} X : m<\omega\} = \{\langle xz,yz\rangle : \langle x,y\rangle \in X,\ |z|<\omega\}\ .$$

LEMMA 1. *The following hold for any relations X and Y on ${}^{[0,\omega)}U$.*

(a) $\overset{[0,\omega)}{\longrightarrow} X$ *coincides with the closure of X under* $\overset{1}{\to}$. *Also,* $\overset{[0,\omega)}{\longrightarrow} X = \bigcup\{\overset{[0,\omega)}{\longrightarrow} X' : X' \subseteq X,\ \overline{\overline{X'}} = 1\}$.

(b) *For any n,* $\overset{\omega}{\to}(\overset{n}{\to} X) = \overset{\omega}{\to} X$. *Hence* $\overset{\omega}{\to}(\overset{[0,\omega)}{\longrightarrow} X) = \overset{\omega}{\to} X$.

(c) $\overset{n}{\to}(X \circ Y) = (\overset{n}{\to} X)\circ(\overset{n}{\to} Y)$, *for any* $n \geq 0$.

Proof. (a) Let W be the closure of X under $\overset{1}{\to}$. For any $m<\omega$ there holds $\overset{m+1}{\longrightarrow} X = \overset{1}{\to}(\overset{m}{\to} X)$. There follows by induction that every $\overset{m}{\to} X$ is included in W. Hence $\overset{[0,\omega)}{\longrightarrow} X = \bigcup\{\overset{m}{\to} X : m<\omega\}$ is included in W. Now consider any $\langle v,w\rangle$ in W. Then either $\langle v,w\rangle$ is in X or $\langle v,w\rangle = \langle v'z', w'z'\rangle$ for some v',w',z' such that $\langle v',w'\rangle$ is in W and $|z'|=1$. There follows by induction that $\langle v,w\rangle = \langle xz,yz\rangle$ for some x,y,z such that $\langle x,y\rangle \in X$ and $|z|<\omega$. Hence $\langle v,w\rangle$ is in $\overset{[0,\omega)}{\longrightarrow} X$. Since this holds for every $\langle v, w\rangle$ in W, therefore $W \subseteq \xrightarrow{[0,\omega)} X$. Thus, $\xrightarrow{[0,\omega)} X$ and the closure W of X under $\xrightarrow{1}$ coincide. The second assertion then follows from the fact that, for any relation Y on ${}^{[0,\omega)}U$, $\xrightarrow{1} Y = \bigcup\{\xrightarrow{1} Y' : Y' \subseteq Y,\ \overline{\overline{Y'}} = 1\}$.

(b) Consider any $m < \omega$. First assume that $\langle v, w\rangle$ is in $\xrightarrow{\omega} X$. Then there are x, y and z such that $\langle x, y\rangle \in X$, $|z| = \omega$, and $\langle xz, yz\rangle = \langle v, w\rangle$. Let $z = z'z''$, where $|z'| = m$ and $|z''| = \omega$. Then $\langle xz', yz'\rangle$ is in $\xrightarrow{m} X$ and hence $\langle v, w\rangle = \langle xz, yz\rangle = \langle xz'z'', yz'z''\rangle$ is in $\xrightarrow{\omega} (\xrightarrow{m} X)$. Now assume that $\langle v, w\rangle$ is in $\xrightarrow{\omega} (\xrightarrow{m} X)$. Then there are x, y and z such that $\langle x, y\rangle \in \xrightarrow{m} X$, $|z| = \omega$, and $\langle xz, yz\rangle = \langle v, w\rangle$. Since $\langle x, y\rangle \in \xrightarrow{m} X$, there are x', y' and z' such that $\langle x', y'\rangle \in X$, $|z'| = m$, and $\langle x'z', y'z'\rangle = \langle x, y\rangle$. Then $\langle v, w\rangle = \langle xz, yz\rangle = \langle x'z'z, y'z'z\rangle$ is in $\xrightarrow{\omega} X$. There now follows that $\xrightarrow{\omega} (\xrightarrow{m} X) = \xrightarrow{\omega} X$. Since this holds for any $m < \omega$ therefore
$\bigcup\{\xrightarrow{\omega} (\xrightarrow{m} X) : m < \omega\} = \xrightarrow{\omega} X$. Since
$\xrightarrow{\omega} (\xrightarrow{[0,\omega)} X) = \xrightarrow{\omega} (\bigcup\{\xrightarrow{m} X : m < \omega\}) = \bigcup\{\xrightarrow{\omega} (\xrightarrow{m} X) : m < \omega\}$,
there follows that $\xrightarrow{\omega} (\xrightarrow{[0,\omega)} X) = \xrightarrow{\omega} X$.

(c) Consider any $\langle v, x\rangle$ in $\xrightarrow{1} (X \circ Y)$. Then $\langle v, x\rangle = \langle v'z, x'z\rangle$, where $\langle v', x'\rangle$ is in $X \circ Y$ and $|z| = 1$. Then, for some w', $\langle v', w'\rangle$ is in X and $\langle w', x'\rangle$ is in Y. Then $\langle v, w'z\rangle = \langle v'z, w'z\rangle$ is in $\xrightarrow{1} X$ and $\langle w'z, x\rangle = \langle w'z, x'z\rangle$ is in $\xrightarrow{1} Y$. Hence $\langle v, x\rangle$ is in $(\xrightarrow{1} X) \circ (\xrightarrow{1} Y)$. This proves that $\xrightarrow{1} (X \circ Y) \subseteq (\xrightarrow{1} X) \circ (\xrightarrow{1} Y)$. By, roughly speaking, inverting the argument just given one proves that $(\xrightarrow{1} X) \circ (\xrightarrow{1} Y) \subseteq \xrightarrow{1} (X \circ Y)$. Hence $(\xrightarrow{1} X) \circ (\xrightarrow{1} Y) = \xrightarrow{1} (X \circ Y)$.

Since this holds for any X and Y, therefore, for any $\xrightarrow{n} X$ and $\xrightarrow{n} Y$, $\xrightarrow{1} (\xrightarrow{n} X) \circ \xrightarrow{1} (\xrightarrow{n} Y) = \xrightarrow{1} (\xrightarrow{n} X \circ \xrightarrow{n} Y)$. From the hypothesis of induction that $\xrightarrow{n} X \circ \xrightarrow{n} Y = \xrightarrow{n} (X \circ Y)$, there then follows that $\xrightarrow{1} (\xrightarrow{n} X) \circ \xrightarrow{1} (\xrightarrow{n} Y) = \xrightarrow{1} (\xrightarrow{n} (X \circ Y))$ and hence that $\xrightarrow{n+1} X \circ \xrightarrow{n+1} Y = \xrightarrow{n+1} (X \circ Y)$. Since $\xrightarrow{0} W = W$ for any W, there now follows by induction that, for any $n \geq 0$, $\xrightarrow{n} (X \circ Y) = \xrightarrow{n} X \circ \xrightarrow{n} Y$. □

For many relations W on ${}^{[0,\omega)}U$, the union $\bigcup\{\xrightarrow{m} W : m < \omega\}$ is not a union of pairwise disjoint sets. There may be overlap between different members $\xrightarrow{m} W$ and $\xrightarrow{n} W$ of $\{\xrightarrow{m} W : m < \omega\}$. We now turn to certain ones among the relations W on ${}^{[0,\omega)}U$ for which there is no such overlap.

Given any $s < \omega$ and $t < \omega$, a relation W on ${}^{[0,\omega)}U$ shall be ***of rank*** $\langle s, t\rangle$ if and only if

$$W \subseteq {}^sU \times {}^tU = \{\langle x, y\rangle : |x| = s,\ |y| = t\}\ .$$

A relation W on ${}^{[0,\omega)}U$ shall be ***ranked*** if and only if, for some $s < \omega$ and $t < \omega$, W is of rank $\langle s, t\rangle$. Thus, while $\emptyset$ is of any rank $\langle s, t\rangle$, any $W \neq \emptyset$ that is ranked is of a unique rank $\langle s, t\rangle$. We let Rk be the set of relations on ${}^{[0,\omega)}U$ that are ranked.

Thus

$$Rk = \bigcup\{\{W \subseteq {}^sU \times {}^tU\} : s < \omega,\ t < \omega\}\ .$$

Consider any W in Rk and let X be the closure $\xrightarrow{[0,\omega)} W$ of W under $\xrightarrow{1}$. If $W = \emptyset$ then, trivially, there is no overlap between different members of $\{\xrightarrow{m} W : m < \omega\}$. Now assume that $W \neq \emptyset$ and hence, for a unique $\langle s,t\rangle$, W is of rank $\langle s,t\rangle$. Then, for any $m < \omega$, $\xrightarrow{m} W$ is of rank $\langle s{+}m, t{+}m\rangle$. Hence, again, distinct members of $\{\xrightarrow{m} W : m < \omega\}$ are disjoint.

We are now ready to define the two kinds of relations that form the topic of this chapter. First, consider any relation X on ${}^{[0,\omega)}U$. Then X shall be ***uniform on*** ${}^{[0,\omega)}U$ if and only if it is the closure $\xrightarrow{[0,\omega)} W$ under $\xrightarrow{1}$ of a relation W that is ranked. We let ${}_{[0,\omega)}Un$ be the set of all relations that are uniform on ${}^{[0,\omega)}U$. Thus,

$$\begin{aligned} {}_{[0,\omega)}Un &= \{\xrightarrow{[0,\omega)} W : W \in Rk\} \\ &= \bigcup\{\xrightarrow{[0,\omega)} W : W \subseteq {}^sU \times {}^tU \text{ for some } s \text{ and } t\}\ . \end{aligned}$$

Now consider any relation X on ${}^{\omega}U$. Then X shall be ***uniform on*** ${}^{\omega}U$ if and only if it is the ω-prolongation $\xrightarrow{\omega} W$ of a relation W that is ranked. We let ${}_{[\omega]}Un$ be the set of all relations that are uniform on ${}^{\omega}U$. Thus,

$$\begin{aligned} {}_{[\omega]}Un &= \{\xrightarrow{\omega} W : W \in Rk\} \\ &= \bigcup\{\xrightarrow{\omega} W : W \subseteq {}^sU \times {}^tU \text{ for some } s \text{ and } t\}\ . \end{aligned}$$

There follows from Lemma 1(b) that ${}_{[\omega]}Un$ is related to ${}_{[0,\omega]}Un$ as follows. (Henceforth this relationship will be taken for granted.)

$${}_{[\omega]}Un = \{\xrightarrow{\omega} X : X \in {}_{[0,\omega)}Un\}\ .$$

We shall say that X is ***uniform*** if and only if either $X \in {}_{[0,\omega)}Un$ or $X \in {}_{[\omega]}Un$. From the context it may then sometimes follow that X is, in fact, in ${}_{[0,\omega)}Un$ or, in fact, in ${}_{[\omega]}Un$. Also, X shall be ***evenly uniform*** if and only if $X = \xrightarrow{[0,\omega)} W$ or $X = \xrightarrow{\omega} W$, for some W that, for some $s < \omega$, is of rank $\langle s,s\rangle$.

We will now begin to discuss uniformity on ${}^{[0,\omega)}U$. Uniformity on ${}^{\omega}U$ will be taken up later on. Consider any X in ${}_{[0,\omega]}Un$ such that $X \neq \emptyset$. Then $X = \xrightarrow{[0,\omega)} W$ for some W in Rk such that $W \neq \emptyset$. Since $W \neq \emptyset$, therefore W is of a unique rank $\langle s,t\rangle$. Further, since

$$X \;=\; \xrightarrow{[0,\omega)} W \;=\; \bigcup\{\langle xz, yz\rangle : \langle xy\rangle \in W,\ |z| < \omega\}\ ,$$

there follows that

$$\begin{aligned} W &= \{\langle x,y\rangle : |x|=s,\ |y|=t,\ \langle xz,yz\rangle \in X \text{ for some } z \in {}^{[0,\omega)}U\} \\ &= \{\langle x,y\rangle : |x|=s,\ |y|=t,\ \langle xz,yz\rangle \in X \text{ for every } z \in {}^{[0,\omega)}U\}\ . \end{aligned}$$

Thus, $X = \overset{[0,\omega)}{\longrightarrow} W$ for a unique $W \neq \emptyset$ in Rk. This W shall be the ***initial section*** of X.

We now classify the members X of ${}_{[0,\omega)}Un$ as follows. If $X \neq \emptyset$, then ***the grade of*** X shall be the rank of the initial section W of X. Further, we choose an object $\dot{0}$ that is not in $\omega \times \omega$, and hence differs from the grade $\langle s,t\rangle$ of any $X \neq \emptyset$ in ${}_{[0,\omega)}Un$, and let $\dot{0}$ be ***the grade of*** $\emptyset$. There follows that every X in ${}_{[0,\omega)}Un$ has a unique grade in $(\omega \times \omega) \cup \{\dot{0}\}$. This grade will be denoted by $gd\ X$. For any X that is not in ${}_{[0,\omega)}Un$, gd shall be undefined. Thus, ${}_{[0,\omega)}Un$ is the set of those relations X on ${}^{[0,\omega)}U$ that have a grade $gd\ X$.

We now turn briefly to the relations on ${}^{[0,\omega)}U$ that were discussed in Chapter I.

LEMMA 2. (a) *Each of the relations* $p_{(i,j)}$, $\dot{q}_i$, q_i r_i, e_{ij}, $\iota_{(j/\lambda)}$, $s_{\lambda,j}$ *is in* ${}_{[0,\omega)}Un$. *Specifically, the grade of* $\dot{q}_i$ *is* $\langle i+1,i\rangle$, *that of* q_i *and that of* $s_{\lambda,i}$ *is* $\langle i,i+1\rangle$, *that of* r_i *is* $\langle i+1,i+2\rangle$, *and if* $i \le j$, *then each of* $p_{(i,j)}$, e_{ij}, $\iota_{(j/\lambda)}$ *has grade* $\langle j+1,j+1\rangle$.

(b) *The grade of* $c_i = \dot{q}_i q_i$ *is* $\langle i+1,i+1\rangle$ *and that of* $q_i\dot{q}_i = {}_{[i,\omega)}\overset{\circ}{1}$ *is* $\langle i,i\rangle$.

(c) *Each of the relations* $p_{(i,j)}$, e_{ij}, $\iota_{(j/\lambda)}$, c_i, ${}_{[i,\omega)}\overset{\circ}{1}$ *is evenly uniform.*

(d) *For any* $i<\omega$ *and* $j<\omega$, *the following relation has grade* $\langle i,j\rangle$:

$$\dot{q}^i q^j = (q_0^{\smile})^i \circ q_0^j = \{\langle xz,yz\rangle : |x|=i,\ |y|=j,\ |z|<\omega\} = \overset{[0,\omega)}{\longrightarrow} ({}^iU \times {}^jU)$$

Proof. Let $W = \{\langle xz,x\rangle : |x|=i,\ |z|=1\}$. Then $\dot{q}_i = \overset{[0,\omega)}{\longrightarrow} W$. Since W is of rank $\langle i+1,i\rangle$, therefore $\dot{q}_i$ has grade $\langle i+1,i\rangle$. For the other relations in part (a) or part (b) the proof is similar. Part (c) follows from parts (a) and (b). The equality in part (d) follows from $q^j = (\dot{q}^{\smile})^j$ and from

$$\dot{q}^i = \{\langle xz,z\rangle : |x|=i,\ 0\le|z|<\omega\} = \overset{[0,\omega)}{\longrightarrow} ({}^iU \times {}^0U)\ . \qquad \square$$

Closely related to the partial functions $\dot{q}_i$ and p_i on ${}^{[0,\omega)}U$ are ${}_{[0,i]}\overset{\circ}{1} \cup \dot{q}_i$ and ${}_{[0,i+1]}\overset{\circ}{1} \cup p_i$. These have the advantage of being total functions on ${}^{[0,\omega)}U$. This advantage seems to be outweighed by their failure of being in ${}_{[0,\omega)}Un$. When ${}_{[0,i]}\overset{\circ}{1}$, ${}_{[0,i+1]}\overset{\circ}{1}$, $\cup$, and $\overline{2}$ are available, then ${}_{[0,i]}\overset{\circ}{1} \cup \dot{q}_i$ and $\dot{q}_i$ are, of course, interdefinable, and so are ${}_{[o,i+1]}\overset{\circ}{1} \cup p_i$ and p_i.

Among relations on ${}^{[0,\omega)}U$ that are less closely related to relations in ${}_{[0,\omega)}Un$ are the following five, each of which is invariant. The first two are permutations of ${}^{[0,\omega)}U$ that are length-preserving.

$$\overset{\leftrightarrow}{t} \quad = \quad \{\langle x,y\rangle : 0 \le |x| = |y| < \omega,\ y_m = x_{(|x|-1)-m} \text{ if } m < |x|\}$$

inversion

$$t^{\nwarrow} \quad = \quad \{\langle\emptyset,\emptyset\rangle\} \cup \{\langle zx, xz\rangle : 0 \le |x| < \omega, |z| = 1\}$$

counterclockwise rotation

$$\overset{\frown}{e} \quad = \quad \{\langle x,x\rangle : 1 \le x < \omega,\ x_0 = x_{|x|-1}\}$$

left-right identity

$$q^{\#} = q \circ t^{\nwarrow} \quad = \quad \{\langle x, xz\rangle : 0 \le |x| < \omega,\ |z| = 1\}$$

right-end insertion

$$\dot{q}^{\#} = q^{\#\smile} \quad = \quad \{\langle xz, x\rangle : 0 \le |x| < \omega,\ |z| = 1\}$$

right-end excision

Let $S^{\#}$ be the closure under $\circ$ of $\{q, \dot{q}, q^{\#}, \dot{q}^{\#}, \hat{e}\}$ and $O^{\#}$ the closure under $\circ$ of $\{q^*, \dot{q}^*, q^{\#*}, \dot{q}^{\#*}, \overset{\frown}{e}^*\}$. In [Mo], Monk considers the clone that is generated by $\{{}^{[0,\omega)}U, \bar{2}, \cup\} \cup O^{\#}$. From his Theorem 1.1 there follows that this clone properly includes the clone generated by the set G_3 of our Theorem I.6 and hence the clone generated by $\{{}^{[0,\omega)}U, P_{\bar{2}}, \cup\} \cup O_{pqe}$.

In contrast, probably none of the relations ${}_{[n+2,\omega)}\overset{\circ}{1}$ is in $S^{\#}$, so that $S^{\#}$ probably does not include S_{pqe}. In fact, we do not know whether there exists a set of relations on ${}^{[0,\omega)}U$ that is finite, such that the closure of this set under $\circ$ and $\smile$ includes S_{pqe}.

Since $\langle S^{\#}, \circ\rangle$ and $\langle O^{\#}, \circ\rangle$ are finitely generated, this shows that use of some relations R on ${}^{[0,\omega)}U$ that are not in ${}_{[0,\omega)}Un$, and of operations R^* they induce, may have certain advantages. There may also be disadvantages. In particular, one can no longer exploit the close relationship between R and $\overset{\omega}{\rightarrow} R$ that holds when R is in ${}_{[0,\omega)}Un$.

THEOREM 3. *Let X be any relation on ${}^{[0,\omega)}U$ such that $X \neq \emptyset$. Then for any s and t in ω, each of the following seven conditions is equivalent to the condition that X has grade $\langle s,t\rangle$. (If $X = \emptyset$, then each of these conditions is trivially satisfied.)*

(i) $X = \{\langle wz, xz\rangle : |w| = s,\ |x| = t,\ |z| < \omega,\ \langle w,x\rangle \in X\}$

(ii) (α) $X \subseteq \dot{q}_0^s \circ q_0^t = \{\langle wz, xz\rangle : |w| = s,\ |x| = t,\ |z| < \omega\}$, *and*

(β) *If* $|w| = s$, $|x| = t$, $|z| < \omega$, *and* $|z'| < \omega$, *then* $\langle wz, xz\rangle \in X$ *if and only if* $\langle wz', xz'\rangle \in X$.

(iii) (α) $X \subseteq \dot{q}_0^s \circ q_0^t = \{\langle wz, xz\rangle : |w| = s,\ |x| = t,\ |z| < \omega\}$, *and*

(β) *If* $|w| = s$, $|x| = t$, $|y| = 1$, *and* $|z| < \omega$, *then* $\langle wyz, xyz\rangle \in X$ *if and only if* $\langle wz, xz\rangle \in X$.

(iv) ${}_{[s+n,\omega)}\mathring{1} \circ X = \stackrel{n}{\rightarrow} X = X \circ {}_{[t+n,\omega)}\mathring{1}$, *for any* n *in* ω.

(v) ${}_{[s+n,\omega)}\mathring{1} \circ X = \stackrel{n}{\rightarrow} X = X \circ {}_{[t+n,\omega)}\mathring{1}$, *if* $n \in \{0, 1\}$.

(vi) $({}_{[s+n,\omega)}\mathring{1} \circ X) \cup (X \circ {}_{[t+n,\omega)}\mathring{1}) \subseteq \stackrel{1}{\rightarrow} X \subseteq$
$X \subseteq ({}_{[s+n,\omega)}\mathring{1} \circ X) \cap (X \circ {}_{[t+n,\omega)}\mathring{1})$.

(vii) $X \subseteq \dot{q}_0^s \circ q_0^t$ *and* $\dot{q}_s \circ X = X \circ \dot{q}_t$.

Proof. Assume throughout that $X \neq \emptyset$. Then the equivalence of (i) to $gd\ X = \langle s, t\rangle$ follows from the definitions involved. So does the equivalence of (ii). From (iii)(β) and induction on the length of z there follows:

$$\text{If } |w| = s \text{ and } |x| = t \text{ then, for any } z,$$
$$\langle wz, xz\rangle \in X \text{ if and only if } \langle w, x\rangle \in X.$$

From this and (iii)(α) there follows (i). Conversely, (iii)(α) and (iii)(β) follow from (i).

To show that $gd\ X = \langle s, t\rangle$ implies (iv), we assume that $gd\ X = \langle s, t\rangle$ and let W be the initial section of X. Consider any $n < \omega$. Then

$$\begin{aligned} \stackrel{n}{\rightarrow} X &= \stackrel{n}{\rightarrow} (\bigcup\{\stackrel{m}{\rightarrow} W : m < \omega\}) \\ &= \bigcup\{\stackrel{n}{\rightarrow}(\stackrel{m}{\rightarrow} W) : m < \omega\} = \bigcup\{\stackrel{m}{\rightarrow}(\stackrel{n}{\rightarrow} W) : m < \omega\} \\ &= \stackrel{[0,\omega)}{\longrightarrow} (\stackrel{n}{\rightarrow} W)\ . \end{aligned}$$

First, consider any $\langle v, w\rangle$ in $\stackrel{n}{\rightarrow} X = \stackrel{[0,\omega)}{\longrightarrow}(\stackrel{n}{\rightarrow} W)$. Then $\langle v, w\rangle = \langle xzz', yzz'\rangle$, where $\langle x, y\rangle \in W$ and hence $|x| = s$ and $|y| = t$, $|z| = n$, and $|z'| < \omega$. Since $\langle x, y\rangle$ is in W, therefore, $\langle v, w\rangle = \langle xzz', yzz'\rangle$ is in $X = \stackrel{[0,\omega)}{\longrightarrow} W$. Since $|xz| = s + n$ and hence $\langle xzz', yzz'\rangle$ is in ${}_{[s+n,\omega)}\mathring{1}$, there follows that $\langle v, w\rangle$ is in ${}_{[s+n,\omega)}\mathring{1} \circ X$. Since $|yz| = t + n$ and hence $\langle yzz', yzz'\rangle$ is in ${}_{[t+n,\omega)}\mathring{1}$, there also follows that $\langle v, w\rangle$ is in $X \circ {}_{[t+n,\omega)}\mathring{1}$. This shows that both $\stackrel{n}{\rightarrow} X \subseteq {}_{[s+n,\omega)}\mathring{1} \circ X$ and $\stackrel{n}{\rightarrow} X \subseteq X \circ {}_{[s+n,\omega)}\mathring{1}$. Now consider any $\langle v, w\rangle$ in ${}_{[s+n,\omega)}\mathring{1} \circ X$. Then $\langle v, v\rangle$ is in ${}_{[s+n,\omega)}\mathring{1}$ and hence $|v| \geq s + n$. Also $\langle v, w\rangle$ is in $X = \stackrel{[0,\omega)}{\longrightarrow} W$ and hence $\langle v, w\rangle = \langle xz, yz\rangle$, where $\langle x, y\rangle$ is in W and $|z| < \omega$. Since $\langle x, y\rangle$ is in W, therefore $|x| = s$. Since $v = xz$ and $|v| \geq s + n$, therefore $z = z'z''$, where $|z'| = n$ and $|z''| < \omega$. Then $\langle xz', yz'\rangle$ is in $\overset{n}{\to} W$ and hence $\langle v, w\rangle = \langle xz'z'', yz'z''\rangle$ is in $\overset{[0,\omega)}{\longrightarrow} (\overset{n}{\to} W) = \overset{n}{\to} X$. This shows that ${}_{[s+n,\omega)}\mathring{1} \circ X \subseteq \overset{n}{\to} X$. A similar argument shows that $X \circ {}_{[t+n,\omega)}\mathring{1} \subseteq \overset{n}{\to} X$. There now follows that ${}_{[s+n,\omega)}\mathring{1} \circ X = \overset{n}{\to} X$ and that $\overset{n}{\to} X = X \circ {}_{[t+n,\omega)}\mathring{1}$. Since this holds for any n, therefore $gd\ X = \langle s, t\rangle$ implies (iv).

To prove the converse implication, we now assume (iv) and let $W = X \cap ({}^sU \times {}^tU)$. First, consider any $\langle v, w\rangle$ in $\overset{[0,\omega)}{\longrightarrow} W$. Then, for some $n < \omega$, $\langle v, w\rangle$ is in $\overset{n}{\to} W$ and hence, since $W \subseteq X$, is also in $\overset{n}{\to} X$. Then, by (iv), $\langle v, w\rangle$ is in ${}_{[s+n,\omega)}\mathring{1} \circ X$ and hence also in X. There follows that $\overset{[0,\omega)}{\longrightarrow} W \subseteq X$. Now consider any $\langle v, w\rangle$ in X. From (iv) for the case where $n = 0$ there follows that ${}_{[s,\omega)}\mathring{1} \circ X = X = X \circ {}_{[t,\omega)}\mathring{1}$. Hence $|v| = s + m$ for some $m \geq 0$ and $|w| = t + n$ for some $n \geq 0$. Then $\langle v, w\rangle$ is in ${}_{[s+m,\omega)}\mathring{1} \circ X$ and in $X \circ {}_{[t+n,\omega)}\mathring{1}$ and hence, by (iv), is in $\overset{m}{\to} X$ and in $\overset{n}{\to} X$. From $\langle v, w\rangle \in \overset{m}{\to} X$ there follows that $\langle v, w\rangle = \langle xz, yz\rangle$, where $\langle x, y\rangle \in X$, $|z| = m$, and hence $|x| = |v| - m = s$ and $|y| = |w| - m = t + n - m$. Now $m > n$ would imply that $|y| < t$. Since $\langle x, y\rangle \in X$ and $X = X \circ {}_{[t,\omega)}\mathring{1}$, this is impossible. Hence $m \leq n$. A similar argument shows that $n \leq m$. Hence $n = m$ and $|y| = t$. There now follows that $\langle x, y\rangle$ is in $X \cap ({}^sU \times {}^tU) = W$ and that $\langle v, w\rangle = \langle xz, yz\rangle$ is in $\overset{m}{\to} W$. We have thus shown that any $\langle v, w\rangle$ in X is in some $\overset{m}{\to} W$ and hence is in $\overset{[0,\omega)}{\longrightarrow} W$. Thus $X \subseteq \overset{[0,\omega)}{\longrightarrow} W$. Since also $\overset{[0,\omega)}{\longrightarrow} W \subseteq X$, therefore $X = \overset{[0,\omega)}{\longrightarrow} W$. Since W is of rank $\langle s, t\rangle$ and $X \neq \emptyset$, therefore $gd\ X = \langle s, t\rangle$. This shows that (iv) implies that $gd\ X = \langle s, t\rangle$ and concludes the proof that the two conditions are equivalent.

Evidently, (iv) implies (v). We now assume the following part of (v): ${}_{[s+1,\omega)}\mathring{1} \circ X = \overset{1}{\to} X$. Also, we let $n \geq 0$ and as inductive hypothesis assume that ${}_{[s+n+1)}\mathring{1} \circ X = \overset{n+1}{\longrightarrow} X$. Then, using this hypothesis for the second equality below, Lemma 1(c) for the third, and the above part of (v) for the fourth, one obtains:

$$\begin{aligned}\overset{n+2}{\longrightarrow} X &= \overset{1}{\to} (\overset{n+1}{\longrightarrow} X) = \overset{1}{\to} ({}_{[s+n+1)}\mathring{1} \circ X) = \overset{1}{\to} ({}_{[s+n+1)}\mathring{1}) \circ \overset{1}{\to} X \\ &= {}_{[s+n+2,\omega)}\mathring{1} \circ ({}_{[s+1,\omega)}\mathring{1} \circ X) = {}_{[s+n+2,\omega)}\mathring{1} \circ X\,.\end{aligned}$$

There follows by induction that, for any $n \geq 0$, ${}_{[s+n+1)}\mathring{1} \circ X = \overset{n+1}{\longrightarrow} X$. A similar argument shows that from the assumption that $\overset{1}{\to} X = X \circ {}_{[t+1,\omega)}\mathring{1}$ there follows that, for any $n \geq 0$, $\overset{n+1}{\longrightarrow} X = X \circ {}_{[t+n+1,\omega)}\mathring{1}$. There now follows that (v) implies (iv).

From (v) for $n = 1$, there follows the first inclusion in (vi). Since ${}_{[s+1,\omega)}\overset{\circ}{1} \circ X \subseteq X$ therefore also from (v), for $n = 1$, there follows the second inclusion in (vi). From (v), for $n = 0$, there follows the third inclusion in (vi). Thus, (v) implies (vi).

Since ${}_{[s,\omega)}\overset{\circ}{1} \circ X \subseteq X$ and $X \circ {}_{[t,\omega)}\overset{\circ}{1} \subseteq X$ always hold, therefore the third inclusion in (vi) implies that ${}_{[s,\omega)}\overset{\circ}{1} \circ X = X = X \circ {}_{[t,\omega)}\overset{\circ}{1}$ and hence implies (v), for $n = 0$. In turn, $X = {}_{[s,\omega)}\overset{\circ}{1} \circ X$ implies that $\overset{1}{\to} X = \overset{1}{\to}({}_{[s,\omega)}\overset{\circ}{1} \circ X) = {}_{[s+1,\omega)}\overset{\circ}{1} \circ \overset{1}{\to} X$. From this and the second inclusion $\overset{1}{\to} X \subseteq X$ in (vi), there follows that $\overset{1}{\to} X \subseteq {}_{[s+1,\omega)}\overset{\circ}{1} \circ X$. Further, from the first inclusion in (vi) there follows that ${}_{[s+1,\omega)}\overset{\circ}{1} \circ X \subseteq \overset{1}{\to} X$. Hence the three inclusions in (vi) together imply that ${}_{[s+1,\omega)}\overset{\circ}{1} \circ X = \overset{1}{\to} X$. They similarly imply that $\overset{1}{\to} X = X \circ {}_{[t+1,\omega)}\overset{\circ}{1}$. Thus, (vi) also implies (v), for $n = 1$. This completes the proof that (vi) implies (v) and thereby the proof that they are equivalent.

To prove that (vii) is equivalent to (iii), and hence to $gd\ X = \langle s, t\rangle$, it suffices to show that if $X \subseteq \dot{q}_0^s \circ q_0^t$, then $\dot{q}_s \circ X = X \circ \dot{q}_t$ is equivalent to (iii)(β). For this purpose we first prove the following.

SUBLEMMA. *Assume that* $X \subseteq \dot{q}_0^s \circ q_0^t$. *Then:*
(a) $\dot{q}_s \circ X = \{\langle wyz, xz\rangle : |w| = s, |x| = t, |y| = 1, |z| < \omega,$ and $\langle wz, xz\rangle \in X\}$.
(b) $X \circ \dot{q}_t = \{\langle wyz, xz\rangle : |w| = s, |x| = t, |y| = 1, |z| < \omega,$ and $\langle wyz, xyz\rangle \in X\}$.

Proof. (a) For any X, there follows from the definition of $\dot{q}_s$ and that of $\circ$ that the set on the right is included in $\dot{q}_s \circ X$. To obtain the converse inclusion, assume that $X \subseteq \dot{q}_0^s \circ q_0^t$ and consider any $\langle v, v''\rangle$ in $\dot{q}_s \circ X$. Then, for some v', $\langle v, v'\rangle$ is in $\dot{q}_s$ and $\langle v', v''\rangle$ is in X. Since $X \subseteq \dot{q}_0^s \circ q_0^t$, there are w, x, z such that $|w| = s$, $|x| = t$, $|z| < \omega$, and $\langle v', v''\rangle = \langle wz, xz\rangle$. Since $\langle v, v'\rangle$ is in $\dot{q}_0^s$ therefore there is a y such that $|y| = 1$ and $\langle v, v'\rangle = \langle wyz, wz\rangle$, and hence such that $\langle v, v''\rangle = \langle wyz, xz\rangle$. This shows that the converse inclusion also holds.

(b) For any X, there follows from the definition of $\circ$ and that of $\dot{q}_t$ that the set on the right is included in $X \circ \dot{q}_t$. To obtain the converse inclusion, assume that $X \subseteq \dot{q}_0^s \circ q_0^t$ and consider any $\langle v, v''\rangle$ in $X \circ \dot{q}_t$. Then $\langle v'', v\rangle$ is in $q_t \circ X^{\smile}$ and hence, for some v', $\langle v'', v'\rangle$ is in q_t and $\langle v', v\rangle$ is in $X^{\smile}$. Since $\langle v'', v'\rangle$ is in q_t, there are x, y, z such that $|x| = t$, $|y| = 1$, $|z| < \omega$, and $\langle v'', v'\rangle = \langle xz, xyz\rangle$. Since $\langle v', v\rangle$ is in $X^{\smile}$ and since $X^{\smile} \subseteq \dot{q}_0^t \circ q_0^s$ therefore there is a w such that $|w| = s$ and $\langle v', v\rangle = \langle xyz, wyz\rangle$, and hence such that $\langle v, v''\rangle = \langle wyz, xz\rangle$. Thus, the converse inclusion also holds and the Sublemma has been proved.

Assume both $X \subseteq \dot{q}_0^s \circ q_0^t$ and (iii)(β). Consider any w, x, y, z such that $|w| = s$, $|x| = t$, $|y| = 1$, and $|z| < \omega$. Suppose that $\langle wyz, xz \rangle$ is in $\dot{q}_s \circ X$. Then, by Sublemma (a), $\langle wx, xz \rangle$ is in X. Then, by (iii)(β), $\langle wyz, xyz \rangle$ is in X and hence, by Sublemma (b), $\langle wyz, xz \rangle$ is in $X \circ \dot{q}_t$. There follows that $\dot{q}_s \circ X \subseteq X \circ \dot{q}_t$. Using Sublemma (b), then (iii)(β), and then Sublemma (a), one similarly shows that $X \circ \dot{q}_t \subseteq \dot{q}_s \circ X$. There follows that $\dot{q}_s \circ X = X \circ \dot{q}_t$.

Assume both $X \subseteq \dot{q}_0^s \circ q_0^t$ and $\dot{q}_s \circ X = X \circ \dot{q}_t$. Again consider any w, x, y, z such that $|w| = s$, $|x| = t$, $|y| = 1$, and $|z| < \omega$. Suppose that $\langle wyz, xyz \rangle$ is in X. Then, by Sublemma (b), $\langle wyz, xz \rangle$ is in $X \circ \dot{q}_t$. Since $X \circ \dot{q}_t = \dot{q}_s \circ X$, therefore $\langle wyz, xz \rangle$ is also in $\dot{q}_s \circ X$. Then, by Sublemma (a), $\langle wz, xz \rangle$ is in X. Using Sublemma (a), then $\dot{q}_s \circ X = X \circ \dot{q}_t$, and then Sublemma (b), one similarly shows that if $\langle wz, xz \rangle$ is in X, then $\langle wyz, xyz \rangle$ is in X. Thus there holds (iii)(β). This concludes the proof that if $X \subseteq \dot{q}_0^s \circ q_0^t$, then $\dot{q}_s \circ X = X \circ \dot{q}_t$ is equivalent to (iii)(β) and thereby the proof that (vii) is equivalent to (iii). □

In the conditions (iv), (v) and (vi) in Theorem 3, the operation $\xrightarrow{1}$ on the set $\{X : X \subseteq {}^{[0,\omega)}U \times {}^{[0,\omega)}U\}$ of all relations on ${}^{[0,\omega)}U$ plays an important role. By removing from $\xrightarrow{1}$ the pair $\langle \emptyset, \emptyset \rangle$ and thus forming $\xrightarrow{1} \cap - \{\langle \emptyset, \emptyset \rangle\}$, one obtains an operation on $\{X : \emptyset \neq X \subseteq {}^{[0,\omega)}U \times {}^{[0,\omega)}U\}$. This latter operation belongs to a class of unary operations which will now be described.

Consider any partial function g on a set A. As usual, for any $x \in A$, let $g^0(x) = x$ and for $0 \le n < \omega$ and $x \in Do\ g^n$ let $g^{n+1}(x) = g(g^n(x))$. We shall say that g is ***striate*** if and only if $g \neq \emptyset$, g is one-one, and, for every x in $Do\ g$, $\{w\colon x = g^n(w) \text{ for some } n < \omega\}$ is finite and $\{y\colon y = g^n(x) \text{ for some } n < \omega\}$ is infinite. Thus, as the referee suggested, one may think of g as striate if and only if $\langle A \cap Do\ g, g \rangle$ is isomorphic to a disjoint union of copies of $\langle \omega, sc \rangle$, where sc is the successor function on ω.

A unary partial function g on a set A shall be ***near-striate*** if and only if, there is some g' that is striate and some z that is not in $Do\ g'$ such that $g = g' \cup \{\langle z, z \rangle\}$. Thus, if g is near-striate, then $g(z) = z$ holds for exactly one z. This z shall be ***the fixed point of*** g.

Assume that g is striate or near-striate. Then the following relation is a partial ordering on $Do\ g$ and, in fact, is the least quasi-ordering that includes g:

$$\{\langle x, y \rangle \colon x \in Do\ g,\ y = g^n(x) \text{ for some } n < \omega\}$$

We call it the ***ordering (of*** $Do\ g$***) generated by*** g and denote it by $\preceq_g$, with the subscript omitted if understood. There is a least relation that includes g and is symmetric and transitive. We call it ***the equivalence relation generated by*** g and denote it by $\approx_g$, again with the subscript omitted if understood. Since g is one-one, there follows:

$$x \approx y \text{ if and only if either } x \preceq y \text{ or } y \preceq x.$$

For any $x \in Do\ g$, we let $x/{\approx}$ be the set $\{y \colon y \approx x\}$. There follows that every $x/{\approx}$ contains an element w such that $w \preceq y$ for every y in $x/{\approx}$. This w satisfies the following condition:

$$\text{Either } \{w\} = (x/{\approx}) \cap -(Rg\ g) \text{ or } w \text{ is the fixed point of } g.$$

These facts about operations that are striate or near-striate may be used tacitly.

We now turn briefly to the set of all relations on ${}^{[0,\omega)}U$. In view of part (a) of the following theorem, any relation X on ${}^{[0,\omega)}U$ such that either $X = \emptyset$ or $X \notin Rg \overset{1}{\rightarrow}$ shall be ***a prototype*** and also, more specifically, ***the prototype for*** $\{\overset{n}{\rightarrow} X : n < \omega\}$. Note that each of the relations ${}_{[0,\omega)}\overset{\circ}{1}$, $\dot{q}_i$, q_i, and also $p_{(i,j)}$, $e_{i,j}$, where $i \neq j$, is a prototype but that ${}_{[i+1,\omega)}\overset{\circ}{1}$, $p_{(i,i)}$, $e_{i,i}$ are not.

THEOREM 4. *Let* $A = \{X : X \subseteq {}^{[0,\omega)}U \times {}^{[0,\omega)}U\}$.

(a) $\overset{1}{\rightarrow} \cap -\{\langle\emptyset,\emptyset\rangle\}$ *is an operation on* $A \cap -\{\emptyset\}$ *that is striate. Also* $\overset{1}{\rightarrow}$ *is an operation on* A *that is near-striate and has* $\emptyset$ *as its fixed point.*

(b) $\overset{1}{\rightarrow}$ *is an endomorphism of* $\langle A, \circ, {}^{\smile}, \emptyset\rangle$ *such that for any* W *and* X *in* A, $W \subseteq X$ *if and only if* $\overset{1}{\rightarrow} W \subseteq \overset{1}{\rightarrow} X$.

(c) *For any* X *in* A, ${}_{[0,\omega)}\overset{\circ}{1} \circ X = X = X \circ {}_{[0,\omega)}\overset{\circ}{1}$ *and* $\emptyset \circ X = \emptyset = X \circ \emptyset$. *Also, for any* $m < \omega$, $\overset{1}{\rightarrow}(\overset{m}{\rightarrow}({}_{[0,\omega)}\overset{\circ}{1})) = {}_{[m+1,\omega)}\overset{\circ}{1}$ *is properly included in* $\overset{m}{\rightarrow}({}_{[0,\omega)}\overset{\circ}{1}) = {}_{[m,\omega)}\overset{\circ}{1}$. *Hence* $m < n$ *implies that* ${}_{[n,\omega)}\overset{\circ}{1} \not\subseteq {}_{[m,\omega)}\overset{\circ}{1}$ *and* $m \leq n$ *implies that* ${}_{[m,\omega)}\overset{\circ}{1} \circ {}_{[n,\omega)}\overset{\circ}{1} = {}_{[n,\omega)}\overset{\circ}{1} = {}_{[n,\omega)}\overset{\circ}{1} \circ {}_{[m,\omega)}\overset{\circ}{1}$.

(d) *For any* X *in* A *there are at most one* m *such that* ${}_{[m,\omega)}\overset{\circ}{1} \circ X = X$ *and* ${}_{[m+1,\omega)}\overset{\circ}{1} \circ X \neq X$ *and at most one* n *such that* $X \circ {}_{[n,\omega)}\overset{\circ}{1} = X$ *and* $X \circ {}_{[n+1,\omega)}\overset{\circ}{1} \neq X$.

Proof. (a) For any W in A, $\overset{1}{\rightarrow} W = \{\langle xz, yz\rangle : \langle x,y\rangle \in W,\ |z| = 1\}$. Assume that $\overset{1}{\rightarrow} W \neq \overset{1}{\rightarrow} X$. First suppose that $\overset{1}{\rightarrow} W \not\subseteq \overset{1}{\rightarrow} X$. Then, for some x, y, z such that $|z| = 1$, $\langle xz, yz\rangle$ is in $\overset{1}{\rightarrow} W$ but not in $\overset{1}{\rightarrow} X$ and hence $\langle x, y\rangle$ is in W but not in X. Hence $W \not\subseteq X$. Similarly, if $\overset{1}{\rightarrow} X \not\subseteq \overset{1}{\rightarrow} W$, then $X \not\subseteq W$. There follows that $\overset{1}{\rightarrow}$ is one-one.

Consider any X in $A \cap -\{\emptyset\}$. There is a unique s_0, $0 \leq s_0 < \omega$, such that some $\langle x_0, y_0\rangle$ is in X, $|x_0| = s_0$, and $|x| \geq s_0$ for any $\langle x, y\rangle \in X$. Now, for any x, y, z

in ${}^{[0,\omega)}U$, $\langle xz, yz\rangle = \langle x_0, y_0\rangle$ implies that $|z| \leq s_0$. Hence $\overset{n}{\to} W = X$ implies that $n \leq s_0$. Since $\overset{1}{\to}$ is one-one, there follows that $\{W\colon \overset{n}{\to} W = X\}$ contains at most s_0 elements. Now let $Y = \overset{n}{\to} X$, where $n < \omega$. Let $z \in {}^nU$. Then $\langle x_0z, y_0z\rangle$ is in Y and, for any $\langle x, y\rangle$ in Y, $|x_0z| = s_0 + n \leq |x|$. There follows that, if $n \neq n'$, then $\overset{n}{\to} X \neq \overset{n'}{\to} X$. Hence the set $\{Y\colon Y = \overset{n}{\to} X \text{ for some } n < \omega\}$ is infinite. Since this holds for any X in $A \cap -\{\emptyset\}$, therefore $\overset{1}{\to} \cap -\{\langle\emptyset,\emptyset\rangle\}$ is striate. Since $\overset{1}{\to} X = \emptyset$ if and only if $X = \emptyset$, there follows that $\overset{1}{\to}$ is near-striate, with $\emptyset$ as its fixed point.

(b) By Lemma 1(c), $\overset{1}{\to}$ is an endomorphism of $\langle A, \circ\rangle$. It is also an endomorphism of $\langle A, {}^{\smile}\rangle$, since the following equalities hold for any $W \in A$:

$$\begin{aligned}
(\overset{1}{\to} W)^{\smile} &= \{\langle xz, yz\rangle : \langle x, y\rangle \in W,\ |z|=1\}^{\smile}\\
&= \{\langle yz, xz\rangle : \langle x, y\rangle \in W,\ |z|=1\}\\
&= \{\langle yz, xz\rangle : \langle y, x\rangle \in W^{\smile},\ |z|=1\} = \overset{1}{\to}(W^{\smile})\ .
\end{aligned}$$

Further, $\overset{1}{\to} \emptyset = \emptyset$.

Assume that $W \subseteq X$. Suppose that $\langle xz, yz\rangle \in \overset{1}{\to} W$, where $|z| = 1$. Then $\langle x, y\rangle \in W$, hence $\langle x, y\rangle \in X$, and hence $\langle xz, yz\rangle \in \overset{1}{\to} X$. Thus, $W \subseteq X$ implies that $\overset{1}{\to} W \subseteq \overset{1}{\to} X$. Now assume that $\overset{1}{\to} W \subseteq \overset{1}{\to} X$. Suppose that $\langle x, y\rangle \in W$. Then, for any z such that $|z|=1$, there holds $\langle xz, yz\rangle \in \overset{1}{\to} W$, hence $\langle xz, yz\rangle \in \overset{1}{\to} X$, and hence $\langle x, y\rangle \in X$. Thus $\overset{1}{\to} W \subseteq \overset{1}{\to} X$ implies that $W \subseteq X$.

(c) The first assertion is obvious and the other two assertions follow from

$$\overset{m}{\to}({}_{[0,\omega)}\overset{\circ}{1}) = {}_{[m,\omega)}\overset{\circ}{1} = \{\langle x, x\rangle : x \in {}^{[0,\omega)}U,\ m \leq |x| < \omega\}\ .$$

(d) This follows from part (c). □

Recall that if X is uniform on ${}^{[0,\omega)}U$, then X has a unique grade in $(\omega\times\omega)\cup\{\dot{0}\}$, where $\dot{0}$ is not in $\omega\times\omega$. For computing the grade $gd(R\circ R')$ of the relative product $R \circ R'$ of relations R and R' that have a grade, we now introduce the following binary operation $\odot$ on $(\omega\times\omega)\cup\{\dot{0}\}$, where $\langle s, t\rangle$ and $\langle s', t'\rangle$ are arbitrary elements of $\omega\times\omega$:

$$\begin{aligned}
\langle s, t\rangle \odot \langle s', t'\rangle &= \langle s - t + \max(t, s'),\ t' - s' + \max(t, s')\rangle\\
\langle s, t\rangle \odot \dot{0} &= \dot{0} = \dot{0} \odot \langle s, t\rangle \quad \text{and} \quad \dot{0} \odot \dot{0} = \dot{0}\ .
\end{aligned}$$

Since $\omega\times\omega$ is closed under $\odot$, therefore $\langle \omega\times\omega, \odot\rangle$ is a subalgebra of $\langle(\omega\times\omega)\cup\{\dot{0}\}, \odot\rangle$. As one can verify, $\odot$ is associative in each of those two algebras. As is done in the literature of the theory of semigroups (cf. [Ho]), we will call $\langle\omega\times\omega, \odot\rangle$ ***the***

bicyclic semigroup. Also, $\langle(\omega \times \omega) \cup \{\dot{0}\}, \odot\rangle$ shall be ***the bicyclic semigroup with zero.***

We let $\overset{\bullet}{\vee}$ and $\overset{\bullet}{\rightarrow}$ be the unary operation on $(\omega \times \omega) \cup \{\dot{0}\}$ that satisfies the following condition, respectively, where $\langle s, t\rangle$ is any element of $\omega \times \omega$:

$$\begin{aligned} \langle s, t\rangle^{\overset{\bullet}{\vee}} &= \langle t, s\rangle \quad \text{and} \quad \dot{0}^{\overset{\bullet}{\vee}} = \dot{0} \\ \overset{\bullet}{\rightarrow} \langle s, t\rangle &= \langle s+1, t+1\rangle \quad \text{and} \quad \overset{\bullet}{\rightarrow} \dot{0} = \dot{0} \end{aligned}$$

There follows that $\overset{\bullet}{\rightarrow} \cap - \{\langle\dot{0}, \dot{0}\rangle\}$ is striate and that $\overset{\bullet}{\rightarrow}$ is near-striate, having $\dot{0}$ as its fixed point.

We let $\overset{\bullet}{\geq}$ be the binary relation on $(\omega \times \omega) \cup \{\dot{0}\}$ that satisfies the following conditions, where $\langle s, t\rangle$ and $\langle s', t'\rangle$ are arbitrary elements of $\omega \times \omega$:

$$\langle s, t\rangle \overset{\bullet}{\geq} \langle s', t'\rangle \text{ if and only if } \langle s, t\rangle = \langle s'+n, t'+n\rangle \quad \text{for some } n \geq 0.$$
$$\dot{0} \overset{\bullet}{\geq} \langle s, t\rangle \text{ and } \dot{0} \overset{\bullet}{\geq} \dot{0}\,.$$

There follows that $\overset{\bullet}{\geq}$ is an ordering of $(\omega \times \omega) \cup \{\dot{0}\}$. Since an element w of $(\omega \times \omega) \cup \{\dot{0}\}$ satisfies $w \odot w = w$, if and only if either $w = \dot{0}$ or $w = \langle s, s\rangle$ for some s, there also follows that, for any x and y in $(\omega \times \omega) \cup \{\dot{0}\}$ there holds:

$$x \overset{\bullet}{\geq} y \ \text{ if and only if } \ x = y \odot w \ \text{ for some } w \text{ such that } \ w \odot w = w.$$

We let $\overset{\bullet}{\leq}$ be the converse of $\overset{\bullet}{\geq}$. There follows that, if $\overset{\bullet}{\precsim}$ is the ordering generated by $\overset{\bullet}{\rightarrow}$, then $\overset{\bullet}{\leq}$ is the union $\overset{\bullet}{\precsim} \cup \{\langle\langle s, t\rangle, \dot{0}\rangle : s \in \omega,\ t \in \omega\}$. Also, for any $\langle s, t\rangle$ in $\omega \times \omega$, the restriction of $\overset{\bullet}{\leq}$ to $\{\langle s', t'\rangle : \langle s', t'\rangle \overset{\bullet}{\leq} \langle s, t\rangle\} \cup \{\langle s', t'\rangle : \langle s, t\rangle \overset{\bullet}{\leq} \langle s', t'\rangle\}$ is a linear ordering of order type ω.

LEMMA 5. *Assume that X and Y are in ${}_{[0,\omega)}Un$ and hence have a grade.*

(a) *$X \circ Y$ is in ${}_{[0,\omega)}Un$. Specifically, if $X \circ Y \neq \emptyset$, $gd\ X = \langle s, t\rangle$, and $gd\ Y = \langle s', t'\rangle$, then $gd(X \circ Y) = \langle s, t\rangle \odot \langle s', t'\rangle$ and if $X \circ Y = \emptyset$, then $gd(X \circ Y) = \dot{0}$.*

(b) *$X^{\smile}$ is in ${}_{[0,\omega)}Un$ and $gd(X^{\smile}) = (gd\ X)^{\overset{\bullet}{\vee}}$.*

(c) *$\overset{1}{\rightarrow} X$ is in ${}_{[0,\omega)}Un$ and $gd(\overset{1}{\rightarrow} X) = \overset{\bullet}{\rightarrow}(gd\ X)$.*

(d) *If $\overset{1}{\rightarrow} W = X$, then W is in ${}_{[0,\omega)}Un$ and $\overset{\bullet}{\rightarrow}(gd\ W) = gd\ X$.*

(e) *If $X \subseteq Y$ then $gd\ X \overset{\bullet}{\geq} gd\ Y$.*

Proof. (a) If $X \circ Y = \emptyset$ then $X \circ Y$ is in ${}_{[0,\omega)}Un$ and $gd(X \circ Y) = \dot{0}$. Now assume that $X \circ Y \neq \emptyset$. Then $X \neq \emptyset$ and $gd\ X = \langle s, t\rangle$ for some $s < \omega$ and $t < \omega$, and $Y \neq \emptyset$ and $gd\ Y = \langle s', t'\rangle$ for some $s' < \omega$ and $t' < \omega$. By the part of Theorem

3 that relates *gd* to condition (iv), there hold:

$$ {}_{[s+m,\omega)}\overset{\circ}{1} \circ X = \overset{m}{\rightarrow} X = X \circ {}_{[t+m,\omega)}\overset{\circ}{1}, \quad \text{for any } m. $$

$$ {}_{[s'+m',\omega)}\overset{\circ}{1} \circ Y = \overset{m'}{\longrightarrow} Y = Y \circ {}_{[t'+m',\omega)}\overset{\circ}{1}, \quad \text{for any } m'. $$

First, suppose that $t' \leq s'$. Consider any n. Then ${}_{[t+n,\omega)}\overset{\circ}{1} \circ {}_{[s'+n,\omega)}\overset{\circ}{1} = {}_{[s'+n,\omega)}\overset{\circ}{1}$. Then from the equalities in the above two sequences there follow the equalities in the next two sequences. From these, there follow those in the third sequence below.

$$ \begin{aligned} \overset{n}{\rightarrow}(X \circ Y) &= \overset{n}{\rightarrow} X \circ \overset{n}{\rightarrow} Y = X \circ {}_{[t+n,\omega)}\overset{\circ}{1} \circ {}_{[s'+n,\omega)}\overset{\circ}{1} \circ Y \\ &= X \circ {}_{[s'+n,\omega)}\overset{\circ}{1} \circ Y\ . \\ {}_{[s+(s'-t)+n,\omega)}\overset{\circ}{1} \circ (X \circ Y) &= X \circ {}_{[t+(s'-t)+n,\omega)}\overset{\circ}{1} \circ Y \\ &= X \circ {}_{[s'+n,\omega)}\overset{\circ}{1} \circ Y = (X \circ Y) \circ {}_{[t'+n,\omega)}\overset{\circ}{1}\ . \\ \overset{n}{\rightarrow}(X \circ Y) &= {}_{[s+(s'-t)+n,\omega)}\overset{\circ}{1} \circ (X \circ Y) = (X \circ Y) \circ {}_{[t'+n,\omega)}\overset{\circ}{1}\ . \end{aligned} $$

Since the equalities in the last sequence hold for any n and since $X \circ Y \neq \emptyset$, there follows from the part of Theorem 3 that relates *gd* to condition (iv) that $X \circ Y$ has grade $\langle s + (s' - t), t' \rangle$. Now suppose that $t \leq s'$. Then a similar argument shows that $X \circ Y$ has grade $\langle s, t' + (t - s) \rangle$. In either case, $X \circ Y$ has grade $\langle s, t \rangle \odot \langle s', t' \rangle$.

(b) Assume that ${}_{[s+n,\omega)}\overset{\circ}{1} \circ X = \overset{n}{\rightarrow} X = X \circ {}_{[t+n,\omega)}\overset{\circ}{1}$. Then ${}_{[t+n,\omega)}\overset{\circ}{1} \circ X^{\smile} = (X \circ {}_{[t+n,\omega)}\overset{\circ}{1})^{\smile} = (\overset{n}{\rightarrow} X)^{\smile} = \overset{n}{\rightarrow}(X^{\smile})$, and $(X \circ {}_{[t+n,\omega)}\overset{\circ}{1})^{\smile} = ({}_{[s+n,\omega)}\overset{\circ}{1} \circ X)^{\smile} = X^{\smile} \circ {}_{[s+n,\omega)}\overset{\circ}{1}$. There follows that, if $gd\ X = \langle s, t \rangle$, then $gd(X^{\smile}) = \langle t, s \rangle = \langle s, t \rangle^{\overset{\bullet}{\vee}}$. Also, if $gd\ X = \dot{0}$ and hence $X = \emptyset$, then $X^{\smile} = \emptyset$ and hence $gd(X^{\smile}) = \dot{0} = (gd\ X)^{\overset{\bullet}{\vee}}$.

(c) Assume that ${}_{[s+n,\omega)}\overset{\circ}{1} \circ X = \overset{n}{\rightarrow} X = X \circ {}_{[t+n,\omega)}\overset{\circ}{1}$. Then there hold:

$$ \begin{aligned} {}_{[(s+1)+n,\omega)}\overset{\circ}{1} \circ \overset{1}{\rightarrow} X &= \overset{1}{\rightarrow} ({}_{[s,\omega)}\overset{\circ}{1}) \circ (\overset{1}{\rightarrow} X) \\ &= \overset{1}{\rightarrow} ({}_{[s+n,\omega)}\overset{\circ}{1} \circ X) = \overset{1}{\rightarrow}(\overset{n}{\rightarrow} X) = \overset{n+1}{\longrightarrow} X\ . \\ \overset{1}{\rightarrow}(\overset{n}{\rightarrow} X) &= \overset{1}{\rightarrow}(X \circ {}_{[t+n,\omega)}\overset{\circ}{1}) = (\overset{1}{\rightarrow} X) \circ \overset{1}{\rightarrow}({}_{[t+n,\omega)}\overset{\circ}{1}) \\ &= \overset{1}{\rightarrow} X \circ {}_{[(t+1)+n,\omega)}\overset{\circ}{1}\ . \end{aligned} $$

There follows that if $gd\ X = \langle s, t \rangle$ then $gd(\overset{1}{\rightarrow} X) = \langle s+1, t+1 \rangle = \overset{\bullet}{\rightarrow}(gd\ X)$. Also, if $gd\ X = \dot{0}$ and hence $X = \emptyset$, then $gd(\overset{1}{\rightarrow} X) = gd\ \emptyset = \dot{0} = \overset{\bullet}{\rightarrow} \dot{0}$.

(d) Assume that $\overset{1}{\rightarrow} W = X$. If $X = \emptyset$, then $W = \emptyset$, hence W is in ${}_{[0,\omega)}Un$ and $gd\ W = \dot{0}$. Now assume that $X \neq \emptyset$ and hence $gd\ X$ is in $\omega \times \omega$. Since

$X = \overset{1}{\to} W$, therefore $gd\ X$ differs from every $\langle 0, t' \rangle$ and from every $\langle s', 0 \rangle$. Hence $gd\ X = \langle s+1, t+1 \rangle$ for some $s < \omega$ and $t < \omega$. Let Q be the initial section of X, so that $Q = X \cap ({}^{s+1}U \times {}^{t+1}U)$ and $X = \bigcup \{ \overset{n}{\to} Q : n < \omega \}$. Let

$$P = \{ \langle x, y \rangle : |x| = s,\ |y| = t, \text{ and, for some } z,\ |z| = 1 \text{ and } \langle xz, yz \rangle \in Q \} \ .$$

Then $Q \subseteq \overset{1}{\to} P$. Conversely, since $\overset{1}{\to} P \subseteq \overset{1}{\to} W = X$ and $\overset{1}{\to} P \subseteq {}^{s+1}U \times {}^{t+1}U$, therefore $\overset{1}{\to} P \subseteq Q$. Hence $Q = \overset{1}{\to} P$. There follows that $X = \bigcup \{ \overset{n+1}{\longrightarrow} P : n < \omega \}$.

Consider any $\langle v, w \rangle$ in W. For any z such that $|z| = 1$, $\langle vz, wz \rangle$ is in $\overset{1}{\to} W = X$ and hence, for some $n < \omega$, in $\overset{n+1}{\longrightarrow} P$. Then $\langle v, w \rangle$ is in $\overset{n}{\to} P$. There follows that $W \subseteq \bigcup \{ \overset{n}{\to} P : n < \omega \}$. Now consider any $n < \omega$. Since $\overset{1}{\to} (\overset{n}{\to} P) = \overset{n+1}{\longrightarrow} P \subseteq X = \overset{1}{\to} W$, therefore $\overset{n}{\to} P \subseteq W$. There follows that $\bigcup \{ \overset{n}{\to} P : n < \omega \} \subseteq W$. Thus $W = \bigcup \{ \overset{n}{\to} P : n < \omega \}$ and hence W is in ${}_{[0,\omega)}Un$. Also, $gd\ W = \langle s, t \rangle$ and hence $\overset{\bullet}{\to} (gd\ W) = \langle s+1, t+1 \rangle = gd\ X$.

(e) Assume that $X \subseteq Y$. If $X = \emptyset$ then $gd\ X = \dot{0}$ and hence $gd\ X \ \succeq \ gd\ Y$. Now assume that $X \neq \emptyset$ and hence $Y \neq \emptyset$. Then for some s, t, s' and t', $gd\ X = \langle s, t \rangle$ and $gd\ Y = \langle s', t' \rangle$. Let W be the initial section of Y. Then W is of rank $\langle s', t' \rangle$. Since $gd\ X = \langle s, t \rangle$, therefore X, and hence also Y, contains a pair $\langle x, y \rangle$ such that $|x| = s$ and $|y| = t$. Since $Y = \overset{[0,\omega)}{\longrightarrow} W$, therefore $\langle x, y \rangle = \langle vz, wz \rangle$ for some v, w, z such that $\langle v, w \rangle$ is in W and $|z| = n$ for some n. There follows that $|v| = s - n$ and $|w| = t - n$ and that W is of rank $\langle s-n, t-n \rangle$. Therefore $\langle s', t' \rangle = \langle s{-}n, t{-}n \rangle$ and hence $\langle s, t \rangle = \langle s'{+}n, t'{+}n \rangle$. There follows that $gd\ X = \langle s, t \rangle \ \succeq \ \langle s', t' \rangle = gd\ Y$. □

LEMMA 6. *Assume that* $gd\ X = \langle s, t \rangle$ *and* $gd\ Y = \langle s', t' \rangle$. *Let* $\precsim$ *be the ordering that is generated by* $\overset{1}{\to}$.

(a) *Consider any* $m < \omega$. *If* $t \leq s'$ *and* $m \leq s' - t$, *then* $\overset{m}{\to} X \circ Y = X \circ Y$. *If* $t \leq s'$, $m \geq s' - t$, *and* $r = m - (s' - t)$, *then* $\overset{m}{\to} X \circ Y = \overset{r}{\to} (X \circ Y)$. *If* $t \geq s'$, *then* $\overset{m}{\to} X \circ Y = \overset{m}{\to} (X \circ Y)$. *Thus, in any case,* $X \circ Y \precsim \overset{m}{\to} X \circ Y$.

(b) *Consider any* $n < \omega$. *If* $s' \leq t$ *and* $n \leq t - s'$, *then* $X \circ \overset{n}{\to} Y = X \circ Y$. *If* $s' \leq t$, $n \geq t - s'$, *and* $r = n - (t - s')$, *then* $X \circ \overset{n}{\to} Y = \overset{r}{\to} (X \circ Y)$. *If* $s' \geq t$, *then* $X \circ \overset{n}{\to} Y = \overset{n}{\to} (X \circ Y)$. *Thus, in any case,* $X \circ Y \precsim X \circ \overset{n}{\to} Y$.

(c) *If* $X \precsim X'$ *and* $Y \precsim Y'$, *then* $X \circ Y \precsim X' \circ Y'$.

Proof. (a) Suppose that $t \leq s'$ and $m \leq s' - t$. Then

$$\overset{m}{\to} X \circ Y = X \circ {}_{[t+m,\omega)}\mathring{1} \circ {}_{[s',\omega)}\mathring{1} \circ Y = X \circ {}_{[s',\omega)}\mathring{1} \circ Y = X \circ Y \ .$$

Suppose that $t \leq s'$, $m \geq s' - t$, and $r = m - (s' - t)$. Then

$$\begin{aligned} \overset{m}{\rightarrow} X \circ Y &= X \circ {}_{[t+m,\omega)}\overset{\circ}{1} \circ Y \\ &= X \circ {}_{[t+r,\omega)}\overset{\circ}{1} \circ {}_{[t+m,\omega)}\overset{\circ}{1} \circ Y \\ &= X \circ {}_{[t+r,\omega)}\overset{\circ}{1} \circ {}_{[s'+r,\omega)}\overset{\circ}{1} \circ Y \\ &= \overset{r}{\rightarrow} X \circ \overset{r}{\rightarrow} Y = \overset{r}{\rightarrow}(X \circ Y) . \end{aligned}$$

Suppose that $t \geq s'$. Then

$$\begin{aligned} \overset{m}{\rightarrow} X \circ Y &= X \circ {}_{[t+m,\omega)}\overset{\circ}{1} \circ Y = X \circ {}_{[t+m,\omega)}\overset{\circ}{1} \circ {}_{[s'+m,\omega)}\overset{\circ}{1} \circ Y \\ &= \overset{m}{\rightarrow} X \circ \overset{m}{\rightarrow} Y = \overset{m}{\rightarrow}(X \circ Y) . \end{aligned}$$

(b) The proof is similar.

(c) Assume that $X \precsim X'$ and $Y \precsim Y'$. Then, for some m and n, $\overset{m}{\rightarrow} X = X'$ and $\overset{n}{\rightarrow} Y = Y'$. Then, by part (a), $X \circ Y \precsim \overset{m}{\rightarrow} X \circ Y = X' \circ Y$. By Lemma 5(c), $gd\ X'$ is in $\omega \times \omega$. Hence, by part (b), $X' \circ Y \precsim X' \circ \overset{n}{\rightarrow} Y = X' \circ Y'$. There follows that $X \circ Y \precsim X' \circ Y'$. □

Some further conventions or terminology will be useful. Let $0 \leq k < \omega$ and let f be a k-ary operation on a set $A \neq \emptyset$. Then $\langle A, f, \leq \rangle$ shall be an ***ordered algebra*** if and only if $\leq$ is an ordering of A and, for every $x_0, \ldots, x_{k-1}, y_0, \ldots, y_{k-1}$ in A, if $x_i \leq y_i$ for every $i < k$, then $f(x_0, \ldots, x_{k-1}) \leq f(y_0, \ldots, y_{k-1})$. Also, if $0 \leq n < \omega$, then $\langle A, f_0, \ldots, f_{(n-1)}, \leq \rangle$ shall be an ***ordered algebra*** if and only if $\langle A, f_m, \leq \rangle$ is an ordered algebra for every $m < n$.

Given any $\langle s, t \rangle$ and $\langle s', t' \rangle$ in $\omega \times \omega$, we shall say that $\langle s, t \rangle$ is ***attuned to*** $\langle s', t' \rangle$ if and only if $t = s'$.

By Lemma 5, the restriction to ${}_{[0,\omega)}Un$ of the operation $\circ$, $\smile$, or $\overset{1}{\rightarrow}$ on $\{X : X \subseteq {}^{[0,\omega)}U \times {}^{[0,\omega)}U\}$ is an operation on ${}_{[0,\omega)}Un$. From now on, we will likewise use $\circ$, $\smile$, or $\overset{1}{\rightarrow}$ for this restriction, and also for other restrictions under consideration. When it should matter, which kind of operations is intended should be apparent from the context.

THEOREM 7. (a) *${}_{[0,\omega)}Un$ is closed under $\circ$, $\smile$, and $\overset{1}{\rightarrow}$, and $\langle {}_{[0,\omega)}Un, \circ, \smile, \overset{1}{\rightarrow}, \subseteq \rangle$ is an ordered algebra. Also, ${}_{[0,\omega)}Un$ includes S_{pqei} and hence contains every ${}_{[s,\omega)}\overset{\circ}{1} = q_s \circ \dot{q}_s$ and every $\dot{q}_0^s \circ q_0^t$.*

(b) *For every X in ${}_{[0,\omega)}Un$ there is at least one $\dot{q}_0^s \circ q_0^t$ such that $X \subseteq \dot{q}_0^s \circ q_0^t$. Moreover, for any relations X and Y on ${}^{[s,\omega)}U$, if $X \subseteq \dot{q}_0^s \circ q_0^t$, $Y \subseteq \dot{q}_0^{s'} \circ q_0^{t'}$, and $s - t \neq s' - t'$, then $X \cap Y = \emptyset$. Also, the set of maximal elements of $\langle {}_{[0,\omega)}Un, \subseteq \rangle$*

is $\{\dot{q}_0^s \circ q_0^t : s < \omega,\ t < \omega\}$, *if the universe* U *satisfies* $\overset{=}{U} \geq 2$, *and* $\{\dot{q}_0^s : s < \omega\} \cup \{q_0^t : t < \omega\}$ *if* $\overset{=}{U} = 1$.

(c) *The operation* $\overset{1}{\rightarrow}$ *on* ${}_{[0,\omega)}Un$ *is near-striate and properly embeds* $\langle {}_{[0,\omega)}Un, \circ, {}^{\smile}, \overset{1}{\rightarrow}\rangle$ *into itself. Also, if* $\overset{1}{\precsim}$ *is the ordering generated by* $\overset{1}{\rightarrow}$, *then* $\langle {}_{[0,\omega)}Un, \circ, {}^{\smile}, \overset{1}{\rightarrow}, \overset{1}{\precsim}\rangle$ *is an ordered algebra. Further, if* $\approx$ *is the equivalence relation generated by* $\overset{1}{\rightarrow}$, *then* $\approx$ *is a congruence of* $\langle {}_{[0,\omega)}Un, \circ, {}^{\smile}, \overset{1}{\rightarrow}\rangle$.

(d) *If* $\emptyset \neq X \in {}_{[0,\omega)}Un$, *then* $\overset{1}{\rightarrow} X \subseteq X \not\subseteq \overset{1}{\rightarrow} X$. *Also,* $\overset{1}{\rightarrow} \emptyset = \emptyset$.

(e) *Let* $X \in {}_{[0,\omega)}Un$. *Then each of the conditions* (i),(ii),(iii) *below is equivalent to* $gd\ X = \langle s, t\rangle$.

(i) ${}_{[s,\omega)}\overset{\circ}{1} \circ X = X = X \circ {}_{[t,\omega)}\overset{\circ}{1}$ *and* ${}_{[s+1,\omega)}\overset{\circ}{1} \circ X \neq X \neq X \circ {}_{[t+1,\omega)}\overset{\circ}{1}$.

(ii) $X \subseteq \dot{q}_0^s \circ q_0^t$ *and* $X \not\subseteq \dot{q}_0^{s+1} \circ q_0^{t+1}$.

(iii) $X \subseteq \dot{q}_0^s \circ q_0^t$ *and* $\dot{q}_s \circ X = X \circ \dot{q}_t$.

(f) *Assume that* $gd\ X = \langle s, t\rangle$. *Then* $\overset{1}{\rightarrow} X = {}_{[s+1,\omega)}\overset{\circ}{1} \circ X = X \circ {}_{[t+1,\omega)}\overset{\circ}{1}$. *Also, if* $\overset{1}{\rightarrow} W = X$, *then* $W = q_{s-1} \circ X \circ q_{t-1}^{\smile}$.

(g) *For any* X *and* Y *in* ${}_{[0,\omega)}Un \cap -\{\emptyset\}$, $X \circ Y = X' \circ Y'$ *for some* X' *and* Y' *such that* $gd\ X'$ *is attuned to* $gd\ Y'$. *Specifically, if* $gd\ X = \langle s, t\rangle$ *and* $gd\ Y = \langle s', t'\rangle$, *then* $t \geq s'$ *implies that* $X \circ Y = X \circ (Y \circ {}_{[t'+t-s',\,\omega)}\overset{\circ}{1})$, *where* $gd(Y \circ {}_{[t'+t-s',\,\omega)}\overset{\circ}{1}) = \langle t, t'+t-s'\rangle$, *while* $t \leq s'$ *implies that* $X \circ Y = ({}_{[s+s'-t,\omega)}\overset{\circ}{1} \circ X) \circ Y$, *where* $gd({}_{[s+s'-t,\omega)}\overset{\circ}{1} \circ X) = \langle s + s' - t, s'\rangle$.

Proof. (a) By Lemma 4(a), 4(b), or 4(c), respectively, ${}_{[0,\omega)}Un$ is closed under $\circ$, ${}^{\smile}$, and $\overset{1}{\rightarrow}$. Then, by Lemma 2, S_{pqei} is included in ${}_{[0,\omega)}Un$. Then also by the set-theoretic definition of $\circ$, ${}^{\smile}$, or $\overset{1}{\rightarrow}$, respectively, $\langle {}_{[0,\omega)}Un, \circ, \subseteq\rangle$, $\langle {}_{[0,\omega)}Un, {}^{\smile}, \subseteq\rangle$, and $\langle {}_{[0,\omega)}Un, \overset{1}{\rightarrow}, \subseteq\rangle$ are an ordered algebra, and hence so is $\langle {}_{[0,\omega)}Un, \circ, {}^{\smile}, \overset{1}{\rightarrow}, \subseteq\rangle$.

(b) By the definition of gd, or by Theorem 3(b), if $gd\ X = \langle s, t\rangle$, then $X \subseteq \dot{q}_0^s \circ q_0^t$. Also $\emptyset \subseteq \dot{q}_0^s \circ q_0^t$. Hence every element of ${}_{[0,\omega)}Un$ is included in some $\dot{q}_0^s \circ q_0^t$.

Let X and Y be any relations on ${}^{[0,\omega)}U$ such that $X \subseteq \dot{q}_0^s \circ q_0^t$ and $Y \subseteq \dot{q}_0^{s'} \circ q_0^{t'}$. Then for every $\langle v, w\rangle$ in X there is some m such that $|v| = s + m$ and $|w| = t + m$. Also, for every $\langle v, w\rangle$ in Y there is some n such that $|v| = s' + n$ and $|w| = t' + n$. There follows that if $s - t \neq s' - t'$, and hence if $\langle s, t\rangle$ differs from every $\langle s' + m, t' + m\rangle$ and $\langle s', t'\rangle$ differs from every $\langle s + n, t + n\rangle$, then X and Y are disjoint.

Since every element of ${}_{[0,\omega)}Un$ is included in some $\dot{q}_0^s \circ q_0^t$, therefore every maximal element of $\langle {}_{[0,\omega)}Un, \subseteq\rangle$ is an element $\dot{q}_0^s \circ q_0^t$. Now consider any $\dot{q}_0^s \circ q_0^t$ and any Y in ${}_{[0,\omega)}Un$ such that $Y \neq \dot{q}_0^s \circ q_0^t$. If $Y = \emptyset$, then $\dot{q}_0^s \circ q_0^t \not\subseteq Y$. Now let $Y \neq \emptyset$ and let $gd\ Y = \langle s', t'\rangle$. If $\langle s, t\rangle$ differs from every $\langle s' + m, t' + m\rangle$ or $\langle s', t'\rangle$ differs from every $\langle s+n, t+n\rangle$, then $\dot{q}_0^s \circ q_0^t$ and Y are disjoint and hence $\dot{q}_0^s \circ q_0^t \not\subseteq Y$. Now suppose that $\langle s', t'\rangle = \langle s+n, t+n\rangle$ for some $n > 0$. Then every pair $\langle v, w\rangle$ such that $|v| = s$ and $|w| = t$ is in $\dot{q}_0^s \circ q_0^t$ but not in Y and hence again $\dot{q}_0^s \circ q_0^t \not\subseteq Y$. Next suppose that $\langle s', t'\rangle = \langle s, t\rangle$. Then, by the definition of gd, $Y \subseteq \dot{q}_0^s \circ q_0^t$. Since $Y \neq \dot{q}_0^s \circ q_0^t$, therefore again $\dot{q}_0^s \circ q_0^t \not\subseteq Y$. Now suppose that $\langle s'+m, t'+m\rangle = \langle s, t\rangle$ for some $m > 0$ and assume, in addition, that $\overline{\overline{U}} \geq 2$. Then there are z and z' in ${}^mU$ such that $z \neq z'$. If $|x| = s'$ and $|y| = t'$, then $\langle xz, yz'\rangle$ is not in $\dot{q}_0^{s'} \circ q_0^{t'}$ and hence not in Y but it is in $\dot{q}_0^s \circ q_0^t$. Hence again $\dot{q}_0^s \circ q_0^t \not\subseteq Y$. Thus, if $\overline{\overline{U}} \geq 2$ then, in all cases, where $Y \in {}_{[0,\omega)}Un$ and $Y \not\subseteq \dot{q}_0^s \circ q^t$ there holds $\dot{q}_0^s \circ q_0^t \not\subseteq Y$. Hence if $\overline{\overline{U}} \geq 2$, then every $\dot{q}_0^s \circ q_0^t$ is a maximal element of $\langle {}_{[0,\omega)}Un, \subseteq\rangle$.

Now assume that $\overline{\overline{U}} = 1$. Then, for every $r < \omega$, there is exactly one v in ${}^rU$. There follows that every element of ${}_{[0,\omega)}Un$ is an element $\dot{q}_0^{s'} \circ q_0^{t'}$. From what was shown earlier there follows that if $\langle s, t\rangle$ differs from each $\langle s'+m, t'+m\rangle$ then $\dot{q}_0^s \circ q_0^t \not\subseteq \dot{q}_0^{s'} \circ q_0^{t'}$. Now assume that $\langle s, t\rangle = \langle s'+m, t'+m\rangle$. Since $\overline{\overline{U}} = 1$, therefore, for every r, the unique $\langle v, w\rangle$ in $\dot{q}_0^s \circ q_0^t$ such that $|v| = s+r$ and $|w| = t+r$ is also in $\dot{q}_0^{s'} \circ q_0^{t'}$. Hence $\dot{q}_0^s \circ q_0^t \subseteq \dot{q}_0^{s'} \circ q_0^{t'}$. Moreover, as we saw earlier, if $m > 0$, then $\dot{q}_0^{s'} \circ q_0^{t'} \not\subseteq \dot{q}_0^s \circ q_0^t$. There follows that if $s \geq t$, then the unique maximal element of $\langle {}_{[0,\omega)}Un, \subseteq\rangle$ that includes $\dot{q}_0^s \circ q_0^t$ is the element $(\dot{q})^{s-t}$ while, if $t \geq s$, then the unique maximal element of $\langle {}_{[0,\omega)}Un, \subseteq\rangle$ that includes $\dot{q}_0^s \circ q_0^t$ is the element $(\dot{q})^{t-s}$.

(c) By part (a), by Theorem 4, and since $\emptyset$ is in ${}_{[0,\omega)}Un$, $\xrightarrow{1}$ is an operation on ${}_{[0,\omega)}Un$ that is near-striate and properly embeds $\langle {}_{[0,\omega)}Un, \circ, {}^{\smile}\rangle$ into itself. Also, trivially, $\xrightarrow{1}$ is an embedding of $\langle {}_{[0,\omega)}Un, \xrightarrow{1}\rangle$ into itself.

Let $\precsim$ be the ordering generated by $\xrightarrow{1}$. Then $X \precsim Y$ implies that $\xrightarrow{1} X \precsim \xrightarrow{1} Y$. Since $\xrightarrow{1}(X^{\smile}) = (\xrightarrow{1} X)^{\smile}$, therefore $X \precsim Y$ implies that $X^{\smile} \precsim Y^{\smile}$. Now assume that $X \precsim Y$ and $X' \precsim Y'$. If $X = \emptyset$, then $Y = \emptyset$, hence both $X \circ X' = \emptyset$ and $Y \circ Y' = \emptyset$, and hence $X \circ X' \precsim Y \circ Y'$. Similarly, if $X' = \emptyset$, then $X \circ X' \precsim Y \circ Y'$. Now suppose that $X \neq \emptyset$ and $X' \neq \emptyset$. Then, by Lemma 6(c), again $X \circ X' \precsim Y \circ Y'$. There now follows that $\langle {}_{[0,\omega)}Un, \circ, {}^{\smile}, \xrightarrow{1}, \precsim\rangle$ is an ordered algebra.

Let $\approx$ be the equivalence relation generated by $\xrightarrow{1}$. Assume that $X \approx Y$ and $X' \approx Y'$. Since $\xrightarrow{1}$ is near-striate therefore, for some W and W', both $W \precsim X$ and $W \precsim Y$ and both $W' \precsim X'$ and $W' \precsim Y'$. Since $\langle {}_{[0,\omega)}Un, \circ, {}^{\smile}, \precsim\rangle$ is an ordered algebra, therefore $W \circ W' \precsim X \circ X'$ and $W \circ W' \precsim Y \circ Y'$. Hence, either $X \circ X' \precsim Y \circ Y'$ or $Y \circ Y' \precsim X \circ X'$. In either case, $X \circ X' \approx Y \circ Y'$. Similarly, if $X \approx Y$, then $X^{\smile} \approx Y^{\smile}$. Evidently, if $X \approx Y$, then $\xrightarrow{1} X \approx \xrightarrow{1} Y$. Thus, $\approx$ is a congruence on $\langle {}_{[0,\omega)}Un, \circ, {}^{\smile}, \xrightarrow{1}\rangle$.

(d) Assume that $\emptyset \neq X \in {}_{[0,\omega)}Un$. Then, for some s and t, X has grade $\langle s,t\rangle$. By Theorem 3, $X = {}_{[s,\omega)}\overset{\circ}{1} \circ X$ and $\overset{1}{\to} X = {}_{[s+1,\omega)}\overset{\circ}{1} \circ X$. Since ${}_{[s+1,\omega)}\overset{\circ}{1} \subseteq {}_{[s,\omega)}\overset{\circ}{1}$, therefore $\overset{1}{\to} X \subseteq X$. Since $\overset{1}{\to}$ is near-striate and $X \neq \emptyset$, therefore $\overset{1}{\to} X \neq X$. There follows that $X \not\subseteq \overset{1}{\to} X$.

(e) Assume that $gd\ X = \langle s,t\rangle$. We now slightly modify the argument just used. By Theorem 3, ${}_{[s,\omega)}\overset{\circ}{1} \circ X = X = X \circ {}_{[t,\omega)}\overset{\circ}{1}$ and ${}_{[s+1,\omega)}\overset{\circ}{1} \circ X = \overset{1}{\to} X = X \circ {}_{[t+1,\omega)}\overset{\circ}{1}$. Since $\overset{1}{\to}$ is near-striate and $X \neq \emptyset$, therefore $\overset{1}{\to} X \neq X$. There follows that ${}_{[s+1,\omega)}\overset{\circ}{1} \circ X \neq X \neq X \circ {}_{[t+1,\omega)}\overset{\circ}{1}$ and hence that X satisfies (i).

Assume that X is in ${}_{[0,\omega)}Un$ and that ${}^{[s,\omega)}\overset{\circ}{1} \circ X = X = X \circ {}_{[t,\omega)}\overset{\circ}{1}$ and ${}_{[s+1,\omega)}\overset{\circ}{1} \circ X \neq X \neq {}_{[t+1,\omega)}X$. Since $X \neq \emptyset$, therefore X has some grade $\langle s',t'\rangle$. Hence, as we just saw, ${}_{[s',\omega)}\overset{\circ}{1} \circ X = X \circ {}_{[s',\omega)}\overset{\circ}{1}$ and ${}_{[s'+1,\omega)}\overset{\circ}{1} \circ X \neq X \neq X \circ {}_{[t'+1,\omega)}\overset{\circ}{1}$. Since by Theorem 4(d), there are at most one s' and at most one t' that satisfy these conditions, therefore $s = s'$, $t = t'$ and $gd\ X = \langle s,t\rangle$. There now follows that, for $X \in {}_{[0,\omega)}Un$, $gd\ X = \langle s,t\rangle$ is equivalent to (i).

Let X be any relation on ${}^{[0,\omega)}U$ such that $X \neq \emptyset$. Assume that both $X \subseteq \dot{q}_0^s \circ q_0^t$ and $X \subseteq \dot{q}_0^{s'} \circ q_0^{t'}$. Then, by part (b), either $\langle s,t\rangle = \langle s'+m, t'+m\rangle$ for some m or $\langle s',t'\rangle = \langle s+n, t+n\rangle$ for some n. Hence from $X \not\subseteq \dot{q}_0^{s+1} \circ q_0^{t+1}$ as a further assumption there follows that $\langle s,t\rangle = \langle s'+m, t'+m\rangle$ for some m. Likewise, from $X \not\subseteq \dot{q}_0^{s'} \circ q_0^{t'}$ as a further assumption there follows that $\langle s',t'\rangle = \langle s+n, t+n\rangle$ for some n. This shows that there is at most one $\langle s,t\rangle$ such that both $X \subseteq \dot{q}_0^s \circ q_0^t$ and $X \not\subseteq \dot{q}_0^{s+1} \circ q_0^{t+1}$.

Assume that $gd\ X = \langle s,t\rangle$. By Theorem 3, $X \subseteq \dot{q}_0^s \circ q_0^t$. Also, as we saw earlier, ${}_{[s+1,\omega)}\overset{\circ}{1} \circ X \neq X$. From this there follows that $X \not\subseteq \dot{q}_0^{s+1} \circ q_0^{t+1}$.

Assume that X is in ${}_{[0,\omega)}Un$ and that $X \subseteq \dot{q}_0^s \circ q_0^t$ and $X \not\subseteq \dot{q}_0^{s+1} \circ q_0^{t+1}$. Since $X \neq \emptyset$, therefore X has some grade $\langle s',t'\rangle$. Hence, as we just saw, $X \subseteq \dot{q}_0^{s'} \circ q_0^{t'}$ and $X \not\subseteq \dot{q}_0^{s'+1} \circ q_0^{t'+1}$. Since there is at most one $\langle s',t'\rangle$ satisfying this condition, therefore $\langle s,t\rangle = \langle s',t'\rangle$ and hence $gd\ X = \langle s,t\rangle$. There now follows that, for $X \in {}_{[0,\omega)}Un$, $gd\ X = \langle s,t\rangle$ is equivalent to (ii).

By the relevant part of Theorem 3, for $X \neq \emptyset$, $gd\ X = \langle s,t\rangle$ is equivalent to (iii).

(f) The first assertion follows from the relevant part of Theorem 3. Now assume that $\overset{1}{\to} W = X$ and $gd\ X = \langle s,t\rangle$. Let Q be the initial section of X. Then $Q = \{\langle wy, xy\rangle : |w| = s-1,\ |x| = t-1,\ |y| = 1,\ \langle w,x\rangle \in W\}$. There follows the second equality below. The third follows from the part of Theorem 3 that concerns condition (i).

$$
\begin{aligned}
X &= \overset{[0,\omega)}{\longrightarrow} Q \\
&= \{\langle wyz, xyz\rangle : |w|=s-1,\ |x|=t-1,\ |y|=1,\ \langle w,x\rangle \in W\} \\
&= \langle wyz, xyz\rangle : |w|=s-1,\ |x|=t-1,\ |y|=1,\ \langle wz,xz\rangle \in W\}\ .
\end{aligned}
$$

According to their definitions, q_{s-1} and $q_{t-1}^{\smile}$ satisfy the following equality respectively:

$$
\begin{aligned}
q_{s-1} &= \{\langle wz, wyz\rangle : |w|=s-1,\ |y|=1,\ |z|<\omega\}\ , \\
q_{t-1}^{\smile} &= \{\langle xyz, xz\rangle : |x|=t-1,\ |y|=1,\ |z|<\omega\}\ .
\end{aligned}
$$

From the three equalities just given there follows:

$$
\begin{aligned}
q_{s-1}\circ X \circ q_{t-1}^{\smile} &= \{\langle wz, xz\rangle : |w|=s-1,\ |x|=t-1,\ |z|<\omega, \\
&\qquad \langle wyz, xyz\rangle \in X \text{ for some } y \in {}^{1}U\} = W\ .
\end{aligned}
$$

(g) Assume that X and Y are in ${}_{[0,\omega)}Un \cap -\{\emptyset\}$ and that $gd\ X = \langle s,t\rangle$ and $gd\ Y = \langle s',t'\rangle$. First, also assume that $t \geq s'$. Then, for any s'', $s' \leq s'' < t$ implies that $X \circ ({}_{[s'']}\mathring{1} \circ Y) = \emptyset$. There follows that $X \circ Y = X \circ ({}_{[t,\omega)}\mathring{1} \circ Y)$. Since $gd\ Y = \langle s',t'\rangle$, therefore ${}_{[t,\omega)}\mathring{1} \circ Y = Y \circ {}_{[t'+t-s',\omega)}\mathring{1}$ and $gd(Y \circ {}_{[t'+t-s',\omega)}\mathring{1}) = \langle t, t'+t-s'\rangle$. There follows that $X \circ Y = X \circ (Y \circ {}_{[t'+t-s',\omega)}\mathring{1})$ and that $gd\ X = \langle s,t\rangle$ is attuned to $gd(Y \circ {}_{[t'+t-s',\omega)}\mathring{1})$. Now assume that $t \leq s'$. Then a similar argument shows that $X \circ Y = (X \circ {}_{[s',\omega)}\mathring{1}) \circ Y = ({}_{[s+s'-t,\omega)}\mathring{1} \circ X) \circ Y$, and that $gd({}_{[s+s'-t,\omega)}\mathring{1} \circ X) = \langle s+s'-t, s'\rangle$ and hence is attuned to $gd\ Y = \langle s',t'\rangle$. □

In contrast to Theorem 7(c), if A is the set of all relations on ${}^{[0,\omega)}U$ and if $\precsim$ and $\approx$ are the ordering and equivalence relation, respectively, that is generated by the operation $\overset{1}{\rightarrow}$ on A, then $\langle A, \circ, {}^{\smile}, \precsim\rangle$ is not an ordered algebra and $\approx$ is not a congruence of $\langle A, \circ, {}^{\smile}\rangle$. To show this we consider the following relation X on ${}^{[0,\omega)}U$:

$$
\begin{aligned}
X &= e_{01} \cup {}_{[3,\omega)}\mathring{1} \\
&= \{\langle x,x\rangle : 2 \leq |x| < \omega,\ x_0 = x_1\} \cup \{\langle x,x\rangle : 3 \leq |x| < \omega\} \\
&= \{\langle x,x\rangle : |x| = 2,\ x_0 = x_1\} \cup \{\langle x,x\rangle : 3 \leq |x| < \omega\}\ .
\end{aligned}
$$

Evidently, $X \precsim X$ and $X \approx X$. Also ${}_{[2,\omega)}\mathring{1} \precsim {}_{[3,\omega)}\mathring{1}$ and ${}_{[2,\omega)}\mathring{1} \approx {}_{[3,\omega)}\mathring{1}$. Further, $X \circ {}_{[2,\omega)}\mathring{1} = X$ and $X \circ {}_{[3,\omega)}\mathring{1} = {}_{[3,\omega)}\mathring{1}$. However, neither $X \precsim {}_{[3,\omega)}\mathring{1}$ nor $X \approx {}_{[3,\omega)}\mathring{1}$.

Theorem 7(e), like Theorem 3, gives criteria for the condition that $gd\ X = \langle s,t\rangle$. Both theorems give the following criterion: $X \subseteq \dot{q}_0^s \circ q_0^t$ and $\dot{q}_s \circ X = X \circ q_t$. The other structural criteria given by Theorem 3 apply to the algebra $\langle A, \circ, {}^\smile, \xrightarrow{1}, {}_{[0,\omega)}\mathring{1}\rangle$ or structure $\langle A, \circ, {}^\smile, \xrightarrow{1}, {}_{[0,\omega)}\mathring{1}, \subseteq\rangle$, where A is the set of all relations on ${}_{[0,\omega)}U$. The two other criteria given by Theorem 7(e), namely (i) and (ii), apply to the algebra $\langle {}_{[0,\omega)}Un, \circ, {}_{[s,\omega)}\mathring{1}\rangle_{s<\omega}$ or structure $\langle {}_{[0,\omega)}Un, \circ, \dot{q}_0, q_0, \subseteq\rangle$, respectively, and do not involve $\xrightarrow{1}$.

Without the further assumption that X is in ${}_{[0,\omega)}Un$, neither (i) nor (ii) implies that $gd\ X = \langle s,t\rangle$. Any $X \neq \emptyset$ that is of rank $\langle s,t\rangle$ satisfies both (i) and (ii) but not $gd\ X = \langle s,t\rangle$.

Theorems 7(e) and 7(f) together show that $\xrightarrow{1}$ can be defined in $\langle {}_{[0,\omega)}Un, \circ, {}_{[s,\omega)}\mathring{1}\rangle_{s<\omega}$. One uses an infinitary language to describe, for each of the pairs $\langle s,t\rangle$ in $\omega\times\omega$, the behavior of $\xrightarrow{1}$ on $\{X : gd\ X = \langle s,t\rangle\}$.

We now turn to the set ${}_{[\omega]}Un$ of uniform relations on ${}^\omega U$. Recall that ${}_{[\omega]}Un = \{\xrightarrow{\omega} R : R \in Rk\} = \{\xrightarrow{\omega} W : W \in {}_{[0,\omega)}Un\}$.

Given any relation X on ${}^\omega U$, we shall say that $\langle s,t\rangle$ is ***a grade of*** X if and only if $X = \xrightarrow{\omega} W$ for some W in ${}_{[\omega]}Un$ whose grade is $\langle s,t\rangle$. Also, $\emptyset = \xrightarrow{\omega}\emptyset$ shall have ***grade*** $\dot{0}$. No other relation on ${}^\omega U$ shall have a grade. Thus, a relation on ${}^\omega U$ has a grade if and only if it is in ${}_{[\omega]}Un$.

Theorem 8(a) below gives a set-theoretic characterization of grade that does not refer to Rk or to ${}_{[\omega]}Un$. It is an analogue of the part of Theorem 3 that involves condition (ii). Whereas later parts of Theorem 3 give algebraic or structural characterizations of grade, Theorem 8 contains no such characterization.

THEOREM 8. *Let X be any relation on ${}^\omega U$.*

(a) *X has grade $\langle s,t\rangle$ if and only if $X \neq \emptyset$ and the following hold:*

(α) *$X \subseteq \xrightarrow{\omega}(\dot{q}_0^s \circ q_0^t) = \{\langle wz, xz\rangle : |w|=s,\ |x|=t,\ |z|=\omega\}$, and*

(β) *If $|w|=s,\ |x|=t,\ |z|=|z'|=\omega$, then $\langle wz, xz\rangle \in X$ if and only if $\langle wz', xz'\rangle \in X$.*

(b) *If X has grade $\langle s,t\rangle$ then, for any n, X also has grade $\langle s+n, t+n\rangle$. Also, assuming that $\overline{\overline{U}} \geq 2$, if X has grade $\langle s,t\rangle$ and $s'-t' \neq s-t$, then X does not have grade $\langle s',t'\rangle$. There follows that, if $\overline{\overline{U}} \geq 2$ then, for any X in ${}_{[\omega]}Un \cap -\{\emptyset\}$, there is a unique $\langle s,t\rangle$ in $\omega\times\omega$ such that the set of grades of X is $\{\langle s+n, t+n\rangle : n<\omega\}$.*

Proof. (a) This follows from ${}_{[\omega]}Un = \{\xrightarrow{\omega} W : W \in {}_{[0,\omega)}Un\}$ and the part of Theorem 3 that concerns (ii).

(b) Assume that X has grade $\langle s,t\rangle$. Then $X = \overset{\omega}{\rightarrow} W$ for some W in ${}_{[0,\omega)}Un$ such that $gd\ W = \langle s,t\rangle$. Let R be the initial section of W, so that R is of rank $\langle s,t\rangle$ and $W = \bigcup\{\overset{n}{\rightarrow} R : n < \omega\}$. Using Lemma 1(b) for the third equality one obtains:

$$\overset{\omega}{\rightarrow}(\overset{1}{\rightarrow} W) = \overset{\omega}{\rightarrow}(\overset{1}{\rightarrow} \bigcup\{\overset{n}{\rightarrow} R : n < \omega\}) = \bigcup\{\overset{\omega}{\rightarrow}(\overset{1}{\rightarrow}(\overset{n}{\rightarrow} R)) : n < \omega\}$$
$$= \bigcup\{\overset{\omega}{\rightarrow}(\overset{n}{\rightarrow} R) : n < \omega\} = \overset{\omega}{\rightarrow}(\bigcup \overset{n}{\rightarrow} R : n < \omega\}) = \overset{\omega}{\rightarrow} W = X\,.$$

Since $gd\ W = \langle s,t\rangle$, therefore, by Lemma 5(c), $gd(\overset{1}{\rightarrow} W) = \langle s+1, t+1\rangle$. There follows that X also has grade $\langle s+1, t+1\rangle$. By induction, X also has grade $\langle s+n, t+n\rangle$.

Assume that $\overline{\overline{U}} \geq 2$. Also assume that X has grade $\langle s,t\rangle$ and that $s'-t' \neq s-t$. For proof by contradiction, suppose that $\langle s',t'\rangle$ also is a grade of X. Then, as we just saw, for any n and n', $\langle s+n, t+n\rangle$ and $\langle s'+n', t'+n'\rangle$ also are grades of X. Choose n and n' such that $s+n = s'+n'$. Since $s'-t' \neq s-t$, there follows that $(s+n)-(t+n) \neq (s'+n')-(t'+n')$ and hence that either $t+n > t'+n'$ or $t+n < t'+n'$. Suppose that $t+n > t'+n'$. Let $r = (t+n)-(t'+n')$. Since X has grade $\langle s+n, t+n\rangle$, there are w, x, y such that $|w| = s+n = s'+n'$, $|x| = t'+n'$, $|y| = r$, and hence $|xy| = t+n$, and such that, for any z in ${}^{\omega}U$, $\langle wz, xyz\rangle$ is in X. Since $\overline{\overline{U}} \geq 2$, there is some y' such that $|y'| = r$ and $y' \neq y$. Then, for any z' in ${}^{\omega}U$, $\langle wy'z', xyy'z'\rangle$ is in X. On the other hand, since $y'z' \neq yy'z'$ and since X has grade $\langle s'+n, t'+n\rangle$, there follows from condition (α) in part (a) that $\langle wy'z', xyy'z\rangle$ is not in X. If $t+n < t'+n'$, then a similar argument again leads to a contradiction. There now follows that, if $s'-t' \neq s-t$, then $\langle s',t'\rangle$ is not a grade of X.

Again assume that $\overline{\overline{U}} \geq 2$. Also assume that X is in ${}_{[\omega]}Un \cap -\{\emptyset\}$ and hence $X = \overset{\omega}{\rightarrow} W$ for some $W \neq \emptyset$ in ${}_{[0,\omega)}Un$. Let $\langle s,t\rangle$ be the grade of W. Since $\overset{1}{\rightarrow}$ is near-striate, there is a unique W_0 and a unique $m \geq 0$ such that $W = \overset{m}{\rightarrow} W_0$ and $W_0 \notin Rg \overset{1}{\rightarrow}$. Let $s_0 = s-m$ and $t_0 = t-m$. Then W_0 is in ${}_{[0,\omega)}Un$ and $gd\ W_0 = \langle s_0, t_0\rangle$. Also, for any $n \geq 0$, $\overset{\omega}{\rightarrow}(\overset{n}{\rightarrow} W_0) = \overset{\omega}{\rightarrow}(\overset{m}{\rightarrow} W_0) = \overset{\omega}{\rightarrow} W = X$. There follows that the set of those grades $\langle s',t'\rangle$ of X such that $s'-t' = s-t$ is the set $\{\langle s_0+n, t_0+n\rangle : n < \omega\}$. Since $s'-t' \neq s-t$ implies that $\langle s',t'\rangle$ is not a grade of X, therefore $\{\langle s_0+n, t_0+n\rangle : n < \omega\}$ is the set of all grades of X. □

Assume that $\overline{\overline{U}} \geq 2$ and that X is in ${}_{[\omega]}Un$. We define $lgd\ X$, ***the least grade of*** X, as follows: If $X = \emptyset$, then $lgd\ X = \dot{0}$, and if $X \neq \emptyset$ and $\{\langle s+n, t+n\rangle : n < \omega\}$

is the set of grades of X, then $lgd\ X = \langle s,t\rangle$. If $\overline{\overline{U}} = 1$ or if X is a relation on ${}^{\omega}U$ that is not in ${}_{[\omega]}Un$, then lgd shall be undefined.

The following relations in ${}_{[\omega]}Un$ will serve to illustrate these notions.

$$\begin{aligned}
\overset{\omega}{\to} p_{(i,j)} &= \{\langle x,y\rangle : |x|=|y|=\omega,\ x_i=y_j,\ x_j=y_i,\ x_k=y_k \text{ if } k\notin\{i,j\}\}\\
\overset{\omega}{\to} e_{ij} &= \{\langle x,x\rangle : |x|=\omega,\ x_i=x_j\}\\
\overset{\omega}{\to} \dot{q}_i &= \{\langle x,y\rangle : |x|=|y|=\omega,\ y_n=x_n \text{ if } n<i,\ y_n=x_{n+1} \text{ if } n\geq i\}\\
&= \{\langle vwz, vz\rangle : |v|=i,\ |w|=1,\ |z|=\omega\}\\
\overset{\omega}{\to} q_i &= \overset{\omega}{\to}(\dot{q}_i^{\smile}) = (\overset{\omega}{\to}\dot{q}_i)^{\smile}\\
\overset{\omega}{\to} \iota_{(j/\lambda)} &= \{\langle x,x\rangle : |x|=\omega,\ x_j=u_\lambda\}
\end{aligned}$$

By Lemma 2(a), the grade of $\dot{q}_i$ is $\langle i+1,i\rangle$, that of q_i is $\langle i,i+1\rangle$ and, if $i\leq j$, then that of $p_{(i,j)}$, that of e_{ij}, and that of $\iota_{(j/\lambda)}$ is $\langle j+1,j+1\rangle$. Moreover, none of these relations is in $Rg\overset{1}{\to}$. There follows that, if $\overline{\overline{U}}\geq 2$, then the least grade of $\overset{\omega}{\to}\dot{q}_i$ is $\langle i+1,i\rangle$, that of $\overset{\omega}{\to}q_i$ is $\langle i,i+1\rangle$ and, if $i\leq j$, then that of $\overset{\omega}{\to}p_{(i,j)}$, $\overset{\omega}{\to}e_{ij}$, and $\overset{\omega}{\to}i_{(j/\lambda)}$ is $\langle j+1,j+1\rangle$. Also, for example, the set of grades of $\overset{\omega}{\to}\dot{q}_i$ is $\{\langle i+1+n, i+n\rangle : n<\omega\}$.

LEMMA 9. (a) *If Q and R are of the same rank, then $\overset{\omega}{\to}Q\subseteq\overset{\omega}{\to}R$ implies that $Q\subseteq R$. Also, if X and Y are in ${}_{[0,\omega)}Un$ and have the same grade, then $\overset{\omega}{\to}X\subseteq\overset{\omega}{\to}Y$ implies that $X\subseteq Y$.*

(b) *If Q is of rank $\langle r,s\rangle$ and R is of rank $\langle s,t\rangle$, then $\overset{\omega}{\to}(Q\circ R)=\overset{\omega}{\to}Q\circ\overset{\omega}{\to}R$.*

(c) *For any X and Y in ${}_{[0,\omega)}Un$, $\overset{\omega}{\to}(X\circ Y)=\overset{\omega}{\to}X\circ\overset{\omega}{\to}Y$.*

(d) *If $\overline{\overline{U}}\geq 2$, then the following two relations coincide:*

$$\{\langle X,Y\rangle : X\in{}_{[0,\omega)}Un,\ Y\in{}_{[0,\omega)}Un,\ \overset{\omega}{\to}X=\overset{\omega}{\to}Y\}$$
$$\{\langle X,Y\rangle : X\in{}_{[0,\omega)}Un,\ Y\in{}_{[0,\omega)}Un,\ \overset{m}{\to}X=\overset{n}{\to}Y \textit{ for some } m \textit{ and } n\}$$

Proof. (a) Assume that Q and R both are of rank $\langle s,t\rangle$ and that $\overset{\omega}{\to}Q\subseteq\overset{\omega}{\to}R$. Consider any $\langle x,y\rangle$ in Q. Then $|x|=s$ and $|y|=t$. Choose any z such that $|z|=\omega$. Then $\langle xz,yz\rangle$ is in $\overset{\omega}{\to}Q$ and hence in $\overset{\omega}{\to}R$. Since $|x|=s$ and $|y|=t$, there follows that $\langle x,y\rangle$ is in R. There follows that $Q\subseteq R$.

Now assume that X and Y are in ${}_{[0,\omega)}Un$, both have grade $\langle s,t\rangle$, and $\overset{\omega}{\to}X\subseteq\overset{\omega}{\to}Y$. Let Q and R be the initial segment of X or of Y, respectively. Then Q and R both have rank $\langle s,t\rangle$ and, by Lemma 1(b), $\overset{\omega}{\to}Q=\overset{\omega}{\to}X$ and $\overset{\omega}{\to}R=\overset{\omega}{\to}Y$. Since

$\xrightarrow{\omega} X \subseteq \xrightarrow{\omega} Y$, therefore $\xrightarrow{\omega} Q \subseteq \xrightarrow{\omega} R$. Then $Q \subseteq R$, as was shown above. There follows that $X = \xrightarrow{[0,\omega)} Q \subseteq \xrightarrow{[0,\omega)} R = Y$.

(b) Assume that Q is of rank $\langle r, s\rangle$ and R is of rank $\langle s, t\rangle$. Consider any $\langle v, v''\rangle$ in $\xrightarrow{\omega} Q \circ \xrightarrow{\omega} R$. Then there is some v' such that $\langle v, v'\rangle$ is in $\xrightarrow{\omega} Q$ and $\langle v', v''\rangle$ is in $\xrightarrow{\omega} R$. Let $v' = xz$, where $|x| = s$ and $|z| = \omega$. Since Q is of rank $\langle r, s\rangle$ and $\langle v, v'\rangle = \langle v, xz\rangle$ is in $\xrightarrow{\omega} Q$, therefore $v = wz$ for some w such that $\langle w, x\rangle$ is in Q. Since R is of rank $\langle s, t\rangle$ and $\langle v', v''\rangle = \langle xz, v''\rangle$ is in $\xrightarrow{\omega} R$, therefore $v'' = yz$ for some y such that $\langle x, y\rangle$ is in R. Since $\langle w, x\rangle$ is in Q and $\langle x, y\rangle$ is in R, therefore $\langle w, y\rangle$ is in $Q \circ R$ and hence $\langle v, v''\rangle = \langle wz, yz\rangle$ is in $\xrightarrow{\omega}(Q \circ R)$. There now follows that $\xrightarrow{\omega} Q \circ \xrightarrow{\omega} R \subseteq \xrightarrow{\omega}(Q \circ R)$. Since $\xrightarrow{\omega}(Q' \circ R') \subseteq \xrightarrow{\omega} Q' \circ \xrightarrow{\omega} R'$ holds for any relations Q' and R' on ${}^{[0,\omega)}U$, as is easy to show, there now follows that $\xrightarrow{\omega}(Q \circ R) = \xrightarrow{\omega} Q \circ \xrightarrow{\omega} R$.

(c) We first consider the case where X and Y are in ${}_{[0,\omega)}Un \cap -\{\emptyset\}$ and where, moreover, $gd\ X$ is attuned to $gd\ Y$, so that, for some r, s and t, $gd\ X = \langle r, s\rangle$ and $gd\ Y = \langle s, t\rangle$. We let Q be the initial section of X and R the initial section of Y. Then $X = \bigcup\{\xrightarrow{m} Q : m < \omega\}$ and $Y = \bigcup\{\xrightarrow{n} R : n < \omega\}$. Since $\xrightarrow{m} Q$ is of rank $\langle r+m, s+m\rangle$ and $\xrightarrow{n} R$ is of rank $\langle s+n, t+n\rangle$ therefore, if $m \neq n$, then $\xrightarrow{m} Q \circ \xrightarrow{n} R = \emptyset$. There follows:

$$\begin{aligned} X \circ Y &= \bigcup\{\xrightarrow{m} Q : m < \omega\} \circ \bigcup\{\xrightarrow{n} R : n < \omega\} \\ &= \bigcup\{\xrightarrow{m} Q \circ \xrightarrow{n} R : m < \omega,\ n < \omega\} \\ &= \bigcup\{\xrightarrow{n} Q \circ \xrightarrow{n} R : n < \omega\} = \bigcup\{\xrightarrow{n}(Q \circ R) : n < \omega\}\ , \end{aligned}$$

the last equality following from Lemma 1(c). By Lemma 1(b),

$$\xrightarrow{\omega}(X \circ Y) = \xrightarrow{\omega}(\bigcup\{\xrightarrow{n}(Q \circ R) : n < \omega\}) = \xrightarrow{\omega}(Q \circ R).$$

Also by Lemma 1(b),

$$\begin{aligned} \xrightarrow{\omega} X &= \xrightarrow{\omega}(\bigcup\{\xrightarrow{m} Q : m < \omega\}) = \xrightarrow{\omega} Q \text{ and} \\ \xrightarrow{\omega} Y &= \xrightarrow{\omega}(\bigcup\{\xrightarrow{n} R : n < \omega\}) = \xrightarrow{\omega} R. \end{aligned}$$

There follows by part (b) that $\xrightarrow{\omega}(X \circ Y) = \xrightarrow{\omega} X \circ \xrightarrow{\omega} Y$.

Now consider any X and Y in ${}_{[0,\omega)}Un \cap -\{\emptyset\}$. Then, by Theorem 7(g), for some k and k', $X \circ Y = (\xrightarrow{k} X) \circ (\xrightarrow{k'} Y)$ and $gd(\xrightarrow{k} X)$ is attuned to $gd(\xrightarrow{k'} Y)$. By what has just been proved, $\xrightarrow{\omega}((\xrightarrow{k} X) \circ (\xrightarrow{k'} Y)) = \xrightarrow{\omega}(\xrightarrow{k} X) \circ \xrightarrow{\omega}(\xrightarrow{k'} Y)$ and hence $\xrightarrow{\omega}(X \circ Y) = \xrightarrow{\omega}(\xrightarrow{k} X) \circ (\xrightarrow{k'} Y)$. Since, by Lemma 1(b), $\xrightarrow{\omega}(\xrightarrow{k} X) = \xrightarrow{\omega} X$ and $\xrightarrow{\omega}(\xrightarrow{k'} Y) = \xrightarrow{\omega} Y$, there follows that again $\xrightarrow{\omega}(X \circ Y) = \xrightarrow{\omega} X \circ \xrightarrow{\omega} Y$. Finally, if either $X = \emptyset$ or $Y = \emptyset$, then $\xrightarrow{\omega}(X \circ Y) = \emptyset = \xrightarrow{\omega} X \circ \xrightarrow{\omega} Y$.

(d) Consider any X and Y in ${}_{[0,\omega)}Un$. First assume that $\overset{m}{\rightarrow} X = \overset{n}{\rightarrow} Y$ for some m and n. Then, by Lemma 1(b),

$$\overset{\omega}{\rightarrow} X \;=\; \overset{\omega}{\rightarrow}(\overset{m}{\rightarrow} X) \;=\; \overset{\omega}{\rightarrow}(\overset{n}{\rightarrow} Y) \;=\; \overset{\omega}{\rightarrow} Y\,.$$

Now assume that $\overline{\overline{U}} \geq 2$ and that $\overset{\omega}{\rightarrow} X = \overset{\omega}{\rightarrow} Y$. If $\overset{\omega}{\rightarrow} X = \emptyset$ and hence $\overset{\omega}{\rightarrow} Y = \emptyset$, then $X = \emptyset$ and $Y = \emptyset$, and hence $\overset{m}{\rightarrow} X = \emptyset = \overset{n}{\rightarrow} Y$ for some m and n. Now assume that $\overset{\omega}{\rightarrow} X = \overset{\omega}{\rightarrow} Y \neq \emptyset$. Then, by Lemma 8(b), $\overset{\omega}{\rightarrow} X$ and hence $\overset{\omega}{\rightarrow} Y$ have a least grade $\langle s,t\rangle$ and the set of their grades is $\{\langle s+n, t+n\rangle : n < \omega\}$. Hence, for some m, the grade of X is $\langle s+m, t+m\rangle$ and, for some n, the grade of Y is $\langle s+n, t+n\rangle$. Suppose, first that $m \leq n$. Let $k = n - m$. Then the grade of $\overset{k}{\rightarrow} X$ is $\langle s+n, t+n\rangle$ and $\overset{\omega}{\rightarrow}(\overset{k}{\rightarrow} X) = \overset{\omega}{\rightarrow} X = \overset{\omega}{\rightarrow} Y$. Then, by part (a), $\overset{k}{\rightarrow} X = Y$. Now suppose that $n \geq m$. Let $k = m - n$. Then, by a similar argument, $\overset{k}{\rightarrow} Y = X$. Thus, in either case, there are m' and n' such that $\overset{m'}{\rightarrow} X = \overset{n'}{\rightarrow} Y$. □

The condition that Q and R are of the same rank or that X and Y have the same grade cannot be omitted from Lemma 9(a). For example,

$$\overset{\omega}{\rightarrow}({}_{[s]}\mathring{1}) = \overset{\omega}{\rightarrow}({}_{[s+1]}\mathring{1}) \quad\text{and}\quad \overset{\omega}{\rightarrow}({}_{[s,\omega)}\mathring{1}) = \overset{\omega}{\rightarrow}({}_{[s+1,\omega)}\mathring{1})\,.$$

The condition on the ranks of Q and R in Lemma 9(b) also cannot be omitted. For example, let $Q = {}^{r}U \times {}^{s}U$ and $R = {}^{s+1}U \times {}^{t+1}U$. Then $Q \circ R = \emptyset$ and hence $\overset{\omega}{\rightarrow}(Q \circ R) = \emptyset$, while $\overset{\omega}{\rightarrow} Q = \overset{\omega}{\rightarrow}(\dot{q}_0^{r} \circ q_0^{s})$ and $\overset{\omega}{\rightarrow}(\dot{q}_0^{s} \circ q_0^{t}) \subseteq \overset{\omega}{\rightarrow}(\dot{q}_0^{s+1} \circ q_0^{t+1}) = \overset{\omega}{\rightarrow} R$ and hence $\emptyset \neq \overset{\omega}{\rightarrow}(\dot{q}_0^{r} \circ q_0^{t}) \subseteq \overset{\omega}{\rightarrow} Q \circ \overset{\omega}{\rightarrow} R$.

By the example just given, there are relations X and Y on ${}^{[0,\omega)}U$ such that $\overset{\omega}{\rightarrow} X \circ \overset{\omega}{\rightarrow} Y \not\subseteq \overset{\omega}{\rightarrow}(X \circ Y)$. Thus the condition that X and Y are in ${}_{[0,\omega)}Un$ cannot be omitted from Lemma 9(c). (Note, however, that $\overset{\omega}{\rightarrow}(X \circ Y) \subseteq \overset{\omega}{\rightarrow} X \circ \overset{\omega}{\rightarrow} Y$ holds for any relations X and Y on ${}^{[0,\omega)}U$.)

Suppose that $\overline{\overline{U}} = 1$. Then for any X and Y in ${}_{[0,\omega)}Un$ such that $X \neq \emptyset$ and $Y \neq \emptyset$ there holds $\overset{\omega}{\rightarrow} X = \overset{\omega}{\rightarrow} Y = {}^{\omega}U$. Now suppose that $gd\ X = \langle s,t\rangle$, $gd\ Y = \langle s',t'\rangle$ and $\langle s,t\rangle \neq \langle s',t'\rangle$. Consider any m and n. Then $gd(\overset{m}{\rightarrow} X) = \langle s+m, t+m\rangle$, $gd(\overset{n}{\rightarrow} Y) = \langle s'+n, t'+n\rangle$, and $(s+m)-(t+m) \neq (s'+n)-(t'+n)$. Then, by Theorem 7(b), $\overset{m}{\rightarrow} X \cap \overset{n}{\rightarrow} Y = \emptyset$ and hence $\overset{m}{\rightarrow} X \neq \overset{n}{\rightarrow} Y$. Hence the condition that $\overline{\overline{U}} \geq 2$ cannot be omitted from Lemma 9(d).

THEOREM 10. (a) *${}_{[\omega]}Un$ is closed under $\circ$ and $\smile$, and $\langle {}_{[\omega]}Un, \circ, \smile, \subseteq\rangle$ is an ordered algebra. Also, ${}_{[\omega]}Un$ includes $\{\overset{\omega}{\rightarrow} X : X \in S_{pqei}\}$ and contains $\overset{\omega}{\rightarrow}(q_0 \circ \dot{q}_0) = {}_{\omega}\mathring{1}$ and every $\overset{\omega}{\rightarrow}(\dot{q}_0^{s} \circ q_0^{t}) = (\overset{\omega}{\rightarrow} \dot{q}_0^{s}) \circ (\overset{\omega}{\rightarrow} q_0^{t})$.*

(b) *For every s and t, $\overset{\omega}{\rightarrow}(\dot{q}_0^s \circ q_0^t) \subseteq \overset{\omega}{\rightarrow}(\dot{q}_0^{s+1} \circ q_0^{t+1})$, the inclusion being proper whenever $\overline{\overline{U}} \geq 2$. Also, for every X in ${}_{[\omega]}Un$, $X \subseteq \overset{\omega}{\rightarrow}(\dot{q}_0^s \circ q_0^t)$ for at least one s and t. Moreover, if $\overline{\overline{U}} \geq 2$ then, for any relations X and Y on ${}^{\omega}U$, if $X \subseteq \overset{\omega}{\rightarrow}(\dot{q}_0^s \circ q_0^t)$, $Y \subseteq \overset{\omega}{\rightarrow}(\dot{q}_0^{s'} \circ q_0^{t'})$, and $s-t \neq s'-t'$, then $W \in {}_{[\omega]}Un$, $W \subseteq X$, and $W \subseteq Y$ together imply that $W = \emptyset$. There follows that if $\overline{\overline{U}} \geq 2$, $X \subseteq \overset{\omega}{\rightarrow}(\dot{q}_0^s \circ q_0^t)$, $Y \subseteq \overset{\omega}{\rightarrow}(\dot{q}_0^{s'} \circ q_0^{t'})$, $X \in {}_{[\omega]}Un$, and $X \neq \emptyset$, then $X \subseteq Y$ implies that $s - t = s' - t'$.*

(c) *Assume that X is in ${}_{[\omega]}Un$ and $X \neq \emptyset$. Then $\langle s, t\rangle$ is a grade of X if and only if $X \subseteq \overset{\omega}{\rightarrow}(\dot{q}_0^s \circ q_0^t)$ and $(\overset{\omega}{\rightarrow} \dot{q}_s) \circ X = X \circ (\overset{\omega}{\rightarrow} \dot{q}_t)$. Hence, if $\overline{\overline{U}} \geq 2$, then $lgd\, X = \langle s, t\rangle$ if and only if $X \subseteq \overset{\omega}{\rightarrow}(\dot{q}_0^s \circ q_0^t)$, $(\overset{\omega}{\rightarrow} \dot{q}_s) \circ X = X \circ (\overset{\omega}{\rightarrow} \dot{q}_t)$, and, if neither $s=0$ nor $t=0$, then either $X \not\subseteq \overset{\omega}{\rightarrow}(\dot{q}_0^{s-1} \circ q_0^{t-1})$ or $(\overset{\omega}{\rightarrow} \dot{q}_{s-1}) \circ X \neq X \circ (\overset{\omega}{\rightarrow} \dot{q}_{t-1})$.*

Proof. (a) Since ${}_{[\omega]}Un = \{\overset{\omega}{\rightarrow} X : X \in {}_{[0,\omega)}Un\}$ therefore, by Theorem 7(a) and Lemma 9(c), ${}_{[\omega]}Un$ is closed under $\circ$. Now consider any relation X on $[0,\omega)U$, any x and y in ${}^{[0,\omega)}U$, and any z in ${}^{\omega}U$. Then the following are equivalent to each other: $\langle xz, yz\rangle \in \overset{\omega}{\rightarrow}(X^{\smile})$; $\langle x, y\rangle \in X^{\smile}$; $\langle y, x\rangle \in X$; $\langle yz, xz\rangle \in \overset{\omega}{\rightarrow} X$; $\langle xz, yz\rangle \in (\overset{\omega}{\rightarrow}(X^{\smile})$. Hence $(\overset{\omega}{\rightarrow} X^{\smile}) = \overset{\omega}{\rightarrow}(X^{\smile})$. Since this holds, in particular, for any X in ${}_{[0,\omega)}Un$ and since ${}_{[0,\omega)}Un$ is closed under $^{\smile}$, there follows that ${}_{[\omega]}Un$ is closed under $^{\smile}$. Hence by the set-theoretic definition of $\circ$ and of $^{\smile}$, $\langle {}_{[\omega]}Un, \circ, {}^{\smile}, \subseteq\rangle$ is an ordered algebra. Also, by Lemma 2, ${}_{[\omega]}Un$ includes $\{\overset{\omega}{\rightarrow} X : X \in S_{pqei}\}$. Hence ${}_{[\omega]}Un$ contains $\overset{\omega}{\rightarrow}(q_0 \circ q_0^{\smile}) = {}_{[\omega]}\overset{\circ}{1}$. Also, since $\overset{\omega}{\rightarrow}(\dot{q}_s \circ q_t) = (\overset{\omega}{\rightarrow} \dot{q}_0) \circ (\overset{\omega}{\rightarrow} q_t)$, by Lemma 9(c), therefore ${}_{[\omega]}Un$ also contains every $\overset{\omega}{\rightarrow}(\dot{q}_0) \circ (\overset{\omega}{\rightarrow} \dot{q}_t)$.

(b) Since, by Theorem 7(b), every W in ${}_{[0,\omega)}Un$ is included in at least one $\dot{q}_s \circ q_t$ and since ${}_{[\omega]}Un = \{\overset{\omega}{\rightarrow} W : W \in {}_{[0,\omega)}Un\}$, therefore every X in ${}_{[\omega]}Un$ is included in at least one $\overset{\omega}{\rightarrow}(\dot{q}_s \circ q_t)$.

Since $\overset{1}{\rightarrow}(\dot{q}_s \circ q_t) \subseteq \dot{q}_0^{s+1} \circ q_0^{t+1}$ and $\overset{\omega}{\rightarrow}(\dot{q}_0^s \circ q_0^t) = \overset{\omega}{\rightarrow}(\overset{1}{\rightarrow} \dot{q}_0^s \circ q_0^t))$, therefore $\overset{\omega}{\rightarrow}(\dot{q}_0^s \circ q_0^t) \subseteq \overset{\omega}{\rightarrow}(\dot{q}_0^{s+1} \circ q_0^{t+1})$. Now assume that $\overline{\overline{U}} \geq 2$. Given any s and t, consider any w, x, y, y', z such that $|w| = s$, $|x| = t$, $|y| = |y'| = 1$, $|z| = \omega$, and $y \neq y'$. Then $\langle wyz, xy'z\rangle$ is in $\overset{\omega}{\rightarrow}(\dot{q}_{s+1} \circ q_{t+1})$ but not in $\overset{\omega}{\rightarrow}(\dot{q}_s \circ q_t)$. Hence $\overset{\omega}{\rightarrow}(\dot{q}_{s+1} \circ q_{t+1}) \not\subseteq \overset{\omega}{\rightarrow}(\dot{q}_s \circ q_t)$.

Consider any v in ${}^{\omega}U$. Given any $r > 0$ and any $n_0 \geq 0$, we shall say that v is ***r-periodic from n_0 on*** if and only if $v_n = v_{n+r}$ for every $n \geq n_0$. Also, v shall be ***eventually periodic*** if and only if, for at least one $r > 0$ and at least one $n_0 \geq 0$, v is r-periodic from n_0 on. Whenever $\overline{\overline{U}} \geq 2$, then there are v in ${}^{\omega}U$ that are not eventually periodic.

Consider any W in ${}_{[\omega]}Un$ such that $W \neq \emptyset$. Then, for some s and t, W has grade $\langle s, t\rangle$. By Theorem 8, there are w and x such that $|w| = s$, $|x| = t$, and, for any z, if $|z| = \omega$, then $\langle wz, xz \rangle$ is in W. Now assume that $\overline{\overline{U}} \geq 2$. Then there are z in ${}^{\omega}U$ such that z is not eventually periodic. There follows that if $\overline{\overline{U}} \geq 2$, then W contains pairs $\langle wz, xz \rangle$ such that none of z, wz, xz is eventually periodic.

Next, consider any relations X and Y on ${}^{\omega}U$ such that, for some $r > 0$ and some s and t, $X \subseteq \overset{\omega}{\rightarrow} (\dot{q}_0^s \circ q_0^{t+r})$ and $Y \subseteq \overset{\omega}{\rightarrow} (\dot{q}_0^s \circ q_0^t)$. Consider any $\langle v, w \rangle$ in $X \cap Y$. Since $X \subseteq \overset{\omega}{\rightarrow} (\dot{q}_0^s \circ q_0^{t+r})$, therefore $v_{s+m} = w_{(t+r)+m}$ for any m. Since $Y \subseteq \overset{\omega}{\rightarrow} (\dot{q}_0^s \circ q_0^t)$, therefore $w_{t+m} = v_{s+m}$ for any m. There follows that, for any m,

$$\begin{aligned} v_{s+m} &= w_{(t+r)+m} = w_{t+(r+m)} = v_{s+(r+m)} = v_{s+(m+r)} \\ &= w_{(t+r)+(m+r)} = w_{((t+r)+m)+r} \,. \end{aligned}$$

Hence v is r-periodic from s on and w is r-periodic from $t + r$ on. A similar argument shows that if, for some $r > 0$ and some s and t, $X \subseteq \overset{\omega}{\rightarrow} (\dot{q}_0^s \circ q_0^t)$ and $Y \subseteq \overset{\omega}{\rightarrow} (\dot{q}_0^s \circ q_0^{t+r})$, then any $\langle v, w \rangle$ in $X \cap Y$ is such that v is r-periodic from $s+r$ on and w is r-periodic from t on. In either case, v and w both are eventually periodic.

Now let X and Y be any relations on ${}^{\omega}U$ such that, for some s, t, s', and t', $X \subseteq \overset{\omega}{\rightarrow} (\dot{q}_0^s \circ q_0^t)$, $Y \subseteq \overset{\omega}{\rightarrow} (\dot{q}_0^{s'} \circ q_0^{t'})$, and $s - t \neq s' - t'$. First suppose that $s - t > s' - t'$. Let $r = (s - t) - (s' - t')$. Since $X \subseteq \overset{\omega}{\rightarrow} (\dot{q}_0^{s+m} \circ q_0^{t+m})$ for any m, and since $Y \subseteq \overset{\omega}{\rightarrow} (\dot{q}_0^{s'+n} \circ q_0^{t'_n})$ for any n, there follows that, for some s'' and t'', $X \subseteq \overset{\omega}{\rightarrow} (\dot{q}_0^{s''} \circ q_0^{t''+r})$ and $Y \subseteq \overset{\omega}{\rightarrow} (\dot{q}_0^{s''} \circ q_0^{t''})$. Then, by what was shown above, any $\langle v, w \rangle$ in $X \cap Y$ is such that both v and w are eventually periodic. Now suppose that $s - t < s' - t'$. Then a similar argument shows that any $\langle v, w \rangle$ in $X \cap Y$ is such that both v and w are eventually periodic.

Let X, Y, s, t, s', t' be as above and let $W \subseteq X \cap Y$, so that, if $\langle v, w \rangle$ is in W, then both v and w are eventually periodic. Now also assume that $\overline{\overline{U}} \geq 2$. Then, as was shown earlier, any W' in ${}_{[\omega]}Un$ such that $W' \neq \emptyset$ contains pairs $\langle v, w \rangle$ such that neither v nor w is eventually periodic. There follows that, if W is in ${}_{[\omega]}Un$, then $W = \emptyset$. This proves the third assertion of Theorem 10(b). It implies, for the case where $W = X$, the fourth assertion.

(c) The proof of Theorem 3 contains an argument that, if $X \subseteq \overset{\omega}{\rightarrow} (\dot{q}_0^s \circ q_0^t)$, then $\dot{q}_s \circ X = X \circ \dot{q}_t$ is equivalent to condition (iii)(β) in Theorem 3. A similar argument shows that if, $X \subseteq \overset{\omega}{\rightarrow} (\dot{q}_0^s \circ q_0^t)$, then $\overset{\omega}{\rightarrow} (\dot{q}_s \circ X) = \overset{\omega}{\rightarrow} (X \circ \dot{q}_t)$ is equivalent to the following condition:

$$\begin{aligned} &\text{If } |w| = s, \ |x| = t, \ |y| = 1, \ \text{ and } |z| = \omega, \\ &\text{then } \langle wyz, xyz \rangle \in X \text{ if and only if } \langle wz, xz \rangle \in X. \end{aligned}$$

Since this condition is implied by the condition (β) in Theorem 8(a), there follows from Theorem 8(a) that if X is a relation on ${}^{\omega}U$ that has grade $\langle s,t\rangle$, then $X \subseteq \overset{\omega}{\rightarrow}(\dot{q}_0^s \circ q_0^t)$ and $\overset{\omega}{\rightarrow}(\dot{q}_0^s \circ X) = \overset{\omega}{\rightarrow}(X \circ \dot{q}_t)$.

Assume that $X \subseteq \overset{\omega}{\rightarrow}(\dot{q}_0^s \circ q_0^t)$ and $\overset{\omega}{\rightarrow}(\dot{q}_s \circ X) = \overset{\omega}{\rightarrow}(X \circ \dot{q}_t)$. Then X satisfies the above condition and hence, by induction, also the following condition:

(i) If $|w|=s$, $|x|=t$, $|y| < \omega$, and $|z| = \omega$, then $\langle wyz, xyz\rangle \in X$ if and only if $\langle wz, xz\rangle \in X$.

Now further assume that $\overset{=}{U} \geq 2$, X is in ${}_{[\omega]}Un$, and $X \neq \emptyset$. Then X has some grade $\langle s',t'\rangle$. Moreover, since $X \subseteq \overset{\omega}{\rightarrow}(\dot{q}_0^s \circ q_0^t)$, there follows from part (b) that $s'-t' = s-t$. Then, by Theorem 8(b), there is some n such that X has grade $\langle s+n, t+n\rangle$. Then, by Theorem 8(a), there holds condition (ii)′ below, from which there follows (ii).

(ii)′ If $|w|=s$, $|x|=t$, $|y| = |y'| = n$, and $|z| = |z'| = \omega$, then $\langle wyz, xy'z\rangle \in X$ if and only if $\langle wyz', xy'z'\rangle \in X$.

(ii) If $|w|=s$, $|x|=t$, $|y| = n$, and $|z| = |z'| = \omega$, then $\langle wyz, xyz\rangle \in X$ if and only if $\langle wyz', xyz'\rangle \in X$.

We will now show that (i) and (ii) together imply

(iii) If $|w|=s$, $|x|=t$, and $|z| = |z'| = \omega$, then $\langle wz, xz\rangle \in X$ if and only if $\langle wz', xz'\rangle \in X$.

Assume (i) and (ii). Consider any w, x, z, z' such that $|w| = s$, $|x| = t$, and $|z| = |z'| = \omega$. Assume that $\langle wz, xz\rangle$ is in X. Choose any y such that $|y| = n$. Then, by (i), $\langle wyz, xyz\rangle$ is in X. Then, by (ii), $\langle wyz', xyz'\rangle$ is in X. Then, by (i), $\langle wz', xz'\rangle$ is in X. Similarly, if $\langle wz', xz'\rangle$ is in X, then $\langle wz, xz\rangle$ is in X. There now follows that (iii).

By Theorem 8(a), if $X \neq \emptyset$, then (iii) and $X \subseteq \overset{\omega}{\rightarrow}(\dot{q}_0^s \circ q_0^t)$ together imply that X has grade $\langle s,t\rangle$. Hence we now have proved that, if $\overset{=}{U} \geq 2$ and $X \neq \emptyset$, then the condition that $X \in {}_{[\omega]}Un$, $\overset{\omega}{\rightarrow}(\dot{q}_s \circ X) = \overset{\omega}{\rightarrow}(X \circ \dot{q}_t)$, and $X \subseteq \overset{\omega}{\rightarrow}(\dot{q}_0^s \circ q_0^t)$ implies the condition that X has grade $\langle s,t\rangle$ and hence, by what was shown earlier, is equivalent to it. Since the two conditions are also equivalent if $\overset{=}{U} = 1$ and $X \neq \emptyset$, there now follows the first assertion in Theorem 10(c). From it, Theorem 8(b), and the definition of lgd, there follows the second assertion. □

The pair $\langle x, x\rangle$ such that $|x| = \omega$ and $x_i = x_j$ for every i and j belongs to every $\overset{\omega}{\rightarrow}(\dot{q}_0^s \circ q_0^t)$. Hence, for any s, t, s' and t', the intersection $\overset{\omega}{\rightarrow}(\dot{q}_0^s \circ q_0^t) \cap \overset{\omega}{\rightarrow}(\dot{q}_0^{s'} \circ q_0^{t'})$ is nonempty, but if $s - t \neq s' - t'$ then, by Theorem 10(b), it is not in ${}_{[\omega]}Un$. Thus, ${}_{[\omega]}Un$ is not closed under $\cap$.

Note that, in contrast, ${}_{[0,\omega)}Un$ is closed under $\cap$. This can be seen as follows. If $X \cap Y = \emptyset$ then $X \cap Y$ is in ${}_{[0,\omega)}Un$. Now assume that $X \cap Y \neq \emptyset$, $gd\ X = \langle s, t\rangle$, and $gd\ Y = \langle s', t'\rangle$. Let Q and R be the initial section of X or of Y respectively. Then $t = s' + r$ implies that $X \cap Y = \overset{[0,\omega)}{\longrightarrow}(Q \cap \overset{r}{\rightarrow} R)$, while $s' = t + r'$ implies that $X \cap Y = \overset{[0,\omega)}{\longrightarrow}(\overset{r'}{\rightarrow} Q \cap R)$.

The assumption that X is in ${}_{[\omega]}Un$ cannot be omitted from Theorem 10(c). For example, given any s and t, choose any z such that $|z| = \omega$ and let $X_z = \{\langle wyz, xyz\rangle : |w| = s,\ |x| = t,\ |y| < \omega\}$. Then $X_z \subseteq \overset{\omega}{\rightarrow}(\dot{q}_0^s \circ q_0^t)$ and X_z satisfies condition (i) above and hence also $\overset{\omega}{\rightarrow}(\dot{q}_s \circ X) = \overset{\omega}{\rightarrow}(X \circ \dot{q}_t)$. However, if $\overline{\overline{U}} = 2$, so that there are z' such that $|z'| = \omega$ and $z' \neq z$, then X_z does not satisfy condition (iii) above and hence does not have grade $\langle s, t\rangle$.

We will now give two illustrations of Theorem 10(c). The reader may wish to skip these. First, we let

$$X = \overset{\omega}{\rightarrow} \dot{q}_i = \{\langle vwz, vz\rangle : |v| = i,\ |w| = 1,\ |z| = \omega\}\ .$$

Consider any $n \geq 0$. Then X is included in

$$\overset{\omega}{\rightarrow}(\dot{q}_0^{i+1+n} \circ q_0^{i+n}) = \{\langle vwxz, vyz\rangle : |v| = i,\ |w| = 1,\ |x| = |y| = n,\ |z| = \omega\}.$$

Also, there hold:

$$\begin{aligned}
&(\overset{\omega}{\rightarrow} \dot{q}_{i+1+n}) \circ (\overset{\omega}{\rightarrow} \dot{q}_i)\\
&\quad = \{\langle vwxyz, vwxz\rangle : |v| = i,\ |w| = 1,\ |x| = n, |y| = 1,\ |z| = \omega\} \circ (\overset{\omega}{\rightarrow} \dot{q}_i)\\
&\quad = \{\langle vwxyz, vxz\rangle : |v| = i,\ |w| = 1,\ |x| = n,\ |y| = 1,\ |z| = \omega\}\ ,\\
&(\overset{\omega}{\rightarrow} \dot{q}_i) \circ (\overset{\omega}{\rightarrow} \dot{q}_{i+n})\\
&\quad = \{\langle vwxyz, vxyz\rangle : |v| = i,\ |w| = 1,\ |x| = n, |y| = 1,\ |z| = \omega\} \circ (\overset{\omega}{\rightarrow} \dot{q}_{i+n})\\
&\quad = \{\langle vwxyz, vxz\rangle : |v| = i,\ |w| = 1,\ |x| = n, |y| = 1,\ |z| = \omega\}\ .
\end{aligned}$$

Hence $(\overset{\omega}{\rightarrow} \dot{q}_{i+1+n}) \circ (\overset{\omega}{\rightarrow} \dot{q}_i) = (\overset{\omega}{\rightarrow} \dot{q}_i) \circ (\overset{\omega}{\rightarrow} \dot{q}_{i+n})$. There follows that $X = \overset{\omega}{\rightarrow} \dot{q}_i$ has grade $\langle i + 1 + n, i + n\rangle$. Now also assume that $\overline{\overline{U}} \geq 2$ and that $i > 0$. Then

$$\overset{\omega}{\rightarrow}(\dot{q}_0^i \circ q_0^{i-1}) = \{\langle v'w'z', v'z'\rangle : |v'| = i - 1,\ |w'| = 1,\ |z'| = \omega\}\ .$$

Since $\overline{\overline{U}} \geq 2$, this set contains pairs $\langle v'w'z', v'z'\rangle$ such that $w'_0 \neq z'_0$ and hence such that $(v'w'z')_i = w' \neq z'_0 = (v'z')_i$. Any such pair differs from every $\langle vwz, vz\rangle$ such that $|v|=i$, $|w|=1$, $|z| = \omega$ and hence differs from every pair in $\stackrel{\omega}{\rightarrow} \dot{q}_i$. There follows that $\stackrel{\omega}{\rightarrow} (\dot{q}_0^i \circ q_0^{i-1}) \not\subseteq \stackrel{\omega}{\rightarrow} \dot{q}_i$ and hence that $\langle i+1, i\rangle$ is the least grade of $\stackrel{\omega}{\rightarrow} \dot{q}_i$.

For another illustration of Theorem 10(c), consider any j and λ such that $j < \omega$ and $\lambda < \kappa = \overline{\overline{U}}$ and let

$$X = \stackrel{\omega}{\rightarrow} \iota_{(j/\lambda)} = \{\langle v^\frown\langle u_\lambda\rangle^\frown z, v^\frown\langle u_\lambda\rangle^\frown z\rangle : |v|=j, |z|=\omega\} .$$

Consider any $n < \omega$. Since $\stackrel{\omega}{\rightarrow} \iota_{(j/\lambda)} \subseteq {}_\omega\overset{\circ}{1}$, therefore $\stackrel{\omega}{\rightarrow} \iota_{(j/\lambda)} \subseteq \stackrel{\omega}{\rightarrow} (\dot{q}_0^n \circ q_0^n)$. Also, since

$$\stackrel{\omega}{\rightarrow} \dot{q}_{j+1+n} = \{\langle vwxyz; vwxz\rangle : |v|=j,\ |w|=|y|=1,\ |x|=n, |z|=\omega\} ,$$

therefore

$$\begin{aligned}
&(\stackrel{\omega}{\rightarrow} (\dot{q}_{j+1+n}) \circ (\stackrel{\omega}{\rightarrow} \iota_{(j/\lambda)}))\\
&\quad = \{\langle v^\frown\langle u_\lambda\rangle^\frown xyz, v^\frown\langle u_\lambda\rangle^\frown xz\rangle : |v|=j,\ |x|=n,\ |y|=1,\ |z|=\omega\} ,\\
&\quad = (\stackrel{\omega}{\rightarrow} \iota_{(j/\lambda)}) \circ (\stackrel{\omega}{\rightarrow} \dot{q}_{j+1+n}) .
\end{aligned}$$

From Theorem 10(c) there now follows that $\stackrel{\omega}{\rightarrow} \iota_{(j/\lambda)}$ has grade $\langle j+1+n, j+1+n\rangle$. Since

$$\stackrel{\omega}{\rightarrow} \dot{q}_j \equiv \{\langle vwyz, vyz\rangle : |v|=j,\ |w|=|y|=1,\ |z|<\omega\} ,$$

therefore

$$\begin{aligned}
&(\stackrel{\omega}{\rightarrow} \dot{q}_j) \circ (\stackrel{\omega}{\rightarrow} \iota_{(j/\lambda)}))\\
&\quad = \{\langle vwyz, v^\frown\langle u_\lambda\rangle^\frown z\rangle : |v|=j,\ |w|=|y|=1,\ |z|=\omega\} ,\\
&(\stackrel{\omega}{\rightarrow} \iota_{(j/\lambda)})) \circ (\stackrel{\omega}{\rightarrow} \dot{q}_j)\\
&\quad = \{v^\frown\langle u_\lambda\rangle^\frown yz, v^\frown\langle u_\lambda\rangle^\frown yz\rangle : |v|=j,\ |y|=1,\ |z|=\omega\} \circ (\stackrel{\omega}{\rightarrow} \dot{q}_j)\\
&\quad = \{\langle v^\frown\langle u_\lambda\rangle^\frown yz, v^\frown yz\rangle : |v|=j,\ |y|=1,\ |z|=\omega\} .
\end{aligned}$$

Hence, if $\overline{\overline{U}} \geq 2$, then $(\stackrel{\omega}{\rightarrow} \dot{q}_j) \circ (\stackrel{\omega}{\rightarrow} \iota_{(j/\lambda)})) \neq (\stackrel{\omega}{\rightarrow} \iota_{(j/\lambda)})) \circ (\stackrel{\omega}{\rightarrow} \dot{q}_j)$ and hence, according to Theorem 10(c), $lgd(\stackrel{\omega}{\rightarrow} \iota_{(j/\lambda)}) = \langle j+1, j+1\rangle$.

We now turn to relationships between ${}_{[0,\omega)}Un$ and ${}_{[\omega]}Un$. Henceforth, when it is clear from the context that this is intended, the restriction to ${}_{[0,\omega)}Un$ of the function $\stackrel{\omega}{\rightarrow}$ whose domain is the set of all relations on ${}^{[0,\omega)}U$ will be also denoted by $\stackrel{\omega}{\rightarrow}$.

THEOREM 11. (a) $\overset{\omega}{\to}$ *is a homomorphism of* $\langle {}_{[0,\omega)}Un, \circ, {}^{\smile}, \overset{1}{\to}\rangle$ *onto* $\langle {}_{[\omega]}Un, \circ, {}^{\smile}, id\rangle$, *where id is the identity operation on* ${}_{[\omega]}Un$.

(b) *Assume that* $\overline{\overline{U}} \geq 2$. *Then the kernel* $\{\langle X, Y\rangle : X \in {}_{[0,\omega)}Un,\ Y \in {}_{[0,\omega)}Un,\ \overset{\omega}{\to} X = \overset{\omega}{\to} Y\}$ *of* $\overset{\omega}{\to}$ *coincides with the equivalence relation* $\approx$ *on* ${}_{[0,\omega)}Un$ *that is generated by* $\overset{1}{\to}$. *Hence* $\langle {}_{[\omega]}Un, \circ, {}^{\smile}, id\rangle$ *is isomorphic to the quotient algebra* $\langle {}_{[0,\omega)}Un, \circ, {}^{\smile}, \overset{1}{\to}\rangle / \approx$.

(c) *For any* X *and* Y *in* ${}_{[0,\omega)}Un$, $X \subseteq Y$ *implies that* $\overset{\omega}{\to} X \subseteq \overset{\omega}{\to} Y$ *and, if* $\overline{\overline{U}} \geq 2$, *and* $X \neq \emptyset$, *then* $\overset{\omega}{\to} X \subseteq \overset{\omega}{\to} Y$ *implies that, if gd* $X = \langle s, t\rangle$ *and gd* $Y = \langle s', t'\rangle$, *then* $s - t = s' - t'$ *and either both* $s \leq s'$ *and* ${}_{[s',\omega)}\overset{\circ}{1} \circ X \subseteq Y$ *or both* $s \geq s'$ *and* $X \subseteq Y$.

Proof. (a) By Lemma 9(b) and Lemma 1(b), $\overset{\omega}{\to}$ is a homomorphism of $\langle {}_{[0,\omega)}Un, \circ\rangle$ onto $\langle {}_{[\omega]}Un, \circ\rangle$. By Lemma 1(b) and an observation in the proof of Theorem 10(a), $\overset{\omega}{\to}$ is a homomorphism of $\langle {}_{[0,\omega)}Un, {}^{\smile}\rangle$ onto $\langle {}_{[\omega]}Un, {}^{\smile}\rangle$.

(b) Assume that $\overline{\overline{U}} \geq 2$. Then, by Lemma 9(d), $\approx$ coincides with the kernel of $\overset{\omega}{\to}$. There follows that $\langle {}_{[\omega]}Un, \circ, {}^{\smile}, id\rangle$ is isomorphic to $\langle {}_{[0,\omega)}Un, \circ, {}^{\smile}, \overset{1}{\to}\rangle / \approx$.

(c) Consider any X And Y in ${}_{[0,\omega)}Un$. Evidently, $X \subseteq Y$ implies that $\overset{\omega}{\to} X \subseteq \overset{\omega}{\to} Y$. Now assume that $\overline{\overline{U}} \geq 2$, $\overset{\omega}{\to} X \neq \emptyset$, and $\overset{\omega}{\to} X \subseteq \overset{\omega}{\to} Y$. Then, for some $\langle s, t\rangle$, both $\overset{\omega}{\to} X$ and X have grade $\langle s, t\rangle$ and, for some $\langle s', t'\rangle$, both $\overset{\omega}{\to} Y$ and Y have grade $\langle s', t'\rangle$. Since $\emptyset \neq \overset{\omega}{\to} X \subseteq \overset{\omega}{\to} Y$, there follows from Theorem 10(b) that $s - t = s' - t'$.

Suppose that $s \leq s'$. Let $r = s' - s$. By Lemma 1(b), $\overset{\omega}{\to}(\overset{r}{\to} X) = \overset{\omega}{\to} X$ and hence $\overset{\omega}{\to}(\overset{r}{\to} X) \subseteq \overset{\omega}{\to} Y$. Since $\overset{r}{\to} X$ and Y both have grade $\langle s', t'\rangle$, there follows from Lemma 8(a) that $\overset{r}{\to} X \subseteq Y$. Since X has grade $\langle s, t\rangle$ and $r = s' - s$, therefore, by Theorem 7(f), $\overset{r}{\to} X = {}_{[s',\omega)}\overset{\circ}{1} \circ X$. Hence ${}_{[s',\omega)}\overset{\circ}{1} \circ X \subseteq Y$.

Suppose that $s \geq s'$. Then, an argument similar to the one just given shows that $X \subseteq {}_{[s,\omega)}\overset{\circ}{1} \circ Y$. Since ${}_{[s,\omega)}\overset{\circ}{1} \circ Y \subseteq Y$, therefore $X \subseteq Y$. □

Henceforth we let ${}_{[\omega]}S_{pqei}$ be the closure under $\circ$ of the union of the following four sets:

$$\{\overset{\omega}{\to} p_i : i, j < \omega\},\ \{\overset{\omega}{\to} q_i : i < \omega\} \cup \{\overset{\omega}{\to} \dot{q}_i : i < \omega\}\ ,$$
$$\{\overset{\omega}{\to} e_{ij} : i, j < \omega\},\ \text{and}\ \{\overset{\omega}{\to} \iota_{(j/\lambda)} : \lambda < \kappa,\ j < \omega\}.$$

In terms of the appropriate ones of these four sets we similarly define ${}_{[\omega]}S_p$, ${}_{[\omega]}S_q$, ${}_{[\omega]}S_e, \ldots, {}_{[\omega]}S_{pqe}, \ldots$. Since $\overset{\omega}{\to} W$ is in ${}_{[\omega]}Un$ for every W in ${}_{[0,\omega)}Un$, therefore, by Lemma 2(a), each of the above four sets is a subset of ${}_{[\omega]}Un$. Hence, by Theorem

10(a), each of the sets ${}_{[\omega]}S_p, {}_{[\omega]}S_q, {}_{[\omega]}S_e, \ldots, {}_{[\omega]}S_{pqei}$ also is a subset of ${}_{[\omega]}Un$. Since, for any $i < \omega$, $\overset{\omega}{\to} e_{00} = \overset{\omega}{\to} e_{ii}$, and hence $\{\overset{\omega}{\to} e_{ij} : i, j < \omega\} = \{\overset{\omega}{\to} e_{ij} : \{i, j\} \neq \{\emptyset\}\}$, there follows from Theorem 11(a) that each of these sets is the direct image under $\overset{\omega}{\to}$ of $S_p, S_q, S_e, \ldots, S_{pqei}$, respectively, and also, for example, that $\langle {}_{[\omega]}S_{pqei}, \circ, \smile, id\rangle$ is the homomorphic image under $\overset{\omega}{\to}$ of $\langle {}_{[\omega]}S_{pqei}, \circ, \smile, \overset{1}{\to}\rangle$. These observations will sometimes be used tacitly.

Next, we describe another homomorphism of $\langle {}_{[0,\omega)}Un, \circ, \smile, \overset{1}{\to}\rangle$. Also, we relate it to $\overset{\omega}{\to}$.

THEOREM 12. (a) *gd is a homomorphism of $\langle {}_{[0,\omega)}Un, \circ, \smile, \overset{1}{\to}\rangle$ onto $\langle (\omega \times \omega) \cup \{\dot{0}\}, \odot, \overset{\bullet}{\vee}, \overset{\bullet}{\to}\rangle$ such that for any X and Y in ${}_{[0,\omega)}Un$, $X \subseteq Y$ implies that $gd\ X \geq gd\ Y$.*

(b) *For any X and Y in ${}_{[0,\omega)}Un$, if both $\overset{\omega}{\to} X = \overset{\omega}{\to} Y$ and $gd\ X = gd\ Y$, then $X = Y$.*

Proof. (a) This follows from the definition of $\odot$, $\overset{\bullet}{\vee}$, $\overset{\bullet}{\to}$, $\underset{=}{\geq}$ and from Lemma 5.

(b) Assume that $\overset{\omega}{\to} X = \overset{\omega}{\to} Y$. First, also assume that $\overline{\overline{U}} \geq 2$. Then, by Theorem 11(b), $X \approx Y$. Then either $\overset{m}{\to} X = Y$, for some m, or $X = \overset{n}{\to} Y$, for some n. Suppose that $gd\ X = gd\ Y \neq \dot{0}$. Then $X \neq \emptyset$ and $Y \neq \emptyset$. Moreover, if $m \geq 1$, then $gd(\overset{m}{\to} X) \neq gd\ X$, hence $gd(\overset{m}{\to} X) \neq gd\ Y$, and hence $\overset{m}{\to} X \neq Y$. Also, if $n \geq 1$, then $gd(\overset{n}{\to} Y) \neq gd\ Y$, hence $gd(\overset{n}{\to} Y) \neq gd\ X$, and hence $\overset{n}{\to} Y \neq X$. There follows that either $\overset{0}{\to} X = Y$ or $X = \overset{0}{\to} Y$, and hence that $X = Y$. Now suppose that $gd\ X = gd\ Y = \dot{0}$. Then $X = \emptyset$ and $Y = \emptyset$, and hence again $X = Y$. Now assume that $\overline{\overline{U}} = 1$. Then, for any $\langle s, t\rangle$ there is exactly one W in ${}_{[0,\omega)}Un$ such that $gd\ W = \langle s, t\rangle$. Hence $gd\ X = gd\ Y = \langle s, t\rangle$ implies that $X = Y$. Also, if $gd\ X$ and $gd\ Y$ are not in $\omega \times \omega$ then $X = \emptyset = Y$. □

By definition S_{pqe} and S_{pqei} are closed under $\circ$ and $\smile$. By Lemma 2, both are included in ${}_{[0,\omega)}Un$. Since both contain every ${}_{[n,\omega)}\overset{\circ}{1}$, q_s, and $q_t^{\smile}$ therefore, by Theorem 7(f), both are closed under $\overset{1}{\to}$ and under the converse of $\overset{1}{\to}$. Hence both are also closed under the equivalence relation $\approx$ that is generated by $\overset{1}{\to}$. Hence S_{pqe} and S_{pqei} are among those sets A that are the universe of a substructure $\mathbf{A}$ of $\langle {}_{[0,\omega)}Un, \circ, \smile, \overset{1}{\to}, \subseteq\rangle$ to which there applies the following theorem.

THEOREM 13. *Assume that $\overline{\overline{U}} \geq 2$. Let $\mathbf{A}$ be any substructure of $\langle {}_{[0,\omega)}Un, \circ, \smile, \overset{1}{\to}, \subseteq\rangle$ and let A be the universe of $\mathbf{A}$. Let $\mathbf{B}$ be the substructure of $\langle {}_{[\omega]}Un, \circ, \smile, id, \subseteq\rangle$ whose universe is $\{\overset{\omega}{\to} X : X \in A\}$, let $\mathbf{C}$ be the substructure of $\langle (\omega\times\omega)\cup\{\dot{0}\}, \odot, \overset{\bullet}{\vee}, \overset{\bullet}{\to}$*

$, \geq \rangle$ *whose universe is* $\{gd\ X : X \in A\}$, *and let* h *be the function whose domain is* A *such that, for any* X *in* A, $h(X) = \langle \overset{\omega}{\rightarrow} X, gd\ X \rangle$. *Then* h *is an embedding of* $\mathbf{A}$ *into the direct product of* $\mathbf{B} \times \mathbf{C}$. *Moreover, if* A *is closed under the equivalence relation* $\approx$ *that is generated by* $\overset{1}{\rightarrow}$, *then* $\emptyset \notin A$ *implies that* $Rg\ h = D$ *while* $\emptyset \in A$ *implies that* $Rg\ h = D \cup \{\langle \emptyset, \dot{0} \rangle\}$, *where* D *is the following set:*

$$\{\langle \overset{\omega}{\rightarrow} W, \langle s+n, t+n \rangle \rangle : W \in A \cap -(Rg \overset{1}{\rightarrow}), \langle s,t \rangle = gd\ W,\ 0 \leq n < \omega\}.$$

Proof. Let $\mathbf{A}'$ be the reduct $\langle A, \circ, ^{\smile}, \overset{1}{\rightarrow} \rangle$ of $\mathbf{A}$, $\mathbf{B}'$ the reduct $\langle B, \circ, ^{\smile}, id \rangle$ of $\mathbf{B}$, and $\mathbf{C}'$ the reduct $\langle (\omega \times \omega) \cup \{\dot{0}\}, \odot, \overset{\bullet}{^{\vee}}, \overset{\bullet}{\rightarrow} \rangle$ of $\mathbf{C}$. Then, by Theorems 11(a) and 12(a), h is a homomorphism of $\mathbf{A}'$ into $\mathbf{B}' \times \mathbf{C}'$. By Theorem 12(b), h is one-one. There follows (cf. [BS], Theorem 7.15) that h is an embedding of $\mathbf{A}'$ into $\mathbf{B}' \times \mathbf{C}'$.

Now assume that $\overset{=}{U} \geq 2$ and consider any X and Y in A. First suppose that $X \subseteq Y$. Then, evidently, $\overset{\omega}{\rightarrow} X \subseteq \overset{\omega}{\rightarrow} Y$ and, by Theorem 12(a), $gd\ X \ \geq gd\ Y$. Now suppose that $\overset{\omega}{\rightarrow} X \subseteq \overset{\omega}{\rightarrow} Y$ and $gd\ X \ \geq gd\ Y$. If $\overset{\omega}{\rightarrow} X = \emptyset$ then $X = \emptyset$ and hence $X \subseteq Y$. Now suppose that $\overset{\omega}{\rightarrow} X \neq \emptyset$ and hence $\overset{\omega}{\rightarrow} Y \neq \emptyset$. Then, for some $\langle s,t \rangle$, $\overset{\omega}{\rightarrow} X$ and X have grade $\langle s,t \rangle$ and, for some $\langle s',t' \rangle$, $\overset{\omega}{\rightarrow} Y$ and Y have grade $\langle s',t' \rangle$. Moreover, by Theorem 10(b), $s - t = s' - t'$. Since $gd\ X \ \geq gd\ Y$, therefore $s \geq s'$. Then, by Theorem 11(c), $X \subseteq Y$. Thus, $X \subseteq Y$ if and only if both $\overset{\omega}{\rightarrow} X \subseteq \overset{\omega}{\rightarrow} Y$ and $gd\ X \geq gd\ Y$. There now follows that, if $\overset{=}{U} \geq 2$, then h is an embedding of $\mathbf{A}$ into $\mathbf{B} \times \mathbf{C}$.

Assume that A is closed under $\approx$. First, consider any X in A such that $X \neq \emptyset$. Then $X = \overset{n}{\rightarrow} W$ for some W in $A \cap -(Rg \overset{1}{\rightarrow})$ and some $n \geq 0$. Let $gd\ W = \langle s,t \rangle$. Then $gd\ X = \langle s+n, t+n \rangle$. Since $\overset{\omega}{\rightarrow} X = \overset{\omega}{\rightarrow} W$, there follows that $h(X) = \langle \overset{\omega}{\rightarrow} W, \langle s{+}n, t{+}n \rangle \rangle$. This shows that $\{h(X) : X \in A,\ X \neq \emptyset\} \subseteq D$. Since $\emptyset \in A$ implies that $h(\emptyset) = \langle \emptyset, \dot{0} \rangle$, there follows that, if $\emptyset \in A$, then $\{h(X) : X \in A\} \subseteq D \cup \{\langle \emptyset, \dot{0} \rangle\}$.

Now consider any W, s, t, and n such that $W \in A \cap -(Rg \overset{1}{\rightarrow})$, $\langle s,t \rangle = gd\ W$, and $0 \leq n < \omega$. Let $X = \overset{n}{\rightarrow} W$. Then $\overset{\omega}{\rightarrow} X = \overset{\omega}{\rightarrow} W$ and $gd\ X = \langle s+n, t+n \rangle$. Hence $h(X) = \langle \overset{\omega}{\rightarrow} W, \langle s+n, t+n \rangle \rangle$. Since $X \neq \emptyset$, there follows that $D \subseteq \{h(X) : X \in A,\ X \neq \emptyset\}$ and hence that, if $\emptyset \in A$, then $D \cup \{\langle \emptyset, \dot{0} \rangle\} \subseteq \{h(X) : X \in A\}$. □

The assumption that $\overset{=}{U} \geq 2$ cannot be omitted from Theorem 13. Whenever $s - t \neq s' - t'$, then $(\dot{q}_0^s \circ q_0^t) \cap (\dot{q}_0^{s'} \circ q_0^{t'}) = \emptyset$ and hence $\dot{q}_0^s \circ q_0^t \not\subseteq \dot{q}_0^{s'} \circ q_0^{t'}$. Yet, when $\overset{=}{U} = 1$, then $\overset{\omega}{\rightarrow} (\dot{q}_0^s \circ q_0^t) = {}_{[\omega]}\overset{\circ}{1} = \overset{\omega}{\rightarrow} (\dot{q}_0^{s'} \circ q_0^{t'})$. Hence, although h is an embedding of $\mathbf{A}'$ into $\mathbf{B}' \times \mathbf{C}'$, where $\mathbf{A}', \mathbf{B}', \mathbf{C}'$ are the reducts described in the above proof, if $\overset{=}{U} = 1$, then h is not an embedding of $\mathbf{A}$ into $\mathbf{B} \times \mathbf{C}$.

Let $\mathbf{B}$ be any substructure of $\langle {}_{[\omega]}Un, \circ, ^{\smile}, id, \subseteq \rangle$ and let A be the inverse image $\{W \in {}_{[0,\omega)}Un : \stackrel{\omega}{\to} W \in B\}$ of the universe B of $\mathbf{B}$ under $\stackrel{\omega}{\to}$. Then A is the universe of a substructure $\mathbf{A}$ of $\langle {}_{[0,\omega)}Un, \circ, ^{\smile}, \stackrel{1}{\to}, \subseteq \rangle$ whose homomorphic image is $\mathbf{B}$. The next theorem shows that, in some cases, $\mathbf{B}$ has structural features that allow one to construct an isomorphic image of $\mathbf{A}$.

THEOREM 14. *Assume that* $\overline{\overline{U}} \geq 2$. *Let* $\mathbf{B}$ *be any substructure of* $\langle {}_{[\omega]}Un, \circ, ^{\smile}, id, \subseteq \rangle$ *such that every* $\stackrel{\omega}{\to} \dot{q}_s$ *and every* $\stackrel{\omega}{\to} q_s$ *is in the universe* B *of* $\mathbf{B}$. *For any* $X \neq \emptyset$ *in* B, *let* $lgd\ X$ *be the least grade of* X *(as defined after the proof of Theorem 8), let*

$$E = \{\langle X, \langle s+n, t+n \rangle \rangle : X \in B,\ \langle s, t \rangle = lgd\ X,\ 0 \leq n < \omega\}\ ,$$

and let $E' = E \cup \{\langle \emptyset, \dot{0} \rangle\}$, *if* $\emptyset \in B$, *and* $E' = E$ *if* $\emptyset \notin B$. *Then the substructure of* $\langle {}_{[0,\omega)}Un, \circ, ^{\smile}, \stackrel{1}{\to}, \subseteq \rangle$ *whose universe* A *is the inverse image* $\{W \in {}_{[0,\omega)}Un : \stackrel{\omega}{\to} W \in B\}$ *of* B *under* $\stackrel{\omega}{\to}$ *is isomorphic to the substructure of* $\mathbf{B} \times \langle (\omega \times \omega) \cup \{\dot{0}\}, \odot, \stackrel{\bullet}{\vee}, \stackrel{\bullet}{\to}, \geq \rangle$ *whose universe is* E'.

Proof. Let A be the inverse image of B under $\stackrel{\omega}{\to}$. Since $\overline{\overline{U}} \geq 2$, therefore, by Theorem 11(b), $\approx$ coincides with the kernel of $\stackrel{\omega}{\to}$. Hence A is closed under $\approx$.

Consider any $\langle X, \langle s, t \rangle \rangle$ such that $X \in B$ and $lgd\ X = \langle s, t \rangle$. Then $X \neq \emptyset$ and $X = \stackrel{\omega}{\to} W'$ for some $W' \neq \emptyset$ in A. Since A is closed under $\approx$, therefore there is some W in $A \cap -(Rg \stackrel{1}{\to})$ such that $W' \approx W$. Since $lgd\ X = \langle s, t \rangle$ therefore $gd\ W = \langle s, t \rangle$. Since $\stackrel{\omega}{\to} W' = X$ and $W \approx W'$, therefore $\stackrel{\omega}{\to} W = X$. This shows that if $\langle X, \langle s, t \rangle \rangle$ is in E it is also in the set D of Theorem 13. There follows that $E \subseteq D$. Now consider any W in $A \cap -(Rg \stackrel{1}{\to})$. Let $gd\ W = \langle s, t \rangle$ and $X = \stackrel{\omega}{\to} W$. Then X is in B and $lgd\ X = \langle s, t \rangle$. This shows that if $\langle \stackrel{\omega}{\to} W, \langle s, t \rangle \rangle$ is in D it is also in E. There follows that $D \subseteq E$ and hence that $D = E$. The function h described in Theorem 13 is therefore a mapping of A onto E'. By Theorem 13, it is an isomorphism. □

CHAPTER III

Diagonal Relations

We now turn to a second class of relations between sequences. It also includes S_{pqe} and ${}_{[\omega]}S_{pqe} = \{\overset{\omega}{\to} W : W \in S_{pqe}\}$.

The class is again a counterpart for dimension 2 of a certain class of unary sets X of sequences. There is a close connection between this class of unary sets and logic with equality. Each unary set X in the class can be obtained by choosing an ordinal γ such that $0 \le \gamma \le \omega$ and a binary relation N on $\gamma = \{k : k < \gamma\}$ and then forming the following set, where $xx^{\smile} = x \circ x^{\smile}$.

$$\begin{aligned} X = X_{\langle \gamma, N \rangle} &= \{x : |x| = \gamma,\ x_i = x_j \text{ if } \langle i, j \rangle \in N\} \\ &= \{x : |x| = \gamma,\ N \subseteq xx^{\smile}\}\ . \end{aligned}$$

The same set X is obtained when one lets M be the equivalence relation that is generated by $N \cup \{\langle k, k \rangle : k < \gamma\}$ and then lets

$$X = X_M = \{x : |x| = Do\ M,\ M \subseteq xx^{\smile}\}\ .$$

If X is thus related to M, then one may say of M that it ***determines*** X. If X is thus determined by some equivalence relation M then, in line with usage in algebraic logic, one may say of X that it is a ***diagonal set.***

A 2-dimensional analogue of this notion can be described more simply by first introducing some conventions. Henceforth, unless we say otherwise, we let $M, N, \ldots$ be relations on ω, and we let $0 \le \gamma \le \omega$, $0 \le \delta \le \omega, \ldots$. For any γ, we let $\overset{\circ}{1}_\gamma$ be the identity relation $\{\langle k, k \rangle : k < \gamma\}$ on γ. For any M and any γ, we let $eq_\gamma M$ be the equivalence relation that is generated by $\overset{\circ}{1}_\gamma \cup (M \cap (\gamma \times \gamma))$. There follows that $M = eq_\gamma M$ if and only if M is symmetric and transitive and $Do\ M = \gamma$.

A $\langle \gamma, \delta \rangle$ ***triple*** shall be any $\langle M_0, M_1, M_2 \rangle$ for which the following holds:

$$M_0 = eq_\gamma M_0\ , \qquad M_1 \subseteq \gamma \times \delta\ , \quad \text{and} \quad M_2 = eq_\delta M_2\ .$$

A triple $\langle M_0, M_1, M_2 \rangle$ shall be ***framing*** if and only if it is a $\langle \gamma, \delta \rangle$ triple for at least one $\gamma \le \omega$ and $\delta \le \omega$, and hence for exactly one γ and δ. It shall be ***finite*** if and only if $M_0 \cup M_1 \cup M_2$ is a finite set of pairs. Hence it is finite if and only if it is a $\langle \gamma, \delta \rangle$ triple such that $\gamma = s$ and $\delta = t$ for some $s < \omega$ and $t < \omega$.

Let $\langle M_0, M_1, M_2 \rangle$ be framing. It shall ***determine*** the following set of pairs of sequences for which we now introduce two symbolizations, so that we will be able to omit $\langle\ \rangle$ whenever it is convenient.

$$
\begin{aligned}
[\langle M_0, M_1, M_2 \rangle] &= [M_0, M_1, M_2] \\
&= \{\langle x, y \rangle : |x| = Do\ M_0,\ \ |y| = Do\ M_2,\ \ x_i = x_j \\
&\quad \text{if } \langle i, j \rangle \in M_0,\ \ x_i = y_j \text{if } \langle i, j \rangle \in M_1, \\
&\quad y_i = y_j \text{ if } \langle i, j \rangle \in M_2\} \\
&= \{\langle x, y \rangle : |x| = Do\ M_0,\ \ |y| = Do\ M_2, \\
&\quad M_0 \subseteq xx^{\smile},\ \ M_1 \subseteq xy^{\smile},\ \ M_2 \subseteq yy^{\smile}\}
\end{aligned}
$$

Finite framing triples will also be used in a second way. Let $0 \leq s < \omega$ and $0 \leq t < \omega$ and let $\langle M_0, M_1, M_2 \rangle = \langle eq_s M_0, M_1, eq_t M_2 \rangle$ be an $\langle s, t \rangle$ triple. Then it shall ***mediately determine*** the following set of pairs of sequences, for which we now introduce two symbolizations:

$$
\begin{aligned}
[\![\langle M_0, M_1, M_2 \rangle]\!] &= [\![M_0, M_1, M_2]\!] = \overset{[0,\omega)}{\longrightarrow} [M_0, M_1, M_2] \\
&= \{\langle xz, yz \rangle : |x| = s,\ \ |y| = t,\ \ |z| < \omega,\ \ x_i = x_j \text{ if} \\
&\quad \langle i, j \rangle \in M_0, x_i = y_j \text{ if } \langle i, j \rangle \in M_1, y_i = y_j \text{ if } \langle i, j \rangle \in M_2\} \\
&= \{\langle xz, yz \rangle : |x| = Do\ M_0,\ |y| = Do\ M_2,\ |z| < \omega, \\
&\quad M_0 \subseteq xx^{\smile},\ M_1 \subseteq xy^{\smile},\ M_2 \subseteq yy^{\smile}\}.
\end{aligned}
$$

Henceforth, use of $[M_0, M_1, M_2]$ and of $[\![M_0, M_1, M_2]\!]$ will tacitly assume that $\langle M_0, M_1, M_2 \rangle$ is framing or, respectively, framing and finite.

A ***diagonal relation*** shall be any relation that is either determined or mediately determined by some framing triple. Thus, every diagonal relation W belongs to exactly one of the following two classes: Either $W = [\![M_0, M_1, M_2]\!]$ for some framing triple that is finite or $W = [M_0, M_1, M_2]$ for some $\langle \gamma, \delta \rangle$ triple such that $0 \leq \gamma \leq \omega$ and $0 \leq \delta \leq \omega$.

Every W in the first of these two classes is a diagonal relation that is uniform on ${}^{[0,\omega)}U$. For this class we will henceforth use the notation that is given below. For the other class there is a natural partition into subclasses. For any one of these subclasses we will use the notation that is also given below. For certain ones of these subclasses we will also use an alternative briefer notation, also given below.

$$\begin{aligned} {}_{[0,\omega)}D &= \{[\![M_0, M_1, M_2]\!] : \langle M_0, M_1, M_2\rangle \text{ is framing and finite}\}. \\ {}_{\langle\gamma,\delta\rangle}D &= \{[M_0, M_1, M_2] : \langle M_0, M_1, M_2\rangle \text{ is a } \langle\gamma,\delta\rangle \text{ triple}\}. \\ {}_{[\gamma]}D = {}_{\langle\gamma,\gamma\rangle}D &= \{[M_0, M_1, M_2] : \langle M_0, M_1, M_2\rangle \text{ is a } \langle\gamma,\gamma\rangle \text{ triple}\}. \end{aligned}$$

A diagonal relation in ${}_{[\omega]}D$ need not be uniform on ${}^{\omega}U$. For example, if $\overline{\overline{U}} \geq 2$, then the following relation is not uniform on ${}^{\omega}U$:

$$\{\langle x, x\rangle : |x| = \omega,\ x_i = x_j \text{ for every } i \text{ and } j\}\ .$$

This relation is determined by $\langle \mathring{1}_\omega, \omega \times \omega, \mathring{1}_\omega\rangle$, by $\langle \omega \times \omega, \mathring{1}_\omega, \mathring{1}_\omega\rangle$, and by various other $\langle\omega,\omega\rangle$ triples, and hence is in ${}_{[\omega]}D$.

As is easy to see, every diagonal relation W is invariant or logical in the sense of Chapter I. (In fact, according to Theorem 16 in [Cr 74b], every W satisfies a certain stronger condition of invariance.) There follows that, if $\overline{\overline{U}} = \kappa \geq 2$, then none of the relations $\iota_{(j/\lambda)}$ or $s_{\lambda,j}$ of Chapter I is a diagonal relation in the present sense.

The following generalization of the notion of a diagonal relation may turn out to be of interest but will not be further investigated here. Let R be one of the relations $\overleftrightarrow{t}$, $t^{\nwarrow}$, $\widehat{e}$, $q^{\#}$, $q^{\#\smile}$ briefly considered after Lemma II.2. Then R is not in ${}_{[0,\omega)}Un$ and hence not in ${}_{[0,\omega)}D$. However, there is some $s < \omega$ and $t < \omega$ such that R is the disjoint union $\bigcup\{R \cap ({}^{s+n}U \times {}^{t+n}U) : n < \omega\}$, where, for each $n < \omega$, $R \cap ({}^{s+n}U \times {}^{t+n}U)$ is in ${}_{\langle s+n, t+n\rangle}D$. For example,

$$q^{\#} = \bigcup\{\{\langle x, xz\rangle : |x| = n,\ |z| = 1\} : n < \omega\} = \bigcup\{[\mathring{1}_n, \mathring{1}_{n1}\mathring{1}_{n+1}] : n < \omega\}$$

Every diagonal relation is nonempty. First, consider any W in ${}_{\langle\gamma,\delta\rangle}D$, where $\gamma \leq \omega$ and $\delta \leq \omega$. Then W contains every pair $\langle x, y\rangle$ in ${}^{\gamma}U \times {}^{\delta}U$ such that $x_i = x_j$ if $i \leq j < \gamma$, $x_i = y_j$ if $i < \gamma$ and $j < \delta$, and $y_i = y_j$ if $i \leq j < \delta$. Now consider any W in ${}_{[0,\omega)}D$. Then, by definition, W is a relation that is uniform on ${}^{[0,\omega)}U$ whose initial section is in some ${}_{\langle s,t\rangle}D$ and hence is nonempty. There follows that W also is nonempty.

There now follows that for every diagonal relation W exactly one of the following holds:

W is in ${}_{[0,\omega)}D$ and has a unique grade $\langle s, t\rangle$.
W is of a unique rank $\langle s, t\rangle$.
W is a relation on ${}^{\omega}U$.

There is a unique $r < \omega$ such that either
$W \subseteq {}^rU \times {}^\omega U$ or $W \subseteq {}^\omega U \times {}^rU$.

Although diagonal relations of the fourth kind may be of little interest, it seems simpler not to exclude them.

Let $\langle M_0, M_1, M_2\rangle$ be framing. Then $M_0{}^\smile = M_0$, $M_2{}^\smile = M_2$, and $\langle M_2, M_1{}^\smile, M_0\rangle$ is also framing. Henceforth we let $\langle M_0, M_1, M_2\rangle^\smile = \langle M_2, M_1{}^\smile, M_0\rangle$. Hence the following equalities hold for $[\;\;]$ and analogous equalities hold for $[\![\;\;]\!]$. These will sometimes be used tacitly.

$$[\langle M_0, M_1, M_2\rangle]^\smile = [\langle M_0, M_1, M_2\rangle^\smile] = [\langle M_2, M_1{}^\smile, M_0\rangle] \ .$$

The following three conditions on triples $\langle M_0, M_1, M_2\rangle$, of which (i) is expressed in two ways, will also play a role. If M_0 and M_2 are given then, in a certain respect, (ii) maximizes M_1 while (iii) minimizes M_1. If $\langle M_0, M_1, M_2\rangle$ is framing then, in (ii), one may use $=$ in place of $\subseteq$. For convenience here, and at times elsewhere, we use iMj instead of $\langle i, j\rangle \in M$.

(i) If iM_1j and $i'M_1j'$, then iM_0i' if and only if jM_2j'.
$M_1M_2M_1{}^\smile \subseteq M_0$ and $M_1{}^\smile M_0M_1 \subseteq M_2$.

(ii) $M_0M_1 \subseteq M_1$ and $M_1M_2 \subseteq M_1$.

(iii) If iM_1j and $i'M_1j'$, then iM_0i' and jM_2j' together imply that $i = i'$ and $j = j'$.

A triple $\langle M_0, M_1, M_2\rangle$ shall be ***precanonical*** if and only if it is framing and satisfies condition (i) above. It shall be ***canonical*** if and only if it is framing and satisfies both (i) and (ii). It shall be ***trim*** if and only if it is framing and satisfies condition (iii).

Parts (b),(e),(f), and (g) of the following lemma may sometimes be used tacitly.

LEMMA 1. (a) *If $\langle M_0, M_1, M_2\rangle$ is precanonical, then $\langle M_0, M_0M_1M_2, M_2\rangle$ is canonical.*

(b) *If $\langle M_0, M_1, M_2\rangle$ is canonical, precanonical, or trim, then so is $\langle M_0, M_1, M_2\rangle^\smile$.*

(c) *If $\langle M_0, M_1, M_2\rangle$ is canonical, then $M_1M_1{}^\smile M_1 = M_1$.*

(d) *If $\langle M_0, M_1, M_2\rangle$ is precanonical and $M_2 \subseteq N_2 = eq_\delta N_2$, then $\langle M_0 \cup M_1N_2M_1{}^\smile, M_1, N_2\rangle$ is precanonical.*

(e) *If $\langle M_0, M_1, M_2\rangle$ is precanonical and $N_1 \subseteq M_1$, then $\langle M_0, N_1, M_2\rangle$ is precanonical.*

(f) *If $\langle M_0, M_1, M_2\rangle$ is trim and $N_1 \subseteq M_1$, then $\langle M_0, N_1, M_2\rangle$ is trim.*

(g) *If $\langle M_0, M_1, M_2\rangle$ is precanonical and trim, then M_1 is one-one.*

Proof. (a) Assume that $\langle M_0, M_1, M_2\rangle$ is precanonical. Then, for some $\gamma \leq \omega$ and $\delta \leq \omega$, $M_0 = eq_\gamma M_0$, $M_1 \subseteq \gamma \times \delta$, and $M_2 = eq_\delta M_2$. Also, $M_1 M_2 M_1{}^{\smile} \subseteq M_0$ and $M_1{}^{\smile} M_0 M_1 \subseteq M_2$. There follows that

$$\begin{aligned}(M_0M_1M_2)M_2(M_0M_1M_2)^{\smile} &= M_0M_1M_2M_2M_2{}^{\smile}M_1{}^{\smile}M_0{}^{\smile}\\ &= M_0(M_1M_2M_1{}^{\smile})M_0 \subseteq M_0M_0M_0 = M_0\ .\end{aligned}$$

Similarly, $(M_0M_1M_2)^{\smile}M_0(M_0M_1M_2)^{\smile} \subseteq M_2$. Also $M_0(M_0M_1M_2) \subseteq M_0M_1M_2$ and $(M_0M_1M_2)M_2 \subseteq M_0M_1M_2$. Further, $\langle M_0, M_0M_1M_2, M_2\rangle$ is a $\langle\gamma,\delta\rangle$ triple and hence is framing. There now follows that
$\langle M_0, M_0M_1M_2, M_2\rangle$ is canonical.

(c) The following inclusions hold because $\langle M_0, M_1, M_2\rangle$ is framing, satisfies (i), or satisfies (ii), respectively:

$$M_1M_1{}^{\smile}M_1 \subseteq M_1M_2M_1{}^{\smile}M_1 \subseteq M_0M_1 \subseteq M_1\ .$$

The converse inclusion $M_1 \subseteq M_1M_1{}^{\smile}M_1$ holds for any relation M_1.

(d) Evidently, $M_1N_2M_1{}^{\smile} \subseteq M_0 \cup M_1N_2M_1{}^{\smile}$. Since $M_1{}^{\smile}M_1 \subseteq M_1{}^{\smile}M_0M_1 \subseteq M_2 \subseteq N_2 = eq_\delta N_2$, therefore

$$\begin{aligned}&M_1{}^{\smile}(M_0 \cup M_1N_2M_1{}^{\smile})M_1\\ &\quad= M_1{}^{\smile}M_0M_1 \cup M_1{}^{\smile}M_1N_2M_1{}^{\smile}M_1 \subseteq N_2\ .\end{aligned}$$

□

The two conditions that $\langle M_0, M_1, M_2\rangle$ is precanonical and that M_1 is one-one, do not together imply that $\langle M_0, M_1, M_2\rangle$ is trim. For example, let $M_0 = M_2 = \{\langle 0,1\rangle, \langle 1,0\rangle\} \cup \mathring{1}_\omega$ and let $M_1 = \mathring{1}_\omega$. Then $\langle M_0, M_1, M_2\rangle$ is precanonical and M_1 is one-one but $\langle M_0, M_1, M_2\rangle$ is not trim, since the following all hold: $0M_10$, $1M_11$, $0M_01$, $0M_21$, and $0 \neq 1$.

THEOREM 2. *Let $\langle M_0, M_1, M_2\rangle$ be any $\langle\gamma,\delta\rangle$ triple. Let $N_0 = eq_\gamma(M_0 \cup M_1M_2M_1{}^{\smile})$ and $N_2 = eq_\delta(M_1{}^{\smile}M_0M_1 \cup M_2)$. Then $\langle N_0, M_1, N_2\rangle$ is a $\langle\gamma,\delta\rangle$ triple that is precanonical, $\langle N_0, N_0M_1N_2, N_2\rangle$ is a $\langle\gamma,\delta\rangle$ triple that is canonical, and there exists an N_1 such that $N_1 \subseteq M_1$, $N_0N_1N_2 = N_0M_1N_2$, and $\langle N_0, N_1, N_2\rangle$ is trim. Moreover, the following equalities hold and, if $\langle M_0, M_1, M_2\rangle$ is finite, then similar equalities hold with $[\![\ ]\!]$ in place of $[\ ]$:*

$$[M_0, M_1, M_2] = [N_0, M_1, N_2] = [N_0, N_0M_1N_2, N_2] = [N_0, N_1, N_2]\ .$$

Proof. We form the following relations:

$$
\begin{aligned}
M_0' &= \{\langle i,j\rangle \in M_0 : \langle i,k\rangle \in M_0 \text{ for some } k \in Do\ M_1\}\\
M_2' &= \{\langle i,j\rangle \in M_2 : \langle i,k\rangle \in M_2 \text{ for some } k \in Rg\ M_1\}\\
N_0' &= \bigcup_{1\le n<\omega} (M_0'M_1M_2'M_1{}^{\smile}M_0')^n\\
N_2' &= \bigcup_{1\le n<\omega} (M_2'M_1{}^{\smile}M_0'M_1M_2')^n
\end{aligned}
$$

Then $M_1M_2M_1{}^{\smile} = M_1M_2'M_1{}^{\smile}$. Also $N_2' = eq_\delta(M_2'M_1{}^{\smile}M_0'M_1M_2')$ and $Do\ N_2' = Do\ M_2'$. There follows that $M_1N_2M_1{}^{\smile} = M_1N_2'M_1{}^{\smile}$. Similarly, $M_1{}^{\smile}N_0M_1 = M_1{}^{\smile}N_0'M_1$.

For proving that $M_1N_2M_1{}^{\smile} \subseteq N_0$, it therefore suffices to prove that $M_1N_2'M_1{}^{\smile} \subseteq N_0'$. Since $\circ$ distributes over $\bigcup$, it therefore suffices to prove the following inclusion:

$$\bigcup_{1\le n<\omega} M_1(M_2'M_1{}^{\smile}M_0'M_1M_2')^nM_1{}^{\smile} \subseteq \bigcup_{1\le n<\omega} (M_0'M_1M_2'M_1{}^{\smile}M_0')^n\ .$$

For this purpose, we first prove by induction that, for any $n \ge 1$, there holds the following equality:

$$M_0'M_1(M_2'M_1{}^{\smile}M_0'M_1M_2')^n = (M_0'M_1M_2'M_1{}^{\smile}M_0')^nM_1M_2'\ .$$

For the case where $n = 1$, this holds by associativity. Now let $n \ge 1$ and assume as inductive hypothesis that the equality holds for n. Then, first using the inductive hypothesis and then $M_0' = M_0'M_0'$, $M_2'M_2' = M_2'$, and associativity one obtains

$$
\begin{aligned}
&M_0'M_1(M_2'M_1{}^{\smile}M_0'M_1M_2')^n(M_2'M_1{}^{\smile}M_0'M_1M_2')\\
&= (M_0'M_1M_2'M_1{}^{\smile}M_0')^nM_1M_2'(M_2'M_1{}^{\smile}M_0'M_1M_2')\\
&= (M_0'M_1M_2'M_1{}^{\smile}M_0')^n(M_0'M_1M_2'M_1{}^{\smile}M_0')M_1M_2'\ .
\end{aligned}
$$

Let any $n \ge 1$ be given. Using for the second step the inductive result just derived, and elsewhere using $M_1 \subseteq M_0'M_1$, $M_0'^{\smile} = M_0'$, $M_0'M_0' = M_0'$, and associativity one obtains:

$$
\begin{aligned}
&M_1(M_2'M_1{}^{\smile}M_0'M_1M_2')^nM_1{}^{\smile}\\
&\subseteq M_0'M_1(M_2'M_1{}^{\smile}M_0'M_1M_2')^nM_1{}^{\smile}M_0'\\
&= (M_0'M_1M_2'M_1{}^{\smile}M_0')^nM_1M_2'M_1{}^{\smile}M_0'\\
&= (M_0'M_1M_2'M_1{}^{\smile}M_0')^nM_0'(M_1M_2'M_1{}^{\smile}M_0')\\
&= (M_0'M_1M_2'M_1{}^{\smile}M_0')^{n+1}\ .
\end{aligned}
$$

This proves the desired inclusion and hence proves that $M_1N_2M_1^{\smile} \subseteq N_0$. The proof that $M_1^{\smile}N_0M_1 \subseteq N_2$ is similar. There follows that $\langle N_0, M_1, N_2\rangle$ is precanonical. Then, by Lemma 1(a), $\langle N_0, N_0M_1N_2, N_2\rangle$ is canonical.

By induction, if M_1 is finite, and by the axiom of choice, if M_1 is infinite, there are relations N_1 such that $N_1 \subseteq M_1$, $N_0N_1N_2 = N_0M_1N_2$, and $\langle N_0, N_1, N_2\rangle$ is trim.

Since $N_0N_1N_2 = N_0M_1N_2$, therefore

$$[N_0, N_0N_1N_2, N_2] = [N_0, N_0M_1N_2, N_2] \ .$$

Since $M_1 \subseteq N_0M_1N_2$, $M_0 \subseteq N_0$, and $M_2 \subseteq N_2$, therefore

$$[N_0, N_0M_1N_2, N_2] \subseteq [N_0, M_1N_2] \subseteq [M_0, M_1, M_2] \ .$$

To prove that the third of these relations is included in the first and hence that all three relations coincide, we let $|x| = \gamma$ and $|y| = \delta$ and assume that $\langle x, y\rangle \in [M_0, M_1, M_2]$. Then $M_0 \subseteq xx^{\smile}$, $M_1 \subseteq xy^{\smile}$, and $M_2 \subseteq yy^{\smile}$. Since $y^{\smile}y$ is included in the identity relation on U, there follows:

$$M_1M_2M_1^{\smile} \subseteq (xy^{\smile})(yy^{\smile})(xy^{\smile})^{\smile} = x(y^{\smile}y)(y^{\smile}y)x^{\smile} \subseteq xx^{\smile} \ .$$

From this inclusion, from $M_0 \subseteq xx^{\smile}$, and from $eq_\gamma(xx^{\smile}) = xx^{\smile}$, there follows that $N_0 \subseteq xx^{\smile}$. Similarly, $N_2 \subseteq yy^{\smile}$. Since $xx^{\smile}x = x$ and $yy^{\smile}y = y$ and hence $xy^{\smile} = xx^{\smile}xy^{\smile}yy^{\smile}$, there now follows that $N_0M_1N_2 \subseteq xy^{\smile}$. Thus $\langle x, y\rangle \in [N_0, N_0M_1N_2, N_2]$. This shows that $[M_0, M_1, M_2] \subseteq [N_0, N_0M_1M_2, N_2]$.

This concludes the proof of the four equalities for $[\ \ ]$. For finite $\langle M_0, M_1, M_2\rangle$ there then follow the desired equalities for $[\![\ \ ]\!]$. □

Part (a) of the following theorem may sometimes be used tacitly.

THEOREM 3. *Assume that for some $\gamma \le \omega$ and $\delta \le \omega$ both $\langle M_0, M_1, M_2\rangle$ and $\langle N_0, N_1, N_2\rangle$ are a $\langle\gamma, \delta\rangle$ triple. Then the following assertions hold and, if $\gamma < \omega$ and $\delta < \omega$, they also hold when $[\ \ ]$ is replaced by $[\![\ \ ]\!]$.*

(a) *If $N_0 \subseteq M_0$, $N_1 \subseteq M_1$, and $N_2 \subseteq M_2$, then $[M_0, M_1, M_2] \subseteq [N_0, N_1, N_2]$.*

(b) *Also assume that $\overline{\overline{U}} \ge 2$. If $\langle M_0, M_1, M_2\rangle$ is canonical, then $[M_0, M_1, M_2] \subseteq [N_0, N_1, N_2]$ if and only if $N_0 \subseteq M_0$, $N_1 \subseteq M_1$, and $N_2 \subseteq M_2$. Hence if $\langle M_0, M_1, M_2\rangle$ and $\langle N_0, N_1, N_2\rangle$ both are canonical, then $[M_0, M_1, M_2] = [N_0, N_1, N_2]$ if and only if $M_0 = N_0$, $M_1 = N_1$, and $M_2 = N_2$.*

(c) *Again assume that* $\overline{\overline{U}} \geq 2$. *If* $\langle M_0, M_1, M_2\rangle$ *is precanonical, then* $[M_0, M_1, M_2] \subseteq [N_0, N_1, N_2]$ *implies that* $N_0 \subseteq M_0$ *and* $N_2 \subseteq M_2$. *Also, if* $\langle M_0, M_1, M_2\rangle$ *is precanonical and trim, then* $[M_0, M_1, M_2] = [N_0, N_1, N_2]$ *implies that, if* $N_1 \subseteq M_1$, *then* $M_1 \subseteq N_1$ *and hence* $M_1 = N_1$.

(d) *Assume that* $\langle M_0, M_1, M_2\rangle$ *is precanonical, that* $N_0 \subseteq M_0$, $N_1 \subseteq M_1$, *and* $N_2 \subseteq M_2$, *and hence* $[M_0, M_1, M_2] \subseteq [N_0, N_1, N_2]$, *and that* N_1 *is one-one. Then there is an* M_1' *such that* M_1' *is one-one,* $N_1 \subseteq M_1' \subseteq M_1$, *and* $M_1 \subseteq M_0 M_1' M_2$, *so that* $[M_0, M_1', M_2] = [M_0, M_1, M_2]$ *and hence* $[M_0, M_1', M_2] \subseteq [N_0, N_1, N_2]$.

Proof. (a) This follows from the definitions involved.

(b) Let $\langle M_0, M_1, M_2\rangle$ be any $\langle\gamma, \delta\rangle$ triple that is canonical. Given any $i_0 < \gamma$, let $I = \{i < \gamma : i_0 M_0 i\}$, $J = \{j < \delta : i_0 M_1 j\}$, $I' = \gamma \mathbin{\tilde{-}} I$, and $J' = \delta \mathbin{\tilde{-}} J$. Then

$$\begin{aligned} M_0 &= (M_0 \cap {}^2I) \cup (M_0 \cap {}^2I')\ , \\ M_1 &= (M_1 \cap (I \times J)) \cup (M_1 \cap (I' \times J')) \text{ and} \\ M_2 &= (M_2 \cap {}^2J) \cup (M_2 \cap {}^2J')\ . \end{aligned}$$

Now also assume that $\overline{\overline{U}} = \kappa \geq 2$ and let $\lambda < \mu < \kappa$. Let x and y satisfy the following conditions, respectively:

$$\begin{aligned} &|x| = \gamma, \quad x_n = \lambda \ \text{ if } n \in I, \quad x_n = \mu \ \text{ if } n \in I'. \\ &|y| = \delta, \quad y_n = \lambda \ \text{ if } n \in J, \quad y_n = \mu \ \text{ if } n \in J'. \end{aligned}$$

Then $M_0 \cap {}^2I \subseteq xx^{\smile}$ and $M_0 \cap {}^2I' \subseteq xx^{\smile}$. Hence $M_0 \subseteq xx^{\smile}$. Similarly $M_1 \subseteq xy^{\smile}$ and $M_2 \subseteq yy^{\smile}$. Hence $\langle x, y\rangle \in [M_0, M_1, M_2]$.

Let $\langle N_0, N_1, N_2\rangle$ be any $\langle\gamma, \delta\rangle$ triple such that either $N_0 \not\subseteq M_0$ or $N_1 \not\subseteq M_1$ or $N_2 \not\subseteq M_2$. First, suppose that $\langle i, j\rangle \in N_1$ and $\langle i, j\rangle \notin M_1$. Then for $i_0 = i$, we let $I, I', J, J', |x|$, and $|y|$ be as above. Then $\langle i, j\rangle = \langle i_0, j\rangle \notin xy^{\smile}$, hence $N_1 \not\subseteq xy^{\smile}$, and hence $\langle x, y\rangle \notin [N_0, N_1, N_2]$. Since $\langle x, y\rangle \in [M_0, M_1, M_2]$, there follows that $[M_0, M_1, M_2] \not\subseteq [N_0, N_1, N_2]$. Next, suppose that $\langle i, i'\rangle \in N_0$ and $\langle i, i'\rangle \notin M_0$. Then, for $i_0 = i$, we let $I, I', J, J', |x|$, and $|y|$ be as above. Then $\langle i, i'\rangle = \langle i_0, i'\rangle \notin xx^{\smile}$, hence $N_0 \not\subseteq xx^{\smile}$, and hence $\langle x, y\rangle \notin [N_0, N_1, N_2]$. Hence, again $[M_0, M_1, M_2] \notin [N_0, N_1, N_2]$. Now suppose that $\langle j, j'\rangle \in N_2$ and $\langle j, j'\rangle \notin M_2$. Then using an argument similar to the one just used, we construct an $\langle x, y\rangle$ such that $\langle x, y\rangle \in [M_0, M_1, M_2]$ but such that $\langle j, j'\rangle \notin yy^{\smile}$, hence $N_2 \not\subseteq yy^{\smile}$, and hence $\langle x, y\rangle \notin [N_0, N_1, N_2]$. Therefore again $[M_0, M_1, M_2] \not\subseteq [N_0, N_1, N_2]$. This concludes the proof that if $[M_0, M_1, M_2] \subseteq [N_0, N_1, N_2]$ then $N_0 \subseteq M_0$, $N_1 \subseteq M_1$, and $N_2 \subseteq M_2$. Conversely, if these three inclusions hold then, by part (a),

$[M_0, M_1, M_2] \subseteq [N_0, N_1, N_2]$. This proves the first assertion. The second assertion follows from it.

(c) Assume that $\overline{\overline{U}} \geq 2$, $\langle M_0, M_1, M_2\rangle$ is precanonical and $[M_0, M_1, M_2] \subseteq [N_0, N_1, N_2]$. By Theorem 2, there are $\langle\gamma, \delta\rangle$ triples $\langle M_0, M_1', M_2\rangle$ and $\langle N_0', N_1', N_2'\rangle$ that are canonical such that $[M_0, M_1, M_2] = [M_0, M_1', M_2]$, $[N_0, N_1, N_2] = [N_0', N_1', N_2']$ and, moreover, $N_0 \subseteq N_0'$ and $N_2 \subseteq N_2'$. Since $[M_0, M_1, M_2] \subseteq [N_0, N_1, N_2]$, therefore $[M_0, M_1', M_2] \subseteq [N_0', N_1', N_2']$. Then, by part (b), $N_0' \subseteq M_0$ and $N_2' \subseteq N_2$. Hence $N_0 \subseteq N_0' \subseteq M_0$ and $N_2 \subseteq N_2' \subseteq M_2$. This proves the first assertion.

For proving the second assertion we now assume, in addition, that $\langle M_0, M_1, M_2\rangle$ is trim. Suppose that $N_1 \subseteq M_1$ and $M_1 \not\subseteq N_1$. Then, for some $i < \gamma$ and $j < \delta$, $\langle i, j\rangle \in M_1$ and $\langle i, j\rangle \notin N_1$. Since $\langle M_0, M_1, M_2\rangle$ is trim, therefore there is no pair $\langle i', j'\rangle$ other than $\langle i, j\rangle$ such that $\langle i', i\rangle \in M_0$, $\langle i', j'\rangle \in M_1$ and $\langle j', j\rangle \in M_2$. Since $\langle M_0, M_1, M_2\rangle$ is precanonical and since $\langle i, j\rangle \in M_1$ therefore, if $\langle j', j\rangle \in M_2$ and $\langle i', i\rangle \notin M_0$ then $\langle i', j'\rangle \notin M_1$. There follows that there is no pair $\langle i', j'\rangle$ other than $\langle i, j\rangle$ such that $\langle i', j'\rangle \in M_1$ and $\langle j', j\rangle \in M_2$. Since $N_1 \subseteq M_1$ and $N_2 \subseteq M_2$ therefore there is no pair $\langle i', j'\rangle$ other than $\langle i, j\rangle$ such that $\langle i', j'\rangle \in N_1$ and $\langle j', j\rangle \in N_2$. Since $\langle i, j\rangle \notin N_1$ there follows that if $\langle j', j\rangle \in N_2$ then there is no i' such that $\langle i', j'\rangle \in N_1$.

We now choose any $\langle x, y\rangle$ in $[M_0, M_1, M_2]$ and any u in $U \cap -\{y_j\}$ and let z be the sequence in ${}^{\delta}U$ such that the following holds for any $n < \delta$:

If $\langle j, n\rangle \notin N_2$ then $z_n = y_n$ and if $\langle j, n\rangle \in N_2$ then $z_n = u$.

Consider any $\langle m, n\rangle$ in N_2. If $\langle j, n\rangle \in N_2$, then $\langle j, m\rangle \in N_2$ and $z_m = u = z_n$. Now suppose that $\langle j, n\rangle \notin N_2$, hence $\langle j, m\rangle \notin N_2$, and hence $z_n = y_n$ and $z_m = y_m$. Since $N_2 \subseteq M_2 \subseteq yy^{\smile}$, therefore $y_m = y_n$ and hence again $z_m = z_n$. Thus $N_2 \subseteq zz^{\smile}$.

Now consider any $\langle i', j'\rangle$ in N_1. Then, as we saw above, $\langle j, j'\rangle \notin N_2$. Hence $z_{j'} = y_{j'}$, by the definition of z. Since $\langle i', j'\rangle \in N_1 \subseteq M_1$ and since $\langle x, y\rangle$ is in $[M_0, M_1, M_2]$, therefore $x_{i'} = y_{j'}$. Hence $x_{i'} = z_{j'}$. Since this holds for any $\langle i', j'\rangle$ in N_1, therefore $N_1 \subseteq xz^{\smile}$. Since also $N_0 \subseteq M_0 \subseteq xx^{\smile}$, there now follows that $\langle x, z\rangle \in [N_0, N_1, N_2]$. In contrast, since $z_j = u \neq y_j$ and since $\langle i, j\rangle \in M_1$ and hence $x_i = y_j$, therefore $x_i \neq z_j$, hence $M_1 \not\subseteq xz^{\smile}$, and hence $\langle xz\rangle \not\subseteq [M_0, M_1, M_2]$. Thus $[N_0, N_1, N_2] \not\subseteq [M_0, M_1, M_2]$. From our assumptions and from the supposition that $N_1 \subseteq M_1$ and $M_1 \not\subseteq N_1$ we have thus been led to the conclusion that $[N_0, N_1, N_2] \not\subseteq [M_0, M_1, M_2]$. Hence our assumptions together with $[M_0, M_1, M_2] = [N_0, N_1, N_2]$ imply that, if $N_1 \subseteq M_1$, then $M_1 \subseteq N_1$ and hence $M_1 = N_1$. This proves the second assertion.

(d) Assume that $\langle M_0, M_1, M_2\rangle$ is precanonical, that $N_0 \subseteq M_0$, $N_1 \subseteq M_1$, and $N_2 \subseteq M_2$, and that N_1 is one-one. Since $\langle M_0, M_1, M_2\rangle$ is precanonical, therefore M_1 can be partitioned into a class $\{M_{1,\nu} : \nu < \mu\}$ of non-empty subrelations of M_1 such that pairs $\langle i, j\rangle$ and $\langle i', j'\rangle$ in M_1 belong to the same $M_{1,\nu}$ if and only if both $\langle i, i'\rangle$ is in M_0 and $\langle j, j'\rangle$ is in M_2. The class $\{M_{1,\nu} : \nu < \mu\}$ shall be indexed so that, for some λ, $0 \leq \lambda \leq \mu$, if $\nu < \lambda$, then $N_1 \cap M_{1,\nu} = \emptyset$ and, if $\lambda \leq \nu < \mu$, then $N_1 \cap M_{1,\nu} \neq \emptyset$. For every $\nu < \lambda$, we choose one pair $\langle i_\nu, j_\nu\rangle$ in $M_{1,\nu}$ and we let M_1'' be the union $\bigcup\{\langle i_\nu, j_\nu\rangle : \nu, \lambda\}$ of the chosen pairs. We let $M_1' = M_1'' \cup N_1$.

Evidently, $N_1 \subseteq M_1'$. Also, since both $M_1'' \subseteq M_1$ and $N_1 \subseteq M_1$, therefore $M_1' \subseteq M_1$. Consider any $\nu < \lambda$. Since $N_1 \subseteq M_1$, therefore $Do\ N_1 \cap Do\ M_{1,\nu} \neq \emptyset$ would imply that $N_1 \cap M_{1,\nu} \neq \emptyset$. Hence $Do\ N_1 \cap Do\ M_{1,\nu} = \emptyset$. Similarly, $Rg\ N_1 \cap Rg\ M_{1,\nu} = \emptyset$. There follows that $Do\ M_1' \cap Do\ N_1 = \emptyset$ and that $Rg\ M_1' \cap Rg\ N_1 = \emptyset$. Since M_1'' is one-one and N_1 is one-one, there now follows that $M_1' = M_1'' \cup N_1$ is one-one.

Consider any $\langle i, j\rangle$ in $M_1 = \bigcup\{M_{1,\nu} : \nu < \mu\}$. Suppose that $\langle i, j\rangle$ is in $M_{1,\nu}$ for some $\nu < \lambda$. Then $\langle i, i_\nu\rangle$ is in M_0 and $\langle j, j_\nu\rangle$ is in M_2. Hence $\langle i, j\rangle$ is in $M_0 M_1'' M_2$ and hence in $M_0 M_1' M_2$. Now suppose that $\langle i, j\rangle$ is in $M_{1,\nu}$ for some ν such that $\lambda \leq \nu < \mu$. Then there is some $\langle i', j'\rangle$ in N_1 such that $\langle i, i'\rangle$ is in M_0 and $\langle j, j'\rangle$ is in M_2. Then $\langle i, j\rangle$ is in $M_0 N_1 M_2$ and hence again in $N_0 M_1' M_2$. Thus, $M_1 \subseteq M_0 M_1' M_2$ and hence $[M_0, M_0 M_1' M_2, M_2] \subseteq [M_0, M_1, M_2]$. Since $M_0 \subseteq xx^{\smile}$, $M_1' \subseteq xy^{\smile}$, and $M_2 \subseteq yy^{\smile}$ together imply that $M_0 M_1' M_2 \subseteq xy^{\smile}$, therefore $[M_0 M_1' M_2] = [M_0, M_0 M_1' M_2, M_2]$. Hence $[M_0, M_1', M_2] \subseteq [M_0, M_1, M_2]$. Since $M_1' \subseteq M_1$, therefore $[M_0, M_1, M_2] \subseteq [M_0, M_1', M_2]$. There follows that $[M_0, M_1', M_2] = [M_0, M_1, M_2]$. Since $N_0 \subseteq M_0$, $N_1 \subseteq M_1$, $N_2 \subseteq M_2$, there follows that $[M_0, M_1', M_2] \subseteq [N_0, N_1, N_2]$. □

The assumption that $\overline{\overline{U}} \geq 2$ cannot be omitted from Theorem 3(b) or 3(c). If $\overline{\overline{U}} = 1$, then $[M_0, M_1, M_2] = {}_{[\gamma]}\overset{\circ}{1}$ for every $\langle \gamma, \gamma\rangle$ triple $\langle M_0, M_1, M_2\rangle$.

Assume that $\langle M_0, M_1, M_2\rangle$ and $\langle M_0', M_1', M_2'\rangle$ both are precanonical and trim, $M_0 = M_0'$, $M_2 = M_2'$, and $M_0 M_1 M_2 = M_0 M_1' M_2$. Then $\langle M_0, M_1, M_2\rangle$ shall be ***earlier than*** $\langle M_0', M_1', M_2'\rangle = \langle M_0, M_1', M_2\rangle$ if and only if, whenever iM_0i', iM_1j, and $i'M_1'j'$ (and hence also jM_2j'), then both $i \leq i'$ and $j \leq j'$. From Lemma 1(a) and Theorem 3(c) there follows that, if $\overline{\overline{U}} \geq 2$, then for any $\langle \gamma, \delta\rangle$ triple $\langle N_0, N_1, N_2\rangle$ the set of those $\langle \gamma, \delta\rangle$ triples $\langle M_0, M_1, M_2\rangle$ such that $[M_0, M_1, M_2] = [N_0, N_1, N_2]$ and such that $\langle M_0, M_1, M_2\rangle$ is precanonical and trim contains an earliest element.

A consequence of Theorems 2 and 3 is the following.

THEOREM 4. *Assume that* $\overline{\overline{U}} \geq 2$.

(a) *The restriction of* $[\ \ ]$ *to the set of those* $\langle\gamma,\delta\rangle$ *triples that are canonical is a one-one correspondence between this set and* ${}_{\langle\gamma,\delta\rangle}D$.

(b) *The restriction of* $[\![\ \ ]\!]$ *to the set of those* $\langle\gamma,\delta\rangle$ *triples that are finite and canonical is a one-one correspondence between this set and* ${}_{[0,\omega)}D$. □

A triple shall be ***special*** if and only if it is a framing triple of the form $\langle \mathring{1}_\gamma, M, \mathring{1}_\delta\rangle$. Framing triples of this kind will play a major role.

For special triples we shall henceforth use the following conventions. We let ${}_\gamma\langle M\rangle_\delta = \langle \mathring{1}_\gamma, M, \mathring{1}_\delta\rangle$ and ${}_\gamma[M]_\delta = [\mathring{1}_\gamma, M, \mathring{1}_\delta]$. Also, if $\gamma = s < \omega$ and $\delta = t < \omega$, we let ${}_\gamma[\![M]\!]_\delta = {}_s[\![M]\!]_t = [\![\mathring{1}_s, M, \mathring{1}_t]\!]$. When thus using $[\![\ \ ]\!]$, we will tacitly assume that $\gamma < \omega$ and $\delta < \omega$. Also, in all cases we will tacitly assume that $M \subseteq \gamma \times \delta$. Our use of ${}_\gamma[\ \ ]_\delta$, ${}_\gamma[\![\ \ ]\!]_\delta$, and ${}_s[\![\ \ ]\!]_t$ can also be described as follows

$$\begin{aligned} {}_\gamma[M]_\delta &= \{\langle x,y\rangle : |x|=\gamma,\ |y|=\delta,\ M\subseteq xy^{\smile}\} \\ {}_\gamma[\![M]\!]_\delta = {}_s[\![M]\!]_t &= \{\langle xz,yz\rangle : |x|=s,\ |y|=t,\ |z|<\omega,\ M\subseteq xy^{\smile}\}, \\ &\qquad \text{if } \gamma = s < \omega,\ \delta = t < \omega. \end{aligned}$$

We will use γ and δ as subscripts when we wish to include the case where $\gamma = \delta = \omega$ but not to exclude the case where $\gamma < \omega$ or $\delta < \omega$. In cases where $\gamma < \omega$ and $\delta < \omega$ then, for emphasis, we will mostly use s and t.

At times, to avoid clutter, and when it is clear from the context that the left and right subscript should be the same, we will omit the right subscript. Thus, instead of ${}_\gamma[M]_\gamma$ and ${}_s[\![M]\!]_s$, we will write ${}_\gamma[M]$ and ${}_s[\![M]\!]$.

Given any $\gamma \leq \omega$ and $\delta \leq \omega$, we will regard ${}_\gamma[\ \ ]_\delta$ as the function whose domain is the set $\{M : M \subseteq \gamma\times\delta\}$ and whose value, for any M in its domain, is the relation ${}_\gamma[M]_\delta$ in ${}_{\langle\gamma,\delta\rangle}D$ that is shown above. Likewise, if $\gamma = s < \omega$ and $\delta = t < \omega$, we will regard ${}_\gamma[\![\ \ ]\!]_\delta = {}_s[\![\ \ ]\!]_t$ as the function whose domain is the set $\{M : M \subseteq s\times t\}$ and whose value, for any M in its domain, is the relation ${}_s[\![M]\!]_t$ of grade $\langle s,t\rangle$ that is shown above.

In general, the relations M_0 and M_2 in a framing triple $\langle M_0, M_1, M_2\rangle$ serve two purposes. One of these is to impose conditions $x_i = x_j$ or $y_i = y_j$, respectively, on the pairs $\langle x,y\rangle$ that are in $[M_0,M_1,M_2]$ or in $[\![M_0,M_1,M_2]\!]$. The other is to impose conditions on the lengths of x and of y in these pairs $\langle x,y\rangle$. In a special triple ${}_\gamma\langle M\rangle_\delta = \langle \mathring{1}_\gamma, M, \mathring{1}_\delta\rangle$, $\mathring{1}_\gamma$ and $\mathring{1}_\delta$ are used for the second purpose only. They play a role that is far from trivial. For example, suppose that $M \subseteq s\times t$ and that

$s \leq s' < \omega$ and $t \leq t' < \omega$. If $\langle s, t\rangle \neq \langle s', t'\rangle$, then ${}_s[M]_t$ and ${}_{s'}[M]_{t'}$ are disjoint and if $s - t \neq s' - t'$ then ${}_s[\![M]\!]_t$ and ${}_{s'}[\![M]\!]_{t'}$ also are disjoint.

A relation R shall be ***difunctional*** if and only if $R \circ R^{\smile} \circ R \subseteq R$ and hence $R \circ R^{\smile} \circ R = R$. For example, every R that is both symmetric and transitive is difunctional. Also, if R is single-valued, then R is difunctional and hence $R^{\smile}$ is difunctional. Furthermore, if Q and R are single-valued and hence $Q^{\smile} \circ Q$ and $R^{\smile} \circ R$ are a subidentity on *Rg* Q or *Rg* R, respectively, then $Q \circ R^{\smile}$ is difunctional. In contrast, there are relations Q and R that are single-valued and hence difunctional such that $Q^{\smile} \circ R$ is not difunctional. The study of difunctional relations was initiated by J. Riguet in [Ri].

As usual, the ***transitive closure*** of a relation R will be the least relation that includes R and is transitive. We let the ***difunctional closure*** of a relation R, symbolically *dfc* R, be the least relation that includes R and is difunctional. As is well known, and easy to prove, there hold the following equalities relating *dfc* R, R, and the transitive closure of $R^{\smile} \circ R$ or of $R \circ R^{\smile}$, respectively.

$$\mathit{dfc}\ R = R \circ \bigcup_{1 \leq n < \omega} (R^{\smile} \circ R)^n = \bigcup_{1 \leq n < \omega} (R \circ R^{\smile})^n \circ R\,.$$

LEMMA 5. *The following assertions hold and, if $\gamma < \omega$ and $\delta < \omega$, then they also hold when* $[\ \ ]$ *is replaced by* $[\![\ \ ]\!]$.

(a) *If $M \subseteq N$, then ${}_\gamma[N]_\delta \subseteq {}_\gamma[M]_\delta$.*

(b) ${}_\gamma[M]_\delta = {}_\gamma[\mathit{dfc}\ M]_\delta$.

(c) *Assume that $\overline{\overline{U}} \geq 2$. If M is difunctional, then ${}_\gamma[M]_\delta \subseteq {}_\gamma[N]_\delta$ implies that $N \subseteq M$. Hence if M and N both are difunctional then ${}_\gamma[M]_\delta = {}_\gamma[N]_\delta$ if and only if $M = N$.*

Proof. (a) By Theorem 3(a).

(b) Since $M \subseteq \mathit{dfc}\ M$, $\mathit{Do}\ M = \mathit{Do}(\mathit{dfc}\ M)$, and $\mathit{Rg}\ M = \mathit{Rg}(\mathit{dfc}\ M)$, therefore, by part (a), ${}_\gamma[\mathit{dfc}\ M]_\delta \subseteq {}_\gamma[M]_\delta$.

Assume that $\langle x, y\rangle \in {}_\gamma[M]_\delta$. Then $M \subseteq xy^{\smile}$ and $M^{\smile}M \subseteq yx^{\smile}xy^{\smile} = yy^{\smile}$. Thus, from the inductive hypothesis that $M(M^{\smile}M)^n \subseteq xy^{\smile}$ there follows that $M(M^{\smile}M)^{n+1} \subseteq xy^{\smile}$. Hence $\mathit{dfc}\ M = M \circ \bigcup (M^{\smile}M)^n \subseteq xy^{\smile}$ and therefore $\langle x, y\rangle \in {}_\gamma[\mathit{dfc}\ M]_\delta$. Thus ${}_\gamma[M]_\delta \subseteq {}_\gamma[\mathit{dfc}\ M]_\delta$.

(c) Assume that $\overline{\overline{U}} \geq 2$, $M \subseteq \gamma \times \delta$, $N \subseteq \gamma \times \delta$, and M is difunctional. Suppose that $N \not\subseteq M$. Then, by an argument similar to one in the proof of Theorem 3(b), ${}_\gamma[M]_\delta \not\subseteq {}_\gamma[N]_\delta$. Hence, if ${}_\gamma[M]_\delta \subseteq {}_\gamma[N]_\delta$ then $N \subseteq M$. Hence, if M and N both are difunctional, then ${}_\gamma[M]_\delta = {}_\gamma[N]_\delta$ implies that $M = N$. Conversely, by part (a), if $M = N$, then ${}_\gamma[M]_\delta = {}_\gamma[N]_\delta$. □

A diagonal relation W shall be ***special*** if and only if it is either determined or mediately determined by a special triple.

Evidently, a diagonal relation W is special if and only if $W^\smile$ is special. Also, for any $\gamma \leq \omega, \delta \leq \omega$, and $M \subseteq \gamma \times \delta$,
$[\overset{\circ}{1}_\gamma \smile MM^\smile, M, M^\smile M \smile \overset{\circ}{1}_\delta]$ is a diagonal relation that is special since, according to Theorem 2,

$$[\overset{\circ}{1}_\gamma \smile MM^\smile, M, M^\smile M \smile \overset{\circ}{1}_\delta] = [\overset{\circ}{1}_\gamma, M, \overset{\circ}{1}_\delta] = {}_\gamma[M]_\delta \ .$$

If $\overline{\overline{U}} = 1$ then every ${}^\gamma U \times {}^\delta U$ is a singleton and hence every diagonal relation is special.

Assume that $\overline{\overline{U}} \geq 2$. Then the following relation W in ${}_{[\omega]}D \cap {}_{[\omega]}Un$ is not special:

$$\begin{aligned}&[\{\langle 0,1\rangle, \langle 1,0\rangle\} \cup \overset{\circ}{1}_\omega, \{\langle n+2, n\rangle : n < \omega\}, \overset{\circ}{1}_\omega]\\ &\quad = \{\langle xxz, z\rangle : |x| = 1,\ |z| = \omega\} = (\overset{\omega}{\rightarrow} e_{01}) \circ (\overset{\omega}{\rightarrow} (\dot{q}_0^2)) \ .\end{aligned}$$

To show that W differs from every ${}_\omega[M] = {}_\omega[M]_\omega$ and hence is not special we consider any relation M on ω. First suppose that $\langle 0, j\rangle \in M$ for some j. Then $v_0 = w_j$ for every $\langle v, w\rangle$ in ${}_\omega[M]$. Since $\overline{\overline{U}} \geq 2$, there are x and z such that $|x| = 1$, $|z| = \omega$, and $x_0 \neq z_j$. Then $\langle xxz, z\rangle$ is in W but not in ${}_\omega[M]$. Now suppose that there is no j such that $\langle 0, j\rangle \in M$. Since $\overline{\overline{U}} \geq 2$, there are x and y such that $|x| = 1$, $|y| = \omega$, $x_0 \neq y_0$, and $y_i = y_j$ for every i and j. Then $\langle xy, y\rangle$ is in ${}_\omega[M]$ but not in W.

Below we give two further diagonal relations which, if $\overline{\overline{U}} \geq 2$, are not special. This can be seen by an argument similar to the one just given. The two relations are in ${}_{[0,\omega)}D$ and also in S_{qe}.

$$\begin{aligned}&\{\langle xxz, z\rangle : |x| = 1,\ |z| < \omega\} = e_{01}\dot{q}_0^2\\ &\{\langle xxz, yyz\rangle : |x| = 1,\ |y| = 1,\ |z| < \omega\} = e_{01}\dot{q}_0^2 q_0^2 e_{01} = e_{01}c_0c_1e_{01} \ .\end{aligned}$$

Although a diagonal relation is not always special, by the next lemma it is always a product of special diagonal relations. The lemma follows from the definitions involved and may sometimes be used tacitly.

LEMMA 6. *Let $\langle M_0, M_1, M_2\rangle$ be any $\langle \gamma, \delta\rangle$ triple. Then there holds the following equality and, if $\gamma < \omega$ and $\delta < \omega$, also the equality that results from it when* [ ] *is*

replaced by $[\![\ ]\!]$.
$[M_0, M_1, M_2] = {}_\gamma[M_0] \circ {}_\gamma[M_1]_\delta \circ {}_\delta[M_2] = {}_\gamma[M_0]_\gamma \circ {}_\gamma[M_1]_\delta \circ {}_\delta[M_2]_\delta$. □

We now turn to some further conditions on relations. Let $\gamma \leq \omega$ and $\delta \leq \omega$. A relation M on ω shall be $\langle\gamma, \delta\rangle$ ***full*** if and only if $Do\ M = \gamma$ and $Rg\ M = \delta$. Also, a subset W of ${}^\gamma U \times {}^\delta U$ shall be $\langle\gamma, \delta\rangle$ ***full*** if and only if $Do\ W = {}^\gamma U$ and $Rg\ W = {}^\delta U$. If γ and δ are understood, then M and W shall be ***full*** if and only if they are $\langle\gamma, \delta\rangle$ full.

Let M be any relation on ω. Then M shall be ***downward full*** if and only if, for any i and j such that $i < j < \omega$, $j \in Do\ M$ implies that $i \in Do\ M$ and $j \in Rg\ M$ implies that $i \in Rg\ M$.

Let W be uniform on ${}^{[0,\omega)}U$. Then W shall be ***upward full*** if and only if $W \neq \emptyset$ and, for the unique s and t such that W has grade $\langle s, t\rangle$, W satisfies $Do\ W = {}^{[s,\omega)}U$ and $Rg\ W = {}^{[t,\omega)}U$. Hence W is upward full if and only if its initial section is full.

As usual, given any set X, an ***equivalence relation on*** X shall be any relation R such that $Do\ R = X$ and such that R is symmetric and transitive. Also, a ***permutation of*** X shall be any relation R such that $Do\ R = Rg\ R = X$ and R is one-one. Further, a ***subidentity*** on X shall be any relation R that is a subset of the identity relation on X. Note that any relation R of one of these three kinds is difunctional. If X is understood, reference to it may be omitted.

According to the next theorem, diagonal relations W of certain kinds are always special. Moreover, if $\overline{\overline{U}} \geq 2$, then in each case there is a kind of relations M on ω such that between the two kinds of relations there is a one-one correspondence. Further, between the conditions that give rise to the two kinds of relations there holds a certain duality.

THEOREM 7. *For* (a) *and* (b) *below assume that* W *is in* ${}_{\langle\gamma,\delta\rangle}D$ *and for* (c), (d) *and* (e) *below assume that* W *is in* ${}_{[\gamma]}D$. *If* $\overline{\overline{U}} \geq 2$, *then condition* (i) *and condition* (ii) *in each of* (a),...,(e) *below are equivalent to each other. If* $\overline{\overline{U}} = 1$, *then* (i) *is equivalent to the condition that results from* (ii) *when one replaces "exactly" by "at least".*

Moreover, if $\gamma < \omega$ *and* $\delta < \omega$, *then one can replace* $[\ ]$ *by* $[\![\ ]\!]$, *provided that one replaces the assumption that* W *is in* ${}_{\langle\gamma,\delta\rangle}D$ *or in* ${}_{[\gamma]}D$ *by the assumption that* W *is in* $\{\overset{[0,\omega)}{\longrightarrow} X : X \in {}_{\langle\gamma,\delta\rangle}D\}$, *or in* $\{\overset{[0,\omega)}{\longrightarrow} X : X \in {}_{[\gamma]}D\}$, *respectively, and provided that, in* (c) *and* (e), *one replaces* ${}^\gamma U$ *by* ${}^{[\gamma,\omega)}U$ *and, in* (a), *"full" by "upward full".*

(a) (i) *W is full.*

(ii) *$W = {}_{\gamma}[M]_{\delta}$, for exactly one one-one M.*

(b) (i) *W is one-one.*

(ii) *$W = {}_{\gamma}[M]_{\delta}$, for an M that is full (and hence also for exactly one M that is both full and difunctional).*

(c) (i) *W is an equivalence on ${}^{\gamma}U$.*

(ii) *$W = {}_{\gamma}[M]$, for exactly one subidentity M.*

(d) (i) *W is a subidentity.*

(ii) *$W = {}_{\gamma}[M]$, for exactly one equivalence M on γ.*

(e) (i) *W is a permutation of ${}^{\gamma}U$.*

(ii) *$W = {}_{\gamma}[M]$, for exactly one permutation M of γ.*

Proof. The assertions about $\llbracket\ \rrbracket$ follow from the assertions about $[\ ]$. Regarding these, we first consider the case where $\overset{=}{U} = 1$. Then, for any $\gamma \leq \omega$ and $\delta \leq \omega$, there is exactly one W in ${}_{\langle\gamma,\delta\rangle}D$ and, moreover, for every $M \subseteq \gamma \times \delta$, there holds ${}_{\gamma}[M]_{\delta} = W$. Thus, in this case the assertion about $[\ ]$ holds trivially.

For the rest of the proof we therefore assume that $\overset{=}{U} \geq 2$. Then, by Theorem 5(b), for any W in any ${}_{\langle\gamma,\delta\rangle}D$ there is at most one M such that both ${}_{\gamma}[M]_{\delta} = W$ and M is difunctional. Moreover, if M is either one-one or an equivalence then M is difunctional. There follows that condition (ii) in (a),(c),(d), or (e) is equivalent to the condition that results from it when one replaces "exactly" by "at least". Also, by Theorem 5(b) and 5(a), condition (ii) in (b) is equivalent to the condition that results from it when one deletes the parenthesized part. It therefore suffices to prove the equivalence of condition (i) in (a), ... ,(e), respectively, to the condition thus resulting from (ii).

To prove part (a), we first consider any $W = [M_0, M_1, M_2]$ and first suppose that $M_0 \neq \overset{\circ}{1}_{\gamma}$. Then $i < j < \gamma$ and $\langle i, j\rangle \in M_0$ for some i and j. Since $\overset{=}{U} \geq 2$, there are x in ${}^{\gamma}U$ such that $x_i \neq x_j$. Then $M_0 \not\subseteq xx^{\smile}$ and hence $\langle x, y\rangle \notin W$ for every y in ${}^{\delta}U$. There follows that $Do\ W \neq {}^{\gamma}U$. This shows that if $Do\ W = {}^{\gamma}U$ then $M_0 = \overset{\circ}{1}_{\gamma}$. Similarly, if $Rg\ W = {}^{\delta}U$, then $M_2 = \overset{\circ}{1}_{\delta}$. Thus, if W is full then, for some $M \subseteq \gamma \times \delta$, W is the special diagonal relation $[1_{\gamma}, M, 1_{\delta}] = {}_{\gamma}[M]_{\delta}$.

Now let $W = {}_{\gamma}[M]_{\delta}$. Suppose that M is not single-valued. Then $\langle i, j\rangle$ and $\langle i, k\rangle$ both are in M for some i, j, and k such that $j \neq k$. Since $\overset{=}{U} \geq 2$, there are y in ${}^{\delta}U$ such that $y_j \neq y_k$. Consider any x in ${}^{\gamma}U$. Then, either $x_i \neq y_j$ or $x_i \neq y_k$. In either cases, $M \not\subseteq xy^{\smile}$ and hence $\langle x, y\rangle \notin W$. Since this holds for any x in ${}^{\gamma}U$, there follows that $y \notin Rg\ W$ and hence $Rg\ W \neq {}^{\delta}U$. This shows that if *Rg* $W = {}^{\delta}U$ then M is single-valued. Similarly, if *Do* $W = {}^{\gamma}U$, then $M^{\smile}$ is single-valued. There now follows that if W is full then $W = {}_{\gamma}[M]_{\delta}$ for some M that is one-one.

Now assume that $W = {}_{\gamma}[M]_{\delta}$. First, also assume that M is single-valued. Then for any $i < \gamma$ there is at most one j such that $\langle i, j\rangle \in M$. Consider any y in ${}^{\delta}U$. There are $x \in {}^{\gamma}U$ such that, if $i \in$ *Do* M, then $x_i = y_j$ for the unique j such that $\langle i, j\rangle \in M$. Then $M \subseteq xy^{\smile}$, hence $\langle x, y\rangle \in W$, and hence $y \in$ *Rg* W. There follows that if M is single-valued then *Rg* $W = {}^{\delta}U$. Similarly, if $M^{\smile}$ is single-valued then *Do* $W = {}^{\gamma}U$. There now follows that if M is one-one then W is full. This concludes the proof of part (a).

For the proof of part (b), we first consider any W in ${}_{\langle\gamma,\delta\rangle}D$. By Theorems 2 and 3, $W = [M_0, M_1, M_2]$ for a unique $\langle\gamma, \delta\rangle$ triple $\langle M_0, M_1, M_2\rangle$ that is canonical. Suppose that *Do* $M_1 \neq \gamma$. Then there is some $i < \gamma$ such that $\langle i, j\rangle \notin M_1$ for every $j < \delta$. Since $\langle M_0, M_1, M_2\rangle$ is canonical, therefore if $\langle n, i\rangle \in M_0$, then also $\langle n, j\rangle \notin M_1$ for every $j < \delta$. Since $W \neq \emptyset$, there are $x \in {}^{\gamma}U$ and $y \in {}^{\delta}U$ such that $\langle x, y\rangle \in W$. Since $\overline{\overline{U}} \geq 2$, we can choose some u_λ in U such that $u_\lambda \neq x_i$ and then let w be the sequence in ${}^{\gamma}U$ that satisfies the following condition:

$$w_n = u_\lambda \ \text{ if } \ \langle n, i\rangle \in M_0 \ \text{ and } \ w_n = x_n \ \text{ if } \ n < \gamma \ \text{ and } \ \langle n, i\rangle \notin M_0 \ .$$

Since $\langle x, y\rangle \in W$, therefore $M_0 \subseteq xx^{\smile}$ and $M_1 \subseteq xy^{\smile}$. There follows that $M_0 \subseteq ww^{\smile}$ and $M_1 \subseteq wy^{\smile}$. Since $\langle x, y\rangle \in W$ also implies that $M_2 \subseteq yy^{\smile}$, there follows that $\langle w, y\rangle \in W$. Since $w \neq x$, this shows that, if *Do* $M \neq \gamma$, then $W^{\smile}$ is not single-valued. Hence, if $W^{\smile}$ is single-valued, then *Do* $M_1 = \gamma$. Similarly, if W is single-valued, then *Rg* $M_1 = \delta$. Thus we have shown that if W is one-one, and $\langle M_0, M_1, M_2\rangle$ is the canonical triple such that $W = [M_0, M_1, M_2]$, then M_1 is full.

Consider any $\langle\gamma, \delta\rangle$ triple $\langle M_0, M_1, M_2\rangle$ that is canonical such that M_1 is full. Since *Do* $M_1 = \gamma$, therefore $M_0 \subseteq M_0M_1M_1^{\smile}$. Since $\langle M_0, M_1, M_2\rangle$ is canonical and hence $M_0M_1 \subseteq M_1$, there follows that $M_0 \subseteq M_1M_1^{\smile}$. Hence whenever $x \in {}^{\gamma}U$, $y \in {}^{\delta}U$, and $M_1 \subseteq xy^{\smile}$, then $M_0 \subseteq M_1M_1^{\smile} \subseteq xy^{\smile}yx^{\smile} \subseteq xx^{\smile}$. Since this holds, in particular when $xx^{\smile} = \mathring{1}_{\gamma}$, there follows that $[M_0, M_1, M_2] = [\mathring{1}_{\gamma}, M_1, M_2]$. Similarly, since *Rg* $M_1 = \delta$, therefore $[\mathring{1}_{\gamma}, M_1, M_2] = [\mathring{1}_{\gamma}, M_1, \mathring{1}_{\delta}] = {}_{\gamma}[M_1]_{\delta}$. Hence $[M_0, M_1, M_2] = {}_{\gamma}[M]_{\delta}$. From this and the preceding paragraph there now follows that if W is in ${}_{\langle\gamma,\delta\rangle}D$ and is one-one, then $W = {}_{\gamma}[M]_{\delta}$ for some M that is full.

Now assume that $W = {}_{\gamma}[M]_{\delta}$ for some M. First, assume also that *Rg* $M = \delta$. Then, for every $j < \delta$ there is at least one $i < \gamma$ such that $\langle i, j\rangle \in M$. Let $x \in {}^{\gamma}U$. Then there is at most one $y \in {}^{\delta}U$ such that $y_j = x_i$ for every $\langle i, j\rangle$ in M. Thus there is at most one y for which $\langle x, y\rangle$ is in ${}_{\gamma}[M]_{\delta} = W$ and hence such that $M \subseteq xy^{\smile}$. There follows that W is single-valued. Similarly, if $Do\ M = \gamma$, then $W^{\smile}$ is single-valued. Hence, if M is full, then W is one-one. This concludes the proof of part (b).

To prove part (c), we first assume that W is an equivalence on ${}^{\gamma}U$. Since W is full therefore, by part (a), $W = {}_{\gamma}[M]_{\delta}$ for some M that is one-one. Suppose that M were not symmetric. Then, for some i and j, $\langle i, j\rangle \in M$ and $\langle j, i\rangle \notin M$. Let $u_{\lambda} \in U$, $u_{\mu} \in U$, $u_{\lambda} \neq u_{\mu}$, and let $x \in {}^{\gamma}U$ and $y \in {}^{\gamma}U$ be as follows:

$$x_n = u_{\lambda} \text{ for every } n < \gamma, \ y_j = u_{\mu}, \text{ and}$$
$$y_n = u_{\lambda} \ \text{ if } \ n \neq j \text{ and } n < \gamma.$$

Then $\langle x, y\rangle \in W^{\smile}$ and $\langle x, y\rangle \notin W$, so that W would not be an equivalence, contrary to our assumption. Thus M is symmetric. Since M is also one-one, there follows that M is a subidentity. Conversely, if $M \subseteq \overset{\circ}{1}_{\gamma}$, then $W = {}_{\gamma}[M]_{\delta}$ is an equivalence on ${}^{\gamma}U$.

To prove part (d), we first assume that $W \subseteq {}_{[\gamma]}\overset{\circ}{1}$. Since W is one-one therefore, by part (b), there is an M that is full and difunctional such that $W = {}_{\gamma}[M]_{\delta}$. Since $\overline{\overline{U}} \geq 2$, an argument similar to one in the proof of Theorem 3(b) shows that if there were an $i < \gamma$ such that $\langle i, i\rangle \notin M$, then there would be pairs $\langle x, y\rangle$ in W such that $x_i \neq y_i$. There follows that $\overset{\circ}{1}_{\gamma} \subseteq M$. Hence $M^{\smile} \subseteq MM^{\smile}M$ and $MM \subseteq MM^{\smile}M$. Since M is difunctional and hence $MM^{\smile}M \subseteq M$, there follows that M is symmetric and transitive and hence is an equivalence on $Do\ M = \gamma$. Conversely, if $M = eq_{\gamma}M$, then $[M_0, M, M_2] \subseteq {}_{[\gamma]}\overset{\circ}{1}$ for any M_0 and M_2 satisfying $M_0 = eq_{\gamma}M_0$ and $M_2 = eq_{\gamma}M_2$, respectively, so that, in particular, ${}_{\gamma}[M] = [\overset{\circ}{1}_{\gamma}, M, \overset{\circ}{1}_{\gamma}] \subseteq {}_{[\gamma]}\overset{\circ}{1}$.

To prove part (e), we first assume that W is a permutation of ${}^{\gamma}U$. Then, by parts (a) and (b), $W = {}_{\gamma}[M]$ for some M that is both one-one and full, and hence is a permutation of γ. Conversely, if M is a permutation of γ, then ${}_{\gamma}[M]$ is a permutation of ${}^{\gamma}U$. □

The parts of Theorems 7(e), 7(d) and 7(c) that concern $[\![\ ]\!]$ deal with generalizations of notions that were introduced in Chapter I. To describe this for Theorem 7(e), we now, and henceforth, let

$$(i, j) = \{\langle i, j\rangle, \langle j, i\rangle\} \cup \{\langle n, n\rangle : n \in \omega, \ n \neq i, \ n \neq j\} ,$$

so that (i, j) is the permutation of ω that transposes i and j. Then, in the case where $i \leq j$, $p_{(i,j)}$ is the following permutation of ${}^{[j+1,\omega)}U$:

$$p_{(i,j)} = {}_{j+1}[\![(i, j) \cap (\{0, \ldots, j\} \times \{0, \ldots, j\})]\!] .$$

One also sees that if $i \leq j$, then e_{ij} is the subidentity ${}_{j+1}[\![eq_{j+1}\{\langle i,j\rangle\}]\!]$. Further, c_i is the upward full equivalence ${}_{i+1}[\![\mathring{1}_i]\!]$. An example of a relation in ${}_{[\omega]}D$ considered in 7(e), 7(d), and 7(c), respectively is the relation $\overset{\omega}{\rightarrow} p_{(i,j)} = {}_{\omega}[(i,j)]$, $\overset{\omega}{\rightarrow} e_{ij} = {}_{\omega}[eq_{\omega}\{\langle i,j\rangle\}]$, or $\overset{\omega}{\rightarrow} c_i = {}_{\omega}[\mathring{1}_{\omega} \,\overline{2}\, \{\langle i,i\rangle\}]$, respectively.

The diagonal relations considered in Theorems 7(a) and 7(d) will play a major role. Those considered in Theorem 7(e) will also be of some importance. Henceforth, we let:

$$
\begin{aligned}
{}_{\langle\gamma,\delta\rangle}DF &= \{W \in {}_{\langle\gamma,\delta\rangle}D : W \text{ is full}\} = \{{}_{\gamma}[M]_{\delta} : M \text{ is one-one}\} \\
{}_{[\gamma]}DF &= {}_{\langle\gamma,\gamma\rangle}DF \\
{}_{[0,\omega)}DF &= \{W \in {}_{[0,\omega)}D : W \text{ is upward full}\} \\
&= \{{}_{s}[\![M]\!]_{t} : M \text{ is one-one, } s < \omega,\ t < \omega\} \\
{}_{[\gamma]}DI &= \{W \in {}_{[\gamma]}D : W \text{ is a subidentity}\} = \{{}_{\gamma}[M] : M = eq_{\gamma}M\} \\
{}_{[0,\omega)}DI &= \{W \in {}_{[0,\omega)}D : W \text{ is a subidentity}\} \\
&= \{{}_{s}[\![M]\!] : M = eq_{s}M,\ s < \omega\} \\
{}_{[\gamma]}DP &= \{W \in {}_{[\gamma]}D : W \text{ is a permutation of } {}^{\gamma}U\} \\
&= \{{}_{\gamma}[M] : M \text{ is a permutation of } \gamma\} \\
{}_{[0,\omega)}DP &= \{W \in {}_{[0,\omega)}D : W \text{ is an upward full permutation}\} \\
&= \{{}_{s}[\![M]\!] : M \text{ is a permutation of } s,\ s < \omega\}
\end{aligned}
$$

As we saw earlier, a relative product of diagonal relations that are special is not always a diagonal relation that is special. From parts (b) and (c) of the next lemma there follows that, for certain kinds of special diagonal relations, the relative product of two relations of this kind is again a relation of this kind.

Henceforth, for any relations Q and R, we let $Q \bullet R = dfc(Q \circ R)$ and call it ***the difunctional product of*** Q ***and*** R. Thus, $Q \circ R = Q \bullet R$ if and only if $Q \circ R$ is difunctional. The axiomatic study of difunctional products was initiated by D. A. Bredikhin, perhaps in [Br94].

LEMMA 8. *Assertions* (a),(b),(c) *below hold for any* $\beta \leq \omega$, $\gamma \leq \omega$, $\delta \leq \omega$ *and any* $M \subseteq \beta \times \gamma$ *and* $N \subseteq \gamma \times \delta$. *Moreover, if* $\beta < \omega$, $\gamma < \omega$, *and* $\delta < \omega$, *they also hold when* [ ] *is replaced by* $[\![\ ]\!]$.

(a) ${}_{\beta}[M]_{\gamma} \circ {}_{\gamma}[N]_{\delta} \subseteq {}_{\beta}[M \circ N]_{\delta}$.

(b) *If* M *and* N *both are one-one, then* ${}_{\beta}[M]_{\gamma} \circ {}_{\gamma}[N]_{\delta} = {}_{\beta}[M \circ N]_{\delta}$.

(c) *If* $Rg\ M = Do\ N$, *then* ${}_\beta[M]_\gamma \circ {}_\gamma[N]_\delta = {}_\beta[M \circ N]_\delta = {}_\beta[dfc(M \circ N)]_\delta = {}_\beta[M \bullet N]_\delta$. *Hence, in particular, if* $M = eq_\gamma M$ *and* $N = eq_\gamma N$, *then* ${}_\gamma[M] \circ {}_\gamma[N] = {}_\gamma[(M \bullet N)] = {}_\gamma[eq_\gamma(M \cup N)]$.

Proof. Assertions (a),(b),(c) about $[\![\ ]\!]$ follow from those about $[\ ]$. We only prove therefore those about $[\ ]$.

Consider any $\langle x, z\rangle$ in ${}_\beta[M]_\gamma \circ {}_\gamma[N]_\delta$. Then x is in ${}^\beta U$, z is in ${}^\delta U$, and, for some y in ${}^\gamma U$, $\langle x, y\rangle$ is in ${}_\beta[M]_\gamma$ and $\langle y, z\rangle$ is in ${}_\gamma[N]_\delta$. Then $M \subseteq xy^\smile$ and $N \subseteq yz^\smile$. Hence $M \circ N \subseteq xy^\smile yz^\smile \subseteq xz^\smile$. Thus, $\langle x, z\rangle$ is in ${}_\beta[M \circ N]_\delta$. There follows that ${}_\beta[M]_\gamma \circ {}_\gamma[N]_\delta \subseteq {}_\beta[M \circ N]_\delta$.

(b) Assume that M and N both are one-one. Consider any $\langle x, z\rangle$ in ${}_\beta[M \circ N]_\delta$. Then $x \in {}^\beta U$, $z \in {}^\delta U$, and $M \circ N \subseteq xz^\smile$. Let y be any sequence in ${}^\gamma U$ that satisfies the following two conditions:

(i) If $j \in Rg\ M$, then $y_j = x_i$ for the unique i such that iMj.
(ii) If $j \notin M$ and $j \in Do\ N$, then $y_j = z_k$ for the unique k such that jNk.

By (i), $M \subseteq xy^\smile$ and hence $\langle x, y\rangle \in {}_\beta[M]_\gamma$. Consider any j in $Do\ N$. Suppose that j is not in $Rg\ M$. Then, by (ii), $y_j = z_k$ and hence $\langle j, k\rangle \in yz^\smile$. Now suppose that j is in $Rg\ M$ and hence in $Do\ N \cap Rg\ M$. Then there is a unique i satisfying iMj and a unique k satisfying jNk. Then $\langle i, k\rangle \in M \circ N$. Since $M \circ N \subseteq xz^\smile$, therefore $x_i = z_k$. Since, by (i), $y_j = x_i$, therefore again $y_j = z_k$. Thus $y_j = z_k$ for every $\langle j, k\rangle$ in N, hence $N \subseteq yz^\smile$, and hence $\langle y, z\rangle \in {}_\gamma[N]_\delta$. There follows that $\langle x, z\rangle$ is in ${}_\beta[M]_\gamma \circ {}_\gamma[N]_\delta$. Thus, if M and N are one-one, then ${}_\beta[M \circ N]_\delta \subseteq {}_\beta[M]_\gamma \circ {}_\gamma[N]_\delta$. Therefore, by part (a), if M and N are one-one, then ${}_\beta[M]_\gamma \circ {}_\gamma[N]_\delta = {}_\beta[M \circ N]_\delta$.

(c) Assume that $Rg\ M = Do\ N$. Then $MM^\smile \subseteq (MN)(MN)^\smile$ and $N^\smile N \subseteq (MN)^\smile(MN)$. Consider any $\langle x, z\rangle$ in ${}_\beta[M \circ N]_\delta$. Then $x \in {}^\beta U$, $z \in {}^\delta U$, and $MN \subseteq xz^\smile$. Let y be any sequence in ${}^\gamma U$ that satisfies the following condition:

If j is in $Rg\ M = Do\ N$, then $y_j = x_i$ for the least i such that $\langle i, j\rangle \in M$.

Consider any $\langle i', j\rangle$ in M. For the least i such that $\langle i, j\rangle \in M$ there holds $y_j = x_i$. Since $\langle i', j\rangle \in M$ and $\langle i, j\rangle \in M$, therefore $\langle i', i\rangle \in MM^\smile$. Since $MN \subseteq xz^\smile$, therefore $MM^\smile \subseteq (MN)(MN)^\smile \subseteq xz^\smile zx^\smile \subseteq xx^\smile$. Therefore $\langle i', i\rangle \in xx^\smile$ and hence $x_{i'} = x_i$. There now follows that $x_{i'} = y_j$. Since this holds for any $\langle i', j\rangle$ in M, therefore $M \subseteq xy^\smile$ and hence $\langle x, y\rangle \in {}_\beta[M]_\gamma$.

Consider any $\langle \gamma, k\rangle$ in N. Since $Do\ N \subseteq Rg\ M$, therefore $j \in Rg\ M$. Hence, for the least i such that $\langle i, j\rangle \in M$ there holds $y_j = x_i$. Since $\langle i, j\rangle \in M$ and

$\langle j,k\rangle \in N$, therefore $\langle i,k\rangle \in MN$. Since $MN \subseteq xz^{\smile}$, therefore $x_i = z_k$. Since also $y_j = x_i$, there follows that $y_j = z_k$ and hence that $\langle j,k\rangle \in yz^{\smile}$. Since this holds for any $\langle j,k\rangle$ in N, therefore $N \subseteq yz^{\smile}$ and hence $\langle y,z\rangle \in {}_{\gamma}[N]_{\delta}$.

There now follows that $\langle x,z\rangle$ is in ${}_{\beta}[M]_{\gamma} \circ {}_{\gamma}[N]_{\delta}$. Thus, if $Rg\ M = Do\ N$, then any $\langle x,z\rangle$ in ${}_{\beta}[M\circ N]_{\delta}$ is also in ${}_{\beta}[M]_{\gamma}\circ{}_{\gamma}[N]_{\delta}$ and hence ${}_{\beta}[M\circ N]_{\delta} \subseteq {}_{\beta}[M]_{\gamma}\circ{}_{\gamma}[N]_{\delta}$. There follows from part (a) that, if $Rg\ M = Do\ N$, then ${}_{\beta}[M]_{\gamma}\circ{}_{\gamma}[N]_{\delta} = {}_{\beta}[M\circ N]_{\delta}$. By Lemma 5(a), ${}_{\beta}[M\circ N]_{\delta} = {}_{\beta}[dfc(M\circ N)]_{\delta} = {}_{\beta}[M \bullet N]_{\delta}$.

If $M = eq_{\gamma}M$ and $N = eq_{\gamma}N$, then $Rg\ M = \gamma = Do\ N$ and $M \bullet N = eq_{\gamma}(M\circ N)$. There follows the second assertion in (c). □

LEMMA 9. *Let* $s<\omega$, $t<\omega$, $k<\omega$, $M\subseteq s\times t$, *and* $N \subseteq (s+k)\times(t+k)$.

(a) $\overset{k}{\rightarrow}{}_{s}[\![M]\!]_{t} = {}_{[s+k,\omega)}\overset{\circ}{1} \circ {}_{s}[\![M]\!]_{t} = {}_{s}[\![M]\!]_{t} \circ {}_{[t+k,\omega)}\overset{\circ}{1} =$ ${}_{s+k}[\![M \cup \{\langle s+j, t+j\rangle \colon j<k\}]\!]_{t+k}$.

(b) *Assume that* $\overset{k}{\rightarrow} X = {}_{s+k}[\![N]\!]_{t+k}$. *Then* $X = {}_{s}[\![N\cap(s\times t)]\!]_{t}$. *Moreover, if* $\overline{\overline{U}} \geq 2$, *then* $N = (N\cap(s\times t)) \cup \{\langle s+j,t+j\rangle : j<k\}$.

Proof. (a) Since ${}_{s}[\![M]\!]_{t}$ has grade $\langle s,t\rangle$, therefore $\overset{k}{\rightarrow}{}_{s}[\![M]\!]_{t} = {}_{[s+k,\omega)}\overset{\circ}{1}\circ{}_{s}[\![M]\!]_{t} =$ ${}_{s}[\![M]\!]_{t}\circ{}_{[t+k,\omega)}\overset{\circ}{1}$. Consider any $\langle v,w\rangle$ in ${}_{[s+k,\omega)}\overset{\circ}{1}\circ{}_{s}[\![M]\!]_{t}$. Then $\langle v,v\rangle$ is in ${}_{[s+k,\omega)}\overset{\circ}{1}$ and hence, for a unique x and z, $|x|=s$, $k\leq|z|<\omega$, and $v=xz$. Also, $\langle v,w\rangle = \langle xz,w\rangle$ is in ${}_{s}[\![M]\!]_{t}$ and hence, for a unique y, $|y|=t$, $w=yz$, and $M\subseteq xy^{\smile}$. Conversely, if $v=xz$ and $w=yz$, where $|x|=s$, $|y|=t$, and $k\leq|z|<\omega$, and if $M\subseteq xy^{\smile}$, then $\langle v,w\rangle$ is in ${}_{[s+k,\omega)}\overset{\circ}{1}\circ{}_{s}[\![M]\!]_{t}$.

Consider any $\langle v,w\rangle$ in ${}_{s+k}[\![M\cup\{\langle s+j,t+j\rangle : j<k\}]\!]_{t+k}$. Then, for a unique x,y,z',z'',z''' there holds $|x|=s$, $|z'|=k$, $|y|=t$, $|z''|=k$, $0\leq|z'''|<\omega$, $v=xz'z'''$, $w=yz''z'''$, and $M\cup\{\langle s+j,t+j\rangle : j<k\} \subseteq (xz')(yz'')^{\smile}$. Since $M\subseteq(xz')(yz'')$, therefore $M\subseteq xy^{\smile}$. Since $\{\langle s+j,t+j\rangle : j<k\}\subseteq(xz')(yz'')^{\smile}$, therefore $(xz')_{s+j} = (yz'')_{t+j}$ for every $j<k$, hence $z'_j = z''_j$ for every $j<k$, and hence $z'=z''$. Now let $z = z'z''' = z''z'''$. Then $k\leq|z|<\omega$, $v=xz$, and $w=yz$. Conversely, if $v=xz$ and $w=yz$, where $|x|=s$, $|y|=t$, $k\leq|z|<\omega$, and if $M\subseteq xy^{\smile}$, then $M\subseteq(xz)(yz)^{\smile} = vw^{\smile}$ and $\{\langle s{+}j,t{+}j\rangle : j<k\}\subseteq(xz)(yz)^{\smile} = vw^{\smile}$, and hence $\langle v,w\rangle$ is in ${}_{s+k}[\![M\cup\{\langle s+j,t+j\rangle : j<k\}]\!]_{t+k}$. There now follows that ${}_{[s+k,\omega)}\overset{\circ}{1}\circ{}_{s}[\![M]\!]_{t} = {}_{s+k}[\![M\cup\{\langle s+j,t+j\rangle : j<k\}]\!]_{t+k}$.

It is worth noting that if M is one-one or full, then so, respectively, is $M\cup\{\langle s+j,t+j\rangle : j<k\}$. Also, if $s=t$ and $M=eq_sM$, then

$$M\cup\{\langle s+j,s+j\rangle : j<k\} = eq_{s+k}M = \overset{\circ}{1}_{s+k}\cup M\ .$$

(b) Assume that $\overset{k}{\to} X = {}_{s+k}[\![N]\!]_{t+k}$. If $\overline{\overline{U}} = 1$, then $X = {}^{[s,\omega)}U \times {}^{[t,\omega)}U = {}_s[\![M]\!]_t$ for any $M \subseteq s \times t$ and hence, in particular, $X = {}_s[\![N \cap (s \times t)]\!]_t$. For the rest of the proof assume, therefore, that $\overline{\overline{U}} \geq 2$.

By Theorem II.5(d), X has grade $\langle s, t\rangle$ and hence $\overset{k}{\to}$ satisfies the equality that is given below. Also, by the definition of $[\![\ \ ]\!]$, ${}_{s+k}[\![N]\!]_{t+k}$ satisfies the equality that is given underneath.

$$\begin{aligned} \overset{k}{\to} X &= \{\langle vxz, wxz\rangle : |v|=s,\ |w|=t, \\ &\qquad |x|=k,\ |z| < \omega,\ \langle v, w\rangle \in X\}\ , \\ {}_{s+k}[\![N]\!]_{t+k} &= \{\langle vxz, wyz\rangle : |v|=s,\ |w|=t, \\ &\qquad |x| = |y|=k,\ |z| < \omega,\ N \subseteq vx(wy)^{\smile}\}\ . \end{aligned}$$

Suppose that $\{\langle s+j, t+j\rangle : j < k\}$ were not included in N. Then, for some $j < k$, $\langle s+j, t+j\rangle$ would not be in N. Since $\overline{\overline{U}} \geq 2$, there would then be some v, w, x, and y such that $|v| = s$, $|w| = t$, and $|x| = |y| = k$, such that $N \subseteq vx(wy)^{\smile}$ and hence $\langle vx, wy\rangle$ is in ${}_{s+k}[\![N]\!]_{t+k}$, and such that $x_j \neq y_j$ and hence $\langle vx, wy\rangle$ is not in $\overset{k}{\to} X$. Then $\overset{k}{\to} X$ would not be included in ${}_{s+k}[\![N]\!]_{t+k}$. This shows that $\{\langle s+j, t+j\rangle : j < k\} \subseteq N$.

Suppose that there were some $\langle s+i, t+j\rangle$ in N such that $i \neq j$. Since $\overline{\overline{U}} \geq 2$, therefore $i \neq j$ implies that there are x such that $|x| = k$ and $x_i \neq x_j$. Given any x of this kind, any v such that $|v| = s$, and any w such that $|w| = t$, there hold $(vx)_{s+i} \neq (wx)_{t+j}$, hence $\langle s+i, t+j\rangle \notin (vx)(wx)^{\smile}$, hence $N \nsubseteq (vx)(wx)^{\smile}$, and hence $\langle vx, wx\rangle \notin {}_{s+k}[\![N]\!]_{t+k}$. Since $\langle vx, wx\rangle \in \overset{k}{\to} X$ whenever $\langle v, w\rangle \in X$ and $|x| = k$, there would follow that $\overset{k}{\to} X \nsubseteq {}_{s+k}[\![N]\!]_{t+k}$. This shows that, for any i and j, if $\langle s+i, t+j\rangle$ is in N, then $i = j$.

Suppose that there were some $m < s$ and some j such that $\langle m, t+j\rangle \in N$. Consider any $\langle v, w\rangle$ in X, any u in U such that $u \neq v_m$, and any x such that $|x| = k$ and $x_j = u$. Then $\langle vx, wx\rangle$ is in $\overset{k}{\to} X$ but, since $(vx)_m \neq (wx)_{t+j}$, hence $\langle m, t+j\rangle \notin (vx)(wx)^{sm}$, and hence $N \nsubseteq (vx)(wx)^{\smile}$, $\langle vx, wx\rangle$ would not be in ${}_{s+k}[\![N]\!]_{t+k}$. This shows that there are no $m < s$ and $j < k$ such that $\langle m, t+j\rangle \in N$. A similar argument shows that there are no $j < k$ and $n < t$ such that $\langle s+j, n\rangle \in N$.

There now follows that $N = (N \cap (s \times t)) \cup \{\langle s+j, t+j\rangle : j < k\}$. Let $M = N \cap (s \times t)$. Then $M \subseteq s \times t$ and $N = M \cup \{\langle s+j, t+j\rangle : j < k\}$. Hence, by part (a), $\overset{k}{\to} {}_s[\![M]\!]_t = {}_{s+k}[\![N]\!]_{t+k}$. Since $\overset{k}{\to} X = {}_{s+k}[\![N]\!]_{t+k}$, there follows that $X = {}_s[\![M]\!]_t = {}_s[\![N \cap (s \times t)]\!]_t$. □

Assume that $N = (N \cap (s \times t)) \cup \{\langle s+j, t+j\rangle : j < k\}$. Then N is one-one if and only if $N \cap (s \times t)$ is one-one. Also, if $s = t$, then $N = eq_{s+k}N$ if and only if $N \cap (s \times s) = eq_s(N \cap (s \times s))$. Hence there follows from Lemma 9 and from Theorem 7(a) or 7(d), respectively, that ${}_{[0,\omega)}DF$ and ${}_{[0,\omega)}DI$ are closed under $\xrightarrow{1}$ and also under the converse of $\xrightarrow{1}$. Evidently, Lemma 9 also implies that the set $\{{}_s[\![M]\!]_t : s < \omega,\ t < \omega,\ M \subseteq s \times t\}$ of all members of ${}_{[0,\omega)}D$ that are special is closed under $\xrightarrow{1}$ and also under the converse of $\xrightarrow{1}$. Furthermore, since ${}_{[r,\omega)}\mathring{1} \circ {}_s[\![M]\!]_t = {}_s[\![M]\!]_t$ whenever $r \le s$ and ${}_s[\![M]\!]_t \circ {}_{[r,\omega)}\mathring{1} = {}_s[\![M]\!]_t$ whenever $r \le t$, there also follows from Lemma 9 that, for any X in ${}_{[0,\omega)}D$ that is special and for any $r < \omega$, ${}_{[r,\omega)}\mathring{1} \circ X$ and $X \circ {}_{[r,\omega)}\mathring{1}$ also are members of ${}_{[0,\omega)}D$ that are special.

Consider any Q, M and N such that $Q = eq_\gamma Q$, $M \subseteq \beta \times \gamma$, and $N \subseteq \gamma \times \delta$. For any $j < \gamma$, let $j/Q = \{j' : jQj'\}$. Then M and N shall be ***adequately linked with regard to*** Q if and only if, for any $j < \gamma$, $Rg\ M \cap j/Q \neq \emptyset$ and $Do\ N \cap j/Q \neq \emptyset$ together imply that $Rg\ M \cap Do\ N \cap j/Q \neq \emptyset$.

LEMMA 10. (a) *For any Q, M, and N such that $Q = eq_\gamma Q$, $M \subseteq \beta \times \gamma$, and $N \subseteq \gamma \times \delta$, there is an N' such that M and N' are adequately linked with regard to Q and such that $Q \circ N = Q \circ N'$. Moreover, if N is one-one, N' can be chosen to be one-one.*

(b) *Assume that $Q = eq_\gamma Q$, $N \subseteq \gamma \times \delta$, $N' \subseteq \gamma \times \delta$, and $Q \circ N = Q \circ N'$. Then ${}_\gamma[Q] \circ {}_\gamma[N]_\delta = {}_\gamma[Q] \circ {}_\gamma[N']_\delta$. Moreover, if $\gamma < \omega$ and $\delta < \omega$, then also ${}_\gamma[\![Q]\!] \circ {}_\gamma[\![N]\!]_\delta = {}_\gamma[\![Q]\!] \circ {}_\gamma[\![N']\!]_\delta$.*

(c) *Assume that $M \subseteq \beta \times \gamma$ and M is one-one, $Q = eq_\gamma Q$, and $N \subseteq \gamma \times \delta$ and N is one-one. Also assume that M and N are adequately linked with regard to Q. Then there holds the following equality and, if $\beta < \omega$, $\gamma < \omega$, and $\delta < \omega$, also the equality that results from it when $[\ \ ]$ is replaced by $[\![\ \ ]\!]$:*

$$
{}_\beta[M]_\gamma \circ {}_\gamma[Q] \circ {}_\gamma[N]_\delta = {}_\beta[\mathring{1}_\beta \cup MQM^{\smile}] \circ {}_\beta[MN]_\delta \circ {}_\delta[\mathring{1}_\delta \cup N^{\smile}QN] \ .
$$

(d) *Assume that $\langle M_0, M_1, M_2\rangle$ is a $\langle \beta, \gamma\rangle$ triple, $\langle N_0, N_1, N_2\rangle$ is a $\langle \gamma, \delta\rangle$ triple, M_1 and N_1 are both one-one, and $Q = eq_\gamma(M_2 \cup N_0)$. Also assume that M_1 and N_1 are adequately linked with regard to Q. Then there hold the following equalities and, if $\beta < \omega$, $\gamma < \omega$, and $\delta < \omega$, also the equalities that result from these when $[\ \ ]$*

is replaced by $[\![\ ]\!]$:

$$\begin{aligned}
&[M_0, M_1, M_2] \circ [N_0, N_1, N_2] \\
&= {}_\beta[M_0] \circ {}_\beta[M_1]_\gamma \circ {}_\gamma[Q] \circ {}_\gamma[N_1]_\delta \circ {}_\delta[N_2] \\
&= {}_\beta[M_0] \circ {}_\beta[\overset{\circ}{1}_\beta \cup M_1 Q M_1{}^\smile] \circ {}_\beta[M_1 N_1]_\delta \circ {}_\delta[\overset{\circ}{1}_\delta \cup N_1{}^\smile Q N_1] \circ {}_\delta[N_2] \\
&= {}_\beta[eq_\beta(M_0 \cup M_1 Q M_1{}^\smile)] \circ {}_\beta[M_1 N_1]_\delta \circ {}_\delta[eq_\delta(N_1{}^\smile Q N_1 \cup N_2)] \\
&= [eq_\beta(M_0 \cup M_1 Q M_1{}^\smile), M_1 N_1, eq_\delta(N_1{}^\smile Q N_1 \cup N_2)] \ .
\end{aligned}$$

Proof. Since the assertions in (b),(c), or (d) about $[\![\ ]\!]$ follow from those about $[\ ]$, we shall only prove those about $[\ ]$.

(a) We form N' as follows. First, consider any j such that either $Rg\ M \cap j/Q = \emptyset$ or $Do\ N \cap j/Q = \emptyset$ or $Rg\ M \cap Do\ N \cap j/Q \neq \emptyset$. Then, for any j' in j/Q we let $\langle j', k\rangle \in N'$ if and only if $\langle j', k\rangle \in N$. Now consider any j such that $Rg\ M \cap j/Q \neq \emptyset$, $Do\ N \cap j/Q \neq \emptyset$, and $Rg\ M \cap Do\ N \cap j/Q = \emptyset$. Then we choose some j_0 in $Rg\ M \cap j/Q$ and some j_1 in $Do\ N \cap j/Q$. We let $\langle j_0, k\rangle \in N'$ if and only if $\langle j_1, k\rangle \in N$. Also, no $\langle j_1, k\rangle$ shall be in N'. Further, if $j_2 \in j/Q$, $j_2 \neq j_0$, and $j_2 \neq j_1$, then we let $\langle j_2, k\rangle \in N'$ if and only if $\langle j_2, k\rangle \in N$. Then M and N' are adequately linked with regard to Q and $Q \circ N = Q \circ N'$. Also, if N is one-one, so is N'.

(b) Consider any $\langle x, y\rangle$ such that $|x| = \gamma$ and $|y| = \delta$. Since ${}_\gamma[Q]$ is a subidentity, therefore the following are equivalent to each other: $\langle x, y\rangle \in {}_\gamma[Q] \circ {}_\gamma[N]_\delta$; $\langle x, x\rangle \in {}_\gamma[Q]$ and $\langle x, y\rangle \in {}_\gamma[N]_\delta$; $Q \subseteq x x^\smile$ and $N \subseteq x y^\smile$; $QN \subseteq x y^\smile$. Likewise, $\langle x, y\rangle \in {}_\gamma[Q] \circ {}_\gamma[N']_\delta$ is equivalent to $QN' \subseteq x y^\smile$. Hence from $QN = QN'$ there follows that ${}_\gamma[Q] \circ {}_\gamma[N]_\delta = {}_\gamma[Q] \circ {}_\gamma[N']_\delta$.

(c) Assume that $\langle x, z\rangle$ is in ${}_\beta[M]_\gamma \circ {}_\gamma[Q] \circ {}_\gamma[N]_\delta$. Then there is some y such that $\langle x, y\rangle \in {}_\beta[M]_\gamma$, $\langle y, y\rangle \in {}_\gamma[Q]$, and $\langle y, z\rangle \in {}_\gamma[N]_\delta$. Let $\langle i, j\rangle \in MQM^\smile$. Then there are i' and j' such that $\langle i, i'\rangle \in M$, $\langle i', j'\rangle \in Q$, and $\langle j, j'\rangle \in M$. From $\langle x, y\rangle \in {}_\beta[M]_\gamma$ there follows that $x_i = y_{i'}$ and $x_j = y_{j'}$, and from $\langle y, y\rangle \in {}_\gamma[Q]$ there follows that $y_{i'} = y_{j'}$. Hence $x_i = x_j$. There follows that $\langle x, x\rangle \in {}_\beta[\overset{\circ}{1}_\beta \cup MQM^\smile]$. Similarly, $\langle z, z\rangle \in {}_\delta[\overset{\circ}{1}_\delta \cup N^\smile QN]$. Now let $\langle i, k\rangle \in MN$. Then, for some j, $\langle i, j\rangle \in M$ and $\langle j, k\rangle \in N$. From $\langle x, y\rangle \in {}_\beta[M]_\gamma$ there follows that $x_i = y_j$ and from $\langle x, z\rangle \in {}_\gamma[N]_\delta$ there follows that $y_j = z_k$. Hence $x_i = z_k$. There follows that $\langle x, z\rangle \in {}_\beta[MN]_\delta$. There now follows one of the two desired inclusions.

Assume that $\langle x, x\rangle \in {}_\beta[\overset{\circ}{1}_\beta \cup MQM^\smile]$, $\langle x, z\rangle \in {}_\beta[MN]_\delta$, and $\langle z, z\rangle \in {}_\delta[\overset{\circ}{1}_\delta \cup N^\smile QN]$. Also assume that M and N are adequately linked with regard to Q. We will now form as follows a sequence y in ${}^{\gamma}U$ such that, for every $j < \gamma$, $y_{j'} = y_{j''}$ for every j' and j'' in j/Q, and hence such that $\langle y, y\rangle \in {}_{\gamma}[Q]$.

First, consider the case where $Rg\ M \cap j/Q \neq \emptyset$ and $Do\ N \cap j/Q \neq \emptyset$. Since M and N are adequately linked with regard to Q, there are j' in j/Q such that both $j' \in Rg\ M$ and $j' \in Do\ N$. Moreover, since M and N are one-one therefore, for any such j' there is exactly one i' such that $i'Mj'$ and exactly one k' such that $j'Nk'$. Furthermore, since $\langle x, z\rangle \in {}_{\beta}[MN]_{\delta}$ and hence $MN \subseteq xz^{\smile}$ therefore, for any such j', i', k' there holds $x_{i'} = z_{k'}$. We choose one such j' in $Rg\ M \cap Do\ N \cap j/Q$ and, for every j'' in j/Q, and for the unique i' and k' such that $i'Mj'$ or $j'Nk'$, respectively, we let $y_{j''} = x_{i'} = z_{k'}$.

In the case where $Rg\ M \cap j/Q \neq \emptyset$ and $Do\ N \cap j/Q = \emptyset$, we choose one j' in $Rg\ M \cap j/Q$ and, for every j'' in j/Q, and for the unique i' such that $i'Mj'$, we let $y_{j''} = x_{i'}$. In the case where $Do\ N \cap j/Q \neq \emptyset$ and $Rg\ M \cap j/Q = \emptyset$, we choose one j' in $Do\ N \cap j/Q$ and, for every j'' in Q, and for the unique k' such that $j'Nk'$, we let $y_{j''} = z_{k'}$. Finally, in the case where $Rg\ M \cap j/Q = \emptyset$ and $Do\ N \cap j/Q = \emptyset$, we choose some element u of U and, for every j'' in j/Q, we let $y_{j''} = u$.

Consider any $\langle i, j\rangle$ in M. Regardless of whether $Do\ N \neq \emptyset$ or $Do\ N = \emptyset$, let j' be the chosen element of j/Q. Since jQj' and since $\langle x, x\rangle \in {}_{\beta}[\mathring{1}_{\beta} \cup MQM^{\smile}]_{\gamma}$ and hence $MQM^{\smile} \subseteq xx^{\smile}$, there follows that, for the unique i' such that $i'Mj'$ there holds $x_i = x_{i'}$. Since $y_j = y_{j'}$ and since $x_{j'}My_{j'}$ there follows that x_iMy_j. Since this holds for any $\langle i, j\rangle$ in M, therefore $M \subseteq xy^{\smile}$ and hence $\langle x, y\rangle \in {}_{\beta}[M]_{\gamma}$.

A similar argument shows that $\langle y, z\rangle \in {}_{\gamma}[N]_{\delta}$. Since $\langle y, y\rangle \in {}_{\gamma}[Q]$, there now follows that $\langle x, z\rangle \in {}_{\beta}[M]_{\gamma} \circ {}_{\gamma}[Q] \circ {}_{\gamma}[N]_{\delta}$. This concludes the proof of the other desired inclusion.

(d) The first equality follows from $[M_0, M_1, M_2] = {}_{\beta}[M_0] \circ {}_{\beta}[M_1]_{\delta} \circ {}_{\gamma}[M_2]$ and $[N_0, N_1, N_2] = {}_{\gamma}[N_0] \circ {}_{\gamma}[N_1]_{\delta} \circ {}_{\delta}[N_2]$, and by ${}_{\gamma}[M_2] \circ {}_{\gamma}[N_0] = [eq_{\gamma}(M_2 \cup N_0)]$, which hold by Lemma 6 or by Lemma 8(c) respectively. The second, third, and fourth equality follows from part (c), Lemma 8(c), or Lemma 6, respectively. □

The assumption that M and N are adequately linked with regard to Q cannot be omitted from Lemma 10(c), nor can the analogous assumption about M_1, N_1, and Q be omitted from Lemma 10(d). For example, let $\beta = \gamma = \delta = 2$ and let $M = \{\langle 0, 0\rangle\}$, $N = \{\langle 1, 1\rangle\}$, and $Q = \{0, 1\} \times \{0, 1\}$. Then $MQM^{\smile} = \{\langle 0, 0\rangle\}$, hence $\mathring{1}_2 \cup MQM^{\smile} = \mathring{1}_2$, and hence ${}_2[\mathring{1}_2 \cup MQM^{\smile}] = {}_2[\mathring{1}_2] = {}_{[2]}\mathring{1}$. Also $N^{\smile}QN = \{\langle 1, 1\rangle\}$, hence $\mathring{1}_2 \cup N^{\smile}QN = \mathring{1}_2$, and hence ${}_2[\mathring{1}_2 \cup N^{\smile}QN] = {}_{[2]}\mathring{1}$. Further, $MN = \emptyset$ and hence ${}_2[MN] = {}_2[\emptyset] = \{\langle x, y\rangle : |x| = 2,\ |y| = 2\}$. There follows that

${}_2[\mathring{1}_2 \cup MQM^\smile] \circ {}_2[MN] \circ {}_2[\mathring{1}_2 \cup N^\smile QN] = \{\langle x,y\rangle : |x|=2,\ |y|=2\}$. There also hold the following three equalities, from which there follows the fourth:

$$\begin{aligned} {}_2[M] &= {}_2[\{\langle 0,0\rangle\}] = \{\langle x,y\rangle : |x|=|y|=2,\ x_0=y_0\} \\ {}_2[Q] &= {}_2[\{0,1\}\times\{0,1\}] = \{\langle y,y\rangle : |y|=2,\ y_0=y_1\} \\ {}_2[N] &= {}_2[\{\langle 1,1\rangle\}] = \{\langle y,z\rangle : |y|=|z|=2,\ y_1=z_1\} \\ {}_2[M]\circ{}_2[Q]\circ{}_2[N] &= \{\langle x,z\rangle : |x|=|z|=2,\ x_0=z_1\}\ . \end{aligned}$$

There now follows that, if $\overline{\overline{U}} \geq 2$, then

$${}_2[\mathring{1}_2 \cup MQM^\smile] \circ {}_2[MN] \circ {}_2[\mathring{1}_2 \cup N^\smile QN] \not\subseteq {}_2[M]\circ{}_2[Q]\circ{}_2[N]\ .$$

As we saw earlier, the relative product of two special diagonal relations is not always a special diagonal relation. In contrast, by part (a) of the next theorem, ${}_{[0,\omega)}D$ and ${}_{[\gamma]}D$ are closed under relative product.

In the next theorem and its proof there are various similarities between ${}_{[\gamma]}D$ and ${}_{[0,\omega)}D$, but also some differences.

THEOREM 11. *For any $\gamma \leq \omega$ there hold* (a),(b),(c) *below. Moreover, the assertion that results from* (a), (b), *or* (c) *when one replaces ${}_{[\gamma]}D$, ${}_{[\gamma]}DI$, and ${}_{[\gamma]}DF$ by ${}_{[0,\omega)}D$, ${}_{[0,\omega)}DI$, or ${}_{[0,\omega)}DF$, respectively, also holds.*

(a) *${}_{[\gamma]}D$ is closed under $\circ$ and $^\smile$, and $\langle {}_{[\gamma]}D, \circ, {}^\smile, \subseteq\rangle$ is an ordered algebra. Moreover, up to isomorphism, this ordered algebra is the same for every base set U such that $\overline{\overline{U}} \geq 2$.*

(b) *Consider any X in ${}_{[\gamma]}D$. There are X_0 and X_2 in ${}_{[\gamma]}DI$ such that $X = X_0 \circ X_1 \circ X_2$ for some X_1 in ${}_{[\gamma]}DF$ and such that, moreover, if $X \subseteq Y_0 \circ Y_1 \circ Y_2$ where Y_0 and Y_2 are in ${}_{[\gamma]}DI$, then $X_0 \subseteq Y_0$ and $X_2 \subseteq Y_2$. Also, among the X_1 in ${}_{[\gamma]}DF$ that satisfy $X = X_0 \circ X_1 \circ X_2$ there are some that also satisfy the following maximality conditions: Whenever $X = X_0' \circ X_1' \circ X_2'$ where X_0' and X_2' are in ${}_{[\gamma]}DI$ and X_1' is in ${}_{[\gamma]}DF$, then $X_1 \subseteq X_1'$ implies that $X_1' \subseteq X_1$. Further, for any Y_0 and Y_2 in ${}_{[\gamma]}DI$ and any Y_1 in ${}_{[\gamma]}DF$ such that $X \subseteq Y_0 \circ Y_1 \circ Y_2$, there are X_1 in ${}_{[\gamma]}DF$ such that both $X = X_0 \circ X_1 \circ X_2$ and $X_1 \subseteq Y_1$.*

(c) *$X \circ X^\smile \circ X = X$, for every X in ${}_{[\gamma]}D$.*

(d) *${}_{[0,\omega)}D$ is closed under $\xrightarrow{1}$ and under the converse of $\xrightarrow{1}$.*

Proof. (a) Consider any X and Y in ${}_{[\gamma]}D$. By Theorem 2, Lemma 1(g), and Lemma 6, $X = {}_\gamma[M_0]\circ{}_\gamma[M_1]\circ{}_\gamma[M_2]$ for some M_0, M_1, M_2 such that $M_0 = eq_\gamma M_0$, $M_1 \subseteq \gamma\times\gamma$ and M_1 is one-one, and $M_2 = eq_\gamma M_2$. Likewise, $Y = {}_\gamma[N_0]\circ{}_\gamma[N_1]\circ{}_\gamma[N_2]$ such that $N_0 = eq_\gamma N_0$, $N_1 \subseteq \gamma\times\gamma$ and N_1 is one-one, and $N_2 = eq_\gamma N_2$. Moreover, by Lemmas 10(a) and 10(b), we may assume that M_1 and N_1 are adequately linked with regard to $Q = eq_\gamma(M_2 \cup N_0)$. Since by Lemma 10(d),

$$\begin{aligned}&{}_\gamma[M_0] \circ {}_\gamma[M_1] \circ {}_\gamma[M_2] \circ {}_\gamma[N_0] \circ {}_\gamma[N_1] \circ {}_\gamma[N_2] \\ &\quad = {}_\gamma[eq_\gamma(M_0 \cup M_1 Q M_1{}^\smile)] \circ {}_\gamma[M_1 N_1] \circ {}_\gamma[eq_\gamma(N_1{}^\smile Q N_1 \cup N_2] \\ &\quad = [eq_\gamma(M_0 \cup M_1 Q M_1{}^\smile), M_1 N_1, eq_\gamma(N_1{}^\smile Q N_1 \cup N_2)] \ ,\end{aligned}$$

therefore $X \circ Y$ is determined by a $\langle\gamma,\gamma\rangle$ triple and hence is in ${}_{[\gamma]}D$. There follows that ${}_{[\gamma]}D$ is closed under $\circ$. Since $X^\smile = [M_2, M_1{}^\smile, M_0]$, therefore $X^\smile$ is determined by a $\langle\gamma,\gamma\rangle$ triple and hence is in ${}_{[\gamma]}D$. There follows that ${}_{[\gamma]}D$ is closed under $^\smile$. From the set-theoretic definitions of $\circ$, $^\smile$, and $\subseteq$ there follows that $\langle {}_{[\gamma]}D, \circ, {}^\smile, \subseteq\rangle$ is an ordered algebra and, in fact, an ordered semigroup for which $^\smile$ is an involution (and for which ${}_{[\gamma]}\overset{\circ}{1}$ is the identity element).

The above proof that ${}_{[\gamma]}D$ is closed under $\circ$ is intended to bring out how this result can be derived from the relationship between ${}_{[\gamma]}DI$ and ${}_{[\gamma]}DF$ that is described in Lemma 10. A proof that does not consider this relationship but, instead, uses a direct computation, was given in [Cr 74b]. It yields a theorem (Theorem 2) in which ${}_{[\gamma]}DF$ plays no role, but which has certain advantages. It makes the following two assertions: Let $\langle M_0, M_1, M_2\rangle$ and $\langle N_0, N_1, N_2\rangle$ be arbitrary $\langle\gamma,\gamma\rangle$ triples and let $Q = eq_\gamma(M_2 \cup N_0)$. Then

$$\begin{aligned}&[M_0, M_1, M_2] \circ [N_0, N_1, N_2] \\ &\quad = {}_\gamma[M_0 \cup M_1 Q M_1{}^\smile] \circ {}_\gamma[M_1 Q N_1] \circ {}_\gamma[N_1{}^\smile Q N_1 \cup N_2] \\ &\quad = [M_0 \cup M_1 Q M_1{}^\smile, M_1 Q N_1, N_1{}^\smile Q N_1 \cup N_2] \ ,\end{aligned}$$

where, moreover, if $\langle M_0, M_1, M_2\rangle$ and $\langle N_0, N_1, N_2\rangle$ are both canonical, then so is $\langle M_0 \cup M_1 Q M_1{}^\smile, M_1 Q N_1, N_1{}^\smile Q N_1 \cup N_2\rangle$. The middle factor in the second product above is ${}_\gamma[M_1 Q N_1]$ where, in general, $M_1 Q N_1$ is not one-one. In contrast, in our earlier argument we used ${}_\gamma[M_1 N_1]$ as a middle factor, where $M_1 N_1$ is one-one and hence ${}_\gamma[M_1 N_1]$ is in ${}_{[\gamma]}DF$.

Temporarily, we will now also consider use of a set U' other than U as a universe. For any $\langle\gamma,\gamma\rangle$ triple $\langle M_0, M_1, M_2\rangle$, we let $[M_0, M_1, M_2]'$ be defined with respect to U' as $[M_0, M_1, M_2]$ has been defined with respect to U, and we let

$${}_{[\gamma]}D' = \{[M_0, M_1, M_2]' : \langle M_0, M_1, M_2\rangle \text{ is a } \langle\gamma,\gamma\rangle \text{ triple}\} \ .$$

Assume now that both $\overline{\overline{U}} \geq 2$ and $\overline{\overline{U'}} \geq 2$. Then, by Theorem 7, for every X in ${}_{[\gamma]}D$ there is a unique $\langle M_0, M_1, M_2\rangle$ that is canonical such that $X = [M_0, M_1, M_0]$. Also, by the proof of Theorem 4, adapted to ${}_{[\gamma]}D'$, for every X' in ${}_{[\gamma]}D'$ there is a unique $\langle M_0, M_1, M_2\rangle$ that is canonical such that $X' = [M_0, M_1, M_0]'$. Now let h be the mapping of ${}_{[\gamma]}D$ into ${}_{[\gamma]}D'$ such that, for any X in ${}_{[\gamma]}D$, if $\langle M_0, M_1, M_2\rangle$ is canonical and $X = [M_0, M_1, M_0]$, then $h(X) = [M_0, M_1, M_0]'$. Then h is a one-one mapping of ${}_{[\gamma]}D$ onto ${}_{[\gamma]}D'$.

Suppose that $X \in {}_{[\gamma]}D$, $X = [M_0, M_1, M_0]$, and $\langle M_0, M_1, M_0\rangle$ is canonical. Then $X^{\smile} = [M_2, M_1{}^{\smile}, M_0]$, $\langle M_2, M_1{}^{\smile}, M_0\rangle$ is canonical, and $h([M_2, M_1{}^{\smile}, M_0]) = [M_2, M_1{}^{\smile}, M_0]'$. Since $[M_2, M_1{}^{\smile}, M_0]' = [M_0, M_1, M_2]'^{\smile}$, there follows that $h(X^{\smile}) = (h(X))^{\smile}$. Thus h is an isomorphism between $\langle {}_{[\gamma]}D, {}^{\smile}\rangle$ and $\langle {}_{[\gamma]}D', {}^{\smile}\rangle$.

As we saw above, if X and Y are in ${}_{[\gamma]}D$, then $X \circ Y$ is in ${}_{[\gamma]}D$. A similar proof shows that if X' and Y' are in ${}_{[\gamma]}D'$, then $X' \circ Y'$ is in ${}_{[\gamma]}D'$. Since $\overline{\overline{U}} \geq 2$, therefore, by Theorem 4, each of X, Y, $X \circ Y$ is determined by a unique canonical triple. Likewise, by Theorem 4, adapted to ${}_{[\gamma]}D'$, each of X', Y', $X' \circ Y'$ is determined by a unique canonical triple. Now assume that X and X' both are determined by the canonical triple $\langle M_0, M_1, M_2\rangle$ and that Y and Y' both are determined by the canonical triple $\langle N_0, N_1, N_2\rangle$. Then by Theorem 2 of [Cr 74b], and by its adaptation to ${}_{[\gamma]}D'$, if $Q = eq_\gamma(M_2 \cup N_1)$, then both $X \circ Y$ and $X' \circ Y'$ are determined by $\langle M_0 \cup M_1 Q M_1{}^{\smile}, M_1 Q N_1, N_1{}^{\smile} Q N_1 \cup N_2\rangle$. (Another proof that $X \circ Y$ and $X' \circ Y'$ are determined by the same canonical triple can be given by using Theorem 2 and suitably add to it our earlier proof that ${}_{[\gamma]}D$ is closed under $\circ$.) There now follows: If $X' = h(X)$ and $Y' = h(Y)$, then $X' \circ Y' = h(X \circ Y)$. Thus, h is an isomorphism between $\langle {}_{[\gamma]}D, \circ\rangle$ and $\langle {}_{[\gamma]}D', \circ\rangle$.

Assume that X and Y are in ${}_{[\gamma]}D$, $X = [M_0, M_1, M_2]$ and $\langle M_0, M_1, M_2\rangle$ is canonical, and $Y = [N_0, N_1, N_2]$ and $\langle N_0, N_1, N_2\rangle$ is canonical. Suppose that $X \subseteq Y$. Then, by Theorem 3(b), $N_0 \subseteq M_0$, $N_1 \subseteq M_1$, and $N_2 \subseteq M_2$. Then from Theorem 3(b), adapted to ${}_{[\gamma]}D'$, there follows that $[M_0, M_1, M_2]' \subseteq [N_0, N_1, N_2]'$. Hence $h(X) \subseteq h(Y)$. Using first Theorem 3(b) adapted to ${}_{[\gamma]}D'$ and then Theorem 3(b) itself, one sees that, conversely, if $h(X) \subseteq h(Y)$, then $X \subseteq Y$. Thus, h is an isomorphism between $\langle {}_{[\gamma]}D, \subseteq\rangle$ and $\langle {}_{[\gamma]}D', \subseteq\rangle$.

We now turn to ${}_{[0,\omega)}D$. Since ${}_{[0,\omega)}D \subseteq {}_{[0,\omega)}Un \cap -\{\emptyset\}$, therefore every X in ${}_{[0,\omega)}D$ has a unique grade in $\omega \times \omega$. Now consider any X and Y in ${}_{[0,\omega)}Un \cap -\{\emptyset\}$. Let $gd\ X = \langle s, t\rangle$ and $gd\ Y = \langle s', t'\rangle$. By Theorem 2, Lemma 1(g), and Lemma 6, $X = {}_s[\![M_0]\!] \circ {}_s[\![M_1]\!]_t \circ {}_t[\![M_2]\!]$ for some M_0, M_1, M_2 such that $M_0 = eq_s M_0$, $M_1 \subseteq s \times t$ and M_1 is one-one, and $M_2 = eq_t M_2$. Likewise, $Y = {}_{s'}[\![N_0]\!] \circ {}_{s'}[\![N_1]\!]_{t'} \circ {}_{t'}[\![N_2]\!]$ for some N_0, N_1, N_2 such that $N_0 = eq_{s'} N_0$, $N_1 \subseteq s' \times t'$ and N_1 is one-one, and $N_2 = eq_{t'} N_2$.

Suppose that $t \geq s'$. Then, by Theorem II.7(g),

$$\begin{aligned} X \circ Y &= X \circ (Y \circ {}_{[t'+t-s,\omega)}\mathring{1}) \\ &= X \circ {}_{s'}[\![N_0]\!] \circ {}_{s'}[\![N_1]\!]_{t'} \circ {}_{t'}[\![N_2]\!] \circ {}_{[t'+t-s',\omega)}\mathring{1} \,. \end{aligned}$$

Repeated uses of Lemma 9(a) yield:

$$\begin{aligned} &{}_{s'}[\![N_0]\!] \circ {}_{s'}[\![N_1]\!]_{t'} \circ {}_{t'}[\![N_2]\!] \circ {}_{[t'+t-s',\omega)}\mathring{1} \\ &\quad = {}_{[t,\omega)}\mathring{1} \circ {}_{s'}[\![N_0]\!] \circ {}_{[t,\omega)}\mathring{1} \circ {}_{s'}[\![N_1]\!]_{t'} \circ {}_{[t'+t-s',\omega)}\mathring{1} \circ {}_{t'}[\![N_2]\!] \,. \end{aligned}$$

Let $k = t - s'$. Then ${}_{[t,\omega)}\mathring{1} = {}_{[s'+k,\omega)}\mathring{1}$ and ${}_{[t'+t-s',\omega)}\mathring{1} = {}_{[t'+k,\omega)}\mathring{1}$. Since $N_0 = eq_{s'}N_0$, therefore $N_0 \cup \{\langle s'+j, s'+j\rangle : j < k\} = \mathring{1}_t \cup N_0$ and $\mathring{1}_t \cup N_0 = eq_t N_0$. Since $N_2 = eq_{t'}N_2$, therefore $N_2 \cup \{\langle t'+j, t'+j\rangle : j < k\} = \mathring{1}_{t'+k} \cup N_2$ and $\mathring{1}_{t'+k} \cup N_2 = eq_{t'+k}N_2$. Using these facts about N_0 or N_2, respectively, and, in each of the three cases, using Lemma 9(a), one obtains:

$$\begin{aligned} {}_{[t,\omega)}\mathring{1} \circ {}_{s'}[\![N_0]\!] &= {}_{t}[\![N_0]\!] \cup {}_{s'}[\![\{\langle s'+j, s'+j\rangle : j < k\}]\!] \\ &= {}_{t}[\![\mathring{1}_t \cup N_0]\!] \\ {}_{[t,\omega)}\mathring{1} \circ {}_{s'}[\![N_1]\!]_{t'} &= {}_{t}[\![N_1]\!] \cup {}_{s'}[\![\{\langle s'+j, t'+j\rangle : j < k\}]\!]_{t'} \\ {}_{[t'+t-s',\omega)}\mathring{1} \circ {}_{t'}[\![N_2]\!] &= {}_{t'+k}[\![N_2]\!] \cup {}_{t'}[\![\{\langle t'+j, t'+j\rangle : j < k\}]\!] \\ &= {}_{t'+k}[\![\mathring{1}_{t'+k} \cup N_2]\!] \end{aligned}$$

Let $Q = eq_t(M_2 \cup (\mathring{1}_t \cup N_0)) = eq_t(M_2 \cup N_0)$. Since N_1 is one-one, therefore $N_1 \cup \{\langle s'+j, t'+j\rangle : j < k\}$ is one-one. Hence, by Lemmas 10(a) and 10(b), there is some N' such that $N' \subseteq t \times (t'+k)$, N' is one-one, M_1 and N_1' are adequately linked with respect to Q, and

$${}_{t}[\![Q]\!] \circ {}_{t}[\![N_1 \cup \{\langle s'+j, t'+j\rangle : j < k\}]\!]_{t'+k} = {}_{t}[\![Q]\!] \circ {}_{t}[\![N_1']\!]_{t'+k} \,.$$

Using first the definition of Q and Lemma 8(c) and then Lemma 10(c) one obtains:

$$\begin{aligned} &{}_{s}[\![M_1]\!]_{t} \circ {}_{t}[\![M_2]\!] \circ {}_{t}[\![\mathring{1}_t \cup N_0]\!] \circ {}_{t}[\![N_1']\!]_{t'+k} \\ &\quad = {}_{s}[\![M_1]\!]_{t} \circ {}_{t}[\![Q]\!] \circ {}_{t}[\![N_1']\!]_{t'+k} \\ &\quad = {}_{s}[\![\mathring{1}_s \cup M_1 Q M_1^{\smile}]\!] \circ {}_{s}[\![M_1 N_1']\!]_{t'+k} \circ {}_{t'+k}[\![\mathring{1}_{t'+k} \cup N_1'^{\smile} Q N_1']\!] \,. \end{aligned}$$

Finally, first putting together some of the equalities so far obtained, then applying Lemma 8(c), and then applying Lemma 6, one obtains:

$$\begin{aligned} X \circ Y &= {}_s[\![M_0]\!] \circ {}_s[\![\overset{\circ}{1}_s \cup M_1 Q M_1^{\smile}]\!] \circ {}_s[\![M_1 N_1']\!]_{t'+k} \\ &\qquad \circ_{t'+k}[\![\overset{\circ}{1}_{t'+k} \cup N_1'^{\smile} Q N_1']\!] \circ {}_{t'+k}[\![\overset{\circ}{1}_{t'+k} \cup N_2]\!] \\ &= {}_s[\![eq_s(M_0 \cup M_1 Q M_1^{\smile})]\!] \circ {}_s[\![M_1 N_1']\!]_{t'+k} \\ &\qquad \circ_{t'+k}[\![eq_{t'+k}(N'^{\smile} Q N' \cup N_2)]\!] \\ &= [\![eq_s(M_0 \cup M_1 Q M_1^{\smile}), M_1 N_1', eq_{t'+k}(N'^{\smile} Q N' \cup N_2)]\!] \;. \end{aligned}$$

This shows that if $t \geq s'$ and $k = t - s'$, then $X \circ Y$ is mediately determined by some $\langle s, t' + k\rangle$ triple and hence is in ${}_{[0,\omega)}D$. If $t \leq s'$, then a similar argument shows that $X \circ Y$ is in ${}_{[0,\omega)}D$. Thus, ${}_{[0,\omega)}D$ is closed under $\circ$. The rest of the proof that $\langle {}_{[0,\omega)}D, \circ, {}^{\smile}, \subseteq\rangle$ is an ordered algebra and, up to isomorphism, the same for every base set U satisfying $\overline{\overline{U}} \geq 2$, is similar to that for ${}_{[\gamma]}D$.

(b) Consider any X in ${}_{[\gamma]}D$. By Theorem 2, there is a $\langle \gamma, \gamma\rangle$ triple $\langle M_0, M_1, M_2\rangle$ that is precanonical and trim such that $X = [M_0, M_1, M_2]$. By Lemma 6, $[M_0, M_1, M_2] = {}_\gamma[M_0] \circ {}_\gamma[M_1] \circ {}_\gamma[M_2]$. Let $X_0 = {}_\gamma[M_0]$, $X_1 = {}_\gamma[M_1]$, and $X_2 = {}_\gamma[M_2]$. Since $M_0 = eq_\gamma M_0$ and $M_2 = eq_\gamma M_2$ therefore, by Theorem 7(d) and the definition of ${}_{[\gamma]}DI$, X_0 and X_2 are in ${}_{[\gamma]}DI$. By Lemma 1(g), M_1 is one-one and hence, by Theorem 7(a) and the definition of ${}_{[\gamma]}DF$, X_1 is in ${}_{[\gamma]}DF$.

Consider any Y_0 and Y_2 in ${}_{[\gamma]}DI$ and any Y_1 in ${}_{[\gamma]}D$ such that $X \subseteq Y_0 \circ Y_1 \circ Y_2$. Then, by Theorems 7(d) and 7(a), $Y_0 = {}_\gamma[N_0]$ for some N_0 such that $N_0 = eq_\gamma N_0$, $Y_1 = {}_\gamma[N_1]$ for some $N_1 \subseteq \gamma\times\gamma$, and $Y_2 = {}_\gamma[N_2]$ for some N_2 such that $N_2 = eq_\gamma N_2$. By Lemma 6, $Y_0 \circ Y_1 \circ Y_2 = [N_0, N_1, N_2]$. Hence $[M_0, M_1, M_2] \subseteq [N_0, N_1, N_2]$. First assume that $\overline{\overline{U}} \geq 2$. Then, since $\langle M_0, M_1, M_2\rangle$ is precanonical there follows from Theorem 3(c) that $N_0 \subseteq M_0$ and $N_2 \subseteq M_2$. Then ${}_\gamma[M_0] \subseteq {}_\gamma[N_0]$ and ${}_\gamma[M_2] \subseteq {}_\gamma[N_2]$. Hence $X_0 \subseteq Y_0$ and $X_2 \subseteq Y_2$. If $\overline{\overline{U}} = 1$, then $X_0 = Y_0 = X_2 = Y_2$ and hence again $X_0 \subseteq Y_0$ and $X_2 \subseteq Y_2$.

Consider any X_0' in ${}_{[\gamma]}DI$, X_1' in ${}_{[\gamma]}DF$, and X_2' in ${}_{[\gamma]}DI$ such that $X = X_0' \circ X_1' \circ X_2'$. By Theorems 7(d) and 7(a), $X_0' = {}_\gamma[M_0']$ for some M_0' such that $M_0' = eq_\gamma M_0'$, $X_1' = {}_\gamma[M_1']$ for some $M_1' \subseteq \gamma\times\gamma$ that is one-one, and $X_2' = {}_\gamma[M_2']$ for some M_2' such that $M_2' = eq_\gamma M_2'$. By Lemma 6, ${}_\gamma[M_0'] \circ {}_\gamma[M_1'] \circ {}_\gamma[M_2'] = [M_0', M_1', M_2']$. There now follows that $[M_0, M_1, M_2] = [M_0', M_1', M_2']$. Assume now that $\overline{\overline{U}} \geq 2$. Then, since $\langle M_0, M_1, M_2\rangle$ is precanonical there follows from Theorem 3(c) that if $M_1' \subseteq M_1$

then $M_1 \subseteq M_1'$. Since $X_1' = {}_\gamma[M_1']$ and $X_1 = {}_\gamma[M_1]$ there follows that if $X_1 \subseteq X_1'$ then $X_1' \subseteq X_1$. The same conclusion holds trivially if $\overline{\overline{U}} = 1$, since then $X_1 = X_1'$.

Again consider any Y_0 and Y_2 in ${}_{[\gamma]}DI$ and any Y_1 in ${}_{[\gamma]}DF$ such that $Y_0 \circ Y_1 \circ Y_2$. Let N_0, N_1, N_2 be as above. Assume that $\overline{\overline{U}} \geq 2$. By Theorems 2 and 3(b) there is a unique $\langle N_0', N_1', N_2' \rangle$ that is canonical such that $Y_0 \circ Y_1 \circ Y_2 = [N_0, N_1, N_2] = [N_0', N_1', N_2']$ and such that $N_i \subseteq N_i'$ for each $i \leq 2$. By Lemma 1(a), $\langle M_0, M_0M_1M_2, M_2 \rangle$ is canonical and, by Theorem 2, $X = [M_0, M_0M_1M_2, M_2]$. Since $X \subseteq Y_0 \circ Y_1 \circ Y_2$ and since $\overline{\overline{U}} \geq 2$, there follows from Theorem 3(b) that $N_0' \subseteq M_0$, $N_1' \subseteq M_0M_1M_2$, and $N_2' \subseteq M_2$. Since $N_1 \subseteq N_1'$, therefore also $N_1 \subseteq M_0M_1M_2$. Since $\langle M_0, M_0M_1M_2, M_2 \rangle$ is canonical and hence precanonical and since N_1 is one-one, there follows from Theorem 3(d) that there is an M_1' that is one-one such that $N_1 \subseteq M_1' \subseteq M_0M_1M_2$ and $M_0M_1M_2 \subseteq M_0M_1'M_2$. Now let $X_1 = {}_\gamma[M_1']$. Then X_1 is in ${}_{[\gamma]}DF$. Also, since $Y_1 = {}_\gamma[N_1]$ and $N_1 \subseteq M_1'$, therefore $X_1 \subseteq Y_1$. Furthermore, from $M_1' \subseteq M_0M_1M_2$ and $M_0M_1M_2 \subseteq M_0M_1'M_2$ there follows that $[M_0, M_1', M_2] = [M_0, M_0M_1M_2, M_2]$. Since $[M_0, M_0M_1M_2, M_2] = [M_0, M_1, M_2]$, therefore $[M_0, M_1', M_2] = [M_0, M_1, M_2]$. Since $[M_0, M_1', M_2] = {}_\gamma[M_0] \circ {}_\gamma[M_1'] \circ {}_\gamma[M_2] = X_0 \circ X_1 \circ X_2$ and since $[M_0, M_1, M_2] = X$ there follows that $X_0 \circ X_1 \circ X_2 = X$. If $\overline{\overline{U}} = 1$, then ${}_{[\gamma]}D$, ${}_{[\gamma]}DI$ and ${}_{[\gamma]}DF$ are the same singleton, so that trivially there exists an X_1 that satisfies the desired conditions.

We now turn to ${}_{[0,\omega)}D$. Consider any X in ${}_{[0,\omega)}D$ and let $gd\ X = \langle s, t \rangle$. Then, again by Theorem 2 and Lemma 6, $X = {}_s[\![M_0]\!] \circ {}_s[\![M_1]\!]_t \circ {}_t[\![M_2]\!]$ for some $\langle s, t \rangle$ triple $\langle M_0, M_1, M_2 \rangle$ that is precanonical and trim. Let $X_0 = {}_s[\![M_0]\!]$ and $X_2 = {}_t[\![M_2]\!]$. Then X_0 and X_2 are in ${}_{[0,\omega)}DI$. Also ${}_s[\![M_1]\!]_t$ is in ${}_{[0,\omega)}DF$ and hence $X = X_0 \circ X_1' \circ X_2$ for some X_1' in ${}_{[0,\omega)}DF$.

Consider any Y_0 and Y_2 in ${}_{[0,\omega)}DI$ and any Y_1 in ${}_{[0,\omega)}D$ such that $X \subseteq Y_0 \circ Y_1 \circ Y_2$. Then also $X \subseteq ({}_{[s,\omega)}\overset{\circ}{1} \circ Y_0) \circ Y_1 \circ ({}_{[t,\omega)}\overset{\circ}{1} \circ Y_2)$. Then by an argument similar to the one used earlier for the case where $X \in {}_{[\gamma]}D$, $X_0 \subseteq {}_{[s,\omega)}\overset{\circ}{1} \circ Y_0$ and $X_2 \subseteq {}_{[t,\omega)}\overset{\circ}{1} \circ Y_0$. Since ${}_{[s,\omega)}\overset{\circ}{1} \circ Y_0 \subseteq Y_0$ and ${}_{[t,\omega)}\overset{\circ}{1} \circ Y_2 \subseteq Y_2$, there follows that $X_0 \subseteq Y_0$ and $X_2 \subseteq Y_2$.

Next, we return to the minimality condition. We let X_1 be the unique relation in ${}^{[0,\omega)}U \times {}^{[0,\omega)}U$ that is not in $Rg \overset{1}{\rightarrow}$ such that ${}_s[\![M_1]\!]_t = \overset{r}{\rightarrow} X_1$ for some r, and hence for a unique r satisfying $0 \leq r \leq \min(s, t)$. By Lemma 9(b), $X_1 = {}_{s+r}[\![M_1 \cap ((s-r) \times (t-r))]\!]$. Since M_1 is one-one, therefore $M_1 \cap ((s-r) \times (t-r))$ is one-one. Hence X_1 is in ${}_{[0,\omega)}DF$. Since $gd({}_s[\![M_1]\!]_t) = \langle s, t \rangle$, therefore $gd\ X_1 = \langle s - r, t - r \rangle$

and $\overset{r}{\rightarrow} X_1 = {}_{[s,\omega)}\overset{\circ}{1} \circ X_1$. Hence, by Theorem II.7(g),

$$\begin{aligned} X &= {}_s[\![M_0]\!] \circ {}_s[\![M_1]\!]_t \circ {}_t[\![M_2]\!] \\ &= X_0 \circ \overset{r}{\rightarrow} X_1 \circ X_2 \\ &= X_0 \circ ({}_{[s,\omega)}\overset{\circ}{1} \circ X_1) \circ X_2 = X_0 \circ X_1 \circ X_2 \ . \end{aligned}$$

Consider any X_0' in ${}_{[0,\omega)}DI$, X_1' in ${}_{[0,\omega)}DF$, and X_2' in ${}_{[0,\omega)}DI$ such that $X = X_0' \circ X_1' \circ X_2'$. Let $gd\, X_0' = \langle s', s' \rangle$ and $gd\, X_2' = \langle t', t' \rangle$. Since $s' > s$ would imply that, for some s'' and t'', $gd\ X = gd(X_0' \circ X_1' \circ X_2') = \langle s'', t'' \rangle$, $t'' \geq t'$, and $s'' \geq s' > s$, therefore $s' \leq s$. Similarly, $t' \leq t$ and, for some r' and r'', $gd\ X_1' = \langle s - r', t - r'' \rangle$. Since $X_0' \subseteq {}_{[0,\omega)}\overset{\circ}{1}$ and $X_2' \subseteq {}_{[0,\omega)}\overset{\circ}{1}$, therefore $gd(X_0' \circ X_1' \circ X_2') = \langle s'', t'' \rangle$ for some $s'' \geq s - r'$ and $t'' \geq t - r''$ such that $s'' - t'' = (s - r') - (t - r'')$. Since $\langle s'', t'' \rangle = gd(X_0' \circ X_1' \circ X_2') = gd\ X = \langle s, t \rangle$, therefore $(s - r') - (t - r'') = s - t$ and hence $r' - r''$.

Let $X_0'' = {}_{[s,\omega)}\overset{\circ}{1} \circ X_0'$, $X_2'' = {}_{[t,\omega)}\overset{\circ}{1} \circ X_2'$, and $X_1'' = \overset{r'}{\rightarrow} X_1' = {}_{[s,\omega)}\overset{\circ}{1} \circ X_1'$. Then $gd\ X_0'' = \langle s, s \rangle$, $gd\ X_1'' = \langle s, t \rangle$, and $gd\ X_2'' = \langle t, t \rangle$. Also, since $gd\ X = \langle s, t \rangle$ and $X = X_0' \circ X_1' \circ X_2'$, therefore $X = X_0'' \circ X_1'' \circ X_2''$. Now X is also the product ${}_s[\![M_0]\!] \circ {}_s[\![M_1]\!]_t \circ {}_t[\![M_2]\!]$, where $\langle M_0, M_1, M_2 \rangle$ is an $\langle s, t \rangle$ triple that is precanonical and trim. Then, by an argument about $\langle s, t \rangle$ triples similar to the one used earlier about $\langle \gamma, \gamma \rangle$ triples, ${}_s[\![M_1]\!]_t \subseteq X_1''$ implies that $X_1'' \subseteq {}_s[\![M_1]\!]_t$ and hence ${}_s[\![M_1]\!]_t = X_1''$.

Suppose that $X_1 \subseteq X_1'$. Then ${}_{[s,\omega)}\overset{\circ}{1} \circ X_1 \subseteq {}_{[s,\omega)}\overset{\circ}{1} \circ X_1'$ and hence ${}_s[\![M_1]\!]_t \subseteq X_1''$. Then, as we just saw, ${}_s[\![M_1]\!]_t = X_1''$. Since ${}_s[\![M_1]\!]_t = \overset{r}{\rightarrow} X_1$ and $X_1'' = \overset{r'}{\rightarrow} X_1'$, therefore, $\overset{r}{\rightarrow} X_1 = \overset{r'}{\rightarrow} X_1'$. Since $X_1 \notin Rg \overset{1}{\rightarrow}$, there follows that $r \geq r'$, hence that $X_1' = \overset{r-r'}{\longrightarrow} X_1$, and hence, by Theorem II.7(d), that $X_1' \subseteq X_1$.

The proof of the last assertion in part (b) with regard to $[\![\ \ ]\!]$ is similar to the proof given earlier of this assertion regarding ${}_\gamma[\ \ ]$.

(c) Consider any X in ${}_{[\gamma]}D$. By part (b), $X = X_0X_1X_2$ for some X_0 in ${}_{[\gamma]}DI$, X_1 in ${}_{[\gamma]}DF$, and X_2 in ${}_{[\gamma]}DI$. By Theorem 7(a), $X_1 = {}_\gamma[M_1]$ for some M_1 that is one-one. Since M_1 is one-one, therefore $M_1 M_1^{\smile} M_1 = M_1$. Hence by Lemma 8(b),

$${}_\gamma[\![M_1]\!] \circ {}_\gamma[\![M_1^{\smile}]\!] \circ {}_\gamma[\![M_1]\!] = {}_\gamma[\![M_1 M_1^{\smile} M_1]\!] = {}_\gamma[\![M_1]\!] \ .$$

Since ${}_\gamma[\![M_1^{\smile}]\!] = ({}_\gamma[\![M_1]\!])^{\smile}$, there follows that $X_1 X_1^{\smile} X_1 = X_1$. Since X_0 and X_2 are included in ${}_{[\gamma]}\overset{\circ}{1}$, there now follows:

$$\begin{aligned} XX^{\smile}X &= X_0X_1X_2X_2^{\smile}X_1^{\smile}X_0^{\smile}X_0X_1X_2 \subseteq X_0X_1X_1^{\smile}X_1X_2 \\ &= X_0X_1X_2 = X \ . \end{aligned}$$

Since $X \subseteq XX^{\smile}X$ holds for any relation X, there follows that $XX^{\smile}X = X$. If X is in ${}_{[0,\omega)}D$, then a similar argument shows that X is again difunctional.

(d) Consider any X in ${}_{[0,\omega)}D$. By part (b), $X = X_0 \circ X_1 \circ X_2$ for some X_0 in ${}_{[0,\omega)}DI$, X_1 in ${}_{[0,\omega)}DF$, and X_2 in ${}_{[0,\omega)}DI$. As we saw after Lemma 9, $\overset{1}{\rightarrow} X_0$, $\overset{1}{\rightarrow} X_1$, $\overset{1}{\rightarrow} X_2$ also are in ${}_{[0,\omega)}DI$, ${}_{[0,\omega)}DF$, or ${}_{[0,\omega)}DI$, respectively. Since $\overset{1}{\rightarrow} X = \overset{1}{\rightarrow} X_0 \circ \overset{1}{\rightarrow} X_1 \circ \overset{1}{\rightarrow} X_2$ there follows from part (a) that $\overset{1}{\rightarrow} X$ is in ${}_{[0,\omega)}D$.

Now assume that $X \in Rg \overset{1}{\rightarrow}$. Then, for some s and t, X has grade $\langle s+1, t+1\rangle$. Also, the above X_0, X_1, and X_2 can be chosen to have grade $\langle s+1, s+1\rangle$, $\langle s+1, t+1\rangle$, or $\langle t+1, t+1\rangle$, respectively, and hence such that, for some N_0, N_1 and N_2 there hold $N_0 = eq_{s+1}N_0$, $N_1 \subseteq (s+1) \times (t+1)$ and N_1 is one-one, $N_2 = eq_{t+1}N_2$, $X_0 = {}_{s+1}[\![N_0]\!]$, $X_1 = {}_{s+1}[\![N_1]\!]_{t+1}$, and $X_2 = {}_{t+1}[\![N_2]\!]$. First assume that $\overline{\overline{U}} = 1$. Then $X_0 = {}_{[s+1,\omega)}\overset{\circ}{1}$, $X_2 = {}_{[t+1,\omega)}\overset{\circ}{1}$, and $X_1 = \overset{1}{\rightarrow} W$, where W is the unique relation that has grade $\langle s, t\rangle$. Then W is in ${}_{[0,\omega)}D$ and $X = X_0 \circ X_1 \circ X_2 = \overset{1}{\rightarrow} W$. Now assume that $\overline{\overline{U}} \geq 2$. Suppose that X_0 were not in $Rg \overset{1}{\rightarrow}$. Then, by Lemma 9(a), there would be some $\langle i, s\rangle$ in N_0 such that $i < s$ and hence such that $x_i = x_s$ for every $\langle x, x\rangle$ in ${}_s[\![N_0]\!]$. There would follow that $X = [\![N_0, N_1, N_2]\!]$ is not in $Rg \overset{1}{\rightarrow}$. Thus, X_0 is in $Rg \overset{1}{\rightarrow}$. Similarly X_2 is in $Rg \overset{1}{\rightarrow}$. Furthermore, if X_1 were not in $Rg \overset{1}{\rightarrow}$ then, by Lemma 9(a), there would either be some $\langle i, t\rangle$ in N_1 such that $i < s$ and hence such that $x_i = y_t$ for every $\langle x, y\rangle$ in ${}_s[\![N_1]\!]_t$ or some $\langle s, j\rangle$ in N_1 such that $j < t$ and hence such that $x_s = y_j$ for every $\langle x, y\rangle$ in ${}_s[\![N]\!]_t$. In either case there would follow that $X = [\![N_0, N_1, N_2]\!]$ is not in $Rg \overset{1}{\rightarrow}$. Thus X_1 also is in $Rg \overset{1}{\rightarrow}$. Let $X_0 = \overset{1}{\rightarrow} W_0$, $X_1 = \overset{1}{\rightarrow} W_1$, and $X_2 = \overset{1}{\rightarrow} W_2$. By Lemma 9(b), W_0, W_1 and W_2 are in ${}_{[0,\omega)}DI$, ${}_{[0,\omega)}DF$, or ${}_{[0,\omega)}DI$, respectively. Thus, if $W = W_0 \circ W_1 \circ W_2$ then $X = \overset{1}{\rightarrow} W$ and, by part (a), W is in ${}_{[0,\omega)}D$. □

Quite often, the relations X_1 in ${}_{[\gamma]}DF$ that satisfy the condition that both $X = X_0 \circ X_1 \circ X_2$ and $X_1 \subseteq Y_1$ in the third assertion of Theorem 11(b) must differ from those X_1 in ${}_{[\gamma]}DF$ that satisfy the maximality condition in the second assertion. For example, assume that $\overline{\overline{U}} \geq 2$ and let each of X, X_0, X_2 be the relation ${}_2[2 \times 2]$ in ${}_{[2]}DI$. Since $X_0 \circ {}_2[\emptyset] \circ X_2 \neq X$, therefore the only relations X_1 in ${}_{[2]}DF$ such that $X_0 \circ X_1 \circ X_2 = X$ are the following: $X_{1,0} = {}_2[\{\langle 0,0\rangle\}]$, $X_{1,1} = {}_2[\{\langle 1,1\rangle\}]$, and $X_{1,2} = {}_2[\{\langle 0,0\rangle, \langle 1,1\rangle\}]$. Of these, $X_{1,0}$ and $X_{1,1}$, but not $X_{1,2}$, satisfy the maximality condition. Now let Y_0 and Y_2 also be the relation ${}_2[2 \times 2]$ and let $Y_1 = X_{1,2}$. Then Y_0 and Y_2 are in ${}_{[2]}DI$, Y_1 is in ${}_{[2]}DF$, and $X \subseteq Y_0 \circ Y_1 \circ Y_2$, and the only X_1 in ${}_{[2]}DF$ such that both $X = X_0 \circ X_1 \circ X_2$ and $X_1 \subseteq Y_1$ is $X_{1,2}$.

As we mentioned in the proof of Theorem 11(a), $\smile$ is an involution for $\langle {}_{[\gamma]}D, \circ \rangle$. More precisely, for any X and Y in ${}_{[\gamma]}D$, there hold $(X^\smile)^\smile = X$ and $(X \circ Y)^\smile = Y^\smile \circ X^\smile$. By Theorem 11(c), for every X in ${}_{[\gamma]}D$ there also holds $X \circ X^\smile \circ X = X$ and hence $X^\smile = X^\smile \circ X \circ X^\smile$. Hence, for every X in ${}_{[\gamma]}D$, X and $X^\smile$ are inverses of each other in the sense in which, generalizing from group theory, one uses this notion in the theory of semigroups (cf [Ho]). Since, by Theorem 11(c), similar remarks apply to $\langle {}_{[0,\omega)}D, \circ, {}^\smile \rangle$, we shall therefore say of $\langle {}_{[\gamma]}D, \circ, {}^\smile \rangle$ and of $\langle {}_{[0,\omega)}D, \circ, {}^\smile \rangle$ that they are a ***semigroup with an inverse-forming involution.***

Since in an inverse semigroup every element has a unique inverse (cf. [Ho], 5.11), therefore neither $\langle {}_{[0,\omega)}D, \circ, {}^\smile \rangle$ nor $\langle {}_{[\omega]}D, \circ, {}^\smile \rangle$ is an inverse semigroup, as each contains elements that have more than one inverse. For example, as we saw after Theorem 7, each of the relations e_{ij} and c_i introduced in Chapter I is in ${}_{[0,\omega)}D$. Since each of them is idempotent, i.e., satisfies $e_{ij} \circ e_{ij} = e_{ij}$ or $c_i \circ c_i = c_i$, respectively, therefore each of them is an inverse of itself. Also, if $i \neq j$, then c_i and e_{ij} are inverses of each other since then there hold both $c_i \circ e_{ij} \circ c_i = c_i$ and $e_{ij} \circ c_i \circ e_{ij} = e_{ij}$. Thus every c_i and every e_{ij} such that $i \neq j$ has several inverses. For $\gamma \geq 2$ and $\overline{\overline{U}} \geq 2$ this example can be adapted to ${}_{[\gamma]}D$. In particular, one can show that if $\overline{\overline{U}} \geq 2$ and $0 \leq i < j < \omega$, then $\stackrel{\omega}{\to} c_i$ and $\stackrel{\omega}{\to} e_{ij}$ are elements of $\langle {}_{[\omega]}D, \circ, {}^\smile \rangle$ that have several inverses.

The variety of semigroups with an inverse-forming involution has been investigated by several authors, including Nordahl and Scheiblich in [NS]. Apparently, no name for it has been generally accepted. In Theorem 7 of [Br 91], D. A. Bredikhin gives axioms for the variety that is generated by those algebras $\langle S, \circ, {}^\smile, R \rangle$ of binary relations such that every W in S is difunctional and such that R is the unary operation on S which, for every W in S, satisfies

$$R(W) = \{\langle x, x \rangle : \langle x, y \rangle \in W \text{ for some } y\} .$$

In [Br 94], he gives axioms for the class of ordered algebras that can be isomorphically embedded into some $\langle S, \circ, {}^\smile, R, \subseteq \rangle$. (When $S = {}_{[\gamma]}D$ then, by Theorem 11(b) and by Theorem 17(c) below, for any W in ${}_{[\gamma]}D$, $R(W)$ is the least X in ${}_{[\gamma]}DI$ such that $W = X \circ Y$ for some Y in ${}_{[\gamma]}D$.)

If X and Y are relations on ${}^{[0,\omega)}U$ that are upward full then $X^\smile$ and $X \circ Y$ are upward full. Since ${}_{[0,\omega)}D$ is closed under $\circ$ and $\smile$, there follows that $\langle {}_{[0,\omega)}DF, \circ, {}^\smile, \subseteq \rangle$ is a substructure of $\langle {}_{[0,\omega)}D, \circ, {}^\smile, \subseteq \rangle$. Likewise, $\langle {}_{[0,\omega)}DI, \circ, {}^\smile, \subseteq \rangle$ is a substructure of $\langle {}_{[0,\omega)}D, \circ, {}^\smile, \subseteq \rangle$. Theorem 11(b) gives some information on how $\langle {}_{[0,\omega)}D, \circ, {}^\smile, \subseteq \rangle$ is built from these two substructures.

Given any $\gamma \leq \omega$, Theorem 11(b) also relates $\langle {}_{[\gamma]}D, \circ, ^{\smile}, \subseteq\rangle$ to its two substructures $\langle {}_{[\gamma]}DF, \circ, ^{\smile}, \subseteq\rangle$ and $\langle {}_{[\gamma]}DI, \circ, ^{\smile}, \subseteq\rangle$. According to parts (a) and (c) of the next theorem, if $\overline{\overline{U}} \geq 2$, then each of these substructures is isomorphic to a structure whose universe consists of relations that are more elementary than those in ${}_{[\gamma]}DF$ or in ${}_{[\gamma]}DI$, respectively. Part (c) follows from part (b). Note that if $M = eq_\gamma M$ and $N = eq_\gamma N$, then $M \bullet N = dfc(M \circ N) = eq_\gamma(M \cup N)$.

THEOREM 12. (a) *If* $\overline{\overline{U}} \geq 2$, *then* $\langle {}_{[\gamma]}DF, \circ, ^{\smile}, \subseteq\rangle$ *is the isomorphic image under* ${}_{\gamma}[\ \ ]$ *of* $\langle\{M : M \subseteq \gamma \times \gamma,\ M \text{ is one-one}\}, \circ, ^{\smile}, \supseteq\rangle$. *Hence* $\langle {}_{[\gamma]}DF, \circ, ^{\smile}\rangle$ *is an inverse semigroup such that* $X \subseteq Y$ *if and only if* $(Y \circ Y^{\smile}) \circ X = Y$.

(b) *Let* $S = \{X \in {}_{[\gamma]}D : X \text{ is one-one}\}$. *If* $\overline{\overline{U}} \geq 2$, *then* $\langle S, \circ, ^{\smile}, \subseteq\rangle$ *is the isomorphic image under* ${}_{\gamma}[\,]$ *of*

$$\langle\{M : Do\ M = Rg\ M = \gamma,\ MM^{\smile}M = M\}, \bullet, ^{\smile}, \supseteq\rangle\ .$$

Also, evidently, $\langle S, \circ, ^{\smile}\rangle$ *is an inverse semigroup such that* $X \subseteq Y$ *if and only if* $(X \circ X^{\smile}) \circ Y = X$.

(c) *If* $\overline{\overline{U}} \geq 2$, *then* $\langle {}_{[\gamma]}DI, \circ, ^{\smile}, \supseteq\rangle$ *is the isomorphic image under* ${}_{\gamma}[\,]$ *of*

$$\langle\{M : M = eq_\gamma M\}, \bullet, ^{\smile}, \supseteq\rangle\ .$$

Proof. (a) If $\overline{\overline{U}} = 1$, then ${}_{[\gamma]}DF$ is a singleton and the second assertion holds trivially. Now assume that $\overline{\overline{U}} \geq 2$. Then by Lemma 5(c), Theorem 7(a), and Lemma 8(b), $\langle {}_{[\gamma]}DF, \circ, ^{\smile}\rangle$ is the isomorphic image under ${}_{\gamma}[\ \ ]$ of $\langle\{M : M \subseteq \gamma \times \gamma,\ M \text{ is one-one}\}, \circ, ^{\smile}\rangle$. Now consider any X and Y in ${}_{[\gamma]}DF$. Let $X = {}_{\gamma}[M]$ and $Y = {}_{\gamma}[N]$. Since M and N are one-one, therefore $N \circ N^{\smile} = \{\langle n, n\rangle : n \in Do\ N\}$ and $N \subseteq M$ if and only if $(N \circ N^{\smile}) \circ M = N$ (cf. [Ho], p.153). By Lemmas 5(a) and 5(c), $X \subseteq Y$ is equivalent to $N \subseteq M$ and hence to $(N \circ N^{\smile}) \circ M = N$. Since ${}_{\gamma}[\,]$ is an isomorphism between $\langle\{M : M \subseteq \gamma \times \gamma,\ M \text{ is one-one}, \circ, ^{\smile}, \subseteq\rangle$ and $\langle {}_{[\gamma]}DF, \circ, ^{\smile}, \supseteq\rangle$ there follows that $X \subseteq Y$ if and only if $(Y \circ Y^{\smile}) \circ X = Y$. (Note that $N \circ N^{\smile} \subseteq \overset{\circ}{1}_\gamma$, so that, by Theorem 7(c), $Y^{\smile} \circ Y = {}_{\gamma}[N \circ N^{\smile}]$ is an equivalence relation on ${}^{\gamma}U$.)

(b) The first assertion follows from Theorem 7(b) and Lemma 8(c). The second assertion holds because S is a set of one-one relations. □

In [Sch 87] and [Sch 88], B. M. Schein has proved that every inverse semigroup whose universe consists of one-one relations is isomorphic to an inverse semigroup whose universe, for some set V, consists of relations that are both difunctional and full with respect to V. The isomorphism he gives differs from our mapping

${}_{\gamma}[\ ]$. Nevertheless, learning of his result has been very helpful. In particular, it has caused me to investigate the role of ${}_{[\gamma]}DF$ and of ${}_{[0,\omega)}DF$.

Besides the sets ${}_{[\gamma]}DF$ and $\{X \in {}_{[\gamma]}D : X \text{ is one-one}\}$ of Theorem 12, there is a third fairly comprehensive set S of special diagonal relations in ${}_{[\gamma]}D$ for which there holds an isomorphism result with respect to ${}_{\gamma}[\ ]$. Specifically, let

$$\begin{aligned} S &= \{X \in {}_{[\gamma]}D : Do\ X = {}^{\gamma}U,\ X \text{ is single-valued}\}\ , \\ S' &= \{f^{\smile} : Do\ f = \gamma,\ Rg\ f \subseteq \gamma,\ f \text{ is single-valued}\}\ . \end{aligned}$$

One can show that, if $\overline{\overline{U}} \geq 2$, then $\langle S, \circ \rangle$ is the isomorphic image under ${}_{\gamma}[\ ]$ of $\langle S', \circ \rangle$. We shall not make use of this fact. A proof can be obtained from parts of the proofs of Theorems 7(a) and 7(b) and from the description of ${}_{\gamma}[f^{\smile}]_{\delta}$ that is about to follow. The related fact that, for any U, ${}_{\gamma}[\ ]$ is a homomorphism of $\langle \{f : Do\ f = \gamma,\ Rg\ f \subseteq \gamma,\ f \text{ is single-valued}\}, \circ \rangle$ onto $\langle \{X^{\smile} : X \in S\}, \circ \rangle$ underlies the approach in [Ha] to algebraic first-order logic without equality and is therefore in essence well known.

For further work on ${}_{[\omega]}DF$ and ${}_{[0,\omega)}DF$, it will be useful to consider for γ and δ that are appropriate, a certain subclass of ${}_{\langle\gamma,\delta\rangle}DF$. Throughout the following discussion we let $f, g, \ldots$ be relations on ω that are single-valued. We begin by considering any $\gamma \leq \omega$ and $\delta \leq \omega$, and any f such that $Do\ f = \delta$ and $Rg\ f \subseteq \gamma$. We also consider any sequence x of length γ (i.e., any x such that x is single-valued, $|x| = Do\ x = \gamma$, and $Rg\ x \subseteq U$) and let y be the relative product $f \circ x$. Then y is the sequence such that $|y| = \delta$ and

$$y_n = y(n) = (f \circ x)(n) = x(f(n)) = x_{f(n)}\ , \quad \text{for every } n < \delta\ .$$

Now $f^{\smile} = \{\langle f(n), n \rangle : n < \delta\}$. Also, by the definition of ${}_{\gamma}[M]_{\delta}$, for every $M \subseteq \gamma \times \delta$ there holds

$${}_{\gamma}[M]_{\delta} = \{\langle x, y \rangle : |x| = \gamma,\ |y| = \delta,\ x_m = y_n \text{ for every } \langle m, n \rangle \text{ in } M\}\ .$$

Hence, when $M = f^{\smile}$, there holds:

$$\begin{aligned} {}_{\gamma}[f^{\smile}]_{\delta} &= \{\langle x, y \rangle : |x| = \gamma,\ |y| = \delta,\ x_{f(n)} = y_n \text{ for every } n < \delta\} \\ &= \{\langle x, y \rangle : |x| = \gamma,\ |y| = \delta,\ y = f \circ x\} \\ &= \{\langle x, f \circ x \rangle : |x| = \gamma\}\ . \end{aligned}$$

Again assume that $Do\ f = \delta < \omega$ and $Rg\ f \subseteq \gamma \leq \omega$ but assume, in addition, that f is strictly increasing (or, equivalently, that both f and $f^{\smile}$ are order-preserving), so that $\delta \leq \gamma$. Consider any x in ${}^{\gamma}U$. Then $f \circ x = \langle x_{f(0)}, x_{f(1)}, \ldots \rangle$ is a sequence of length δ such that, for any m and n in δ, $m < n$ implies that

$f(m) < f(n)$. Thus $f \circ x$ is a subsequence of x which, moreover, is proper, except when $\gamma = \delta$ and f is the identity function on δ. Since this holds for any x in ${}^{\gamma}U$, therefore ${}_{\gamma}[f^{\smile}]_{\delta} = \{\langle x, f \circ x\rangle : |x| = \gamma\}$ is a function whose value $({}_{\gamma}[f^{\smile}]_{\delta})(x) = f \circ x$, for any argument x, is a subsequence of length δ of x.

A convenient process of forming, for any x in ${}^{\gamma}U$, this subsequence $f \circ x$ is the following. Let $J = \{j \in \gamma : j \notin \mathit{Rg}\ f\} = \gamma \underset{2}{-} \mathit{Rg}\ f$. Consider any sequence $x = \langle x_0, x_1, \ldots\rangle$ in ${}^{\gamma}U$. First, from x remove, simultaneously for every j in J, the term that is at place j in x; then close up so that there are no gaps between terms, thereby forming a sequence of length $\overline{\overline{\gamma \underset{2}{-} J}}$. Since this closing up seems to be the simplest and most natural way of again forming a sequence after the removals, taking it for granted we will call the entire process ***simultaneous excision at the places** j **in** J* or, more briefly, J-***excision***. For $J = \{i\}$, we will also call it an i-***excision***. Evidently, every y in ${}^{\delta}U$ can be obtained from some x in ${}^{\gamma}U$ by J-excision. For the unique f such that f is strictly increasing, $\mathit{Do}\ f = \delta$, and $\mathit{Rg}\ f = \gamma \underset{2}{-} J$, and hence $J = \gamma \underset{2}{-} (\mathit{Rg}\ f)$, the unique function ${}_{\gamma}[f^{\smile}]_{\delta} = \{\langle x, f \circ x\rangle : |x| = \gamma\}$ shall therefore be ***the*** J-***excision from*** ${}^{\gamma}U$ ***onto*** ${}^{\delta}U$.

To give an illustration, we let $s = \gamma = 7$, $J = \{2, 4, 5\}$, and $t = \delta = \overline{\overline{\gamma \underset{2}{-} J}} = 4$. Then the unique f such that f is strictly increasing, $\mathit{Do}\ f = t$, and $\mathit{Rg}\ f = s \underset{2}{-} J$ is the function $\{\langle 0,0\rangle, \langle 1,1\rangle, \langle 2,3\rangle, \langle 3,6\rangle\}$. Consider any $x = \langle x_0, x_1, x_2, x_3, x_4, x_5, x_6\rangle$ in ${}^{s}U = {}^{7}U$. Then

$$ {}_{7}[f^{\smile}]_{4}(x) = f \circ x = \langle x_{f(0)}, x_{f(1)}, x_{f(2)}, x_{f(3)}\rangle = \langle x_0, x_1, x_3, x_6\rangle\ . $$

From x one obtains the above subsequence by simultaneously excising at the places 2,4,5 (and then closing the resulting two gaps of length 1 or length 2, respectively).

If $s < \omega$, $t \leq s$, $\mathit{Do}\ f = t$, $\mathit{Rg}\ f = s$, f is strictly increasing, and $J = s \underset{2}{-} (\mathit{Rg}\ f)$, then ${}_{s}[\![f^{\smile}]\!]_{t} = \bigcup\{\overset{n}{\rightarrow} ({}_{s}[f^{\smile}]_{t}) : n < \omega\}$ is a function whose domain is ${}^{[s,\omega)}U$ and whose range is ${}^{[s,\omega)}U$. It shall be ***the*** J-***excision from*** ${}^{[s,\omega)}U$ ***onto*** ${}^{[t,\omega)}U$.

For example, again let $s = 7$, $J = \{2, 4, 5\}$, $t = 4$, and $f = \{\langle 0,0\rangle, \langle 1,1\rangle, \langle 2,3\rangle, \langle 3,6\rangle\}$. Then given any $n \geq 0$ and any given $x = \langle x_0, x_1, \ldots, x_6\rangle^\frown\langle x_7, \ldots, x_{n-1}\rangle$ in ${}^{7+n}U$, ${}_{7}[\![f^{\smile}]\!]_{4}(x) = \langle x_0, x_1, x_3, x_6\rangle^\frown\langle x_7, \ldots, x_{n-1}\rangle$.

Some further examples will be instructive. Consider any i such that $0 \leq i < \omega$. Let $s = i + 1 = \{0, \ldots i\}$ and $t = i = \{0, \ldots i - 1\}$. Then the unique f such that f is strictly increasing, $\mathit{Do}\ f = i = t$, and $\mathit{Rg}\ f = s \underset{2}{-} \{i\} = i$ is the identity relation $\overset{\circ}{1}_i = \{\langle n, n\rangle : n < i\}$. Hence the i-excision from ${}^{[s,\omega)}U = {}^{[i+1,\omega)}U$ onto ${}^{[t,\omega)}U = {}^{[i,\omega)}U$ is the function ${}_{i+1}[\![(\overset{\circ}{1}_i)^{\smile}]\!]_{i} = {}_{i+1}[\![\overset{\circ}{1}_i]\!]_{i}$. It is the function $\dot{q}_i$ that was introduced in Chapter I.

For any i, $0 \le i < \omega$, there are further i-excisions in ${}_{[0,\omega)}DF$. Given any $k \ge 0$, there is a unique f' such that f' is strictly increasing, $Do\ f' = i+k = \{0,\ldots,i+k-1\}$, and $Rg\ f' = (i+k) \mathbin{\overline{\scriptstyle 2}} \{i\}$. Then the function ${}_{i+k+1}[\![f'^{\smile}]\!]_{i+k}$ is the i-excision from ${}^{[i+k+1,\omega)}U$ onto ${}^{[i+k,\omega)}U$. It is the restriction of $\dot{q}_i = {}_{i+1}[\![\mathring{1}_i]\!]_i$ to the domain ${}^{[i+k+1,\omega)}U$. It also satisfies ${}_{i+k+1}[\![f'^{\smile}]\!]_{i+k} = \overset{k}{\rightarrow} \dot{q}_i$. Note that $\dot{q}_i$ is the prototype of these i-excisions, i.e., the only i-excision in ${}_{[0,\omega)}DF$ that is not in $Rg \overset{1}{\rightarrow}$.

The converse of ${}_{i+1}[\![\mathring{1}_i]\!]_i$ is the relation ${}_i[\![\mathring{1}_i]\!]_{i+1}$, which also is in ${}_{[0,\omega)}DF$. It is the relation $q_i = \dot{q}_i{}^{\smile}$ that also was introduced in Chapter I.

Again let $0 \le i < \omega$. Now let $\gamma = \delta = \omega$. Then the unique f such that f is strictly increasing, $Do\ f = \delta = \omega$, and $Rg\ f = \gamma \mathbin{\overline{\scriptstyle 2}} \{i\} = \omega \mathbin{\overline{\scriptstyle 2}} \{i\}$ is the union $\mathring{1}_i \cup \{\langle n, n+1\rangle : i \le n < \omega\}$. Hence the i-excision from ${}^{\omega}U$ onto ${}^{\omega}U$ is the following function in ${}_{[\omega]}DF$:

$$_{\omega}[\mathring{1}_i \cup \{\langle n+1, n\rangle : i \le n < \omega\}] \ .$$

This function is the same as $\overset{\omega}{\rightarrow} ({}_{i+1}[\![\mathring{1}_i]\!]_i)$ and hence the same as $\overset{\omega}{\rightarrow} \dot{q}_i = \overset{\omega}{\rightarrow} (q^{\smile}{}_i)$. Its converse is the same as $\overset{\omega}{\rightarrow} ({}_i[\![\mathring{1}]\!]_{i+1})$ and hence the same as $\overset{\omega}{\rightarrow} q_i$.

Recall from Chapter I that S_{pqe} is the closure under $\circ$ of the union of the three sets $\{p_i : i < \omega\}$, $\{q_i : i < \omega\} \cup \{q_i{}^{\smile} : i < \omega\}$, and $\{e_{ij} : \{i,j\} \ne \{0\}\}$. As we saw after Theorem 7, every p_i and every e_{ij} is in ${}_{[0,\omega)}D$. As we saw above, every $q_i{}^{\smile} = \dot{q}_i$ and every q_i is in ${}_{[0,\omega)}D$. By Theorem 11(a), ${}_{[0,\omega)}D$ is closed under $\circ$. There now follows that $S_{pqe} \subseteq {}_{[0,\omega)}D$.

Recall from Chapter II that ${}_{[\omega]}S_{pqe}$ is the closure under $\circ$ of the union of the three sets $\{\overset{\omega}{\rightarrow} p_i : i < \omega\}$, $\{\overset{\omega}{\rightarrow} q_i : i < \omega\} \cup \{\overset{\omega}{\rightarrow} q_i{}^{\smile} : i < \omega\}$, and $\{\overset{\omega}{\rightarrow} e_{ij} : i,j < \omega\}$. As we saw after Theorem 7 or above, respectively, each of these is a subset of ${}_{[\omega]}D$. By Theorem 11(a), ${}_{[\omega]}D$ is closed under $\circ$. There now follows that ${}_{[\omega]}S_{pqe} \subseteq {}_{[\omega]}D$.

Let us look once more at excision from ${}^{\gamma}U$ onto ${}^{\delta}U$. Let any $\gamma \le \omega$ and any $\delta \le \gamma$ be given. Then between the set $\{f : Do\ f = \delta,\ Rg\ f \subseteq \gamma,\ f$ is strictly increasing$\}$ and the set $\{K \subseteq \gamma : \overline{\overline{K}} = \delta\}$ there is a simple one-one correspondence. If f is in the first set, then f is one-one and hence the unique $K \subseteq \gamma$ such that $K = Rg\ f$ is in the second set. By induction one shows that if K is in the second set then there is a unique f in the first set such that $K = Rg\ f$. Since complementation with respect to γ is an involution, therefore this complementation is a one-one correspondence between $\{K \subseteq \gamma : \overline{\overline{K}} = \delta\}$ and $\{J \subseteq \gamma : \overline{\overline{\gamma \mathbin{\overline{\scriptstyle 2}} J}} = \delta\}$. Composing these two one-one correspondences, one obtains the following one-one correspondence between $\{f : Do\ f = \delta,\ Rg\ f \subseteq \gamma,\ f$ is strictly increasing$\}$ and $\{J \subseteq \gamma : \overline{\overline{\gamma \underset{2}{-} J}} = \delta\}$: To each f there corresponds $\gamma \underset{2}{-} (Rg\ f)$.

In view of the one-one correspondence just described, we define each of the following sets in two ways. We let $\gamma \leq \omega$ and assume that $\delta \leq \gamma$ since, if $\delta > \gamma$, then there are no f such that $Do\ f = \delta$, $Rg\ f \subseteq \gamma$, and f is strictly increasing. Note that, if $\gamma < \omega$, then the set ${}_{[\gamma]}DE$ defined below is the singleton $\{{}_{[\gamma]}\mathring{1}\}$. Also note that, as always, our use of ${}_{\gamma}[f^{\smile}]_{\delta}$ presupposes that $f^{\smile} \subseteq \gamma \times \delta$ and hence that $Rg\ f \subseteq \gamma$.

$$
\begin{aligned}
{}_{[0,\omega)}DE &= \{{}_{s}[\![f^{\smile}]\!]_{t} : t \leq s < \omega,\ Do\ f = t, \text{ and } f \text{ is strictly increasing}\} \\
&= \{W : \text{for some } t,\ s, \text{ and } J,\ t \leq s < \omega, J \subseteq s,\ \ \overline{\overline{s \underset{2}{-} J}} = t, \\
&\qquad \text{and } W \text{ is the } J\text{-excision from } {}^{[s,\omega)}U \text{ onto } {}^{[t,\omega)}U\}\ . \\
{}_{\langle\gamma,\delta\rangle}DE &= \{{}_{\gamma}[f^{\smile}]_{\delta} : Do\ f = \delta \text{ and } f \text{ is strictly increasing}\} \\
&= \{W : \text{for some } J \subseteq \gamma,\ \overline{\overline{\gamma \underset{2}{-} J}} = \delta \text{ and } W \text{ is the } J\text{-excision} \\
&\qquad \text{from } {}^{\gamma}U \text{ onto } {}^{\delta}U\}\ . \\
{}_{[\gamma]}DE &= {}_{\langle\gamma,\gamma\rangle}DE
\end{aligned}
$$

THEOREM 13. *$\langle {}_{[\omega]}DE, \circ\rangle$ is a subalgebra of $\langle {}_{[\omega]}DF, \circ\rangle$ and, if $\overline{\overline{U}} \geq 2$, is isomorphic under ${}_{\omega}[\ \ ]$ to $\langle\{f^{\smile} : Do\ f = \omega,\ \ f$ is strictly increasing$\}, \circ\rangle$.*

Proof. Since the universe of the latter algebra is closed under $\circ$ therefore, by Lemma 8(b), so is ${}_{[\omega]}DE$. Hence, $\langle {}_{[\omega]}DE, \circ\rangle$ is a subalgebra of $\langle {}_{[\omega]}DF, \circ\rangle$. Then, by Theorem 12(a), if $\overline{\overline{U}} \geq 2$, it is isomorphic under ${}_{\omega}[\ \ ]$ to $\langle\{f^{\smile} : Do\ f = \omega,\ \ f$ is strictly increasing$\}, \circ\rangle$. □

THEOREM 14. *Let $S = \{M \subseteq \omega \times \omega : M$ and $M^{\smile}$ are order-preserving$\} = \{M \subseteq \omega \times \omega : M$ is one-one and strictly increasing$\}$. Also let $T = \{f : Do\ f = \omega,\ f$ is strictly increasing$\}$.*

(a) *If $M \in S$ and $\overline{\overline{M}} = \omega$, then there are exactly one f and g in T such that $M = f^{\smile} \circ g$.*

(b) *Given any $r < \omega$, let $h_r = \mathring{1}_r \cup \{\langle r+n, r+2n\rangle : n < \omega\}$ and $h'_r = \mathring{1}_r \cup \{\langle r+n, r+2n+1\rangle : n < \omega\}$. Then h_r and h'_r are in T and $\mathring{1}_r = h_r \circ h_r^{\prime\smile}$.*

(c) *Assume that $N \in S$ and $\overline{\overline{N}} < \omega$, so that, for some $s < \omega$ and $t < \omega$, $N \subseteq s \times t$. Let $M = N \cup \{\langle s+m, t+m\rangle : m < \omega\}$. Then $M = f^{\smile} \circ g$ for exactly one f and g in T, and $N = \mathring{1}_s \circ f^{\smile} \circ g \circ \mathring{1}_t$.*

(d) *The closure under $\circ$ and ${}^{\smile}$ of T coincides with S.*

(e) *Let* S' *be the closure under* $\circ$ *and* $\smile$ *of* ${}_{[\omega]}DE$. *If* $\overset{=}{U} \geq 2$, *then* $\langle S', \circ, \smile, \subseteq\rangle$ *is isomorphic under* ${}_{\omega}[\ ]$ *to* $\langle S, \circ, \smile, \supseteq\rangle$

Proof. (a) f and g are the unique element of T such that, respectively, $Rg\ f = Do\ f^{\smile} = Do\ M$ and $Rg\ g = Rg\ M$.

(c) This follows from part (a).

(d) Since $T \subseteq S$ and S is closed under $\circ$ and $\smile$, therefore the closure of T under $\circ$ and $\smile$ is included in S. The converse inclusion holds by parts (a) and (c).

(e) Since ${}_{[\omega]}DE \subseteq {}_{[\omega]}DF$, this follows from Theorem 12(a). □

Henceforth, any W in the set S' of Theorem 14(e) shall be an ***excision-insertion product.***

THEOREM 15. *Let* $S = \{M \subseteq \omega \times \omega : M$ *is one-one*$\}$, *let* $T = \{f : Do\ f = \omega$, f *is strictly increasing*$\}$, *let* $P = \{\pi : \pi$ *is a permutation of* $\omega\}$, *and let* S' *be the closure under* $\circ$ *and* $\smile$ *of* $T \cup P$.

(a) *Assume that* $M \in S$ *and* $\overset{=}{M} = \omega$. *Let* f *and* g *be the unique elements of* T *such that, respectively,* $Rg\ f = Do\ M$ *and* $Rg\ g = Rg\ M$. *Let* $\pi \doteq f \circ M \circ g^{\smile}$. *Then* π *is in* P *and* $M = f^{\smile} \circ \pi \circ g$. *Therefore* M *is in* S'.

(b) *Assume that* $N \in S$ *and* $\overset{=}{N} < \omega$, *so that, for some* $s < \omega$ *and* $t < \omega$, $N \subseteq s \times t$. *Let* $M = N \cup \{\langle s+m, t+m\rangle : m < \omega\}$. *Then, for exactly one* f *and* g *in* T, *and for* $\pi = f \circ M \circ g^{\smile}$, *there hold* $Rg\ f = Do\ M$, $Rg\ g = Rg\ M$, $M = f^{\smile} \circ \pi \circ g$, *and hence* $N = \overset{\circ}{1}_s \circ f^{\smile} \circ \pi \circ g \circ \overset{\circ}{1}_t$. *Hence* N *is in* S'.

(c) $S = S'$.

(d) *For any* ${}_{\omega}[M]$ *in* ${}_{[\omega]}DF$ *such that* $\overset{=}{M} = \omega$, *there are exactly one* X_0, X_1, X_2 *in* ${}_{[\omega]}DE$, ${}_{[\omega]}DP$, ${}_{[\omega]}DE$, *respectively, such that* ${}_{\omega}[M] = X_0 \circ X_1 \circ X_2^{\smile}$. *Also, for any* ${}_{\omega}[N]$ *in* ${}_{[\omega]}DF$ *such that* $\overset{=}{N} < \omega$, *there are* X_0, X_1, X_2 *in* ${}_{[\omega]}DE$, ${}_{[\omega]}DP$, ${}_{[\omega]}DE$, *respectively, such that, for some* $s < \omega$ *and* $t < \omega$, ${}_{\omega}[N] = {}_{\omega}[\overset{\circ}{1}_s] \circ X_0 \circ X_1 \circ X_2^{\smile} \circ {}_{\omega}[\overset{\circ}{1}_t]$. *There follows that* ${}_{[\omega]}DF$ *is the closure under* $\circ$ *and* $\smile$ *of* ${}_{[\omega]}DE \cup {}_{[\omega]}DP$.

Proof. (a) Since $Do\ f = \omega$, $Rg\ f = Do\ M$, $Rg\ M = Rg\ g = Do\ g^{\smile}$, and $Rg\ g^{\smile} = \omega$, therefore $Do\ \pi = \omega$ and $Rg\ \pi = \omega$. Since f, M and $g^{\smile}$ are one-one, therefore $\pi = f \circ M \circ g^{\smile}$ is one-one. Hence π is in P. Since $f^{\smile} \circ f$ is the identity relation on $Rg\ f = Do\ M$ and $g^{\smile} \circ g$ is the identity relation on $Rg\ g = Rg\ M$, therefore $f^{\smile} \circ \pi \circ g = f^{\smile} \circ f \circ M \circ g^{\smile} \circ g = M$.

(b) This follows from part (a) and Theorem 14(b).

(c) Since $T \subseteq S$, $P \subseteq S$, and S is closed under $\circ$ and $\smile$, therefore $S' \subseteq S$. By parts (a) and (b), $S \subseteq S'$.

(d) If $\overline{\overline{U}} = 1$, then the first two assertions hold trivially. Now assume that $\overline{\overline{U}} \geq 2$. Then the first assertion follows from part (a), the definition of ${}_{[\omega]}DE$ and ${}_{[\omega]}DP$, and Theorem 12(a). Using these definitions and Theorem 12(a), one obtains the second assertion similarly from part (b). From these there follows the third assertion. □

From certain substructures of $\langle\{M : M \subseteq \omega \times \omega\}, \circ, {}^{\smile}, \supseteq\rangle$ and applications to these of Theorem 12(a), we now turn to a substructure of $\langle\{M : M \subseteq \omega \times \omega\}, \bullet, {}^{\smile}, \supseteq\rangle$ and an application to it of Theorem 12(b). Let N be difunctional. Then N shall satisfy the ***equicardinality condition*** if and only if the following holds for any $\langle i, j\rangle$ in N:

$$\overline{\overline{\{i' : \langle i, i'\rangle \in N \circ N^{\smile}\}}} = \overline{\overline{\{j' : \langle j, j'\rangle \in N^{\smile} \circ N\}}} \; .$$

THEOREM 16. *Let $P = \{\pi : \pi$ is a permutation of $\omega\}$, let $E = \{M : M = eq_\omega M\}$, and let S be the closure of $P \cup E$ under $\bullet$.*

(a) *Assume that M and N are in E and that π and ρ are in P. Then the following hold:*

$$\pi^{\smile} \circ M \circ \pi \text{ and } \pi \circ M \circ \pi^{\smile} \text{ are in } E;$$
$$M \circ \pi = M \bullet \pi = \pi \circ (\pi^{\smile} \circ M \circ \pi);$$
$$\pi \circ M = \pi \bullet M = (\pi \circ M \circ \pi^{\smile}) \circ \pi \; .$$

Also, if either $M \circ \pi = N \circ \rho$ or $\pi \circ M = \rho \circ N$, then $M = N$. Further, $M \circ \pi = M \circ \rho$ is equivalent to $\pi \circ \rho^{\smile} \subseteq M$ and $\pi \circ M = \rho \circ M$ is equivalent to $\rho^{\smile} \circ \pi \subseteq M$.

(b) *For any M and N in E and for any π in P there hold the following:*

$$dfc(M \circ N \circ \pi) = dfc(M \circ N) \circ \pi = (M \bullet N) \circ \pi;$$
$$dfc(\pi \circ M \circ N) = \pi \circ dfc(M \circ N) = \pi \circ (M \bullet N)$$

(c) *The following sets are the same: S, $\{M \circ \pi : M \in E,\ \pi \in P\}$, $\{\pi \circ M : \pi \in P,\ M \in E\}$, and $\{N : Do\ N = Rg\ N = \omega,\ N \circ N^{\smile} \circ N = N$, and N satisfies the equicardinality condition$\}$.*

(d) *Let S' be the closure of ${}_{[\omega]}DP \cup {}_{[\omega]}DI$ under $\circ$. If $\overline{\overline{U}} \geq 2$, then $\langle S', \circ, {}^{\smile}, \subseteq\rangle$ is isomorphic under ${}_{\omega}[\;]$ to $\langle S, \bullet, {}^{\smile}, \supseteq\rangle$. Hence*

$$S' = \{X \circ Y : X \in {}_{[\omega]}DP,\ Y \in {}_{[\omega]}DI\}$$
$$= \{Y \circ X : Y \in {}_{[\omega]}DI,\ X \in {}_{[\omega]}DP\} \; .$$

Proof. (a) Since $\langle i,j\rangle \in \pi^{\smile}\circ M\circ\pi$ if and only if $\langle \pi^{\smile}(i),\pi^{\smile}(j)\rangle \in M$, therefore $\pi^{\smile}\circ M\circ\pi$ is the direct image of M under π. Since M is in E and π is in P, therefore $\pi^{\smile}\circ M\circ\pi$ is in E. Since $\pi\circ M\circ\pi^{\smile}$ is the inverse image of M under π, therefore $\pi\circ M\circ\pi^{\smile}$ likewise is in E.

Since $\pi\in P$, therefore $\pi\circ\pi^{\smile}=\mathring{1}_\omega$. There follows that $M\circ\pi=\pi\circ(\pi^{\smile}\circ M\circ\pi)$. There also follows that $(M\circ\pi)\circ(M\circ\pi)^{\smile}\circ(M\circ\pi)=M\circ\mathring{1}_\omega\circ M^{\smile}\circ M\circ\pi=M\circ\pi$. Hence $M\circ\pi=\mathit{dfc}(M\circ\pi)=M\bullet\pi$. By similar arguments, $\pi\circ M=(\pi\circ M\circ\pi^{\smile})\circ M=\pi\bullet M$.

Assume that $M\circ\pi=N\circ\rho$. Since $(\pi\circ M)\circ(M\circ\pi)^{\smile}=M\circ\mathring{1}_\omega\circ M^{\smile}=M$ and $(N\circ\rho)\circ(N\circ\rho)^{\smile}=N$, there follows that $M=N$. Similarly, if $\pi\circ M=\rho\circ N$, then $M=N$.

First assume that $M\circ\pi=M\circ\rho$. Then $\pi\subseteq M\circ\rho$ and hence $\pi\circ\rho^{\smile}\subseteq M\circ\rho\circ\rho^{\smile}=M$. Now assume that $\pi\circ\rho^{\smile}\subseteq M$. Consider any $\langle i,j\rangle$ in $M\circ\pi$. Then $\langle i,\pi^{\smile}(j)\rangle$ is in M. Let $k=\rho^{\smile}(j)$. Then $\langle\pi^{\smile}(j),k\rangle$ is in $\pi\circ\rho^{\smile}$ and therefore in M. There follows that $\langle i,k\rangle$ is in $M\circ M$ and therefore in M. Since $\langle k,j\rangle\in\rho$, there follows that $\langle i,j\rangle$ is in $M\circ\rho$. Thus, we have shown that if $\pi\circ\rho^{\smile}\subseteq M$ then $M\circ\pi\subseteq M\circ\rho$. Since this holds for any π and ρ in P there follows that if $\rho\circ\pi^{\smile}\subseteq M$ then $M\circ\rho\subseteq M\circ\pi$. Since $\pi\circ\rho^{\smile}\subseteq M$ implies that $\rho\circ\pi^{\smile}\subseteq M^{\smile}=M$, there now follows that $\pi\circ\rho^{\smile}\subseteq M$ implies that $M\circ\pi=M\circ\rho$. This concludes the proof that, for any π and ρ in P, $M\circ\pi=M\circ\rho$ and $\pi\circ\rho^{\smile}\subseteq M$ are equivalent.

There follows that $M\circ\pi^{\smile}=M\circ\rho^{\smile}$ and $\pi^{\smile}\circ\rho=\pi^{\smile}\circ\rho^{\smile\smile}\subseteq M$ are equivalent. Since $M\circ\pi^{\smile}=M\circ\rho^{\smile}$ is equivalent to $(M\circ\pi^{\smile})^{\smile}=(M\circ\rho^{\smile})^{\smile}$ and hence to $\pi\circ M=\rho\circ M$ and since $\pi^{\smile}\circ\rho=M$ and $\rho^{\smile}\circ\pi\subseteq M^{\smile}=M$ are equivalent, there follows that $\pi\circ M=\rho\circ M$ and $\rho^{\smile}\circ\pi\subseteq M$ are equivalent.

(b) Since π, M, N are in P, E, E, respectively, therefore $\pi\circ\pi^{\smile}=\mathring{1}_\omega$, $N\circ N^{\smile}=N$, and $M^{\smile}=M$. There follows that $(M\circ N\circ\pi)\circ(M\circ N\circ\pi)^{\smile}=M\circ N\circ M$. From this there follows by induction

$$((M\circ N\circ\pi)\circ(M\circ N\circ\pi)^{\smile})^n=(M\circ N)^n\circ M,\quad \text{if } n\geq 1.$$

According to a remark when *dfc* was introduced, just before Lemma III.5,

$$\mathit{dfc}(M\circ N\circ\pi)=\bigcup_{1\leq n<\omega}((M\circ N\circ\pi)\circ(M\circ N\circ\pi)^{\smile})^n\circ(M\circ N\circ\pi)\ .$$

There now follows that

$$\begin{aligned} dfc(M \circ N \circ \pi) &= \bigcup_{1 \le n < \omega} ((M \circ N)^n \circ M) \circ (M \circ N \circ \pi) \\ &= \bigcup_{1 \le n < \omega} (M \circ N)^{n+1} \circ \pi \ . \end{aligned}$$

Since M and N are in E, therefore $M \bullet N = dfc(M \cup N) = eq_\omega(M \cup N)$. Hence $M \bullet N$ is the transitive closure $\bigcup_{1 \le n < \omega}(M \circ N)^n = \bigcup_{1 \le n < \omega}(M \circ N)^{n+1}$ of $M \circ N$. Thus,

$$(M \bullet N) \circ \pi = \bigcup_{1 \le n < \omega} (M \circ N)^{n+1} \circ \pi \ .$$

There now follows that $dfc(M \circ N \circ \pi) = (M \bullet N) \circ \pi$. By definition, $(M \bullet N) \circ \pi = dfc(M \circ N) \circ \pi$. The second set of equalities in part (b) can be derived by a similar argument.

(c) By part (a), if $M \in E$ and $\pi \in P$, then $M \circ \pi = M \bullet \pi$. There follows that $\{M \circ \pi : M \in E,\ \pi \in P\} \subseteq S$, where S is the closure under $\bullet$ of $P \cup E$.

To prove that, conversely, every relation R in S also is in $\{M \circ \pi : M \in E,\ \pi \in P\}$ we use induction. If R is in P then, since $R = \overset{\circ}{1}_\omega \circ R$, therefore R is in $\{M \circ \pi : M \in E,\ \pi \in P\}$. If R is in E then, since $R = R \circ \overset{\circ}{1}_\omega$, therefore R is in $\{M \circ \pi : M \in E,\ \pi \in P\}$. Now consider any R and R' in S and assume as inductive hypothesis that $R = M \circ \pi$ and $R' = N \circ \rho$ for some M and N in E and some π and ρ in P. Let $\sigma = \pi \circ \rho$, so that σ is in P. Let $N' = \pi \circ N \circ \pi^{\smile}$ so that, by part (a), N' is in E and hence $M \bullet N'$ also is in E. Then the following equalities hold, the last one by part (b):

$$\begin{aligned} R \bullet R' &= dfc(R \circ R') = dfc((M \circ \pi) \circ (N \circ \rho)) \\ &= dfc(M \circ (\pi \circ N \circ \pi^{\smile}) \circ (\pi \circ \rho)) \\ &= dfc(M \circ N' \circ \sigma) = (M \bullet N') \circ \sigma \ . \end{aligned}$$

Since $M \bullet N'$ is in E and σ is in P there now follows by induction that every R in S is also in $\{M \circ \pi : M \in E,\ \pi \in P\}$. This concludes the proof that $S = \{M \circ \pi : M \in E,\ \pi \in P\}$. The proof that $S = \{\pi \circ M : \pi \in P,\ M \in E\}$ is similar.

Consider any M in E and π in P. Let $N = M \circ \pi$. Since $Do\ M = \omega$ and $Do\ \pi = \omega$, therefore $Do\ N = \omega$. Since $Rg\ \pi = \omega$ and $Rg\ M = \omega$, therefore $Rg\ N = \omega$. Also, $N \circ N^{\smile} = M \circ \pi \circ \pi^{\smile} \circ M^{\smile} = M$ so that $N \circ N^{\smile}$ is in

E, and $N^{\smile} \circ N = \pi^{\smile} \circ M \circ \pi$, so that, by part (a), $N^{\smile} \circ N$ is in E. Further, $N \circ N^{\smile} \circ N = M \circ (M \circ \pi) = M \circ \pi = N$, so that N is difunctional.

Assume that $\langle i, j\rangle$ is in $N = M \circ \pi$. Then $\langle i, \pi^{\smile}(j)\rangle$ is in M. Consider any i' such that $\langle i, i'\rangle$ is in $N \circ N^{\smile} = M$. Then $\langle i, \pi^{\smile}(j)\rangle$ is in M and hence $\langle i', j\rangle$ is in $M \circ \pi$. Then $\langle \pi(i'), j\rangle$ is in $\pi^{\smile} \circ M \circ \pi = N^{\smile} \circ N$. This shows that if i' is in $\{i' : \langle i, i'\rangle \in N \circ N^{\smile}\}$ then $\pi(i')$ is in $\{j' : \langle j, j'\rangle \in N^{\smile} \circ N\}$. Now consider any j' such that $\langle j, j'\rangle$ is in $N^{\smile} \circ N = \pi^{\smile} \circ M \circ \pi$. Then $\langle \pi^{\smile}(j'), j\rangle$ is in $\pi \circ \pi^{\smile} \circ M \circ \pi = M \circ \pi$ and $\langle \pi^{\smile}(j'), \pi^{\smile}(j)\rangle$ is in $M \circ \pi \circ \pi^{\smile} = M$. Since $\langle i, \pi^{\smile}(j)\rangle$ also is in M, therefore $\langle \pi^{\smile}(j'), i\rangle$ is in M. This shows that if j' is in $\{j' : \langle j, j'\rangle \in N^{\smile} \circ N\}$, then $\pi^{\smile}(j')$ is in $\{i' : \langle i, i'\rangle \in N \circ N^{\smile}\}$. There now follows that the restriction of π to the set $\{i' : \langle i, i'\rangle \in N \circ N^{\smile}\}$ is a mapping of this set onto the set $\{j' : \langle j, j'\rangle \in N^{\smile} \circ N\}$. Since π is one-one, therefore the two sets have the same cardinality. This concludes our proof that the set $\{M \circ \pi : M \in E,\ \pi \in P\}$ is included in the set

$$\begin{aligned}&\{N : Do\ N = Rg\ N = \omega,\ N \circ N^{\smile} \circ N = N,\\ &\text{and, for any } \langle i, j\rangle \text{ in } N,\\ &\overline{\overline{\{i' : \langle i, i'\rangle \in N \circ N^{\smile}\}}} = \overline{\overline{\{j' : \langle j, j'\rangle \in N^{\smile} \circ N\}}}\}\end{aligned}$$

To prove that the converse inclusion holds, consider any N in this last set. For any $i < \omega$, choose any one-one mapping π_i of $\{i' : \langle i, i'\rangle \in N \circ N^{\smile}\}$ onto $\{j' : \langle j, j'\rangle \in N^{\smile} \circ N\}$. Let $\pi = \bigcup\{\pi_i : i < \omega\}$. Then π is in P. Also, $N \circ N^{\smile}$ is in E. Further, for any i and j the following conditions are equivalent to each other: $\langle i, j\rangle \in N$, $\langle i, \pi_i{}^{\smile}(j)\rangle \in N \circ N^{\smile}$; $\langle i, j\rangle \in (N \circ N^{\smile}) \circ \pi$. Hence $N = (N \circ N^{\smile}) \circ \pi$ and N is in $\{M \circ \pi : M \in E,\ \pi \circ P\}$.

(d) The first assertion follows from Theorem 12(b), the definition of ${}_{[\omega]}DP$, and that of ${}_{[\omega]}DI$. From it and from the relevant parts of part (c) there follows the second assertion. □

Whenever Y is in ${}_{[\omega]}DI$, then $Y \circ X$ shall be X ***restricted to*** $D \circ Y$. If, moreover, X is in ${}_{[\omega]}DP$, so that $Y \circ X$ is in the set S' of Theorem 16(d), then $Y \circ X$ shall be a ***restricted permutation (in*** ${}_{[\omega]}D$***)***.

Among the algebras considered in Theorem 12(a) is the algebra $\langle\{M \subseteq \omega \times \omega :$ M is one-one$\}, \circ, {}^{\smile}\rangle$, while among the algebras considered in Theorem 12(b) is the algebra $\langle\{N : Do\ N = Rg\ N = \omega,\ NN^{\smile}N = N\}, \bullet, {}^{\smile}\rangle$. In Theorem 14 we considered the following subalgebra of the first of these:

$$\mathbf{S} = \langle\{M \subseteq \omega \times \omega : M \text{ is one-one and strictly increasing}\}, \circ, {}^{\smile}\rangle\ .$$

A certain subuniverse S' of the second of these algebras will be introduced presently. Later on we will show that $\mathbf{S}$ and the algebra $\mathbf{S}'$ whose universe is S' are isomorphic.

Consider any relation N on ω that is difunctional, i.e., satisfies $NN^{\smile}N = N$. Then, for a unique γ, $0 \leq \gamma \leq \omega$, there is a pair $\langle\langle I_r : r < \gamma\rangle, \langle J_r : r < \gamma\rangle\rangle$ of sequences of length γ such that $\langle I_r : r < \gamma\rangle$ and $\langle J_r : r < \gamma\rangle$ are an indexed partition into pairwise disjoint nonempty sets of *Do* N or of *Rg* N, respectively, and such that the following condition is satisfied:

$$N = \bigcup\{I_r \times J_r : r < \gamma\} = \{\langle i, j\rangle : \text{for some } r < \gamma,\ i \in I_r \text{ and } j \in J_r\}\ .$$

Now assume in addition, that $\langle I_r : r < \gamma\rangle$ and $\langle J_r : r < \gamma\rangle$ can be indexed so that they satisfy the following condition:

> If $r < r' < \gamma$, then $i \in I_r$ and $i' \in I_{r'}$ together imply that $i < i'$,
> and $j \in J_r$ and $j' \in J_{r'}$ together imply that $j < j'$.

Then both $\langle\langle I_r : r < \gamma\rangle, \langle J_r : r < \gamma\rangle\rangle$ and the relation $N = \bigcup\{I_r \times J_r : r < \gamma\}$ shall be ***biordered***. Evidently, the assignment to each biordered pair $\langle\langle I_r : r < \gamma\rangle, \langle J_r : r < \gamma\rangle\rangle$ of the relation $N = \bigcup\{I_r \times J_r : r < \gamma\}$ is a one-one correspondence between the set of such pairs and the set of difunctional relations on ω that are biordered.

We let $S' = \{N : Do\, N = Rg\, N = \omega,\ NN^{\smile}N = N,\ N \text{ is biordered}\}$. Consider any N in S'. Let $\langle\langle I_r : r < \gamma\rangle, \langle J_r : r < \gamma\rangle\rangle$ be the corresponding biordered pair of partitions. Then $1 \leq \gamma \leq \omega$ and both $\langle I_r : r < \gamma\rangle$ and $\langle J_r : r < \gamma\rangle$ are a partition of ω into γ (consecutive) intervals. Moreover, if $\gamma < \omega$, then $I_{\gamma-1}$ and $J_{\gamma-1}$ are infinite. Conversely, if $\langle\langle I_r : r < \gamma\rangle, \langle J_r : r < \gamma\rangle\rangle$ is a biordered pair of partitions of ω into intervals, then the corresponding relation N is in S'.

Let K be any subset of ω. There is a unique β, $0 \leq \beta \leq \omega$, and a unique sequence $\langle k_r : r < \beta\rangle$ that is strictly increasing such that K is the range $\{k_r : r < \gamma\}$ of the sequence. If $\beta < \omega$ let $\gamma = \beta + 1$ and if $\beta = \omega$ let $\gamma = \beta = \omega$. Then there is a unique sequence $\langle K_r : r < \gamma\rangle$ of length γ of intervals such that the following conditions are satisfied. If $\beta = 0$ and hence $\gamma = 1$, then $K_0 = \omega$. Now assume that $\beta \neq 0$. Then $K_0 = \{k : 0 \leq k < k_0\}$. Also, for any r such that $1 \leq r < \beta$, $K_r = \{k : k_{r-1} < k \leq k_r\}$. Further, if $\beta < \omega$ and $r = \beta$, then $K_r = K_\beta = \{k : k_{r-1} < k < \omega\}$. Thus, the sequence $\langle K_r : r < \gamma\rangle$ is an indexed partition of ω into intervals that is ***ordered*** in the sense that if $r < r' < \gamma$, $k \in K_r$, and $k' \in K_{r'}$, then $k < k'$. Conversely, given any such $\langle K_r : r < \gamma\rangle$, the set $\{k : \text{for some } r < \gamma,\ k \text{ is the greatest element of } K_r\}$ is the set K. We shall say of K and of $\langle K_r : r < \gamma\rangle$ that they are ***correlated.***

Let M be any relation on ω such that both M and $M^{\smile}$ are order-preserving or, equivalently, such that M is one-one and strictly increasing. Then there is a unique β, $0 \leq \beta \leq \omega$, and a unique sequence $\langle\langle i_r, j_r\rangle : r < \beta\rangle$ of pairs such that both $\langle i_r : r < \beta\rangle$ and $\langle j_r : r < \beta\rangle$ are strictly increasing and such that $M = \{\langle i_r, j_r\rangle : r < \beta\}$. If $\beta < \omega$ let $\gamma = \beta + 1$ and if $\beta = \omega$ let $\gamma = \beta = \omega$. Let $\langle I_r : r < \gamma\rangle$ and $\langle J_r : r < \gamma\rangle$ be the ordered partition of ω into intervals that is correlated with $\{i_r : r < \beta\}$ or with $\{j_r : r < \beta\}$, respectively. Then $\langle\langle I_r : r < \gamma\rangle, \langle J_r : r < \gamma\rangle\rangle$ is a biordered pair of partitions of ω into intervals. From this pair one can recover M by first forming the sets $\{i_r : r < \beta\}$ and $\{j_r : r < \beta\}$ that are correlated with $\langle I_r : r < \gamma\rangle$ and $\langle J_r : r < \gamma\rangle$, respectively, and then forming the set of pairs $\{\langle i_r, j_r\rangle : r < \beta\}$. We shall say of M and $\langle\langle I_r : r < \gamma\rangle, \langle J_r : r < \gamma\rangle\rangle$ that they are ***correlated.***

By composing the correlations just described with the one-one correspondence between the biordered pairs $\langle\langle I_r : r < \gamma\rangle, \langle J_r : r < \gamma\rangle\rangle$ of partitions of ω into intervals and the difunctional relations N on ω that are biordered and full one obtains a one-one correspondence between the following two sets of relations:

$$\begin{aligned} S &= \{M \subseteq \omega \times \omega : M \text{ is one-one and strictly increasing}\} \\ S' &= \{N : \mathit{Do}\ N = \mathit{Rg}\ N = \omega,\ NN^{\smile}N = N,\ N \text{ is biordered }\}. \end{aligned}$$

For any M in S we denote the corresponding N in S' by $M^{\frown+}$. Thus, $M^{\frown+}$ results from M by a certain process of adding pairs or fleshing out.

To be more explicit, let $0 \leq \beta \leq \omega$, let $\langle i_r : r < \beta\rangle$ and $\langle j_j : r < \beta\rangle$ be strictly increasing, let $\langle I_r : r < \beta\rangle$ and $\langle J_r : r < \beta\rangle$ be the sequence of partitions of ω that is correlated with $\langle i_r : r < \beta\rangle$ or $\langle j_r : r < \beta\rangle$, respectively, and let M be the one-one strictly increasing relation on ω that is shown below. Then $M^{\frown+}$ is the full, difunctional, biordered relation on ω that is shown underneath.

$$\begin{aligned} M &= \{\langle i_r, j_r\rangle : r < \beta\} \\ M^{\frown+} &= \bigcup\{I_r \times J_r : r < \beta\} \\ &= \{\langle i, j\rangle : \text{ for some } r < \beta, i \in I_r \text{ and } j \in J_r\} \end{aligned}$$

THEOREM 17. *The assignment* $^{\frown+}$ *to each* M *in* S *of the corresponding* $N = M^{\frown+}$ *in* S' *is an isomorphism between* $\langle S, \circ, {}^{\smile}, \subseteq\rangle$ *and* $\langle S', \bullet, {}^{\smile}, \supseteq\rangle$.

Proof. We just saw that $^{\frown+}$ is a one-one mapping of S onto S'. Now consider any M and M' in S. Let $\langle\langle I_r : r < \gamma\rangle, \langle J_r : r < \gamma\rangle\rangle$ and $\langle\langle I'_r : r < \gamma'\rangle, \langle J'_r : r < \gamma'\rangle\rangle$ be the biordered pair of partitions of ω into intervals that is correlated with M or with M', respectively, and hence corresponds to N or to N', respectively, where

$$\begin{aligned} N &= \widehat{M^+} = \bigcup\{I_r \times J_r : r < \gamma\} \text{ and} \\ N' &= \widehat{(M')^+} = \bigcup\{I'_r \times J'_r : r < \gamma'\} . \end{aligned}$$

With $M^\smile$ there is correlated the pair $\langle\langle J_r < r < \gamma\rangle, \langle I_r : r < \gamma\rangle\rangle$. To this pair there corresponds $N^\smile$. There follows that $\widehat{(M^\smile)^+} = (\widehat{M^+})^\smile$.

Assume that $M \subseteq M'$. Then for each $r < \gamma$ there is some $r' < \gamma'$ such that $I_r \subseteq I'_{r'}$ and $J_r \subseteq J'_{r'}$. There follows that $N' \subseteq N$.

Let $K = Rg\ M \cap Do\ M'$ and let $\langle K_s : s < \delta\rangle$ be the ordered partition of ω into intervals that is correlated with K. Then every J_r, $r < \gamma$, is included in some K_s, $s < \delta$, and every I'_r, $r < \gamma'$, is included in some K_s, $s < \delta$. Moreover, if $\langle K'_s : s < \delta'\rangle$ is any ordered partition of ω into intervals that is similarly related to $\langle J_r : r < \gamma\rangle$ and $\langle I'_r : r < \gamma'\rangle$, then every K_s, $s < \delta$, is included in some K'_s, $s < \delta'$. For every $s < \delta$, let $I''_s = \bigcup\{I_r : r < \gamma,\ J_r \subseteq K_s\}$ and let $J''_s = \bigcup\{J'_r : r < \gamma',\ I'_r \subseteq K_s\}$. Then $\langle\langle I''_s : s < \delta\rangle, \langle J''_s : s < \delta\rangle\rangle$ is a biordered pair of partitions of ω, into intervals. We denote the corresponding difunctional, biordered, and full relation by N''. Thus

$$N'' = \bigcup\{I''_s \times J''_s : s < \delta\} .$$

Consider any $\langle i, j'\rangle$ in $N \circ N'$. Let $i \in I_r$ and $j' \in J'_{r'}$. Then $J_r \cap I'_{r'} \neq \emptyset$. Hence, for some $s < \delta$, $J_r \cup I'_{r'} \subseteq K_s$. Then $i \in I''_s$ and $j' \in J''_s$. Hence $\langle i, j'\rangle \in I''_s \times J''_s \subseteq N''$. There follows that $N \circ N' \subseteq N''$. Hence, since N'' is difunctional and $N \bullet N'$ is the difunctional closure $dfc(N \circ N')$ of $N \circ N'$, therefore $N \bullet N' \subseteq N''$.

Consider any $\langle i, j'\rangle$ in N''. Then for some $s < \delta$, $i \in I''_s$ and $j' \in J''_s$. Also, there are $r < \gamma$ and $r' < \gamma'$ such that $i \in I_r \subseteq I''_s$, $j' \in J'_{r'} \subseteq J''_s$, $J_r \subseteq K_s$, and $I'_{r'} \subseteq K_s$. There follows that, for some n, $1 \leq n < \omega$, there are $J_{r_0}, \ldots, J_{r_n}$ and $I'_{r'_0}, \ldots, I'_{r'_n}$ such that $J_{r_0} = J_r$, $I'_{r'_n} = I'_{r'}$, $J_{r_m} \cap I'_{r'_m} \neq \emptyset$ for each $m \leq n$, and $I'_{r'_m} \cap J'_{r_{m+1}} \neq \emptyset$ for each $m < n$. Consider any $m < n$, any i_m in I_{r_m}, and any i_{m+1} in $I_{r_{m+1}}$. Since $J_{r_m} \cap I'_{r'_m} \neq \emptyset$, therefore there is some j'_m in $J'_{r'_m}$ such that $\langle i_m, j'_m\rangle \in N \circ N'$. Since $I'_{r'_m} \cap J_{r_{m+1}} \neq \emptyset$, therefore $\langle j'_m, i_{m+1}\rangle \in N'^\smile \circ N^\smile$. Hence $\langle i_m, i_{m+1}\rangle$ is in $(N \circ N') \circ (N \circ N')^\smile$. There follows that, for any i_n in $I_{r_n} = I_{r'}$, $\langle i, i_n\rangle$ is in $((N \circ N') \circ (N \circ N')^\smile)^n$. Since j' is in $J'_{r'}$, since $J_{r_n} \cap I'_{r_n} \neq \emptyset$, and since $J'_{r_n} = J'_{r'}$, therefore $\langle i_n, j'\rangle$ is in $N \circ N'$. Hence

$$\langle i, j'\rangle \in ((N \circ N') \circ (N \circ N')^\smile)^n \circ (N \circ N') .$$

Since this holds for any $\langle i, j'\rangle$ in N'' and since, as we saw earlier, $dfc\ R = \left(\bigcup_{1 \leq n < \omega}(R \circ R^\smile)\right) \circ R$ holds for any R, there now follows that $N'' \subseteq$

$dfc(N \circ N') = N \bullet N'$. Together with $N \bullet N' \subseteq N''$, shown earlier, this implies that $N'' = N \bullet N'$.

Since $M \circ M'$ is the set of pairs $\langle i, j' \rangle$ such that, for some k in $K = Rg\ M \cap Do\ M'$, $\langle i, k \rangle \in M$ and $\langle k, j' \rangle \in M'$, therefore the biordered pair of partitions of ω into intervals that is correlated with $M \circ M'$ is the pair $\langle \langle I''_s : s < \delta \rangle, \langle J''_s : s < \delta \rangle \rangle$. Hence, by the definition of $\widehat{+}$, $(M \circ M')^{\widehat{+}}$ is the biordered difunctional relation that corresponds to this pair. In other words, $(M \circ M')^{\widehat{+}} = N''$. There now follows that $(M \circ M')^{\widehat{+}} = N'' = N \bullet N' = M^{\widehat{+}} \bullet (M')^{\widehat{+}}$. □

We now return to ${}_{\langle \gamma, \delta \rangle}D$.

THEOREM 18. *Let $0 \le \gamma \le \omega$ and $0 \le \delta \le \omega$.*

(a) *$\langle {}_{\langle \gamma, \delta \rangle}D, \subseteq \rangle$ is a complete lattice whose greatest element is the set ${}_{\gamma}[\emptyset]_{\delta} = {}^{\gamma}U \times {}^{\delta}U$ and whose least element is the set*

$$\begin{aligned} &[\gamma \times \gamma,\ \gamma \times \delta,\ \delta \times \delta] \\ &\quad = \{\langle x, y \rangle \in {}^{\gamma}U \times {}^{\delta}U : x_i = x_j \text{ if } i \le j < \gamma,\ x_i = y_j \\ &\quad \text{if } i < \gamma \text{ and } j < \delta, \text{ and } y_i = y_j \text{ if } i \le j < \delta\} \ . \end{aligned}$$

(b) *Consider any $W \subseteq {}^{\gamma}U \times {}^{\delta}U$. There is a least X such that $W \subseteq X$ and $X \in {}_{\langle \gamma, \delta \rangle}D$. Specifically, let $\langle M_{W,0}, M_{W,1}, M_{W,2} \rangle = \langle \gamma \times \gamma, \gamma \times \delta, \delta \times \delta \rangle$ if $W = \emptyset$ and, if $W \ne \emptyset$, let*

$$\begin{aligned} M_{W,0} &= \mathring{1}_{\gamma} \smile \bigcap \{x x^{\smile} : \langle x, y \rangle \in W \text{ for some } y\} \ , \\ M_{W,1} &= \bigcap \{x y^{\smile} : \langle x, y \rangle \in W\} \ , \\ M_{W,2} &= \mathring{1}_{\delta} \smile \bigcap \{y y^{\smile} : \langle x, y \rangle \in W \text{ for some } x\} \ . \end{aligned}$$

*Then this least X is the set $[M_{W,0}, M_{W,1}, M_{W,2}]$. We call it **the** $\langle \gamma, \delta \rangle$ **diagonal closure of** W.*

(c) *Consider any $W \subseteq {}^{\gamma}U \times {}^{\delta}U$. Then $W \in {}_{\langle \gamma, \delta \rangle}D$ if and only if $W = [M_{W,0}, M_{W,1}, M_{W,2}] = {}_{\gamma}[M_{W,0}] \circ {}_{\gamma}[M_{W,1}]_{\delta} \circ {}_{\delta}[M_{W,2}]$. Also, ${}_{\gamma}[M_{W,0}] = \{\langle x, x \rangle : x \in Do\ W\}$ and ${}_{\delta}[M_{W,2}] = \{\langle y, y \rangle : y \in Rg\ W\}$.*

Proof. (a) Consider any family $\{\langle M_{0,i}, M_{1,i}, M_{2,i} \rangle : i \in I\}$ of $\langle \gamma, \delta \rangle$ triples. Let $N_0 = eq_{\gamma}(\bigcup \{M_{0,i} : i \in I\})$, $N_1 = \bigcup \{M_{1,i} : i \in I\}$, and $N_2 = eq_{\delta}(\bigcup \{M_{2,i} : i \in I\})$. Then $\langle N_0, N_1, N_2 \rangle$ is a $\langle \gamma, \delta \rangle$ triple. Moreover, from the definition of [ ] there follows:

$$\bigcap \{[M_{0,i}, M_{1,i}, M_{2,i}] : i \in I\} = [N_0, N_1, N_2] \ .$$

Since, evidently, ${}_{\gamma}[\emptyset]_{\delta} = {}^{\gamma}U \times {}^{\delta}U$ is the greatest element of ${}_{\langle\gamma,\delta\rangle}D$, there follows that $\langle {}_{\langle\gamma,\delta\rangle}D, \subseteq \rangle$ is a complete lattice. As one can verify, its least element is $[\gamma \times \gamma, \gamma \times \delta, \delta \times \delta]$.

(b) Consider any $W \subseteq {}^{\gamma}U \times {}^{\delta}U$. Since ${}^{\gamma}U \times {}^{\delta}U$ is in ${}_{\langle\gamma,\delta\rangle}D$, therefore the set $\{X \in {}_{\langle\gamma,\delta\rangle}D : W \subseteq X\}$ is nonempty. By part (a), it has a least element.

From their definitions there follows that $M_{W,0} = eq_{\gamma}M_{W,0}$, $M_{W,1} \subseteq \gamma \times \delta$, and $M_{W,2} = eq_{\delta}M_{W,2}$ and hence that $\langle M_{W,0}, M_{W,1}, M_{W,2}\rangle$ is a $\langle\gamma,\delta\rangle$ triple. Let $X = [M_{W,0}, M_{W,1}, M_{W,0}]$. Then $X \in {}_{\langle\gamma,\delta\rangle}D$. Consider any $\langle v, w\rangle$ in W. Then $M_{W,0} \subseteq vv^{\smile}$, $M_{W,1} \subseteq vw^{\smile}$, and $M_{W,2} \subseteq ww^{\smile}$. Hence $\langle v, w\rangle \in [M_{W,0}, M_{W,1}, M_{W,2}] = X$. There follows that $W \subseteq X$.

Consider any $\langle\gamma,\delta\rangle$ triple $\langle N_0, N_1, N_2\rangle$ such that either $N_0 \not\subseteq M_{W,0}$, or $N_1 \not\subseteq M_{W,1}$, or $N_2 \not\subseteq M_{W,2}$. First, consider the case when $N_0 \not\subseteq M_{W,0}$. Then, since $M_{W,0} = eq_{\gamma}M_{W,0}$, $\langle i, j\rangle \in N_0$ and $\langle i, j\rangle \notin M_{W,0}$, for some $i < j < \gamma$. Since $\langle i, j\rangle \notin M_{W,0}$, there is some $\langle x, y\rangle$ in W such that $\langle i, j\rangle \notin xx^{\smile}$ and hence $x_i \neq x_j$. Since $\langle i, j\rangle \in N_0$, therefore $N_0 \not\subseteq xx^{\smile}$, hence $\langle x, y\rangle \notin [N_0, N_1, N_2]$, and hence $W \not\subseteq [N_0, N_1, N_2]$. In the case where $N_1 \not\subseteq M_{W,1}$ or where $N_2 \not\subseteq M_{W,2}$, there similarly follows that $X \not\subseteq [N_0, N_1, N_2]$. Thus, $W \subseteq [N_0, N_1, N_2]$ implies that $N_i \subseteq M_{W,i}$ for each $i \leq 2$ and hence, by Theorem 3(a), that $X = [M_{W,0}, M_{W,1}, M_{W,2}] \subseteq [N_0, N_1, N_2]$.

(c) This follows from part (b). □

According to our original definition, W is in ${}_{\langle\gamma,\delta\rangle}D$ if and only if $W = [M_0, M_1, M_2]$ for some $\langle\gamma,\delta\rangle$ triple $\langle M_0, M_1, M_2\rangle$. Theorem 18(c) gives a criterion that is more specific and sometimes more useful. In particular, it seems more suitable for establishing non-membership in ${}_{\langle\gamma,\delta\rangle}D$.

It is important to note that, in contrast to Theorem 18(a), neither ${}_{[0,\omega)}D$ nor ${}_{[\omega]}Un \cap {}_{[\omega]}D$ is closed under $\cap$. As to ${}_{[0,\omega)}D$, consider any $\langle s, t\rangle$ and $\langle s', t'\rangle$, and any X and Y in ${}_{[0,\omega)}D$ such that $X \subseteq \dot{q}_0^s \circ q_0^t$ and $Y \subseteq \dot{q}_0^{s'} \circ q_0^{t'}$. If $s - t = s' - t'$ and $s' - s = k \geq 0$, then $X \cap Y = \overset{k}{\rightarrow} X \cap Y$ and therefore is in ${}_{[0,\omega)}D$. If $s - t = s' - t'$ and $s - s' = k \geq 0$, then $X \cap Y = X \cap \overset{k}{\rightarrow} Y$ and therefore again is in ${}_{[0,\omega)}D$. However, if $s - t \neq s' - t'$, then $X \cap Y = \emptyset$ and hence is not in ${}_{[0,\omega)}D$. There follows that ${}_{[0,\omega)}D$ is not closed under $\cap$, but that the closure of ${}_{[0,\omega)}D$ under $\circ$, ${}^{\smile}$, and $\cap$ is ${}_{[0,\omega)}D \cup \{\emptyset\}$.

As was remarked after the proof of Theorem II.10, if $s - t \neq s' - t'$, then $\overset{\omega}{\rightarrow} (\dot{q}_0^s \circ q_0^t) \cap \overset{\omega}{\rightarrow} (\dot{q}_0^{s'} \circ q_0^{t'})$ is nonempty but not in ${}_{[\omega]}Un$. Since $\overset{\omega}{\rightarrow} (\dot{q}_0^s \circ q_0^t)$ and $\overset{\omega}{\rightarrow} (\dot{q}_0^{s'} \circ q_0^{t'})$ are in ${}_{[\omega]}Un \cap {}_{[\omega]}D$, therefore ${}_{[\omega]}Un \cap {}_{[\omega]}D$ is not closed under $\cap$. Since by Theorems 11(a) and 18(a), ${}_{[\omega]}D$ is closed under $\circ$, $\smile$, and $\frown$. Therefore the closure under $\circ$, $\smile$, and $\frown$ of ${}_{[\omega]}Un \cap {}_{[\omega]}D$ is a subset of ${}_{[\omega]}D$. We do not have a characterization of this subset.

CHAPTER IV

Uniform Diagonal Relations and Some Kinds of Bisections or Bisectable Relations

Of the sets ${}_{[0,\omega)}D$ and ${}_{[\omega]}D$ discussed in Chapter III, ${}_{[0,\omega)}D$ is included in the set ${}_{[0,\omega)}Un$ of relations that are uniform on ${}^{[0,\omega)}U$, but ${}_{[\omega]}D$ is not included in the set ${}_{[\omega]}Un$ of relations that are uniform on ${}^{\omega}U$. Henceforth, for brevity, we let ${}_{[\omega]}UD = {}_{[\omega]}Un \cap {}_{[\omega]}D$, ${}_{[\omega]}UDF = {}_{[\omega]}Un \cap {}_{[\omega]}DF$, ${}_{[\omega]}UDP = {}_{[\omega]}Un \cap {}_{[\omega]}DP$, ${}_{[\omega]}UDE = {}_{[\omega]}Un \cap {}_{[\omega]}DE$, and ${}_{[\omega]}UDI = {}_{[\omega]}Un \cap {}_{[\omega]}DI$.

Each of these five sets will be shown to be the image under $\xrightarrow{\omega}$ of the corresponding subset of ${}_{[0,\omega)}D$. We will also characterize these sets in another way. For ${}_{[\omega]}UD$ we will describe the set of those $\langle \omega, \omega \rangle$ triples that determine a member of the set and, for each of the other four sets, we will describe the set of those special $\langle \omega, \omega \rangle$ triples ${}_{\omega}\langle N \rangle = \langle \overset{\circ}{1}_{\omega}, N, \overset{\circ}{1}_{\omega} \rangle$ that determine a member of that set.

For the latter purpose we will make use of the following notion: A relation N on ω shall be ***bisectable*** if and only if, for some $s < \omega$ and $t < \omega$,

$$N = (N \cap (s \times t)) \cup \{\langle s+n, t+n \rangle : n < \omega\} .$$

Thus, if N is bisectable, then N is determined in a simple way by an initial section $N \cap (s \times t)$. Thus, the bisectable relations are those relations on ω that are infinite but, in the sense just indicated, have finite support or are finitary. The set of bisectable relations will be denoted by Bsl.

Specifically, we will show the following:

$$\begin{aligned} {}_{[\omega]}UDF &= \{{}_{\omega}[N] : N \text{ is one-one and bisectable}\} \\ &= \{{}_{\omega}[N] \in {}_{[\omega]}DF : N \in Bsl\} . \end{aligned}$$

For ${}_{[\omega]}UDP$, ${}_{[\omega]}UDE$, and ${}_{[\omega]}UDI$ we will show that it is that subset of ${}_{[\omega]}DP$, ${}_{[\omega]}DE$, or ${}_{[\omega]}DI$, respectively, that is related to it as ${}_{[\omega]}UDF$ is related to ${}_{[\omega]}DF$.

Results from Chapter III will then be used to show that, if $\overline{\overline{U}} \geq 2$, then, for each of these four sets, a certain algebra having this set as universe is isomorphic under the converse of ${}_{\omega}[\ ]$ to an algebra whose universe consists of certain relations on ω.

The latter algebra consists of relations that are more elementary and more easily understood. Below we list, on the left, the four algebras whose universe is ${}_{[\omega]}UDF$, ${}_{[\omega]}UDP$, ${}_{[\omega]}UDE$, or ${}_{[\omega]}UDI$, respectively, and to the right of each the algebra of relations on ω to which, if $\overset{=}{U} \geq 2$, it is isomorphic under ${}_{\omega}[\;]$.

$$\begin{array}{ll} \langle {}_{[\omega]}UDF, \circ, {}^{\smile} \rangle & \langle \{N \in Bsl : N \text{ is one-one}\}, \circ, {}^{\smile} \rangle \\ \langle {}_{[\omega]}UDP, \circ, {}^{\smile} \rangle & \langle \{\pi \in Bsl : \pi \text{ is a permutation of } \omega\}, \circ, {}^{\smile} \rangle \\ \langle {}_{[\omega]}UDE, \circ \rangle & \langle \{f^{\smile} \in Bsl : Do\ f = \omega,\ f \text{ is strictly increasing}\}, \circ \rangle \\ \langle {}_{[\omega]}UDI, \circ \rangle & \langle \{N \in Bsl : N = eq_{\omega} N\}, \bullet \rangle \ . \end{array}$$

For the algebra whose universe is ${}_{[\omega]}UDP$, ${}_{[\omega]}UDE$, ${}_{[\omega]}UDF$, or ${}_{[\omega]}UDI$, respectively, we will obtain a set G_P, G_E, G_F, G_I, respectively, of generators, obtaining first a set of generators for the corresponding algebra of relations on ω. From the definitions involved there will follow that $G_F \subseteq {}_{[\omega]}S_{pq}$ and hence, since ${}_{[\omega]}S_{pq}$ is closed under $\circ$ and ${}^{\smile}$, that ${}_{[\omega]}UDF \subseteq {}_{[\omega]}S_{pq}$. There will likewise follow that $G_I \subseteq {}_{[\omega]}S_e$ and hence that ${}_{[\omega]}UDI \subseteq {}_{[\omega]}S_e$. There will then follow that ${}_{[\omega]}UD \subseteq {}_{[\omega]}S_{pqe}$. Since, as we showed in Chapter II or III, respectively, ${}_{[\omega]}S_{pqe} \subseteq {}_{[\omega]}Un$ and ${}_{[\omega]}S_{pqe} \subseteq {}_{[\omega]}D$, this will conclude our proof that ${}_{[\omega]}S_{pqe} = {}_{[\omega]}Un \cap {}_{[\omega]}D = {}_{[\omega]}UD$.

For dealing with questions about ${}_{[0,\omega)}D$ that are similar to those about ${}_{[\omega]}D$ that have just been discussed, we will make use of the following set:

$$\begin{aligned} Bst \;\; = \;\; & \{\langle N, \langle s,t\rangle\rangle : s<\omega,\ t<\omega, \\ & N = (N \cap (s \times t)) \cup \{\langle s+n, t+n\rangle : n<\omega\}\} \ . \end{aligned}$$

Any $\langle N, \langle s,t\rangle\rangle$ in Bst shall be a ***bisection*** or, more explicitly, a ***bisection of*** N or, fully explicitly, ***the bisection of*** N ***into*** $N \cap (s \times t)$ ***and*** $\{\langle s+n, t+n\rangle : n<\omega\}$. We shall also say that $\langle s,t\rangle$ ***bisects*** N, or is a ***bisector of*** N, if and only if $\langle N, \langle s,t\rangle\rangle$ is in Bst. Thus, N has a bisector, i.e., there is some $\langle s,t\rangle$ such that $\langle N, \langle s,t\rangle\rangle \in Bst$, if and only if $N \in Bsl$. Evidently, whenever N is in Bsl then there is a unique $\langle s_0, t_0\rangle$ such that the set of bisectors of N is $\{\langle s_0+n, t_0+n\rangle : n<\omega\}$. This $\langle s_0, t_0\rangle$ shall be ***the least bisector of*** N.

Recall from Chapter III that the set of triples that are both finite and special is the following set:

$$\begin{aligned} & \{\langle \mathring{1}_s, M, \mathring{1}_t\rangle : s<\omega,\ t<\omega,\ M \subseteq s \times t\} \\ & \quad = \{{}_s\langle M\rangle_t : s<\omega,\ t<\omega,\ M \subseteq s \times t\} \ . \end{aligned}$$

If ${}_s\langle M\rangle_t$ is a triple that is finite and special and if $N = M \cup \{\langle s+n, t+n\rangle : n < \omega\}\}$, then $\langle N, \langle s,t\rangle\rangle$ is in Bst. If $\langle N, \langle s,t\rangle\rangle$ is in Bst and if $M = N \cap (s \times t)$, then ${}_s\langle M\rangle_t$ is a triple that is finite and special. Thus, between Bst and the set of triples that are finite and special there is a simple one-one correspondence.

In the later parts of the present chapter, this correspondence will be used to obtain results that are somewhat analogous to those sketched above. For example, we will show that, if $\overline{\overline{U}} \geq 2$, then $\langle {}_{[0,\omega)}DF, \circ, {}^{\smile}, \overset{1}{\to}\rangle$ is isomorphic to a certain algebra whose universe is $\{\langle N, \langle s,t\rangle\rangle \in Bst\colon N \text{ is one-one}\}$. For each of the algebras concerned we shall again obtain a set of generators. We shall then see the following: $S_p = {}_{[0,\omega)}DP \cap -\{{}_{[0,\omega)}\mathring{1}, {}_{[1,\omega)}\mathring{1}\}$; the closure of $\{q_i^{\smile} : i < \omega\}$ under $\circ$ is ${}_{[0,\omega)}DE \cap -(Rg \overset{1}{\to}) \cap -\{{}_{[0,\omega)}\mathring{1}\}$; S_q is the closure of ${}_{[0,\omega)}DE$ under $\circ$ and ${}^{\smile}$; $S_{pq} = {}_{[0,\omega)}DF$; $S_e = {}_{[0,\omega)}DI \cap - \{{}_{[0,\omega)}\mathring{1}, {}_{[1,\omega)}\mathring{1}\}$; $S_{qe} = S_{pqe} = {}_{[0,\omega)}D$.

THEOREM 1. (a) *Given any $s < \omega$, $t < \omega$, and $\langle s,t\rangle$ triple $\langle M_0, M_1, M_2\rangle$, let $N_0 = M_0 \cup \{\langle s+n, s+n\rangle : n < \omega\}$, $N_1 = M_1 \cup \{\langle s+n, t+n\rangle : n < \omega\}$, and $N_2 = M_2 \cup \{\langle t+n, t+n\rangle : n < \omega\}$. Then $\overset{\omega}{\to} [\![M_0, M_1, M_2]\!] = [N_0, N_1, N_2]$ and $\overset{\omega}{\to} {}_s[\![M_1]\!]_t = {}_\omega[N_1]$.*

(b) *If X is in ${}_{[\omega]}UD$, $X = \overset{\omega}{\to} W$, and $gd\, W = \langle s,t\rangle$, then $({}^sU \times {}^tU) \cap W$ is in ${}_{\langle s,t\rangle}D$.*

(c) *${}_{[\omega]}UD$ is the direct image $\{\overset{\omega}{\to} W : W \in {}_{[0,\omega)}UD\}$ of ${}_{[0,\omega)}D$ under $\overset{\omega}{\to}$. Also, ${}_{[\omega]}UDF$, ${}_{[\omega]}UDP$, ${}_{[\omega]}UDE$, and ${}_{[\omega]}UDI$ is the direct image under $\overset{\omega}{\to}$ of ${}_{[0,\omega)}DF$, ${}_{[0,\omega)}DP$, ${}_{[0,\omega)}DE$, or ${}_{[0,\omega)}DI$, respectively*

Proof. (a) Since $M_0 = eq_s M_0$ and $M_2 = eq_t M_2$, therefore $N_0 = eq_\omega N_0$ and $N_2 = eq_\omega N_2$. Hence $\langle N_0, N_1, N_2\rangle$ is an $\langle \omega, \omega\rangle$ triple.

Consider any x, y, z such that $|x| = s$, $|y| = t$, and $|z| = \omega$. Then $M_0 \subseteq xx^{\smile}$ if and only if $N_0 \subseteq (xz)(xz)^{\smile}$, $M_1 \subseteq xy^{\smile}$ if and only if $N_1 \subseteq (xz)(yz)^{\smile}$, and $M_2 \subseteq yy^{\smile}$ if and only if $N_2 \subseteq (yz)(yz)^{\smile}$. There follows that $\overset{\omega}{\to} [M_0, M_1, M_2] = [N_0, N_1, N_2]$ and that $\overset{\omega}{\to} {}_s[M]_t = {}_\omega[N_1]$. Hence, by Lemma II.1(b), $\overset{\omega}{\to} [\![M_0, M_1, M_2]\!] = [N_0, N_1, N_2]$ and $\overset{\omega}{\to} {}_s[\![M_1]\!]_t = {}_\omega[N_1]$.

(b) Assume that $gd\, W = \langle s,t\rangle$, $W' = ({}^sU \times {}^tU) \cap W$, and $X = \overset{\omega}{\to} W$, and hence, by Lemma II.1(b), $X = \overset{\omega}{\to} W'$. Let $M_{W',0} = \mathring{1}_\omega \cup \bigcap\{xx^{\smile} : \langle x,y\rangle \in W' \text{ for some } y\}$, $M_{W',1} = \{xy^{\smile} : \langle x,y\rangle \in W'\}$, and $M_{W',2} = \mathring{1}_\omega \cup \bigcap\{yy^{\smile} : \langle x,y\rangle \in W' \text{ for some } x\}$. Let $M_{X,0} = \mathring{1}_\omega \cup \bigcap\{vv^{\smile} : \langle v,w\rangle \in X \text{ for some } w\}$, $M_{X,1} = \bigcap\{vw^{\smile} : \langle v,w\rangle \in X\}$, and $M_{X,2} = \mathring{1}_\omega \cup \bigcap\{ww^{\smile} : \langle v,w\rangle \in X \text{ for some } v\}$. Since $X = \overset{\omega}{\to} W'$, therefore

$M_{X,0} = \overset{\circ}{1}_\omega \cup \bigcap\{(xz)(xz)^\smile : |z| = \omega,\ \langle x,y\rangle \in W' \text{ for some } y\}$, $M_{X,1} = \bigcap\{(xz)(yz)^\smile :$ $|z| = \omega,\ \langle x,y\rangle \in W'\}$, and $M_{X,2} = \overset{\circ}{1}_\omega \cup \bigcap\{(yz)(yz)^\smile : \langle x,y\rangle \in W' \text{ for some } x\}$.

Now assume also that W' is not in ${}_{\langle s,t\rangle}D$. Then, by Theorems III.18(b) and III.18(c), W' is a proper subset of $[M_{W',0}, M_{W',1}, M_{W',2}]$. Then there is some $\langle x,y\rangle$ in $[M_{W',0}, M_{W',1}, M_{W',2}]$ which is not in W'. Then, for any z in ${}^\omega U$, $\langle xz, yz\rangle$ is in $[M_{X,0}, M_{X,1}, M_{X,2}]$ but not in $\overset{\omega}{\rightarrow} W' = X$. From Theorem III.18(c) there then follows that X is not in ${}_{[\omega]}D$.

This shows that if X is in ${}_{[\omega]}D$, $X = \overset{\omega}{\rightarrow} W$, and $gd\ W = \langle s,t\rangle$, then $({}^sU \times {}^tU) \cap W$ is in ${}_{\langle s,t\rangle}D$ and hence W is in ${}_{[0,\omega)}D$. There follows that if W is in ${}_{[0,\omega)}Un$ and $\overset{\omega}{\rightarrow} W$ is in ${}_{[\omega]}D$, as well as in ${}_{[\omega]}Un$, then W is in ${}_{[0,\omega)}D$.

(c) By part (a), $\{\overset{\omega}{\rightarrow} W : W \in {}_{[0,\omega)}UD\}$ is included in ${}_{[\omega]}UD$. Hence, by part (b) they coincide.

Consider any W in ${}_{[0,\omega)}Un$. Then W is upward full, an upward full permutation, an upward full excision, or a subidentity if and only if $\overset{\omega}{\rightarrow} W$ is full, a permutation of ${}^\omega U$, an excision from ${}^\omega U$ onto ${}^\omega U$, or a subidentity, respectively. Our second assertion therefore follows from our first. □

THEOREM 2. (a) *Let $\langle N_0, N_1, N_2\rangle$ be any $\langle \omega,\omega\rangle$ triple. If N_0, N_1, and N_2 are bisectable, then $[N_0, N_1, N_2]$ is in ${}_{[\omega]}UD$. Also, if $\overline{\overline{U}} \geq 2$ and $[N_0, N_1, N_2]$ is in ${}_{[\omega]}UD$, then N_0, N_1, and N_2 are bisectable. There follows:*

$${}_{[\omega]}UD = \{[N_0, N_1, N_2] : N_0, N_1 \text{ and } N_2 \text{ are bisectable}\}.$$

(b) ${}_{[\omega]}UDF = \{{}_\omega[N] : N \text{ is one-one and bisectable}\}$.

(c) ${}_{[\omega]}UDP = \{{}_\omega[\pi] : \pi \text{ is a bisectable permutation of } \omega\}$
$= \{{}_\omega[\pi] : \pi \text{ is a permutation of } \omega \text{ and } \{n : \pi(n) \neq n\} \text{ is finite}\}$.

(d) ${}_{[\omega]}UDE = \{{}_\omega[f^\smile] : Do\ f = \omega,\ f \text{ is strictly increasing},$ $f \text{ is bisectable}\} = \{{}_\omega[f^\smile] : Do\ f = \omega,\ f \text{ is strictly increasing},$ $\omega \,\bar{\underset{2}{}}\, (Rg\ f) \text{ is finite}\}$.

(e) ${}_{[\omega]}UDI = \{{}_\omega[N] : N = eq_\omega N \text{ and } N \text{ is bisectable}\}$
$= \{{}_\omega[N] : N = eq_\omega N \text{ and } N \,\bar{\underset{2}{}}\, \overset{\circ}{1}_\omega \text{ is finite}\}$.

Proof. Let $\langle N_0, N_1, N_2\rangle$ be any $\langle\omega,\omega\rangle$ triple. First assume that N_0, N_1, and N_2 are bisectable. By Lemma III.6, $[N_0, N_1, N_2] = {}_\omega[N_0] \circ {}_\omega[N_1] \circ {}_\omega[N_2]$. Since N_0, N_1, N_2 are bisectable and since $N_0 = eq_\omega N_0$ and $N_2 = eq_\omega N_2$, therefore there are s, s', t', t such that $\langle s,s\rangle$ bisects N_0, $\langle s',t'\rangle$ bisects N_1, and $\langle t,t\rangle$ bisects N_2. Let $M_0 = N_0 \cap (s \times s)$, $M_1 = N_1 \cap (s' \times t')$, and $M_2 = N_2 \subseteq (t \times t)$. By Theorem 1(a), $\overset{\omega}{\rightarrow} ({}_s[\![M_0]\!]) = {}_\omega[N_0]$, $\overset{\omega}{\rightarrow} ({}_{s'}[\![M_1]\!]_{t'}) = {}_\omega[N_1]$, and $\overset{\omega}{\rightarrow} ({}_t[\![M_2]\!]) = {}_\omega[N_2]$. Then

$$[N_0, N_1, N_2] = \overset{\omega}{\rightarrow} ({}_s[\![M_0]\!]) \circ \overset{\omega}{\rightarrow} ({}_{s'}[\![M_1]\!]_{t'}) \circ \overset{\omega}{\rightarrow} ({}_t[\![M_2]\!])\ .$$

Let $W = {}_s[\![M_0]\!] \circ {}_{s'}[\![M_1]\!]_{t'} \circ {}_t[\![M_2]\!]$. Then, by Lemma II.9(c),

$$\overset{\omega}{\rightarrow} ({}_s[\![M_0]\!]) \circ \overset{\omega}{\rightarrow} ({}_{s'}[\![M_1]\!]_{t'}) \circ \overset{\omega}{\rightarrow} ({}_t[\![M_2]\!]) = \overset{\omega}{\rightarrow} W \ .$$

There follows that $[N_0, N_1, N_2] = \overset{\omega}{\rightarrow} W$. By Theorem II.10(a), W is in ${}_{[0,\omega)}Un$. Hence $[N_0, N_1, N_2]$ is in ${}_{[\omega]}Un$. Since $[N_0, N_1, N_2]$ also is in ${}_{[\omega]}D$, it therefore is in ${}_{[\omega]}UD$. Thus $\{[N_0, N_1, N_2] : N_0,\ N_1$ and N_2 are bisectable$\} \subseteq {}_{[\omega]}UD$.

Now assume that $\overline{\overline{U}} \geq 2$ and that $[N_0, N_1, N_2]$ is in ${}_{[\omega]}UD$ and hence is in ${}_{[\omega]}Un$. Since $[N_0, N_1, N_2] \neq \emptyset$, therefore $[N_0, N_1, N_2] = \overset{\omega}{\rightarrow} W$ for some W in ${}_{[0,\omega)}Un$ such that $W \neq \emptyset$. Let $gd\ W = \langle s, t\rangle$. Then there is some x in ${}^sU$ and some y in ${}^tU$, such that $\langle x, y\rangle$ is in W and such that $\langle xz, yz\rangle$ is in $[N_0, N_1, N_2]$ for every z in ${}^\omega U$. Since $\langle N_0, N_1, N_2\rangle$ is an $\langle \omega, \omega\rangle$ triple, therefore $N_0 = eq_\omega N_0$ and $N_2 = eq_\omega N_2$. Now suppose that there were some $\langle i, j\rangle$ in N_0 such that $i \neq j$ and $i \geq s$. Since $\overline{\overline{U}} \geq 2$, if $j < s$, then there are z in ${}^\omega U$ such that $x_j \neq z_i$ and if $j \geq s$, then there are z in ${}^\omega U$ such that $z_{j-s} \neq z_{i-s}$. In either case, there would be z in ${}^\omega U$ such that $N_0 \subseteq (xz)(xz)^\smile$ and hence such that $\langle xz, yz\rangle$ is not in $[N_0, N_1, N_2]$. There follows that $N_0 \mathbin{\bar{2}} \overset{\circ}{1}_\omega \subseteq s \times s$. Similarly, $N_2 \mathbin{\bar{2}} \overset{\circ}{1}_\omega \subseteq t \times t$. Finally, since $\overline{\overline{U}} \geq 2$ and $\langle xz, yz\rangle$ is in $[N_0, N_1, N_2]$ for every z in ${}^\omega U$, there follows, by an argument similar to the proof of Lemma III.9(b), that $N_1 = (N_1 \cap (s \times t)) \cup \{\langle s+n, t+n\rangle : n < \omega\}$. This proves the second assertion of part (a). If $\overline{\overline{U}} = 1$, then ${}_{[\omega]}D = {}_{[\omega]}UD = \{{}_{[\omega]}\overset{\circ}{1}\}$. Therefore from the first and the second assertion of part(a) there follows the third.

A consequence of this third assertion is the following: X is a special diagonal relation in ${}_{[\omega]}Un$ if and only if $X = {}_{[\omega]}[N]$ for some N that is bisectable. Part (b) and the first equality in (c), (d), or (e) then follows from the definition of ${}_{[\omega]}DF$, ${}_{[\omega]}DE$ or ${}_{[\omega]}DI$, respectively.

Assume that $Do\ f = \omega$ and that f is strictly increasing. First, suppose that f is bisectable. Let $\langle s, t\rangle$ be a bisector of f. Then $\omega \mathbin{\bar{2}} (Rg\ f)$ is finite and hence is included in some $\{0, \ldots, t-1\}$. Let $r = \omega \mathbin{\bar{2}} (Rg\ f)$. Then $\langle t-r, t\rangle$ is a bisector of f. There now follows the second equality in (d). The proof of the second equality in (c) or in (e), respectively, is likewise straightforward. □

THEOREM 3. *Assume that* $\overline{\overline{U}} \geq 2$. *In* (d) *below, let* S *be the closure under* $\circ$ *and* $^\smile$ *of* ${}_{[\omega]}UDE$. *In* (f) *below, let* S' *be the closure under* $\circ$ *of* ${}_{[\omega]}UDI \cup {}_{[\omega]}UDP$ *and let* S'' *be the closure under* $\bullet$ *of* $\{N \in Bsl : N = eq_\omega N\} \cup \{\pi \in Bsl :$ π *is a permutation of* $\omega\}$. *Then each structure below is isomorphic under* ${}_\omega[\ \ ]$ *to the structure given directly underneath.*

(a) $\langle {}_{[\omega]}UDF, \circ, \smile, \subseteq \rangle$
$\langle \{N \in Bsl : N \text{ is one-one}\}, \circ, \smile, \supseteq \rangle$

(b) $\langle {}_{[\omega]}UDP, \circ, \smile \rangle$
$\langle \{\pi \in Bsl : \pi \text{ is a permutation of } \omega\}, \circ, \smile \rangle$

(c) $\langle {}_{[\omega]}UDE, \circ \rangle$
$\langle \{f^{\smile} \in Bsl : Do\ f = \omega,\ f \text{ is strictly increasing}\}, \circ \rangle$

(d) $\langle S, \circ, \smile, \subseteq \rangle$
$\langle \{N \in Bsl : N \text{ and } N^{\smile} \text{ are order-preserving}\}, \circ, \smile, \supseteq \rangle$

(e) $\langle {}_{[\omega]}UDI, \circ, \subseteq \rangle$
$\langle \{N \in Bsl : N = eq_{\omega} N\}, \bullet, \supseteq \rangle$

(f) $\langle S', \circ, \subseteq \rangle$
$\langle S'', \bullet, \supseteq \rangle$

Proof. Part (a) follows from Theorem III.12(a) and Theorem 2(b). Parts (b) and (c) follow from part (a) and from Theorem 2(c) or 2(d), respectively. Part (d) follows from part (c) and Theorem III.14. Part (e) follows from Theorem III.12(b) and Theorem 2(e). Part (f) follows from Theorem III.16 and Theorems 2(c) and 2(e). □

A certain factorizability of ${}_{[\omega]}D$ using ${}_{[\omega]}DI$ and ${}_{[\omega]}DF$ is asserted by Theorem III.11(b). Also, of a certain subset of ${}_{[\omega]}DF$, a certain factorizability that uses ${}_{[\omega]}DE$ and ${}_{[\omega]}DP$ is asserted by the first sentence in Theorem III.15(d). According to the next theorem, there holds a similar factorizability of ${}_{[\omega]}UD$ using ${}_{[\omega]}UDI$ and ${}_{[\omega]}UDF$ and likewise a factorizability of ${}_{[\omega]}UDF$ that uses ${}_{[\omega]}UDE$ and ${}_{[\omega]}UDP$. Another factorizabilitiy is described in part (c).

THEOREM 4. (a) *Replacement in Theorem* III.11(b) *of* ${}_{[\gamma]}D$, ${}_{[\gamma]}DI$, *and* ${}_{[\gamma]}DF$ *by* ${}_{[\omega]}UD$, ${}_{[\omega]}UDI$, *or* ${}_{[\omega]}UDF$, *respectively, results in a true assertion about* ${}_{[\omega]}UD$.

(b) *For any* X *in* ${}_{[\omega]}UDF$ *there are exactly one* X_0, X_1, X_2 *in* ${}_{[\omega]}UDE$, ${}_{[\omega]}UDP$, *or* ${}_{[\omega]}UDE$, *respectively, such that* $X = X_0 \circ X_1 \circ X_2^{\smile}$.

(c) *Consider any* X *in the closure under* $\circ$ *of* ${}_{[\omega]}UDI \cup {}_{[\omega]}UDP$. *There is exactly one* X_1 *in* ${}_{[\omega]}UDI$ *such that* $X = X_1 \circ X_2$ *for some* X_2 *in* ${}_{[\omega]}UDP$. *Moreover, if* X_2' *is in* ${}_{[\omega]}UDP$ *and* $X_1 \circ X_2 = X_1 \circ X_2'$, *then* $X_1 \subseteq X_2 \circ X_2'^{\smile}$.

Proof. (a) By Theorem III.11(b), Lemma III.6, and Theorems 2(a), 2(b), and 2(e).

(b) By the first sentence in Theorem III.15(d) and Theorems 2(b), 2(c), and 2(d).

(c) By Theorem III.16 and Theorem 3(f). □

For any γ and δ such that $\delta \leq \gamma \leq \omega$, we described, in Chapter III, a one-one correspondence between the two sets $\{f : Do\ f = \delta,\ Rg\ f \subseteq \gamma,\ f \text{ is strictly increasing}\}$ and $\{J \subseteq \gamma : \overline{\overline{\gamma \,\bar{_2}\, J}} = \delta\}$. Under it there corresponds to every J in the second set the unique f in the first set such that $\gamma \,\bar{_2}\, (Rg\ f) = J$.

We will now make use of this correspondence in the case where $\gamma = \delta = \omega$. For this purpose we will, in what follows, tacitly assume that $I, J, K, \ldots$ are subsets of ω and also, for example, will let $-J = \omega \,\bar{_2}\, J$. Thus $\{J \subseteq \omega : \overline{\overline{\omega \,\bar{_2}\, J}} = \omega\} = \{J : \overline{\overline{-J}} = \omega\}$. Also note that $\{f : Do\ f = \omega,\ f \text{ is strictly increasing}\}$, coincides with $\{f : Do\ f = \omega,\ f \text{ is strictly increasing},\ Rg\ f \subseteq \omega\}$ and therefore is in one-one correspondence with $\{J : \overline{\overline{-J}} = \omega\}$.

We will use $\{J : \overline{\overline{-J}} = \omega\}$ to index $\{f : Do\ f = \omega,\ f \text{ is strictly increasing}\}$ as follows: For any J such that $\overline{\overline{-J}} = \omega$ we let f_J be the unique function f such that $Do\ f = \omega$, f is strictly increasing, and $-(Rg\ f) = J$. We will also apply this method of indexing to the smaller set $\{f : Do\ f = \omega, f \text{ is strictly increasing}, \overline{\overline{-(Rg\ f)}} < \omega\}$. The indexing set then is $\{J : \overline{\overline{J}} < \omega\}$.

In the case where J is a singleton $\{i\}$ there holds

$$f_{\{i\}} = \{\langle n, n\rangle : n < i\} \cup \{\langle n, n+1\rangle : n \geq i\}\ .$$

Thus $f_{\{0\}}$ is the successor function on ω. More generally, $f_{\{i\}}$ coincides with $\mathring{1}_i$ on $i = \{n : n < i\}$ and with the successor function on $\{n : n \geq i\}$. Thus, $f_{\{i\}}$ shall be the successor function that ***delays (or procrastinates) until*** i.

THEOREM 5. (a) *Assume that* $\overline{\overline{-I}} = \overline{\overline{-J}} = \omega$. *Let* $K = \{f_J(i) : i \in I\} \cup J$. *Then* $Rg(f_I \circ f_J) = -K$, $\overline{\overline{-K}} = \omega$, *and* $f_I \circ f_J = f_K$.

(b) *Assume that* $J = \{i_0, \ldots i_{n-1}\}$, $n \geq 1$, *and, if* $k < m < n$, *then* $i_k < i_m$. *Then* $f_J = f_{\{i_0\}} \circ \cdots \circ f_{\{i_{n-1}\}}$. *There follows that* $\{f_{\{i\}} : i < \omega\}$ *is a generating set for* $\langle \{f_J : 1 \leq \overline{\overline{J}} < \omega\}, \circ\rangle$ *and, in fact, is the least generating set for it.*

(c) *If* $\overline{\overline{U}} \geq 2$, *then* $\{\xrightarrow{\omega} (q_i{}^{\smile}) : i < \omega\}$ *is a generating set for* $\langle {}_{[\omega]}UDE \cap -\{{}_{[\omega]}\mathring{1}\}, \circ\rangle$.

(d) *The following sets are the same:* $\{N \in Bsl : N$ *and* $N^{\smile}$ *are order-preserving*$\}$; $\{(f_J)^{\smile} \circ f_K : \overline{\overline{J}} < \omega,\ \overline{\overline{K}} < \omega\}$; *the closure of* $\{f_{\{i\}} : i < \omega\}$ *under* $\circ$ *and* $^{\smile}$.

(e) *Each of the following sets coincides with* ${}_{[\omega]}S_q$: *The closure of* ${}_{[\omega]}UDE$ *under* $\circ$ *and* $^{\smile}$; $\{X \circ Y^{\smile} : X$ *and* Y *are in* ${}_{[\omega]}UDE\}$; *the closure of* $\{\overset{\omega}{\rightarrow}(q_i{}^{\smile}) : i < \omega\}$ *under* $\circ$ *and* $^{\smile}$.

Proof. (a) Assume that n is not in $K = (\{f_J(i) : i \in I\} \cup J)$. Then $n \notin J$ and hence $n = f_J(m)$ for some m. Since n is not in $\{f_J(i) : i \in I\}$, therefore m is not in I. Therefore m is in $Rg\ f_I$ and hence n is in $Rg(f_I \circ f_J)$. Thus $-K \subseteq Rg(f_I \circ f_J)$.

Assume that $n \in Rg(f_I \circ f_J)$. Then $n \in Rg\ f_J$ and hence $n \notin J$. Since f_J is one-one, there is a unique m such that $f_J(m) = n$. Moreover, $m \in Rg\ f_I$, hence $m \notin I$, and hence $n \notin \{f_J(i) : i \in I\}$. There now follows that n is not in $\{f_J(i) : i \in I\} \cup J = K$. Thus $Rg(f_I \circ f_J) \subseteq -K$.

There now follows that $Rg(f_I \circ f_J) = -K$. Since f_I and f_J are strictly increasing, therefore $f_I \circ f_J$ is strictly increasing and hence is one-one. Since $\overline{\overline{Do\ f_I}} = \overline{\overline{Do\ f_J}} = \omega$, therefore $Do(f_I \circ f_J) = \omega$. There now follows that $\overline{\overline{Rg(f_I \circ f_J)}} = \overline{\overline{Do(f_I \circ f_J)}} = \omega$ and hence that $\overline{\overline{-K}} = \omega$.

Since $\overline{\overline{-K}} = \omega$ therefore, by definition, f_K is the unique f such that $Do\ f = \omega$, f is strictly increasing, and $-(Rg\ f) = K$. Since $Do(f_I \circ f_J) = \omega$, $f_I \circ f_J$ is strictly increasing, and $-(Rg(f_I \circ f_J)) = K$, therefore $f_I \circ f_J$ also is this unique f. There follows that $f_I \circ f_J = f_K$.

(b) Given any $n \geq 1$, assume as inductive hypothesis that the first assertion holds for any J such that $\overline{\overline{J}} = n$. Consider any $\{i_0, \ldots, i_{n-1}, i_n\}$ such that $k < m \leq n$ implies that $i_k < i_m$. Let $I = \{i_0, \ldots, i_{n-1}\}$ and let $K = \{i_0, \ldots, i_{n-1}, i_n\}$. Then $\{f_{\{i_n\}}(i) : i \in I\} = I$ and hence $K = I \cup \{i_n\} = \{f_{\{i_n\}}(i) : i \in I\} \cup \{i_n\}$. By part (a) there follows that $f_K = f_I \circ f_{\{i_n\}}$. By the inductive hypothesis, $f_I = f_{\{i_0\}} \circ \cdots \circ f_{\{i_{n-1}\}}$. There now follows that $f_K = f_{\{i_0\}} \circ \cdots \circ f_{\{i_{n-1}\}} \circ f_{\{i_n\}}$. The first assertion now follows by induction. From it there follows that $\{f_{\{i\}} : i < \omega\}$ is a generating set for $\langle \{f_J : 1 \leq \overline{\overline{J}} < \omega\}, \circ\rangle$. Since $f_{\{i\}}$ differs from any $f_{\{j\}}$, $j \neq i$, and from any $f_{\{i_0\}} \circ \cdots \circ f_{\{i_n\}}$ such that $n \geq 1$, it is the least generating set.

(c) Since ${}_{\omega}[(f_{\{i\}})^{\smile}] = \overset{\omega}{\rightarrow}(q_i{}^{\smile})$, therefore by Theorem 3(c) and part (b), if $\overline{\overline{U}} \geq 2$, then $\{\overset{\omega}{\rightarrow}(q_i{}^{\smile}) : i < \omega) : i < \omega\}$ is a generating set for $\langle {}_{[\omega]}UDE \cap -\{{}_{[\omega]}\overset{\circ}{1}\}, \circ\rangle$. (Note that if $\overline{\overline{U}} = 1$, then ${}_{[\omega]}UDE = {}_{[\omega]}\overset{\circ}{1}$ and hence ${}_{[\omega]}UDE \cap -\{{}_{[\omega]}\overset{\circ}{1}\}$ is empty.)

(d) Let $S = \{N \in Bsl : N$ and $N^{\smile}$ are order-preserving$\}$. Then a straightforward argument shows that $\{f_J{}^{\smile} \circ f_K : \overline{\overline{J}} < \omega,\ \overline{\overline{K}} < \omega\}$ is included in S. Now consider any N in S. Then $\overline{\overline{N}} = \omega$ and hence, by Theorem III.14(a), $N = f^{\smile} \circ g$ for exactly one f and g such that $Do\ f = Do\ g = \omega$ and such that f and g are strictly increasing. Let $J = -(Do\ f)$ and $K = -(Do\ g)$. Since N is bisectable, therefore, for some $s < \omega$ and $t < \omega$, $\langle s, t \rangle$ bisects N. Then $J \subseteq s$ and $K \subseteq t$, and hence $\overline{\overline{J}} < \omega$ and $\overline{\overline{K}} < \omega$. Then $f = f_J$ and $g = f_K$. Thus $S \subseteq \{f_J{}^{\smile} \circ f_K : \overline{\overline{J}} < \omega,\ \overline{\overline{K}} < \omega\}$. Since $f_{\{i\}} \circ (f_{\{i\}})^{\smile} = \mathring{1}_\omega = f_\emptyset$, there now follows from part (b) that S is the closure under $\circ$ and $^{\smile}$ of $\{f_{\{i\}} : i < \omega\}$.

(e) By its definition (in Chapter II), ${}_{[\omega]}S_q$ is the closure under $\circ$ of $\{\overset{\omega}{\to} q_i : i < \omega\} \cup \{\overset{\omega}{\to} (q_i{}^{\smile}) : i < \omega\}$. Hence it is also the closure under $\circ$ and $^{\smile}$ of $\{\overset{\omega}{\to} (q_i{}^{\smile}) : i < \omega\}$. By part (c), if $\overline{\overline{U}} \geq 2$, then this closure is also the closure under $\circ$ and $^{\smile}$ of ${}_{[\omega]}UDE$. By Theorems 3(d) and III.14(a) this closure, in turn, coincides with $\{X \circ Y^{\smile} : X \text{ and } Y \text{ are in } {}_{[\omega]}UDE\}$. If $\overline{\overline{U}} = 1$, then all the sets concerned trivially coincide. □

Membership in the set ${}_{[\omega]}S_{pqe}$ was defined, in Chapter II (before Theorem II.12), inductively. In contrast, Theorem 2(a) gives an explicit criterion for membership in ${}_{[\omega]}UD$. Part (d) of the next theorem asserts that the two sets coincide. The final assertion in parts (a), (b), or (c), respectively, is of a similar kind. Recall that, by definition, $p_i = p_{(i,i+1)}$.

THEOREM 6. (a) *$\langle\{\pi : \pi$ a permutation of ω, $\overline{\overline{\{n : \pi(n) \neq n\}}} < \omega\}, \circ, {}^{\smile}\rangle$ is generated by $\{(i, i+1) : i < \omega\}$. There follows that $\langle {}_{[\omega]}UDP, \circ, {}^{\smile}\rangle$ is generated by $\{\overset{\omega}{\to} p_i : i < \omega\}$ and that ${}_{[\omega]}UDP = {}_{[\omega]}S_p$.*

(b) *If N is one-one and bisectable, then $N = f_J{}^{\smile} \circ \pi \circ f_K$ for exactly one J, K and π such that $\overline{\overline{J}} < \omega$, $\overline{\overline{K}} < \omega$, π is a permutation of ω, and $\overline{\overline{\{n : \pi(n) \neq n\}}} < \omega$. There follows that $\langle\{N \in Bsl : N$ is one-one$\}, \circ, {}^{\smile}\rangle$ is generated by $\{f_{\{0\}}\} \cup \{(i, i+1) : i < \omega\}$. There also follows that $\langle {}_{[\omega]}UDF, \circ, {}^{\smile}\rangle$ is generated by $\{\overset{\omega}{\to} q_0\} \cup \{\overset{\omega}{\to} p_i : i < \omega\}$ and that ${}_{[\omega]}UDF = {}_{[\omega]}S_{pq}$.*

(c) *$\langle\{N : N = eq_\omega N,\ \overline{\overline{N \underset{2}{-} \mathring{1}_\omega}} < \omega\}, \bullet\rangle$ is generated by $\{eq_\omega\{\langle i, j\rangle\} : i, j < \omega\}$. There follows that $\langle {}_{[\omega]}UDI, \circ\rangle$ is generated by $\{\overset{\omega}{\to} e_{ij} : i, j < \omega\}$ and that ${}_{[\omega]}UDI = {}_{[\omega]}S_e$.*

(d) *${}_{[\omega]}UD = {}_{[\omega]}S_{pqe} = {}_{[\omega]}S_{qe}$.*

Proof. (a) As is well known, every permutation π of ω such that $\pi(n) \neq n$ for only finitely many n is a finite product $\pi_0 \circ \cdots \circ \pi_m$ of cycles and every cycle is a finite product of transpositions (j, k). Further, if $j < k$, then $(j, k+1) = (j, k) \circ (k, k+1) \circ (j, k)$. There follows by induction that if $j < k$ then (j, k) is a product of transpositions in $\{(i,i+1) : j \leq i < k\}$. Hence if $j > k$, then (j,k) is a product of transpositions $(i,i+1)^{\smile} = (i,i+1)$. Further, $(i,i) = \overset{\circ}{1}_\omega = (0,1) \circ (0,1)$. Hence any (j,k), therefore any cycle, and therefore any permutation π of ω such that $\overline{\overline{\{n : \pi(n) \neq n\}}} < \omega$ is in the closure of $\{(i,i+1) : i < \omega\}$ under $\circ$. Moreover, the set of these permutations π is closed under $^{\smile}$. From Theorem 3(b) there now follows that, if $\overline{\overline{U}} \geq 2$, then ${}_{[\omega]}UDP$ is the closure under $\circ$, and hence under $\circ$ and $^{\smile}$, of $\{\overset{\omega}{\rightarrow} p_{(i,i+1)} : i < \omega\}$. The same conclusion holds trivially if $\overline{\overline{U}} = 1$.

(b) The first assertion can be proved by an argument using Theorem III.15(a) that is similar to the argument using Theorem III.14(a) that was used in the proof of Theorem 5(c). The second assertion then follows from Theorem 5(c), part (a), and $f_{\{i+1\}} = f_{\{i\}} \circ (i,i+1)$. The first part of the third assertion then follows from Theorem 3(a) for the case where $\overline{\overline{U}} \geq 2$ and it holds trivially if $\overline{\overline{U}} = 1$. It also follows from Theorem 4(b), Theorem 5(c), and part (a). From it there follows that ${}_{[\omega]}UDF = {}_{[\omega]}S_{pq}$.

(c) By induction on $\overline{\overline{N \mathbin{\bar{2}} \overset{\circ}{1}_\omega}}$ one proves that $\{eq_\omega\{\langle i,j\rangle\} : i,j < \omega\}$ is a set of generators for $\langle\{N, N = eq_\omega N,\ \overline{\overline{N \mathbin{\bar{2}} \overset{\circ}{1}_\omega}} < \omega\}, \bullet\rangle$. Using Theorem 3(e), one can then prove the second assertion.

(d) The first equality follows from parts (b), (c), and Theorem 4(a). The second follows from Theorems I.1(a) and II.11(a). □

In Chapter I, we introduced the equivalence relation $c_i = q_i^{\smile} \circ q_i$ on ${}^{[i+1,\omega)}U$ and the function $r_i = q_i \circ e_{i(i+1)}$ of replication at place i that maps ${}^{[i+1,\omega)}U$ into ${}^{[i+2,\omega)}U$. If one forms their respective ω-prolongation, one obtains the following equivalence relation on ${}^\omega U$ or operation on ${}^\omega U$, respectively.

$$
\begin{aligned}
\overset{\omega}{\rightarrow} c_i &= (\overset{\omega}{\rightarrow} q_i^{\smile}) \circ (\overset{\omega}{\rightarrow} q_i) \\
&= \{\langle x,y\rangle : |x| = |y| = \omega,\ x_n = y_n \text{ if } n \neq i,\ n < \omega\}\ , \\
\overset{\omega}{\rightarrow} r_i &= (\overset{\omega}{\rightarrow} q_i) \circ (\overset{\omega}{\rightarrow} e_{i(i+1)}) \\
&= \{\langle x,y\rangle : |x| = |y| = \omega,\ y_n = x_n \text{ if } n \leq i,\ y_{n+1} = x_n \text{ if } i \leq n < \omega\} \\
&= \{\langle vwz, vwwz\rangle : |v| = i,\ |w| = 1,\ |z| = \omega\}\ .
\end{aligned}
$$

Let ${}_{[\omega]}S_r$ and ${}_{[\omega]}S_{prc}$ be the closure under $\circ$ and $^{\smile}$ of $\{\overset{\omega}{\rightarrow} r_i : i < \omega\}$ or of $\{p_i : i < \omega\} \cup \{\overset{\omega}{\rightarrow} r_i : i < \omega\} \cup \{\overset{\omega}{\rightarrow} c_i : i < \omega\}$, respectively.

We conclude our discussion of ${}_{[\omega]}UD$ with a result about $\langle {}_{[\omega]}S_r, \circ, ^{\smile}, \supseteq \rangle$ and one about ${}_{[\omega]}S_{prc}$.

THEOREM 7. (a) *The mapping* $\overset{\frown}{+}$ *defined just before Theorem III.17 is an isomorphism between the following:*

$$\langle \{M \in Bsl : M \text{ is one-one and strictly increasing }\}, \circ, {}^{\smile}, \subseteq \rangle \ ,$$

$$\langle \{N \in Bsl : Do\ N = Rg\ N = \omega,\ \ NN^{\smile}N = N,\ \ N \text{ is biordered}\}, \bullet, {}^{\smile}, \supseteq \rangle \ .$$

Hence $\langle {}_{[\omega]}S_r, \circ, {}^{\smile}, \supseteq \rangle$ *is isomorphic to* $\langle {}_{[\omega]}S_q, \circ, {}^{\smile}, \supseteq \rangle$.

(b) ${}_{[\omega]}S_{prc} = {}_{[\omega]}S_{qe} = {}_{[\omega]}UD$.

Proof. (a) The first assertion follows from Theorem III.17. The second assertion then follows from Theorem III.12(a) and III.12(b).

(b) One can verify that $\overset{\omega}{\to} r_i = (\overset{\omega}{\to} q_i) \circ \overset{\omega}{\to} e_{i(i+1)}$, $\overset{\omega}{\to} c_i = (\overset{\omega}{\to} q_i{}^{\smile}) \circ (\overset{\omega}{\to} q_i)$, and $\overset{\omega}{\to} p_i = \overset{\omega}{\to} q_i \circ \overset{\omega}{\to} e_{i(i+2)} \circ \overset{\omega}{\to} (q_{i+2}^{\smile})$. Hence ${}_{[\omega]}S_{prc} \subseteq {}_{[\omega]}S_{qe}$. One can verify that $\overset{\omega}{\to} q_i = (\overset{\omega}{\to} r_i) \circ (\overset{\omega}{\to} c_i)$. Hence ${}_{[\omega]}S_q \subseteq {}_{[\omega]}S_{rc}$ and ${}_{[\omega]}S_{qe} \subseteq {}_{[\omega]}S_{prc}$. Thus ${}_{[\omega]}S_{prc} = {}_{[\omega]}S_{qe}$ and therefore, by Theorem 6(d), ${}_{[\omega]}S_{prc} = {}_{[\omega]}UD$. □

Consider, for a moment, the closure S_{prc} under $\circ$ and ${}^{\smile}$ of the set $\{p_i : i < \omega\} \cup \{r_i : i < \omega\} \cup \{c_i : i < \omega\}$ of relations on ${}^{[0,\omega)}U$. As we saw after Theorem I.1, it contains only relations on ${}^{[1,\omega)}U$ and hence, for example, does not contain q_0 or ${}_{[0,\omega)}\overset{\circ}{1}$. Hence, in contrast to Theorem 7(b), S_{prc} is a proper subset of ${}_{[0,\omega)}D$.

From ${}_{[\omega]}UD$ we now turn to ${}_{[0,\omega)}D$. In Theorem 3, several subsets of the set Bsl of bisectable relations were shown to be the universe of a structure that is isomorphic to a structure whose universe is a subset of ${}_{[\omega]}UD$. We will begin by considering a structure whose universe is the set Bst of bisections. Several subsets of Bst will then be shown to be the universe of a structure that is isomorphic to a structure whose universe is a subset of ${}_{[0,\omega)}D$.

A structure that played a role in Chapter II is the structure $\mathbf{B}$ that is given below. $\mathbf{B}$ results from the bicyclic semigroup $\langle \omega \times \omega, \odot \rangle$ by adjoining the unary operations $\overset{\bullet}{\vee}$ and $\overset{\bullet}{\to}$ and the binary relation $\dot{\geq}$ that were defined when the structure was introduced. We shall refer to $\mathbf{B}$ as ***the expanded bicyclic semigroup*** or, more explicitly, as ***the bicyclic semigroup expanded by*** $\overset{\bullet}{\vee}$, $\overset{\bullet}{\to}$ ***and*** $\dot{\geq}$.

$$\mathbf{B} = \langle \omega \times \omega, \odot, \overset{\bullet}{\vee}, \overset{\bullet}{\to}, \dot{\geq} \rangle$$

As one can verify, the reduct $\langle \omega \times \omega, \odot, \overset{\bullet}{\vee} \rangle$ of $\mathbf{B}$ is an inverse semigroup (cf. [Ho], p.145).

Let A be the set $\{N : N \subseteq \omega \times \omega\}$ of all binary relations on ω. Let $\circ$, ${}^{\smile}$, and id be the relative product, the converse, or the identity operation $\{\langle N, N \rangle : N \in A\}$, respectively, and let $N \supseteq N'$ if and only if $N' \subseteq N$. Then the following structure **A** is of the same similarity type as **B**.

$$\mathbf{A} = \langle \{N : N \subseteq \omega \times \omega\}, \circ, {}^{\smile}, \mathrm{id}, \supseteq \rangle \ .$$

For the direct product $\mathbf{C} = \mathbf{A} \times \mathbf{B}$, whose universe is the set $C = \{N : N \subseteq \omega \times \omega\} \times (\omega \times \omega)$, we will use the following notation:

$$\mathbf{C} = \mathbf{A} \times \mathbf{B} = \langle C, \odot, \overset{\bullet}{\vee}, \overset{\bullet}{\rightarrow}, \dot{\geq} \rangle \ .$$

Recall that Bst is the following subset of C:

$$\begin{aligned} Bst &= \{\langle N, \langle s, t\rangle\rangle : s < \omega,\ t < \omega, \\ &\qquad N = (N \cap (s \times t)) \cup \{\langle s+n, t+n\rangle : n < \omega\}\} \ . \end{aligned}$$

THEOREM 8. (a) *Bst is the universe of a substructure of* $\mathbf{C} = \mathbf{A} \times \mathbf{B}$.
(b) *Bsl is the universe of a substructure of* $\mathbf{A}$.

Proof. Assume that $\langle N, \langle s, t\rangle\rangle$ and $\langle N', \langle s', t'\rangle\rangle$ are in Bst. Then $\langle s, t\rangle$ bisects N and $\langle s', t'\rangle$ bisects N'. First suppose that $t \geq s'$. Let $r = t - s'$. Then

$$\langle s, t\rangle \odot \langle s', t'\rangle = \langle s - t + \max(t, s'), t' - s + \max(t, s')\rangle = \langle s, t' + r\rangle \ .$$

Since $\langle s', t'\rangle$ bisects N', therefore $\langle s' + r, t' + r\rangle = \langle t, t' + r\rangle$ also bisects N'. Thus there hold:

$$\begin{aligned} N &= (N \cap (s \times t)) \cup \{\langle s+n, t+n\rangle : n < \omega\} \ , \\ N' &= (N' \cap (t \times (t' + r))) \cup \{\langle t+n, t'+r+n\rangle : n < \omega\} \ . \end{aligned}$$

Since the range of $N \cap (s \times t)$ is disjoint from the domain of $\{\langle t+n, t'+r+n\rangle : n < \omega\}$ and since the range of $\{\langle s+n, t+n\rangle : n < \omega\}$ is disjoint from the domain of $N' \cap (t \times (t' + r))$, therefore $N \circ N'$ is the union of the following two relations:

$$\begin{gathered} \{\langle s+n, t+n\rangle : n < \omega\} \circ \{\langle t+n, t'+r+n\rangle : n < \omega\} \\ = \{\langle s+n, t'+r+n\rangle : n < \omega\} \ , \\ (N \cap (s \times t)) \circ (N' \cap (t \times (t' + r))) \ . \end{gathered}$$

One can verify that, for any relations N and N', if $\langle i, k\rangle$ is in $(N \cap (s \times t)) \circ (N' \cap (t \times (t' + r)))$, then $\langle i, k\rangle$ is in $(N \circ N') \cap (s \times (t' + r))$. Now assume that $\langle i, k\rangle$ is in $(N \circ N') \cap (s \times (t' + r))$. Then $i < s$, $k < t' + r$, and, for some j, $\langle i, j\rangle \in N$ and $\langle j, k\rangle \in N'$. Since $i < s$, $\langle i, j\rangle \in N$, and $\langle s, t\rangle$ bisects N, therefore

$j < t$. Hence $\langle i, j\rangle$ is in $N \cap (s \times t)$, $\langle j, k\rangle$ is in $N' \cap (t \times (t' \times r))$, and $\langle i, k\rangle$ is in $(N \cap (s \times t)) \circ (N' \cap (t \times (t' + r)))$. This shows that

$$(N \cap (s \times t)) \circ (N' \cap (t \times (t' + r))) = (N \circ N') \cap (s \times (t' + r)) \; .$$

There now follows that

$$N \circ N' = ((N \circ N') \cap (s \times (t' + r))) \cup \{\langle s + n, t' + r + n\rangle : n < \omega\} \; .$$

Thus $\langle s, t' + r\rangle = \langle s, t\rangle \odot \langle s', t'\rangle$ is a bisector of $N \circ N'$. In the case where $t' \leq s$, a similar argument shows that $\langle s, t\rangle \odot \langle s', t'\rangle$ again is a bisector of $N \circ N'$. Hence, in either case, $\langle N, \langle s, t\rangle\rangle \odot \langle N', \langle s', t'\rangle\rangle = \langle N \circ N', \langle s, t\rangle \odot \langle s', t'\rangle\rangle$ is in Bst. Hence Bst is closed under $\odot$.

Assume that $\langle N, \langle s, t\rangle\rangle$ is in Bst. Then $\langle s, t\rangle$ is a bisector of N. There follows that $\langle t, s\rangle$ is a bisector of $N^{\smile}$. Since $\langle N, \langle s, t\rangle\rangle^{\overset{\bullet}{\vee}} = \langle N^{\smile}, \langle s, t\rangle^{\overset{\bullet}{\vee}}\rangle = \langle N^{\smile}, \langle t, s\rangle\rangle$, therefore $\langle N, \langle s, t\rangle\rangle^{\overset{\bullet}{\vee}}$ is in Bst.

Assume that $\langle N, \langle s, t\rangle\rangle$ is in Bst. Then $\langle s, t\rangle$ is a bisector of N. Hence $\langle s + 1, t + 1\rangle$ is a bisector of N. Since $\overset{\bullet}{\rightarrow} \langle N, \langle s, t\rangle\rangle = \langle N, \overset{\bullet}{\rightarrow} \langle s, t\rangle\rangle = \langle N, \langle s + 1, t + 1\rangle\rangle$, therefore $\overset{\bullet}{\rightarrow} \langle N, \langle s, t\rangle\rangle$ is in Bst.

This concludes the proof of part (a). Part (b) follows from it but also can be proved by an argument similar to the one just given, but somewhat simpler. □

Henceforth, we let **Bst** be the substructure of $\mathbf{C} = \mathbf{A} \times \mathbf{B}$ described in Theorem 8(a), and **Bsl** the substructure of $\mathbf{A}$ described in Theorem 8(b).

The next lemma is concerned with the following subset of Bst:

$$Bst \cap -(Rg \overset{\bullet}{\rightarrow}) = \{\langle N, \langle s, t\rangle\rangle \in Bst : \langle s, t\rangle \text{ is the least bisector of } N\}$$

LEMMA 9. (a) *Let S be the following set:*

$$S = \{\langle f, \langle s, t\rangle\rangle \in Bst \cap -(Rg \overset{\bullet}{\rightarrow}) : Do\ f = \omega,\ \ f \textit{ is strictly increasing}\} \; .$$

Then $\langle S, \odot\rangle$ is a subalgebra of $\langle Bst, \odot\rangle$. It is isomorphic to the subalgebra of $\langle Bsl, \circ\rangle$ whose universe is

$$\{f \in Bsl : Do\ f = \omega,\ \ f \textit{ is strictly increasing}\} \; .$$

(b) *If S is as in* (a) *and if $\langle f, \langle s, t\rangle\rangle$ and $\langle f', \langle s', t'\rangle\rangle$ are in S, then $\langle f^{\smile}, \langle t, s\rangle\rangle \odot \langle f', \langle s', t'\rangle\rangle$ is in $Bst \cap -(Rg \overset{\bullet}{\rightarrow})$.*

(c) *If $\langle N, \langle s, s\rangle\rangle$ and $\langle N', \langle s', s'\rangle\rangle$ are in $Bst \cap -(Rg \overset{\bullet}{\rightarrow})$ and if $N = eq_\omega N$ and $N' = eq_\omega N'$, then $\langle N \bullet N', \langle s, t\rangle \odot \langle s', t'\rangle\rangle$ is in $Bst \cap -(Rg \overset{\bullet}{\rightarrow})$.*

Proof. (a) Let $\langle f, \langle s,t\rangle\rangle$ and $\langle f', \langle s',t'\rangle\rangle$ be any elements of S. Let I, J, and K be the subset of ω that corresponds to f, f', or $f \circ f'$, respectively, so that $f = f_I$, $f' = f_J$, and $f \circ f' = f_K$. Since f and f' are in Bsl, therefore I and J are finite. By Theorem 5(a), $K = \{f_J(i) : i \in I\} \cup J$. Since f_J is one-one and since $Rg\ f_J$ and J are disjoint, therefore $\overline{\overline{K}} = \overline{\overline{I}} + \overline{\overline{J}}$.

Let $\langle s'',t''\rangle$ be the least bisector of $f_K = f \circ f'$. Then $t'' = \{0,\ldots,t''-1\}$ is the least initial segment of ω such that $K \subseteq t''$. There follows that $\langle s'',t''\rangle = \langle t'' - \overline{\overline{K}}, t''\rangle$. Similarly, since $\langle s,t\rangle$ and $\langle s',t'\rangle$ are the least bisector of f_I or of f_J, respectively, therefore $\langle s,t\rangle = \langle t - \overline{\overline{I}}, t\rangle$ and $\langle s',t'\rangle = \langle t' - \overline{\overline{J}}, t'\rangle$.

Assume that $I = \emptyset$, and hence $f = f_I = \mathring{1}_\omega$ and $\langle s,t\rangle = \langle 0,0\rangle$. Then $f \circ f' = \mathring{1}_\omega \circ f' = f'$ and hence $\langle s'',t''\rangle = \langle s',t'\rangle = \langle 0,0\rangle \odot \langle s',t'\rangle = \langle s,t\rangle \odot \langle s',t'\rangle$. Hence $\langle s,t\rangle \odot \langle s',t'\rangle$ is the least bisector of $f \circ f'$. If $J = \emptyset$, and hence $f' = f_J = \mathring{1}$ and $\langle s',t'\rangle = \langle 0,0\rangle$, then, by a similar argument, $\langle s,t\rangle \odot \langle s',t'\rangle$ is again the least bisector of $f \circ f'$.

For the rest of the proof assume that $I \neq \emptyset$ and $J \neq \emptyset$, and hence also $K \neq \emptyset$. Then the greatest element of I, J, or K is $t-1$, $t'-1$, or $t''-1$, respectively, and also $s + \overline{\overline{I}} - 1$, $s' + \overline{\overline{J}} - 1$, or $s'' + \overline{\overline{K}} - 1$, respectively.

Suppose that $t > s'$ and hence $t-1 \geq s'$. Since $f_J(n) = n + \overline{\overline{J}}$ for any $n \geq s'$, therefore $f_J(t-1) = t-1+\overline{\overline{J}}$. Since $t \geq s'$, therefore $f_J(t-1) > s'-1+\overline{\overline{J}}$ and hence $f_J(t-1)$ is greater than the greatest element of J. Since $K = \{f_J(i) : i \in I\} \cup J$, there follows that the greatest element of K is $t-1+\overline{\overline{J}}$. The least bisector of $f'' = f_K$ therefore is $\langle t+\overline{\overline{J}}-\overline{\overline{K}}, t+\overline{\overline{J}}\rangle = \langle t-\overline{\overline{I}}, t+\overline{\overline{J}}\rangle = \langle s, t+t'-s'\rangle$. Since $t > s'$, therefore $\langle s,t\rangle \odot \langle s',t'\rangle = \langle s, t'+t-s'\rangle$. There now follows that the least bisector of $f \circ f' = f_K$ again is $\langle s,t\rangle \odot \langle s',t'\rangle$.

Suppose that $t \leq s'$. Then $t-1 < s'$. Since $\langle s',t'\rangle$ is a bisector of $f' = f_J$ and hence $f_J(s') = t'$, there follows that $f_J(t-1) \leq t'-1$. Moreover, since $t'-1$ is in J and since J and $Rg\ f_J$ are disjoint, therefore $f_J(t-1) < t'-1$. Since $K = \{f_J(i) : i \in I\} \cup J$, there now follows that the greatest element of K is $t'-1$. The least bisector of $f'' = f_K$ therefore is $\langle t' - \overline{\overline{K}}, t'\rangle = \langle t' - \overline{\overline{I}} - \overline{\overline{J}}, t'\rangle = \langle t' - (t-s) - (t'-s'), t'\rangle = \langle s+s'-t, t'\rangle$. Since $t \leq s'$, therefore $\langle s,t\rangle \odot \langle s',t'\rangle = \langle s+s'-t, t'\rangle$. There now follows that the least bisector of $f \circ f' = f_K$ again is $\langle s,t\rangle \odot \langle s',t'\rangle$.

This concludes the proof that in all cases, if $\langle f, \langle s,t\rangle\rangle$ and $\langle f', \langle s',t'\rangle\rangle$ belong to S, then the least bisector of $f \circ f'$ is $\langle s,t\rangle \odot \langle s',t'\rangle$, so that $\langle f, \langle s,t\rangle\rangle \odot \langle f', \langle s',t'\rangle\rangle = \langle f \circ f', \langle s,t\rangle \odot \langle s',t'\rangle\rangle$ also belongs to S. Thus $\langle S, \odot\rangle$ is a subalgebra of $\langle Bst, \odot\rangle$.

There follows that the projection map that to every pair $\langle f, \langle s,t\rangle\rangle$ in S assigns f is an isomorphism between the two algebras concerned.

(b) Assume that $\langle f, \langle s,t\rangle\rangle$ and $\langle f', \langle s',t'\rangle\rangle$ are in S. Let I and J be the subset of ω such that $f = f_I$ and $f' = f_J$ respectively. Then $\langle f^{\smile}, \langle t,s\rangle\rangle = \langle f_I{}^{\smile}, \langle t, t - \overline{\overline{I}}\rangle\rangle$ and $\langle f', \langle s',t'\rangle\rangle = \langle f_J, \langle t', t - \overline{\overline{J}}, t'\rangle\rangle$. Now $\langle s'', t''\rangle$ is a bisector of $f_I{}^{\smile} \circ f_J$ if and only if $I \subseteq s''$, $J \subseteq t''$, and $t'' - s'' = \overline{\overline{J}} - \overline{\overline{I}}$. Since $\langle t,s\rangle$ is the least bisector of $f_I{}^{\smile}$ and $\langle s',t'\rangle$ is the least bisector of f_J, there follows that $\langle s'',t''\rangle$ is a bisector of $f_I{}^{\smile} \circ f_J$ if and only if $t \leq s''$, $t' \leq t''$, and $t'' = s'' = \overline{\overline{J}} - \overline{\overline{I}}$.

Suppose that $t - \overline{\overline{I}} \geq t' - \overline{\overline{J}}$ and hence $s \geq s'$. Then $t' \leq t - \overline{\overline{I}} + \overline{\overline{J}}$. Since also $t \leq t$ and $(t - \overline{\overline{I}} + \overline{\overline{J}}) - t = \overline{\overline{J}} - \overline{\overline{I}}$, there follows that $\langle t, t - \overline{\overline{I}} + \overline{\overline{J}}\rangle$ is a bisector of $f_I{}^{\smile} \circ f_J$. Since $t = \{0, \ldots, t-1\}$ is the least initial segment of ω that includes I, therefore $\langle t, t - \overline{\overline{I}} + \overline{\overline{J}}\rangle$ is a bisector of $f_I{}^{\smile} \circ f_J$. Since $s \geq s'$, therefore

$$\begin{aligned}
\langle t,s\rangle \odot \langle s',t'\rangle &= \langle t, t' + s - s'\rangle \\
&= \langle t, t' + (t - \overline{\overline{I}}) - (t' - \overline{\overline{J}})\rangle \\
&= \langle t, t - \overline{\overline{I}} + \overline{\overline{J}}\rangle \,.
\end{aligned}$$

Hence the least bisector of $f^{\smile} \circ f' = f_I{}^{\smile} \circ f_J$ is $\langle t,s\rangle \odot \langle s',t'\rangle$. There follows that $\langle f^{\smile}, \langle s,t\rangle\rangle \odot \langle f', \langle s',t'\rangle\rangle$ is in $Bst \cap -(Rg \overset{\bullet}{\rightarrow})$.

Suppose that $t - \overline{\overline{I}} \leq t' - \overline{\overline{J}}$ and hence $s' \leq s$. Then a similar argument shows that the least bisector of $f_I{}^{\smile} \circ f_J$ is $\langle t' - \overline{\overline{J}} + \overline{\overline{I}}, t'\rangle = \langle t,s\rangle \odot \langle s',t'\rangle$ and hence that $\langle f^{\smile}, \langle s,t\rangle\rangle \odot \langle f', \langle s',t'\rangle\rangle$ again is in $Bst \cap -(Rg \overset{\bullet}{\rightarrow})$.

(c) If either $N = \overset{\circ}{1}_\omega$ or $N' = \overset{\circ}{1}_\omega$ this holds trivially. Now assume that $N \neq \overset{\circ}{1}_\omega$ and $N' \neq \overset{\circ}{1}_\omega$. Then, for some $s \geq 2$, the greatest I such that $\langle i,j\rangle \in N$ for some $j \neq i$ is $s-1$. Also, for some $s' \geq 2$, the greatest i' such that $\langle i',j'\rangle \in N'$ for some $j' \neq i'$ is in $s'-1$. There follows that the greatest i'' such that $\langle i'',j''\rangle \in eq(N \cup N')$ for some $j'' \neq i''$ is $\max(s-1, s'-1)$. Hence the least bisector of $eq(N \cup N')$ is $\langle \max(s,s'), \max(s,s')\rangle = \langle s,s\rangle \odot \langle s',s'\rangle$. □

From Theorem 8(a) there follows that if N and N' are in Bsl and if $\langle s,t\rangle$, $\langle s',t'\rangle$, and $\langle s'',t''\rangle$ are the least bisector of N, N', or $N \circ N'$, respectively, then $\langle s'',t''\rangle \leq \langle s,t\rangle \odot \langle s',t\rangle$. Quite often, in circumstances other than those considered in Lemma 9, $\langle s'',t''\rangle \neq \langle s,t\rangle \odot \langle s',t'\rangle$. For example, given any i, let $N = f_{\{i\}}$ and $N' = f_{\{i\}}{}^{\smile}$. Then the least bisector of N is $\langle i, i+1\rangle$, the least bisector of N' is $\langle i+1, i\rangle$, and $\langle i, i+1\rangle \odot \langle i+1, i\rangle = \langle i,i\rangle$. Also, $N \circ N' = f_{\{i\}} \circ f_{\{i\}}{}^{\smile} = \overset{\circ}{1}_\omega$ and its least bisector is $\langle 0,0\rangle$. This differs from $\langle i,i\rangle$ whenever $i \neq 0$. For a second example, let $i < j < k$, let N be the cycle (i, j, k), i.e., the permutation of ω such that iNj, jNk, kNi, and mNm for any $m \notin \{i, j, k\}$, and let N' be the transposition (i, k). Then for both N and N', the least bisector is $\langle k+1, k+1\rangle$. Since $N \circ N' = (i, j)$, the least bisector of $N \circ N'$ is $\langle j+1, j+1\rangle$, which differs from $\langle k+1, k+1\rangle \odot \langle k+1, k+1\rangle$.

As we saw earlier, between the set Bst of bisections $\langle N, \langle s, t\rangle\rangle$ and the set of triples ${}_s\langle M\rangle_t$ that are finite and special there exists a simple one-one correspondence. Under it there corresponds to any $\langle N, \langle s, t\rangle\rangle$ in Bst the triple ${}_s\langle N \cap (s \times t)\rangle_t$. According to Chapter III, if ${}_s\langle M\rangle_t$ is any triple that is finite and special, then the relation mediately determined by ${}_s\langle M\rangle_t$ is the following relation in ${}_{[0,\omega)}D$:

$$ {}_s[\![M]\!]_t = \{\langle xz, yz\rangle : |x| = s,\ |y| = t,\ |z| < \omega,\ M \subseteq xy^{\smile}\} . $$

Given any $\langle N, \langle s,t\rangle\rangle$ in Bst, the relation ***mediately determined by*** $\langle N, \langle s,t\rangle\rangle$ shall be the following relation in ${}_{[0,\omega)}D$, for which it will be convenient to use two notations:

$$ \begin{aligned} [\![\langle N, \langle s, t\rangle\rangle]\!] &= [\![N, \langle s, t\rangle]\!] \\ &= \{\langle xz, yz\rangle : |x| = s,\ |y| = t,\ |z| < \omega,\ N \cap (s \times t) \subseteq xy^{\smile}\} . \end{aligned} $$

Henceforth, use of $[\![\langle N, \langle s, t\rangle\rangle]\!]$ or of $[\![N, \langle s, t\rangle]\!]$ will tacitly assume that $\langle N, \langle s, t\rangle\rangle$ is in Bst. From the definitions involved there follows that any $\langle N, \langle s, t\rangle\rangle$ in Bst and the corresponding ${}_s\langle N \cap (s \times t)\rangle_t$ mediately determine the same special diagonal relation in ${}_{[0,\omega)}D$. In other words, there holds the following equality, of which henceforth we will make free use:

$$ [\![N, \langle s, t\rangle]\!] = {}_s[\![N \cap (s \times t)]\!]_t $$

From Lemma III.5 one readily obtains the following.

LEMMA 10. *Consider any* $\langle N, \langle s, t\rangle\rangle$ *and* $\langle N', \langle s, t\rangle\rangle$ *in* Bst.

(a) *If* $N \subseteq N'$, *then* $[\![N', \langle s, t\rangle]\!] \subseteq [\![N, \langle s, t\rangle]\!]$.

(b) *dfc* $N = (\textit{dfc}(N \cap (s \times t))) \cup \{\langle s + n, t + n\rangle : n < \omega\}$. *Hence* $[\![N, \langle s, t\rangle]\!] = [\![\textit{dfc}\ N, \langle s, t\rangle]\!]$.

(c) *Assume that* $\overset{=}{U} \geq 2$. *If* N *is difunctional, then* $[\![N, \langle s, t\rangle]\!] \subseteq [\![N', \langle s, t\rangle]\!]$ *implies that* $N' \subseteq N$. *Hence if* N *and* N' *both are difunctional, then* $[\![N, \langle s, t\rangle]\!] = [\![N', \langle s, t\rangle]\!]$ *if and only if* $N = N'$. *There follows that* $[\![\ ]\!]$ *is a one-one correspondence between* $\{\langle N, \langle s, t\rangle\rangle \in Bst : N \textit{ is difunctional}\}$ *and* $\{W \in {}_{[0,\omega)}D : W \textit{ is special}\}$. □

In some cases, a restriction of the one-one correspondence in Lemma 10(c) is an isomorphism. The next theorem describes some of these. The relationships between Bst and ${}_{[0,\omega)}D$ it describes are mostly similar to those between Bsl and ${}_{[\omega]}UD$ that were described in Theorem 3.

The first of the two structures in each of the five pairs given below is always either a substructure of **Bst** or a substructure of a reduct of **Bst**.

THEOREM 11. *Assume that* $\overset{=}{U} \geq 2$. *In* (e) *below, let* S *be the closure, under* $\circ$ *and* $\smile$, *either of* ${}_{[0,\omega)}DE \cap -(Rg \overset{1}{\rightarrow})$ *or of* ${}_{[0,\omega)}DE$. *Then the second structure in* (a)... (e) *below is the isomorphic image under* $[\![\ ]\!]$ *of the first.*

(a) $\langle \{\langle N, \langle s,t \rangle\rangle \in Bst : N \text{ is one-one}\}, \odot, \overset{\bullet}{\vee}, \overset{\bullet}{\rightarrow}, \geq \rangle$
$\langle {}_{[0,\omega)}DF, \circ, \smile, \overset{1}{\rightarrow}, \subseteq \rangle$

(b) $\langle \{\langle \pi, \langle s,s \rangle\rangle \in Bst : \pi \text{ is a permutation of } \omega\}, \odot, \overset{\bullet}{\vee}, \overset{\bullet}{\rightarrow}, \geq \rangle$
$\langle {}_{[0,\omega)}DP, \circ, \smile, \overset{1}{\rightarrow}, \subseteq \rangle$

(c) $\langle \{\langle f^{\smile}, \langle s,t \rangle\rangle \in Bst \cap -(Rg \overset{\bullet}{\rightarrow}) :$
$Do\ f = \omega,\ f \text{ is strictly increasing}\}, \odot \rangle$
$\langle {}_{[0,\omega)}DE \cap -(Rg \overset{1}{\rightarrow}), \circ \rangle$

(d) $\langle \{\langle f^{\smile}, \langle s,t \rangle\rangle \in Bst : Do\ f = \omega,\ f \text{ is strictly increasing}\}, \odot, \overset{\bullet}{\rightarrow}, \geq \rangle$
$\langle {}_{[0,\omega)}DE, \circ, \overset{1}{\rightarrow}, \subseteq \rangle$

(e) $\langle \{\langle N, \langle s,t \rangle\rangle \in Bst : N \text{ and } N^{\smile} \text{ are order-preserving}\}, \odot, \overset{\bullet}{\vee}, \overset{\bullet}{\rightarrow}, \geq \rangle$
$\langle S, \circ, \smile, \overset{1}{\rightarrow}, \subseteq \rangle$

Proof. (a) Assume that $\langle N, \langle s,t \rangle\rangle$ and $\langle N', \langle s',t' \rangle\rangle$ are in Bst and that N and N' are one-one. From the definitions involved there follows that to prove the first equality below it suffices to prove the equality that is given underneath.

$$\begin{aligned} [\![\langle N, \langle s,t\rangle\rangle \odot \langle N', \langle s',t'\rangle\rangle]\!] &= [\![N, \langle s,t\rangle]\!] \circ [\![N', \langle s',t'\rangle]\!] \\ [\![N \circ N', \langle s,t\rangle \odot \langle s',t'\rangle]\!] &= {}_s[\![N \cap (s \times t)]\!]_t \circ {}_{s'}[\![N' \cap (s' \times t')]\!]_{t'} \end{aligned}$$

First suppose that $t \geq s'$. Let $r = t - s'$. By Theorems II.7(g) and II.7(f), by Lemma III.9(a), by Lemma III.8(b), and by the proof of Theorem 8(a), respectively:

$$\begin{aligned} &{}_s[\![N \cap (s \times t)]\!]_t \circ {}_{s'}[\![N' \cap (s' \times t')]\!]_{t'} \\ &= {}_s[\![N \cap (s \times t)]\!]_t \circ \overset{r}{\rightarrow} {}_{s'}[\![N' \cap (s' \times t')]\!]_{t'} \\ &= {}_s[\![N \cap (s \times t)]\!]_t \circ {}_t[\![N' \cap (t \times (t'+r))]\!]_{t'+r} \\ &= {}_s[\![(N \cap (s \times t)) \circ (N' \cap (t' \times (t'+r))]\!]_{t'+r} \\ &= {}_s[\![(N \circ N') \cap (s \times (t'+r))]\!]_{t'+r} \end{aligned}$$

Since, by the definitions involved,

$$\begin{aligned} {}_s[\![(N \circ N') \cap (s \times (t'+r))]\!]_{t'+r} &= [\![\langle N \circ N', \langle s, t'+r\rangle\rangle]\!] \\ &= [\![\langle N \circ N', \langle s,t\rangle \odot \langle s',t'\rangle\rangle]\!] \ , \end{aligned}$$

there now follows the desired equality. If $t \leq s'$, then the desired equality follows by a similar argument.

The following equalities follow mostly from the definitions involved:

$$\begin{aligned} [\![\langle N, \langle s,t\rangle\rangle^{\overset{\bullet}{\vee}}]\!] &= [\![\langle N^{\smile}, \langle t,s\rangle\rangle]\!] = {}_t[\![N^{\smile} \cap (t \times s)]\!]_s \\ &= {}_t[\![(N \cap (s \times t))^{\smile}]\!]_s \\ &= ({}_s[\![N \cap (s \times t)]\!]_t)^{\smile} = [\![N, \langle s,t\rangle]\!]^{\smile} \ . \end{aligned}$$

The third equality below follows from Lemma III.9(a), while the others follow from the definitions involved:

$$\begin{aligned} [\![\overset{\bullet}{\rightarrow} \langle N, \langle s,t\rangle\rangle]\!] &= [\![\langle N, \langle s+1, t+1\rangle\rangle]\!] \\ &= {}_{s+1}[\![N \cap ((s+1) \times (t+1))]\!]_{t+1} \\ &= \overset{1}{\longrightarrow} {}_s[\![(N \cap (s \times t))]\!]_t = \overset{1}{\longrightarrow} [\![\langle N, \langle s \times t\rangle\rangle]\!] \ . \end{aligned}$$

Assume that $\langle N, \langle s,t\rangle\rangle \cdot\!\geq \langle N', \langle s',t'\rangle\rangle$. Then $N \supseteq N'$. Also, $\langle s,t\rangle \cdot\!\geq \langle s',t'\rangle$ and hence $\langle s,t\rangle = \langle s'+k, t'+k\rangle$, for some k. Then:

$$\begin{aligned} [\![N, \langle s,t\rangle]\!] &= {}_s[\![N \cap (s \times t)]\!]_t = {}_{s'+k}[\![N \cap ((s' \times k) \times (t'+k))]\!]_{t'+k} \\ &\subseteq {}_{s'}[\![N \cap ((s'+k) \times (t'+k))]\!]_{t'} \subseteq {}_{s'}[\![N' \cap (s' \times t')]\!]_{t'} \\ &= [\![N', \langle s',t'\rangle]\!] \end{aligned}$$

where the second equality holds by Theorems II.7(d) and II.7(f), the first inclusion by Lemma III.5(a), and the second inclusion by $N' \subseteq N$ and Lemma III.5(a). This shows that $\langle N, \langle s,t\rangle\rangle \geq \langle N', \langle s',t'\rangle\rangle$ implies that $[\![\langle N, \langle s,t\rangle\rangle]\!] \subseteq [\![\langle N', \langle s',t'\rangle\rangle]\!]$.

Assume that $[\![\langle N, \langle s,t\rangle\rangle]\!] \subseteq [\![\langle N', \langle s',t'\rangle\rangle]\!]$. Making use also of our assumption that $\overset{=}{U} \geq 2$ and that N is one-one, we now want to show that $\langle s,t\rangle \geq \langle s',t'\rangle$ and that $N \supseteq N'$, and hence that $\langle N, \langle s,t\rangle\rangle \geq \langle N', \langle s',t'\rangle\rangle$. The above inclusion is equivalent to ${}_s[\![N \cap (s \times t)]\!]_t \subseteq {}_{s'}[\![N' \cap (s' \times t')]\!]_{t'}$. From this inclusion and from Lemma II.5(e) there follows that $\langle s,t\rangle \geq \langle s',t'\rangle$ and hence that $\langle s,t\rangle = \langle s'+k, t'+k\rangle$, for some k. There follows that

$$\begin{aligned} &{}_s[\![N \cap (s \times t)]\!]_t \\ &= {}_{[s'+k,\omega)}\overset{\circ}{1} \circ {}_s[\![N \cap (s \times t)]\!]_t \subseteq {}_{[s'+k,\omega)}\overset{\circ}{1} \circ {}_{s'}[\![N' \cap (s' \times t')]\!]_{t'} \ . \end{aligned}$$

By Theorem II.7(f) or Lemma II.9(a), respectively:

$$\begin{aligned}
&{}_{[s'+k,\omega)}\overset{\circ}{1} \circ {}_{s'}[\![N' \cap (s' \times t')]\!]_{t'} \\
&= \overset{k}{\rightarrow}({}_{s'}[\![N' \cap (s' \times t')]\!]_{t'}) \\
&= {}_{s}[\![(N' \cap (s' \times t')) \cup \{\langle s'+n, t'+n\rangle : n<k\}]\!]_{t} \ .
\end{aligned}$$

There now follows that

$${}_{s}[\![N \cap (s \times t)]\!]_{t} \subseteq {}_{s}[\![(N' \cap (s' \times t')) \cup \{\langle s'+n, t'+n\rangle : n < k\}]\!]_{t} \ .$$

Since $\overset{=}{U} \geq 2$ and since N and hence $N \cap (s \times t)$ is one-one, therefore by Lemma III.5(c), this last inclusion implies the following:

$$(N' \cap (s' \times t')) \cup \{\langle s'+n, t'+n\rangle : n < k\} \subseteq N \cap (s \times t) \ .$$

Since $\langle s, t\rangle = \langle s'+k, t'+k\rangle$, since $\langle N', \langle s', t'\rangle\rangle$ is in Bst, and since $\langle N, \langle s, t\rangle\rangle$ is in Bst, therefore the following hold:

$$\begin{aligned}
N' &= (N' \cap (s' \times t')) \cup \{\langle s'+n, t'+n\rangle : n < k\} \\
&\quad \cup \{\langle s+n, t+n\rangle : n < \omega\} \ , \\
N &= (N \cap (s \times t)) \cup \{\langle s+n, t+n\rangle : n < \omega\} \ .
\end{aligned}$$

From these two equalities and from the preceding inclusion there follows that $N \supseteq N'$. This completes our proof that $[\![\ ]\!]$ is an isomorphism between $\langle\{\langle N, \langle s, t\rangle\rangle \in Bst : N \text{ is one-one}\}, \geq\rangle$ and $\langle {}_{[0,\omega)}DF, \subseteq\rangle$ and thereby also our proof that it is an isomorphism between the two ordered algebras concerned.

(b) Assume that $\langle \pi, \langle s, s\rangle\rangle$ is in Bst and that π is a permutation of ω. Then $\pi \cap (s \times s)$ is a permutation of $s = \{0, \ldots, s-1\}$ and $[\![\pi, \langle s, s\rangle]\!] = {}_{s}[\![\pi \cap (s, s)]\!]$ is in ${}_{[0,\omega)}DP$. Now assume that W is in ${}_{[0,\omega)}DP$. Then, for some $s < \omega$ and M, M is a permutation of $\{0, \ldots, s-1\}$ and $W = {}_{s}[\![M]\!]$. Let $\pi = M \cup \{\langle s+n, t+n\rangle : n < \omega\}$. Then $W = [\![\langle \pi, \langle s, s\rangle\rangle]\!]$. There now follows:

$${}_{[0,\omega)}DP = \{[\![\langle \pi, \langle s, s\rangle\rangle]\!] : \langle \pi, \langle s, s\rangle\rangle \in Bst,\ \pi \text{ is a permutation of } \omega\} \ .$$

From this, since ${}_{[0,\omega)}DP$ is closed under $\circ$, $\smile$, and $\overset{1}{\rightarrow}$, and from part (a), there follows part (b).

(c) Assume that $\langle f^{\smile}, \langle s, t\rangle\rangle$ is in Bst and that $Do\ f = \omega$, f is strictly increasing and $\langle s, t\rangle$ is the least bisector of $f^{\smile}$. Let $f' = f \cap (t \times s)$. Then f' is strictly increasing, $Do\ f' = t$ and $Rg\ f' \subseteq s$. Moreover, since $\langle s, t\rangle$ is the least bisector of $f^{\smile}$, therefore either $s = 0$, $f = \overset{\circ}{1}_{\omega}$, and $f' = \emptyset$, or else $s \geq 1$ and $s - 1 \notin Rg\ f'$. There follows that $[\![\langle f^{\smile}, \langle s, t\rangle\rangle]\!] = {}_{s}[\![f'^{\smile}]\!]_{t}$ is the $(s \cap -Rg\ f')$-excision from ${}^{[s,\omega)}U$

onto ${}^{[t,\omega)}U$ and hence is in ${}_{[0,\omega)}DE$. Moreover, either $s = 0$ and $[\![\langle f^{\smile}, \langle s,t\rangle\rangle]\!] = {}_s[\![\emptyset]\!] = {}_{[0,\omega)}\overset{\circ}{1}$, or else $s \geq 1$ and $s-1 \notin Rg\ f'$. In either case, $[\![\langle f^{\smile}, \langle s,t\rangle\rangle]\!] = {}_s[\![f'^{\smile}]\!]_t$ is not in $Rg \overset{1}{\rightarrow}$.

Now assume that W is in ${}_{[0,\omega)}DE$ and W is not in $Rg \overset{1}{\rightarrow}$. If $W = {}_{[0,\omega)}\overset{\circ}{1}$, then $W = [\![\langle \overset{\circ}{1}_\omega, \langle 0,0\rangle\rangle]\!] = [\![\langle \overset{\circ}{1}_\omega{}^{\smile}, \langle 0,0\rangle\rangle]\!]$, where $\overset{\circ}{1}_\omega$ is the function f such that $Do\ f = \omega$, f is strictly increasing, and $\langle 0,0\rangle$ is the least bisector of $f^{\smile}$. Now assume that $W \neq {}_{[0,\omega)}\overset{\circ}{1}$. Then, for some $s \geq 1$ and some $t < s$, W is an excision from ${}^{[s,\omega)}U$ onto ${}^{[t,\omega)}U$. Hence, $W = {}_s[\![f'^{\smile}]\!]_t$ for some f' such that f' is strictly increasing, $Do\ f' = t$, and $Rg\ f' \subseteq s$. Moreover, since $W \notin Rg \overset{1}{\rightarrow}$, therefore $s-1 \notin Rg\ f'$. Let $f = f' \cup \{\langle t+n, s+n\rangle : n \in \omega\}$. Then $\langle s,t\rangle$ is the least bisector of $f^{\smile}$, $Do\ f = \omega$, f is strictly increasing, and $W = [\![\langle f^{\smile}, \langle s,t\rangle\rangle]\!]$.

We have thus shown that ${}_{[0,\omega)}DE \cap -(Rg \overset{1}{\rightarrow})$ is the direct image under $[\![\ \ ]\!]$ of the following subset of Bst:

$$\{\langle f^{\smile}, \langle s,t\rangle\rangle \in Bst : Do\ f = \omega,\ f \text{ is strictly increasing, } \langle s,t\rangle \text{ is the least bisector of } f^{\smile}\}\ .$$

Since, by Lemma 9(a), this subset is closed under $\odot$, there now follows from part (a) that $[\![\ \ ]\!]$ is an isomorphism between the two algebras concerned.

(d) Since the set $\{\langle f^{\smile}, \langle s,t\rangle\rangle : Do\ f = \omega,\ f \text{ is strictly increasing}\}$ is closed under $\odot$ and $\overset{\bullet}{\rightarrow}$, this follows from part (a) and from the following equality, which can be proved by an argument that is similar to the one used for part (c), but somewhat simpler:

$${}_{[0,\omega)}DE = \{[\![f^{\smile}, \langle s,t\rangle\rangle]\!] : Do\ f = \omega,\ f \text{ is strictly increasing}\}\ .$$

(e) For every n, $\dot{q}_n$ is in ${}_{[0,\omega)}DE \cap -(Rg \overset{1}{\rightarrow})$ and hence ${}_{[n,\omega)}\overset{\circ}{1} = q_n q_n{}^{\smile} = \dot{q}_n{}^{\smile}\dot{q}_n$ is in the closure of ${}_{[0,\omega)}DE \cap -(Rg \overset{1}{\rightarrow})$ under $\circ$ and ${}^{\smile}$. For every W' in ${}_{[0,\omega)}DE$ there is some W in ${}_{[0,\omega)}DE \cap -(Rg \overset{1}{\rightarrow})$ and some r such that $W' = \overset{r}{\rightarrow} W$. If W has grade $\langle s,t\rangle$ then, by Theorem II.7(f), $W = {}_{[s+r,\omega)}\overset{\circ}{1} \circ W'$. There now follows that the closure under $\circ$ and ${}^{\smile}$ of ${}_{[0,\omega)}DE \cap -(Rg \overset{1}{\rightarrow})$ and that of ${}_{[0,\omega)}DE$ coincide. Since this closure is closed under $\overset{1}{\rightarrow}$, as well as under $\circ$ and ${}^{\smile}$, there follows from part (a) and Theorem 3(d) that $[\![\ \ ]\!]$ is an isomorphism between the two structures concerned. $\square$

Since $\langle \omega\times\omega, \odot, \overset{\bullet}{\vee} \rangle$ and $\langle \{N \in Bsl : N \text{ is one-one}\}, \circ, {}^{\smile} \rangle$ are inverse semigroups, therefore $\langle \{\langle N, \langle s,t\rangle\rangle \in Bst : N \text{ is one-one}\}, \odot, \overset{\bullet}{\vee} \rangle$ is an inverse semigroup. There follows from Theorem 11(a) that so is $\langle {}_{[0,\omega)}DF, \circ, {}^{\smile} \rangle$.

From Theorem 11(c), Lemma 9(a), and Theorem 3(c), there follows that, if $\overline{\overline{U}} \geq 2$, then $\langle {}_{[0,\omega)}DE \cap -(Rg \overset{1}{\rightarrow}), \circ \rangle$ and $\langle {}_{[\omega]}UDE, \circ \rangle$ are isomorphic.

THEOREM 12. (a) $\langle \{\langle \pi, \langle s,s\rangle\rangle \in Bst : \pi \textit{ is a permutation of } \omega\}, \odot, \overset{\bullet}{\vee} \rangle$ *is generated by* $\{\langle \overset{\circ}{1}_\omega, \langle i,i\rangle\rangle : i \leq 1\} \cup \{\langle (i,i+1), \langle i+2, i+2\rangle\rangle : i < \omega\}$. *There follows that* $\langle {}_{[0,\omega)}DP, \circ, {}^{\smile} \rangle$ *is generated by* $\{{}_{[i,\omega)}\overset{\circ}{1} : i \leq 1\} \cup \{p_i : i < \omega\}$ *and that* $S_p = {}_{[0,\omega)}DP \cap -\{{}_{[i,\omega)}\overset{\circ}{1} : i \leq 1\}$.

(b) $\langle \{\langle N, \langle s,t\rangle\rangle \in Bst : N \textit{ and } N^{\smile} \textit{ are order-preserving}\}, \odot, \overset{\bullet}{\vee} \rangle$ *is generated by* $\{\langle f_{\{i\}}, \langle i, i+1\rangle\rangle : i < \omega\}$. *There follows that* S_q *coincides with the closure under* $\circ$ *and* ${}^{\smile}$ *of* ${}_{[0,\omega)}DE$.

(c) $\langle \{\langle N, \langle s,t\rangle\rangle \in Bst : N \textit{ is one-one}\}, \odot, \overset{\bullet}{\vee} \rangle$ *is generated by*

$$\{(i,i+1), \langle i+2, i+2\rangle\rangle : i < \omega\} \cup \{\langle f_{\{0\}}, \langle 0,1\rangle\rangle\} .$$

There follows that $S_{pq} = {}_{[0,\omega)}DF$.

Proof. (a) Let S, S_0, and G' be the following sets:

$$\begin{aligned} S &= \{\langle \pi, \langle s,s\rangle\rangle \in Bst : \pi \text{ is a permutation of } \omega\} , \\ S_0' &= \{\langle \pi, \langle s,s\rangle\rangle \in Bst : \pi \neq \overset{\circ}{1}_\omega,\ \langle s,s\rangle \text{ is the least bisector of } \pi\} , \\ G' &= \{\langle (i,i+1), \langle i+2, i+2\rangle\rangle : i < \omega\} . \end{aligned}$$

We first want to show that every $\langle \pi, \langle s,s\rangle\rangle$ in S_0' is in the closure of G' under $\odot$ and $\overset{\bullet}{\vee}$. If $\pi = (i,i+1)$ then, for $s = i+2$, $\langle \pi, \langle s,s\rangle\rangle$ is in S_0' and also in G'. If $\pi = (i+1,i)$ then, for $s = i+2$, $\langle \pi, \langle s,s\rangle\rangle$ is in S_0' and also in the closure of G' under $\overset{\bullet}{\vee}$. Now let $j < k-1$ and let $\pi = (j,k)$ so that, for $s = \langle k+1, k+1\rangle$, $\langle \pi, \langle s,s\rangle\rangle$ is in S_0'. Then

$$\begin{aligned} &\langle (j,j+1), \langle j+2, j+2\rangle\rangle \odot \cdots \odot \langle (k-1,k), \langle k+1, k+1\rangle\rangle \\ &\quad = \langle (j,k), \langle k+1, k+1\rangle\rangle . \end{aligned}$$

Hence $\langle \pi, \langle s,s\rangle\rangle$ is in the closure of G' under $\odot$. There follows that $\langle (k,j), \langle k+1, k+1\rangle\rangle = \langle (j,k), \langle k+1, k+1\rangle\rangle^{\overset{\bullet}{\vee}}$ is in the closure of G' under $\odot$ and $\overset{\bullet}{\vee}$. Now for $m \geq 2$ and for distinct $j_0, \ldots, j_m$, let π be the cycle $(j_0, j_1) \circ \cdots \circ (j_{m-1}, j_m) \circ (j_m, j_0)$. Let $s = 1 + \max(j_0, j_1, \ldots, j_m)$, so that $\langle \pi, \langle s,s\rangle\rangle$ is in S_0'. From what was shown above about every $\langle (j,k), \langle k+1, k+1\rangle\rangle$ and every $\langle (k,j), \langle k+1, k+1\rangle\rangle$ such that

$j < k$, there follows that $\langle \pi, \langle s,s\rangle\rangle$ is in the closure of G' under $\odot$ and $\overset{\bullet}{\vee}$. Finally, let $n \geq 2$, for each $n' < n$ let $\pi_{n'}$ be either some (i,j) or (j,i) such that $i \neq j$ or a cycle $(j_0, j_1) \circ \cdots \circ (j_{m-1}, j_m) \circ (j_m, j_0)$ as above. Moreover, for $n' < n'' < n$, let $\pi_{n'}$ and $\pi_{n''}$ be disjoint in the sense that if (j', k') is a factor of $\pi_{n'}$ and (j'', k'') a factor of $\pi_{n''}$, then $\{j', k'\} \cap \{j'', k''\} = \emptyset$. Let

$$s = 1 + \max\{\max\{j', k'\} : (j', k') \text{ is a factor of } \pi_{n'}\} : n' < n\}$$

Then $\langle \pi, \langle s,s\rangle\rangle$ is in S_0' and, from what was shown previously, there follows that $\langle \pi, \langle s,s\rangle\rangle$ is in the closure of G' under $\odot$ and $\overset{\bullet}{\vee}$. This concludes our proof that S_0' is included in this closure.

This closure also contains every

$$\langle \overset{\circ}{1}_\omega, \langle i+2, i+2\rangle\rangle = \langle (i,i+1), \langle i+2, i+2\rangle\rangle \odot \langle (i,i+1), \langle i+2, i+2\rangle\rangle^{\overset{\bullet}{\vee}} ,$$

hence every $\langle \pi, \langle s,s\rangle\rangle \odot \langle \overset{\circ}{1}_\omega, \langle s+n, s+n\rangle\rangle = \langle \pi, \langle s+n, s+n\rangle\rangle$, such that $\langle \pi, \langle s,s\rangle\rangle$ is in S_0', and hence every $\langle \pi, \langle s,s\rangle\rangle$ in $S \cap -\{\langle \overset{\circ}{1}_\omega, \langle i,i\rangle\rangle : i \leq 1\}$. There now follows the first assertion. From Theorem 11(b) there follows that the second assertion holds if $\overline{\overline{U}} \geq 2$. If $\overline{\overline{U}} = 1$, the second assertion holds trivially.

(b) Let S, S' and G be the following sets:

$$\begin{aligned} S &= \{\langle N, \langle s,t\rangle\rangle \in Bst : N \text{ and } N^{\smile} \text{ are order-preserving } \} , \\ S' &= \{\langle N, \langle s,t\rangle\rangle \in Bst \cap -(Rg \overset{\bullet}{\rightarrow}) : N \text{ and } N^{\smile} \text{ are order preserving}\} , \\ G &= \{\langle f_{\{i\}}, \langle i, i+1\rangle\rangle : i < \omega\} . \end{aligned}$$

By Lemma 9(b) and Theorem 5(c), the closure of G under $\odot$ and $\overset{\bullet}{\vee}$ includes S'. It also contains every $\langle f_{\{i\}}, \langle i, i+1\rangle\rangle \odot \langle f_{\{i\}}, \langle i, i+1\rangle\rangle^{\overset{\bullet}{\vee}} = \langle \overset{\circ}{1}_\omega, \langle i,i\rangle\rangle$, hence every $\langle \overset{\circ}{1}_\omega, \langle s+n, s+n\rangle\rangle \odot \langle N, \langle s,t\rangle\rangle$ such that $\langle N, \langle s,t\rangle\rangle$ is in S', and hence every $\langle N, \langle s,t\rangle\rangle$ in S. It therefore coincides with S.

Since $[\![\langle f_{\{i\}}, \langle i, i+1\rangle\rangle]\!] = q_i$ and $[\![\langle f_{\{i\}}, \langle i, i+1\rangle\rangle^{\overset{\bullet}{\vee}}]\!] = q_i^{\smile}$, therefore, by Theorem 11(e), the second assertion holds if $\overline{\overline{U}} \geq 2$. If $\overline{\overline{U}} = 1$, it holds trivially.

(c) The first assertion follows from parts (a) and (b) and from

$$\langle f_{\{i+1\}}, \langle i+1, i+2\rangle\rangle = \langle f_{\{i\}}, \langle i, i+1\rangle\rangle \odot \langle (i,i+1), \langle i+2, i+2\rangle\rangle .$$

By Theorem 11(a), it implies the second assertion. □

THEOREM 13. (a) $S_e = {}_{[0,\omega)}DI \cap -\{{}_{[0,\omega)}\overset{\circ}{1}, {}_{[1,\omega)}\overset{\circ}{1}\}$.

(b) $S_{qe} = S_{pqe} = {}_{[0,\omega)}D$.

Proof. (a) Since ${}_{[0,\omega)}DI = \{{}_s[\![M]\!] : M = eq_sM,\ s < \omega\}$, therefore ${}_{[0,\omega)}DI \cap -(Rg \stackrel{1}{\to})$ is the following set:

$$ {}_{[0,\omega)}\mathring{1} \cup \{{}_s[\![M]\!] : M = eq_sM,\ 2 \leq s < \omega,\ \langle i, s-1\rangle \in M \text{ for some } i < s-1\} $$

Using induction on $M \cap -\mathring{1}_s$ one proves that if ${}_s[\![M]\!]$ is in ${}_{[0,\omega)}DI \cap -(Rg \stackrel{1}{\to})$ and ${}_s[\![M]\!] \neq {}_{[0,\omega)}\mathring{1}$, then ${}_s[\![M]\!]$ is in the closure under $\circ$ of $\{e_{ij} : 0 \leq i < j < \omega\}$. Since $e_{ii} = {}_{[i+1,\omega)}\mathring{1}$, for every $i < \omega$, therefore, for every W in ${}_{[0,\omega)}DI \cap -\{{}_{[0,\omega)}\mathring{1}, {}_{[1,\omega)}\mathring{1}\}$, there is some W' in ${}_{[0,\omega)}DI \cap -(Rg \stackrel{1}{\to})$ and some e_{ii} such that $W = W' \circ e_{ii}$. There follows that ${}_{[0,\omega)}DI \cap -\{{}_{[0,\omega)}\mathring{1}, {}_{[1,\omega)}\mathring{1}\} = S_e$.

(b) By Theorem I.1(a), $S_{qe} = S_{pqe}$. By part (a), Theorem 12(c), and Theorem III.11(a), $S_{pqe} = {}_{[0,\omega)}D$. □

CHAPTER V

Presentation of $\mathbf{S}_q$, $\mathbf{S}_p$, and Related Structures

Throughout this and the next two chapters we shall assume that the base set U satisfies $\overline{\overline{U}} \geq 2$. Then, by Theorem III.11(a), there is up to isomorphism, only one algebra $\mathbf{S}_{pqe} = \langle S_{pqe}, \circ, ^{\smile}\rangle = \langle {}_{[0,\omega)}D, \circ, ^{\smile}\rangle$ and likewise only one ordered algebra $\langle \mathbf{S}_{pqe}, \subseteq\rangle$. In Chapter VII, these two structures will be given a presentation. In preparation, but also because of their intrinsic interest, we will give, in this and the next chapter, presentations of various substructures.

To round out the picture, for each of these various structures we will also give a presentation of its homomorphic image under $\xrightarrow{\omega}$. In each case, the two presentations turn out to be closely related. To further round out the picture, in Theorems 1, 7, 12 below we also give presentations of some further structures.

To save space, we will present ${}_{[\omega]}\mathbf{S}_q$, ${}_{[\omega]}\mathbf{S}_{pq}$, ${}_{[\omega]}\mathbf{S}_{pe}, \ldots$ (but not ${}_{[\omega]}\mathbf{S}_p$) by utilizing our presentation of $\mathbf{S}_q, \mathbf{S}_{pq}, \mathbf{S}_{pe}, \ldots$, respectively. This reverses the order of discovery. In fact, we first found a presentation of ${}_{[\omega]}\mathbf{S}_q$, ${}_{[\omega]}\mathbf{S}_{pq}$, ${}_{[\omega]}\mathbf{S}_{pe}, \ldots$ and then made the changes needed to take grade into account and thereby obtain a presentation of $\mathbf{S}_q, \mathbf{S}_{pq}, \mathbf{S}_{pe}, \ldots$. Helpful for finding a presentation of ${}_{[\omega]}\mathbf{S}_q, {}_{[\omega]}\mathbf{S}_{pq}, {}_{[\omega]}\mathbf{S}_{pe}$ were the following, respectively: Theorem IV.3(d), IV.5(d), and IV.5(e); Theorem IV.3(a) and IV.6(b); Theorem IV.3(f) and III.16.

Each of the algebras $\mathbf{S}_q, \ldots, \mathbf{S}_{pqe}$ and each of its homomorphic images ${}_{[\omega]}\mathbf{S}_q, \ldots, {}_{[\omega]}\mathbf{S}_{pqe}$ will be presented as a semigroup with involution. More precisely, for any nonempty set Γ, let $\mathbf{F}_{SI}(\Gamma)$ be the free semigroup with involution having Γ as a free generating set. Then each of the above algebras $\mathbf{S}$ will be presented by giving a set Γ freely generating $\mathbf{F}_{SI}(\Gamma)$, a homomorphism h of $\mathbf{F}_{SI}(\Gamma)$ onto $\mathbf{S}$, induced by mapping Γ onto the chosen set of generators of $\mathbf{S}$, and a binary relation R on the universe $F_{SI}(\Gamma)$ of $\mathbf{F}_{SI}(\Gamma)$ such that the following holds:

> If $\equiv$ is the congruence relation on $\mathbf{F}_{SI}(\Gamma)$ that is generated by R, then $\equiv$ coincides with the kernel $h \circ h^{\smile}$ of h and hence, if h' is the natural homomorphism of $\mathbf{F}_{SI}(\Gamma)$ onto $\mathbf{F}_{SI}(\Gamma)/\equiv$, and h'' is

the unique map such that $h' \circ h'' = h$, then h'' is an isomorphism between $\mathbf{F}_{SI}(\Gamma)/\!\equiv$ and $\mathbf{S}$.

Each of the corresponding ordered algebras $\langle \mathbf{S}, \subseteq \rangle$ will be presented by giving, in addition to the Γ, h, and R that together present $\mathbf{S}$, a binary relation R' on $F_{SI}(\Gamma)$ such that, if $\leq$ is the least quasiordering of $F_{SI}(\Gamma)$ that includes $R \cup R^{\smile} \cup R'$ and is compatible with $\mathbf{F}_{SI}(\Gamma)$, then $\leq$ coincides with the inverse image $\{\langle x, y \rangle : h(x) \subseteq h(y)\}$ of $\subseteq$ under h. Hence if $\leq /\!\equiv$ is the relation $\{\langle x/\!\equiv, y/\!\equiv \rangle : \langle x, y \rangle \in \leq\}$ on $F_{SI}(\Gamma)/\!\equiv$, then the unique map h'' such that $h' \circ h'' = h$ is an isomorphism between $\langle \mathbf{F}_{SI}(\Gamma)/\!\equiv, \leq /\!\equiv \rangle$ and $\langle \mathbf{S}, \subseteq \rangle$.

As regards the algebras $\mathbf{S}$, our definition of their presentation is the usual one, provided that the class $\mathbf{K}$ of algebras $\mathbf{S}$ that is involved contains algebras $\mathbf{F}$ that are free for $\mathbf{K}$. Our notion of presenting ordered algebras $\langle \mathbf{S}, \subseteq \rangle$ may call for an explanation. If Γ, h and R are related to $\mathbf{S}$ as described above, then one may think of any x in $F_{SI}(\Gamma)$ as denoting the element $h(x)$ of S, of any $\langle x, y \rangle$ in R as the true assertion that $h(x) = h(y)$, and of $\equiv$ as the set of all true assertions of this kind. Thus, as is suggested by Mal'cev in section 11.2 of [Ma], one may think of $\equiv$ as the positive atomic diagram of $\mathbf{S}$ and of R as an axiomatization of this diagram. Similarly, if Γ, h, R and R' are related to $\langle \mathbf{S}, \subseteq \rangle$ as described above, then one may think of any $\langle x, y \rangle$ in $R \cup R^{\smile} \cup R'$ as the true assertion that $h(x) \subseteq h(y)$, and of $\leq$ as the set of all true assertions of this kind. Thus one may think of $\leq$ as the positive atomic diagram of $\langle \mathbf{S}, \subseteq \rangle$ and of $R \cup R^{\smile} \cup R'$, and hence also of $\equiv \cup R'$, as an axiomatization of this diagram. (According to current usage, the set of atomic sentences that are true in $\mathbf{S}$ is ***the positive atomic diagram of*** $\mathbf{S}$, whereas ***the atomic diagram of*** $\mathbf{S}$ is the larger set of those sentences ψ true in $\mathbf{S}$ such that ψ is either an atomic sentence or the negation of an atomic sentence. Although the phrase used in [Ma] is 'atomic diagram', according to current usage the notion employed there is that of positive atomic diagram.)

Suppose that one replaces the above condition that $\leq$ coincides with the inverse image of $\subseteq$ under h by the weaker condition that $\subseteq$ is the direct image $\{\langle h(x), h(y) \rangle : \langle x, y \rangle \in \leq\}$ of $\leq$ under h. Then it will still be the case that the unique map h'' such that $h' \circ h'' = h$ is an isomorphism between $\langle \mathbf{F}_{SI}(\Gamma)/\!\equiv, \leq /\!\equiv \rangle$ and $\langle \mathbf{S}, \subseteq \rangle$. However, there will then be no guarantee that $\leq$ contains all pairs $\langle x, y \rangle$ such that $h(x) \subseteq h(y)$, thus serving as the positive atomic diagram of $\langle \mathbf{S}, \subseteq \rangle$.

Very fortunately, free semigroups with involution turn out to have a representation that is concrete and manageable. Apparently, this was first noted by V. V. Vagner (cf. [Pe], p.54). Our description follows I.10.5 of [Pe]. Let Γ be any nonempty set, let Γ' be a set disjoint from Γ such that $\overline{\overline{\Gamma'}} = \overline{\overline{\Gamma}}$, and let $F_{SI}(\Gamma)$ be the set of all ***nonempty*** finite sequences $x = \langle x_0, \ldots, x_{n-1}\rangle$ such that each x_m belongs to $\Gamma \cup \Gamma'$. Assuming that a one-one mapping g from Γ onto Γ' has been given, we first let $x_m^{-1} = g(x_m)$ if $x_m \in \Gamma$ and $x_m^{-1} = g^{\smile}(x_m)$ if $x_m \in \Gamma'$, and then define a unary operation $^{-1}$ on $F_{SI}(\Gamma)$ as follows:

$$\langle x_0, \ldots, x_{n-1}\rangle^{-1} = \langle x_{n-1}^{-1}, \ldots, x_0^{-1}\rangle\ ,$$
$$\text{for any } \langle x_0, \ldots, x_{n-1}\rangle \text{ in } F_{SI}(\Gamma)\ .$$

Henceforth we will ignore the distinction between an element x_m of $\Gamma \cup \Gamma'$ and the one-letter sequence $\langle x_m\rangle$, and hence also between x_m^{-1} and $\langle x_m\rangle^{-1}$. We may then regard Γ as a set of generators of $\mathbf{F}_{SI}(\Gamma) = \langle F_{SI}(\Gamma), {}^\frown, {}^{-1}\rangle$, where $^\frown$ is the operation of concatenation restricted to the sequences in $F_{SI}(\Gamma)$. As is easily shown (cf. [Pe], p.52), if $\mathbf{S} = \langle S, \circ, {}^\smile\rangle$ is any semigroup with involution, G is a set of generators of $\mathbf{S}$, and h' is a mapping of Γ onto G, then h' can be extended to a homomorphism of $\mathbf{F}_{SI}(\Gamma)$ onto $\mathbf{S}$. In other words, $\mathbf{F}_{SI}(\Gamma)$ is a free algebra in the class SI of all semigroups with involution and has Γ as free generating set.

Henceforth, $\mathbf{F}_{SI}(\Gamma)$ shall be the algebra $\langle F_{SI}(\Gamma), {}^\frown, {}^{-1}\rangle$ just described. When they are understood, we may omit reference to SI and sometimes to Γ and write, more briefly, $\mathbf{F}(\Gamma)$, $\mathbf{F}$, or $\langle F, {}^\frown, {}^{-1}\rangle$. Also, unless we indicate otherwise, a ***word*** shall be any sequence that either belongs to $F_{SI}(\Gamma)$, and hence is nonempty, or else is the empty sequence $\emptyset$.

Recall from Chapter I that q_i is the multivalued i-insertion from ${}^{[i,\omega)}U$ onto ${}^{[i+1,\omega)}U$ and that r_i is the i-replication from ${}^{[i+1,\omega)}U$ into ${}^{[i+2,\omega)}U$. Also recall that S_q and S_r are the closure of $\{q_i : i < \omega\}$ or of $\{r_i : i < \omega\}$, respectively, under $\circ$ and $^\smile$, while T_q and T_r are the closure of $\{q_i : i < \omega\}$ or of $\{r_i : i < \omega\}$, respectively, under $\circ$ only. Before presenting $\mathbf{S}_q = \langle S_q, \circ, {}^\smile\rangle$ and $\mathbf{S}_r = \langle S_q, \circ, {}^\smile\rangle$ as semigroups with involution, we will present $\mathbf{T}_q = \langle T_q, \circ\rangle$ and $\mathbf{T}_r = \langle T_r, \circ\rangle$ as a semigroup. Our presentation will show that they are isomorphic.

From Theorem IV.7(a) there follows that ${}_{[\omega]}\mathbf{T}_q = \langle {}_{[\omega]}T_q, \circ\rangle$ and ${}_{[\omega]}\mathbf{T}_r = \langle {}_{[\omega]}T_r, \circ\rangle$ are isomorphic, where ${}_{[\omega]}T_q$ and ${}_{[\omega]}T_r$ are the closure under $\circ$ of $\{\overset{\omega}{\rightarrow} q_i : i < \omega\}$ or of $\{\overset{\omega}{\rightarrow} r_i : i < \omega\}$, respectively.

Let $\mathbf{T}' = \langle T', \circ\rangle$ and $\mathbf{T}'' = \langle T'', \circ\rangle$, where

$$\begin{aligned} T' &= \{f_J : 1 \leq \overline{\overline{J}} < \omega\} \\ &= \{f \in Bsl : Do\, f = \omega,\ f \text{ is strictly increasing } \}\ , \\ T'' &= \{f_J^{\overset{\frown}{+}} : 1 \leq \overline{\overline{J}} < \omega\} \\ &= \{f^{\smile} \in Bsl : Do\, f = Rg\, f = \omega,\ f \text{ is order-preserving}\}\ . \end{aligned}$$

Since $\circ$ and $\bullet$ coincide on T'', therefore from Theorem III.17 there follows that $\mathbf{T}''$ is isomorphic under $\overset{\frown}{+}$ to $\mathbf{T}'$. By Theorem IV.5(b), $\mathbf{T}'$ is generated by the set $\{f_{\{i\}} : i < \omega\}$ of delaying successor functions. Hence $\mathbf{T}''$ is generated by the set $\{f_{\{i\}}^{\overset{\frown}{+}} : i < \omega\}$. Note that $(f_{\{i\}}^{\overset{\frown}{+}})^{\smile}$ is the function on ω that satisfies the following two conditions and hence may be called ***the predecessor function that delays until*** i:

$$\begin{aligned} (f_{\{i\}}^{\overset{\frown}{+}})^{\smile}(n) &= n \quad \text{if } n \leq i. \\ (f_{\{i\}}^{\overset{\frown}{+}})^{\smile}(n+1) &= n \quad \text{if } n > i. \end{aligned}$$

From Theorem IV.5(b) there follows that ${}_{[\omega]}\mathbf{T}_q$ and $\mathbf{T}'$ are isomorphic. There now follows that ${}_{[\omega]}\mathbf{T}_q$, ${}_{[\omega]}\mathbf{T}_r$, $\mathbf{T}'$ and $\mathbf{T}''$ are isomorphic to each other.

For any set $\Gamma \neq \emptyset$, we let $F_T(\Gamma)$ be the set $\bigcup\{{}^n\Gamma : 1 \leq n < \omega\}$ of nonempty finite sequences $x = \langle x_0, \ldots, x_n\rangle$ such that each x_m belongs to Γ and let $\mathbf{F}_T(\Gamma) = \langle F_T(\Gamma), {}^\frown\rangle$, so that $\mathbf{F}_T(\Gamma)$ is the free semigroup with Γ as free generating set.

We let $\Gamma(\tilde{\mathbf{q}}) = \{\mathbf{q}_i : i < \omega\}$ and $\Gamma(\tilde{\mathbf{r}}) = \{\mathbf{r}_i : i < \omega\}$, where $\mathbf{q}_i \neq \mathbf{q}_j$ and $\mathbf{r}_i \neq \mathbf{r}_j$ if $i \neq j$, and where $\Gamma(\tilde{\mathbf{q}}) \cap \Gamma(\tilde{\mathbf{r}}) = \emptyset$. We let Q_1 and R_1 be the following relation on $F_T(\Gamma(\tilde{\mathbf{q}}))$ or on $F_T(\Gamma(\tilde{\mathbf{r}}))$, respectively.

$$\begin{aligned} Q_1 &= \{\langle \mathbf{q}_i\mathbf{q}_j, \mathbf{q}_{j-1}\mathbf{q}_i\rangle : 0 \leq i < j < \omega\} \\ R_1 &= \{\langle \mathbf{r}_i\mathbf{r}_j, \mathbf{r}_{j-1}\mathbf{r}_i\rangle : 0 \leq i < j < \omega\} \end{aligned}$$

We let $\equiv_{Q_1}$ and $\equiv_{R_1}$ be the congruence relation on $\mathbf{F}_T(\Gamma(\tilde{\mathbf{q}}))$ or on $\mathbf{F}_T(\Gamma(\tilde{\mathbf{r}}))$, respectively, that is generated by Q_1 or by R_1, respectively.

THEOREM 1. *Let* h *be the homomorphism of* $\mathbf{F}_T(\Gamma(\tilde{\mathbf{q}}))$ *onto* $\mathbf{T}_q$ *such that* $h(\mathbf{q}_i) = q_i$ *for each* $i < \omega$ *and let* h' *be the homomorphism of* $\mathbf{F}_T(\Gamma(\tilde{\mathbf{r}}))$ *onto* $\mathbf{T}_r$ *such that* $h'(\mathbf{r}_i) = r_i$ *for each* $i < \omega$. *Then* $\equiv_{Q_1}$ *coincides with* $h \circ h^{\smile}$ *and also with* $(h\circ \overset{\omega}{\rightarrow}) \circ (h\circ \overset{\omega}{\rightarrow})^{\smile}$, *and* $\equiv_{R_1}$ *coincides with* $h' \circ h'^{\smile}$ *and also with* $(h'\circ \overset{\omega}{\rightarrow}) \circ (h'\circ \overset{\omega}{\rightarrow})^{\smile}$. *There follows that* $\mathbf{F}_T(\Gamma(\tilde{\mathbf{q}}))/{\equiv_{Q_1}}$ *is isomorphic to* $\mathbf{T}_q$ *and to* ${}_{[\omega]}\mathbf{T}_q$, *that* $\mathbf{F}_T(\Gamma(\tilde{\mathbf{r}}))/{\equiv_{R_1}}$ *is isomorphic to* $\mathbf{T}_r$ *and to* ${}_{[\omega]}\mathbf{T}_r$, *and that* $\mathbf{T}_q$, ${}_{[\omega]}\mathbf{T}_q$,

$\mathbf{T}_r$, ${}_{[\omega]}\mathbf{T}_r$ *are isomorphic to each other. Furthermore, they are isomorphic to* $\mathbf{T}'$ *and to* $\mathbf{T}''$.

Proof. As one can verify, and as will be shown in the proof of Lemma 2 below, if $0 \leq i < j < \omega$, then $q_i q_j = q_{j-1} q_i$. Hence $Q_1 \subseteq h \circ h^{\smile}$. There follows that $\equiv_{Q_1}$ is included in $h \circ h^{\smile}$ and hence also in $(h\circ \overset{\omega}{\rightarrow}) \circ (h\circ \overset{\omega}{\rightarrow})^{\smile}$.

As one can verify, and as will be shown in the proof of Lemma 2 below, if $0 \leq i < j < \omega$, then $r_i r_j = r_{j-1} r_i$. Hence $R_1 \subseteq h' \circ h'^{\smile}$. There follows that $\equiv_{R_1}$ is included in $h' \circ h'^{\smile}$ and hence also in $(h'\circ \overset{\omega}{\rightarrow}) \circ (h'\circ \overset{\omega}{\rightarrow})^{\smile}$.

To prove that the converse inclusions hold, we begin with a definition. A word $\mathbf{q}_{i_0} \ldots \mathbf{q}_{i_{m-1}}$ in $F_T(\Gamma(\tilde{\mathbf{q}}))$ or a word $\mathbf{r}_{i_0} \ldots \mathbf{r}_{i_{m-1}}$ in $F_T(\Gamma(\tilde{\mathbf{r}}))$ shall be ***strictly ascending*** if and only if, whenever $k < k' < m$, then $i_k < i_{k'}$.

Assume that $x = \mathbf{q}_{i_0} \ldots \mathbf{q}_{i_{m-1}}$ and $y = \mathbf{q}_{j_0} \ldots \mathbf{q}_{j_{n-1}}$ are strictly ascending. Let x' and y' be the unique element of T' such that ${}_{\omega}[x'] = \overset{\omega}{\rightarrow}(h(x))$ and ${}_{\omega}[y'] = \overset{\omega}{\rightarrow}(h(y))$. Then $x' = f_{\{i_0,\ldots,i_{m-1}\}}$ and $y' = f_{\{j_0,\ldots,j_{n-1}\}}$. Now also assume that $\overset{\omega}{\rightarrow}(h(x)) = \overset{\omega}{\rightarrow}(h(y))$. Then $x' = y'$, hence $f_{\{i_0,\ldots,i_{m-1}\}} = f_{\{j_0,\ldots,j_{n-1}\}}$, hence $\{i_0, \ldots, i_{m-1}\} = \{j_0, \ldots, j_{n-1}\}$, and hence x and y are the same word $\mathbf{q}_{i_0} \ldots \mathbf{q}_{i_{m-1}}$. In sum, if x and y are words in $F_T(\Gamma(\tilde{\mathbf{q}}))$ that are strictly ascending, then $\overset{\omega}{\rightarrow}(h(x)) = \overset{\omega}{\rightarrow}(h(y))$ implies that $x = y$.

Assume that $x = \mathbf{r}_{i_0}, \ldots, \mathbf{r}_{i_{m-1}}$ and $y = \mathbf{r}_{j_0}, \ldots, \mathbf{r}_{j_{n-1}}$ are strictly ascending. Let x'' and y'' be the unique element of T'' such that ${}_{\omega}[x''] = \overset{\omega}{\rightarrow}(h'(x))$ and ${}_{\omega}[y''] = \overset{\omega}{\rightarrow}(h'(y))$. Then $x'' = \widehat{(f_{\{i_0,\ldots,i_{m-1}\}})^{+}}$ and $y'' = \widehat{(f_{\{j_0,\ldots,j_{n-1}\}})^{+}}$. Now also assume that $\overset{\omega}{\rightarrow}(h'(x)) = \overset{\omega}{\rightarrow}(h'(y))$. Then $x'' = y''$, hence $\widehat{(f_{\{i_0,\ldots,i_{m-1}\}}))^{+}} = \widehat{(f_{\{j_0,\ldots,j_{n-1}\}}))^{+}}$, hence $\{i_0, \ldots, i_{m-1}\} = \{j_0, \ldots, j_{n-1}\}$, and hence x and y are the same word $\mathbf{r}_{i_0} \ldots \mathbf{r}_{i_{n-1}}$. In sum, if x and y are words in $F_T(\Gamma(\tilde{\mathbf{r}}))$ that are strictly ascending, then $\overset{\omega}{\rightarrow}(h'(x)) = \overset{\omega}{\rightarrow}(h'(y))$ implies that $x = y$.

If in the description of Q_1 given earlier one simultaneously replaces $j-1$ by i and i by j then one sees that $Q_1{}^{\smile}$ can be described as follows:

$$Q_1{}^{\smile} = \{\langle \mathbf{q}_i \mathbf{q}_j, \mathbf{q}_j \mathbf{q}_{i+1} \rangle : i \geq j\} \ .$$

Since $\equiv_{Q_1}$ also is the congruence on $\mathbf{F}_T(\Gamma(\tilde{\mathbf{q}}))$ that is generated by $Q_1{}^{\smile}$, therefore by induction on the length of words one can prove the following: For any word v in $F_T(\Gamma(\tilde{\mathbf{q}}))$ there is some word x in $F_T(\Gamma(\tilde{\mathbf{q}}))$ that is of the same length as v and is strictly ascending such that $v \equiv_{Q_1} x$. By a similar argument a similar conclusion can be proved for $F_T(\Gamma(\tilde{\mathbf{r}}))$.

Consider any v and w in $F_T(\Gamma(\tilde{\mathbf{q}}))$ such that $\overset{\omega}{\rightarrow}(h(v)) = \overset{\omega}{\rightarrow}(h(w))$. There are x and y in $F_T(\Gamma(\tilde{\mathbf{q}}))$ that are strictly ascending such that $v \equiv_{Q_1} x$ and $w \equiv_{Q_1} y$. Since, as we saw, $\equiv_{Q_1}$ is included in $(h \circ \overset{\omega}{\rightarrow}) \circ (h \circ \overset{\omega}{\rightarrow})^{\smile}$, therefore $\overset{\omega}{\rightarrow}(h(x)) = \overset{\omega}{\rightarrow}(h(y))$. As we also saw, there follows that $x = y$. From $x = y$, $v \equiv_{Q_1} x$, and $w \equiv_{Q_1} y$, there follows that $v \equiv_{Q_1} w$. This shows that $(h \circ \overset{\omega}{\rightarrow}) \circ (h \circ \overset{\omega}{\rightarrow})^{\smile}$ is included in $\equiv_{Q_1}$. Hence $h \circ h^{\smile}$ also is included in $\equiv_{Q_1}$.

There now follows that $h \circ h^{\smile}$, $(h \circ \overset{\omega}{\rightarrow}) \circ (h \circ \overset{\omega}{\rightarrow})^{\smile}$, and $\equiv_{Q_1}$ coincide. A similar argument shows that $h' \circ h'^{\smile}$, $(h' \circ \overset{\omega}{\rightarrow}) \circ (h' \circ \overset{\omega}{\rightarrow})^{\smile}$, and $\equiv_{R_1}$ coincide. There follows that $\mathbf{F}_T(\Gamma(\tilde{\mathbf{q}}))/\equiv_{Q_1}$ is isomorphic to $\mathbf{T}_q$ and to ${}_{[\omega]}\mathbf{T}_q$, and hence that $\mathbf{T}_q$ and ${}_{[\omega]}\mathbf{T}_q$ are isomorphic. Also, $\mathbf{F}_T(\Gamma(\tilde{\mathbf{r}}))/\equiv_{R_1}$ is isomorphic to $\mathbf{T}_r$ and to ${}_{[\omega]}\mathbf{T}_r$, and hence $\mathbf{T}_r$ and ${}_{[\omega]}\mathbf{T}_r$ are isomorphic. The assignment to each $\mathbf{q}_i$ of $\mathbf{r}_i$ induces an isomorphism between $\mathbf{F}_T(\Gamma(\tilde{\mathbf{q}}))$ and $\mathbf{F}_T(\Gamma(\tilde{\mathbf{r}}))$, one between $\langle \mathbf{F}_T(\Gamma(\tilde{\mathbf{q}})), Q_1 \rangle$ and $\langle \mathbf{F}_T(\Gamma(\tilde{\mathbf{r}})), R_1 \rangle$, and one between $\langle \mathbf{F}_T(\Gamma(\tilde{\mathbf{q}})), \equiv_{Q_1} \rangle$ and $\langle \mathbf{F}_T(\Gamma(\tilde{\mathbf{r}})), \equiv_{R_1} \rangle$. Therefore $\langle \mathbf{F}_T(\Gamma(\tilde{\mathbf{q}})), \equiv_{Q_1} \rangle$ and $\langle \mathbf{F}_T(\Gamma(\tilde{\mathbf{r}})), \equiv_{R_1} \rangle$ are isomorphic. There now follows that $\mathbf{T}_q$ and ${}_{[\omega]}\mathbf{T}_q$ are isomorphic to $\mathbf{T}_r$ and to ${}_{[\omega]}\mathbf{T}_r$. Finally, as we saw earlier, $\mathbf{T}'$ is isomorphic to ${}_{[\omega]}\mathbf{T}_q$ and $\mathbf{T}''$ is isomorphic to ${}_{[\omega]}\mathbf{T}_r$. Thus, all the semigroups $\mathbf{T}_q$, $\mathbf{T}_r$, ${}_{[\omega]}\mathbf{T}_q$, ${}_{[\omega]}\mathbf{T}_r$, $\mathbf{T}'$, $\mathbf{T}''$ are isomorphic to each other. □

We now turn to the task of giving a presentation of $\mathbf{S}_q = \langle S_q, \circ, {}^{\smile} \rangle$ and one of $\mathbf{S}_r = \langle S_r, \circ, {}^{\smile} \rangle$. These presentations will show that $\mathbf{S}_q$ and $\mathbf{S}_r$ are isomorphic. Again let $\Gamma(\tilde{\mathbf{q}}) = \{\mathbf{q}_i : i < \omega\}$, where $\mathbf{q}_i \neq \mathbf{q}_j$ if $i \neq j$. Also let $= \{\mathbf{q}_i^{-1} : i < \omega\}$ be such that $\mathbf{q}_i^{-1} \neq \mathbf{q}_j^{-1}$ if $i \neq j$, and such that $\{\mathbf{q}_i : i < \omega\} \cap \{\mathbf{q}_i^{-1} : i < \omega\} = \emptyset$. Let

$$F_{SI}(\Gamma(\tilde{\mathbf{q}})) =$$
$$\{x : 1 \leq |x| < \omega,\ x_m \in \{\mathbf{q}_i : i < \omega\} \cup \{\mathbf{q}_i^{-1} : i < \omega\},\ \text{if } m < |x|\}$$
$$\mathbf{F}_{SI}(\Gamma(\tilde{\mathbf{q}})) = \langle F_{SI}(\Gamma(\mathbf{q})), {}^\frown, {}^{-1} \rangle$$

Let $\Gamma(\tilde{\mathbf{r}}) = \{\mathbf{r}_i : i < \omega\}$, where $\mathbf{r}_i \neq \mathbf{r}_j$, if $i \neq j$, and where $\Gamma(\tilde{\mathbf{q}}) \cap \Gamma(\tilde{\mathbf{r}}) = \emptyset$. Define $F_{SI}(\Gamma(\tilde{\mathbf{r}}))$ and $\mathbf{F}_{SI}(\Gamma(\tilde{\mathbf{r}}))$ in terms of $\Gamma(\tilde{\mathbf{r}})$ as $F_{SI}(\Gamma(\tilde{\mathbf{q}}))$ and $\mathbf{F}_{SI}(\Gamma(\tilde{\mathbf{q}}))$ were defined in terms of $\Gamma(\tilde{\mathbf{q}})$. (Thus $F_T(\Gamma(\tilde{\mathbf{q}}))$ is a proper subset of $F_{SI}(\Gamma(\tilde{\mathbf{q}}))$, $\mathbf{F}_T(\Gamma(\tilde{\mathbf{q}}))$ is a proper subreduct of $\mathbf{F}_{SI}(\Gamma(\tilde{\mathbf{q}}))$, and similar remarks apply to $F_T(\Gamma(\tilde{\mathbf{r}}))$ and $\mathbf{F}_T(\Gamma(\mathbf{r}))$.)

We now repeat the definition of Q_1 and let Q_2, Q_3, Q_4 be the relations on $F(\Gamma(\tilde{\mathbf{q}}))$ that are shown below. Also, henceforth R_1, R_2, R_3, R_4 shall be the relation on $F(\Gamma(\tilde{\mathbf{r}}))$ that results from Q_1, Q_2, Q_3, Q_4, respectively by the replacement of each

$\mathbf{q}_k$ by $\mathbf{r}_k$.

$$\begin{aligned} Q_1 &= \{\langle \mathbf{q}_i\mathbf{q}_j, \mathbf{q}_{j-1}\mathbf{q}_i\rangle : i < j < \omega\} \\ Q_2 &= \{\langle \mathbf{q}_i\mathbf{q}_j^{-1}, \mathbf{q}_{j-1}^{-1}\mathbf{q}_i\rangle : i < j < \omega\} \\ Q_3 &= \{\langle \mathbf{q}_i\mathbf{q}_i^{-1}\mathbf{q}_i, \mathbf{q}_i\rangle : i < \omega\} \\ Q_4 &= \{\langle \mathbf{q}_i\mathbf{q}_i^{-1}\mathbf{q}_{i+1}, \mathbf{q}_{i+1}\rangle : i < \omega\} \end{aligned}$$

Until the end of the proof of Theorem 12 we let $\equiv$ be the congruence on $\mathbf{F}(\Gamma(\tilde{\mathbf{q}}))$ that is generated by $Q_1 \cup Q_2 \cup Q_3 \cup Q_4$ and $\equiv'$ the congruence on $\mathbf{F}(\Gamma(\tilde{\mathbf{r}}))$ that is generated by $R_1 \cup R_2 \cup R_3 \cup R_4$. Also, until then, h shall be the homomorphism of $\mathbf{F}(\Gamma(\tilde{\mathbf{q}}))$ onto $\mathbf{S}_q$ such that $h(\mathbf{q}_i) = q_i$ for each $i < \omega$ and h' the homomorphism of $\mathbf{F}(\Gamma(\tilde{\mathbf{r}}))$ onto $\mathbf{S}_r$ such that $h'(\mathbf{r}_i) = r_i$ for each $i < \omega$.

LEMMA 2. *$\equiv$ is included in $h \circ h^{\smile}$ and $\equiv'$ is included in $h' \circ h'^{\smile}$.*

Proof. Since $q_i q_i^{\smile}$ is the identity relation ${}_{[i,\omega)}\mathring{1}$ on *Do* $q_i = {}^{[i,\omega)}U$, there follows that $Q_3 \subseteq h \circ h^{\smile}$. Since *Do* $q_{i+1} = {}^{[i+1,\omega)}U \subseteq {}^{[i,\omega)}U$, there also follows that $Q_4 \subseteq h \circ h^{\smile}$. Since $r_i r_i^{\smile}$ is the identity relation ${}_{[i+1,\omega)}\mathring{1}$ on *Do* $r_i = {}^{[i+1,\omega)}U$, there follows that $R_3 \subseteq h' \circ h'^{\smile}$. Since *Do* $r_{i+1} = {}^{[i+2,\omega)}U \subseteq {}^{[i+1,\omega)}U$, there also follows that $R_4 \subseteq h' \circ h'^{\smile}$.

To verify that Q_1 and Q_2 are included in $h \circ h^{\smile}$, assume that $i < j$. Then the following equalities hold, where v, w, x, y, z are elements of ${}^{[0,\omega)}U$.

$$\begin{aligned} &\text{(i)} & {}_{[j-1,\omega)}\mathring{1} \circ q_i &= \{\langle vxz, vwxz\rangle : \\ &&& \quad |v|=i,\ |x|=j-i-1,\ |w|=1,\ |z| < \omega\} \\ &\text{(ii)} & {}_{[j,\omega)}\mathring{1} \circ q_i &= \{\langle vxyz, vwxyz\rangle : \\ &&& \quad |v|=i,\ |x|=j-i-1,\ |w|=|y|=1,\ |z| < \omega\} \\ &\text{(iii)} & q_{j-1} &= \{\langle vxz, vxyz\rangle : \\ &&& \quad |v|=i,\ |x|=j-i-1,\ |y|=1,\ |z| < \omega\} \\ &\text{(iv)} & q_j &= \{\langle vwxz, vwxyz\rangle : \\ &&& \quad |v|=i,\ |x|=j-i-1,\ |w|=|y|=1,\ |z| < \omega\} \end{aligned}$$

$$\begin{aligned}
\text{(v)}\quad q_{j-1}{}^{\smile} &= \{\langle vxyz, vxz\rangle : \\
&\quad |v|=i,\ |x|=j-i-1,\ |y|=1,\ |z|<\omega\} \\
\text{(vi)}\quad q_j{}^{\smile} &= \{\langle vwxyz, vwxz\rangle : \\
&\quad |v|=i,\ |x|=j-i-1,\ |w|=|y|=1,\ |z|<\omega\}
\end{aligned}$$

Since $q_i \circ q_j = q_i \circ {}_{[j,\omega)}\overset{\circ}{1} \circ q_j = {}_{[j-1,\omega)}\overset{\circ}{1} \circ q_i \circ q_j$, and since $q_{j-1} \circ q_i = q_{j-1} \circ {}_{[j,\omega)}\overset{\circ}{1} \circ q_i$, there follows from (i) and (iv) and from (iii) and (ii), respectively, that each of the relative products $q_i \circ q_j$ and $q_{j-1} \circ q_i$ is the following relation on ${}^{[0,\omega)}U$:

$$\{\langle vxz, vwxyz\rangle : |v|=i,\ |x|=j-i-1,\ |w|=|y|=1,\ |z|<\omega\}\ .$$

Thus $i<j$ implies that $q_i \circ q_j = q_{j-1} \circ q_i$. Hence $Q_1 \subseteq h \circ h^{\smile}$.

Since $q_i \circ q_j{}^{\smile} = q_i \circ {}_{[j+1,\omega)}\overset{\circ}{1} \circ q_j{}^{\smile} = {}_{[j,\omega)}\overset{\circ}{1} \circ q_i \circ q_j{}^{\smile}$ and since $q_{j-1}^{\smile} \circ q_i = q_{j-1}^{\smile} \circ {}_{[j-1,\omega)}\overset{\circ}{1} \circ q_i$, there follows from (ii) and (vi) and from (v) and (i), respectively that each of the relative products $q_i \circ q_j{}^{\smile}$ and $q_{j-1}^{\smile} \circ q_i$ is the following relation on ${}^{[0,\omega)}U$:

$$\{\langle vxyz, vwxz\rangle : |v|=i,\ |x|=j-i-1,\ |w|=|y|=1,\ |z|<\omega\}\ .$$

Thus $i<j$ implies that $q_i \circ q_j{}^{\smile} = q_{j-1}^{\smile} \circ q_i$. Hence $Q_2 \subseteq h \circ h^{\smile}$.

There now follows that $Q_1 \cup Q_2 \cup Q_3 \cup Q_4$ and hence $\equiv$ are included in $h \circ h^{\smile}$.

Since $r_i = q_i e_{i,i+1}$ and since $e_{j,k} = e_{k,j}$, therefore to show that R_1 or R_2, respectively, is included in $h' \circ h'^{\smile}$ it suffices to show the following respectively:

$$\begin{aligned}
&(1)\quad q_i e_{i,i+1} q_j e_{j,j+1} = q_{j-1} e_{j-1,j}\ q_i e_{i,i+1}, \quad \text{if } i<j. \\
&(2)\quad q_i e_{i,i+1} e_{j,j+1} q_j{}^{\smile} = e_{j-1,j}\ q_{j-1}^{\smile} q_i e_{i,i+1}, \quad \text{if } i<j.
\end{aligned}$$

One can verify that (3) below holds, and also that (4) holds and therefore (4)′.

$$\begin{aligned}
&(3)\quad e_{i,i+1} q_j = q_j e_{i,i+1}, \quad \text{if } i<j. \\
&(4)\quad e_{i,i+1} q_j = q_j e_{i+1,i+2}, \quad \text{if } i\geq j. \\
&(4)'\quad e_{j-1,j} q_i = q_i e_{j,j+1}, \quad \text{if } i<j.
\end{aligned}$$

Using (3) or (4)′, respectively, one sees that the following hold:

$$\begin{aligned}
&(5)\quad q_i e_{i,i+1} q_j e_{j,j+1} = q_i q_j e_{i,i+1} e_{j,j+1}, \quad \text{if } i<j. \\
&(6)\quad q_{j-1} e_{j-1,j}\ q_i e_{i,i+1} = q_{j-1} q_i e_{j,j+1} e_{i,i+1}, \quad \text{if } i<j.
\end{aligned}$$

As we verified earlier, if $i<j$, then $q_i q_j = q_{j-1} q_i$. From this, from $e_{k_0k_1} e_{k_2k_3} = e_{k_2k_3} e_{k_0k_1}$, and from (5) and (6) there follows (1).

As we verified earlier, if $i<j$ then $q_{j-1}^{\smile} q_i = q_i q_j^{\smile}$. Hence:

$$(7)\quad e_{j-1,j} q_{j-1}^{\smile} q_i e_{i,i+1} = e_{j-1,j}\ q_i q_j{}^{\smile} e_{i,i+1}, \quad \text{if } i<j.$$

From (4)′, (3) and $e_{k_0k_1} e_{k_2k_3} = e_{k_2k_3} e_{k_0k_1}$ there follows

(8) $e_{j-1}q_iq_j^{\smile}e_{i,i+1} = q_ie_{j,j+1}e_{i,i+1}q_j^{\smile}$, if $i < j$.

From (7) and (8) there follows (2).

From (1) there follows that $R_1 \subseteq h' \circ h'^{\smile}$ and from (2) there follows that $R_2 \subseteq h' \circ h'^{\smile}$. There now follows that $R_1 \cup R_2 \cup R_3 \cup R_4$ and hence $\equiv'$ are included in $h' \circ h'^{\smile}$. □

For showing that $h \circ h^{\smile}$ and $h' \circ h'^{\smile}$ are included in $\equiv$ or in $\equiv'$, respectively, we now prove a series of lemmas. The first two of these concern the behavior of $q_iq_i^{-1}$ with respect to $\equiv$ and that of $r_ir_i^{-1}$ with respect to $\equiv'$. Instead of saying that $x \equiv y$ holds because Q_n is included in $\equiv$ we shall simply say: $x \equiv y$ **by** Q_n. Similar phrases will be used in connection with $\equiv'$ and R_n.

LEMMA 3. *Each of the assertions* (a), ..., (g) *below holds, and so does the assertion* (a)$'$, ..., (g)$'$ *that results from* (a), ..., (g) *respectively by replacing* $\equiv$ *by* $\equiv'$ *and each* $\mathbf{q}_k$ *by* $\mathbf{r}_k$.

(a) $\mathbf{q}_i\mathbf{q}_i^{-1}\mathbf{q}_j \equiv \mathbf{q}_j\mathbf{q}_{i+1}\mathbf{q}_{i+1}^{-1}$, if $i \geq j$.

(b) $\mathbf{q}_i\mathbf{q}_i^{-1}\mathbf{q}_j^{-1} \equiv \mathbf{q}_j^{-1}\mathbf{q}_{i-1}\mathbf{q}_{i-1}^{-1}$, if $i > j$.

(c) $\mathbf{q}_i\mathbf{q}_i^{-1}\mathbf{q}_j \equiv \mathbf{q}_j$, if $i \leq j$.

(d) $\mathbf{q}_i\mathbf{q}_i^{-1}\mathbf{q}_j^{-1} \equiv \mathbf{q}_j^{-1}$, if $i \leq j+1$.

(e) $\mathbf{q}_j\mathbf{q}_i\mathbf{q}_i^{-1} \equiv \mathbf{q}_j$, if $i \leq j+1$.

(f) $\mathbf{q}_j^{-1}\mathbf{q}_i\mathbf{q}_i^{-1} \equiv \mathbf{q}_j^{-1}$, if $i \leq j$.

(g) $\mathbf{q}_i\mathbf{q}_i^{-1}\mathbf{q}_j\mathbf{q}_j^{-1} \equiv \mathbf{q}_{\max(i,j)}\mathbf{q}_{\max(i,j)}^{-1}$.

Proof. (a) Assume that $i \geq j$. Then $j < i+1$ and hence by Q_1 or Q_2, respectively,

$$\mathbf{q}_j\mathbf{q}_{i+1}\mathbf{q}_{i+1}^{-1} \equiv \mathbf{q}_i\mathbf{q}_j\mathbf{q}_{i+1}^{-1} \equiv \mathbf{q}_i\mathbf{q}_i^{-1}\mathbf{q}_j \ .$$

(b) Assume that $i > j$. Then $i-1 \geq j$. Hence, by part (a), $\mathbf{q}_{i-1}\mathbf{q}_{i-1}^{-1}\mathbf{q}_j \equiv \mathbf{q}_j\mathbf{q}_i\mathbf{q}_i^{-1}$. Hence

$$\mathbf{q}_j^{-1}\mathbf{q}_{i-1}\mathbf{q}_{i-1}^{-1} = (\mathbf{q}_{i-1}\mathbf{q}_{i-1}^{-1}\mathbf{q}_j)^{-1} \equiv (\mathbf{q}_j\mathbf{q}_i\mathbf{q}_i^{-1})^{-1} = \mathbf{q}_i\mathbf{q}_i^{-1}\mathbf{q}_j^{-1} \ .$$

(c) Let $\equiv_4$ be the congruence on $\mathbf{F}_{SI}(\Gamma)$ that is generated by Q_4. If $i = j-1$, then $\mathbf{q}_i\mathbf{q}_i^{-1}\mathbf{q}_j \equiv_4 \mathbf{q}_j$. Now let $0 < i < j$ and as inductive hypothesis assume that

$\mathbf{q}_i\mathbf{q}_i^{-1}\mathbf{q}_j \equiv_4 \mathbf{q}_j$. Then by the inductive hypothesis, by Q_4, and by the inductive hypothesis, respectively, there hold

$$\mathbf{q}_i\mathbf{q}_i^{-1}\mathbf{q}_j \equiv_4 \mathbf{q}_{i-1}\mathbf{q}_{i-1}^{-1}\mathbf{q}_i\mathbf{q}_i^{-1}\mathbf{q}_j \equiv_4 \mathbf{q}_i\mathbf{q}_i^{-1}\mathbf{q}_j \equiv_4 \mathbf{q}_j \ .$$

Hence, by induction, $i < j$ implies that $\mathbf{q}_i\mathbf{q}_i^{-1}\mathbf{q}_j \equiv_4 \mathbf{q}_j$. Since, by Q_3, $i = j$ implies that $\mathbf{q}_i\mathbf{q}_i^{-1}\mathbf{q}_j \equiv \mathbf{q}_j$ there now follows that $i \leq j$ implies that $\mathbf{q}_i\mathbf{q}_i^{-1}\mathbf{q}_j \equiv \mathbf{q}_j$.

(f) Assume that $i \leq j$. Then, by part (c), $\mathbf{q}_i\mathbf{q}_i^{-1}\mathbf{q}_j \equiv \mathbf{q}_j$. Hence also

$$\mathbf{q}_j^{-1}\mathbf{q}_i\mathbf{q}_i^{-1} = (\mathbf{q}_i\mathbf{q}_{i-1}^{-1}\mathbf{q}_j)^{-1} \equiv \mathbf{q}_j^{-1} \ .$$

(e) By Q_1, Q_2 or Q_3, respectively,

$$\mathbf{q}_j\mathbf{q}_{j+1}\mathbf{q}_{j+1}^{-1} \equiv \mathbf{q}_j\mathbf{q}_j\mathbf{q}_{j+1}^{-1} \equiv \mathbf{q}_j\mathbf{q}_j^{-1}\mathbf{q}_j \equiv \mathbf{q}_j \ .$$

Hence $i = j+1$ implies that $\mathbf{q}_j\mathbf{q}_i\mathbf{q}_i^{-1} \equiv \mathbf{q}_j$. Now let $0 < i \leq j+1$ and as inductive hypothesis assume that $\mathbf{q}_j\mathbf{q}_i\mathbf{q}_i^{-1} \equiv \mathbf{q}_j$. Then by the inductive hypothesis, part (f), or the inductive hypothesis, respectively,

$$\mathbf{q}_j\mathbf{q}_{i-1}\mathbf{q}_{i-1}^{-1} \equiv \mathbf{q}_j\mathbf{q}_i\mathbf{q}_i^{-1}\mathbf{q}_{i-1}\mathbf{q}_{i-1}^{-1} \equiv \mathbf{q}_j\mathbf{q}_i\mathbf{q}_i^{-1} \equiv \mathbf{q}_j \ .$$

Hence, by induction, $i \leq j+1$ implies that $\mathbf{q}_j, \mathbf{q}_i\mathbf{q}_i^{-1} \equiv \mathbf{q}_j$.

(d) By part (e).

(g) If $i \geq j$, then, by part (f), $\mathbf{q}_i^{-1}\mathbf{q}_j\mathbf{q}_j^{-1} \equiv \mathbf{q}_i^{-1}$. If $i \leq j$, then, by part (c), $\mathbf{q}_i\mathbf{q}_i^{-1}\mathbf{q}_j \equiv \mathbf{q}_j$.

The proof of parts (a)$'$, ..., (g)$'$ is similar to that of (a), ..., (g), respectively. □

Let w be any element of $F(\Gamma(\tilde{\mathbf{q}}))$ or of $F(\Gamma(\tilde{\mathbf{r}}))$. The ***grade associated with*** w shall be the grade $gd(h(w))$ of $h(w)$ or the grade $gd(h'(w))$ of $h'(w)$, respectively. It will be denoted by $gd\ w$. Note that if w is in $F(\Gamma(\tilde{\mathbf{r}}))$ and $gd\ w = \langle s, t\rangle$ then $s \geq 1$ and $t \geq 1$.

The following consequence of the definition just given and of Lemma II.5(a) will be used tacitly:

$$\begin{aligned} &gd\ w^\frown x = gd\ w \odot gd\ x \ , && \text{where} \\ &\langle s,t\rangle \odot \langle s',t'\rangle = \langle s, t'+t-s'\rangle \ , && \text{if } t \geq s' \ , \\ &\langle s,t\rangle \odot \langle s',t'\rangle = \langle s+s'-t, t'\rangle \ , && \text{if } t \leq s' \ . \end{aligned}$$

Parts (a) and (a)$'$ of the next lemma generalize parts (a), (a)$'$, (b), and (b)$'$ of Lemma 3. Also, parts (b), (b)$'$, (c), and (c)$'$ generalize parts (c), ..., (g) and (c)$'$, ..., (g)$'$ of Lemma 3. Parts (a) and (a)$'$ of Lemma 4 are related to Theorem II.7(f), while parts (b), (b)$'$, (c), and (c)$'$ are related to Theorem II.7(e).

LEMMA 4. *Assume that* $x \in F(\Gamma(\tilde{\mathbf{q}}))$ *and* $gd\ x = \langle s,t\rangle$. *Also assume that* $x' \in F(\Gamma(\tilde{\mathbf{r}}))$ *and* $gd\ x' = \langle s+1, t+1\rangle$.

(a) $\mathbf{q}_{s+k}\mathbf{q}_{s+k}^{-1}x \equiv x\mathbf{q}_{t+k}\mathbf{q}_{t+k}^{-1}$, *if* $k < \omega$.

(a)′ $\mathbf{r}_{s+k}\mathbf{r}_{s+k}^{-1}x' \equiv' x'\mathbf{r}_{t+k}\mathbf{r}_{t+k}^{-1}$, *if* $k < \omega$.

(b) $\mathbf{q}_i\mathbf{q}_i^{-1}x \equiv x$, *if* $i \leq s$.

(b)′ $\mathbf{r}_i\mathbf{r}_i^{-1}x' \equiv' x'$, *if* $i \leq s$.

(c) $x\mathbf{q}_i\mathbf{q}_i^{-1} \equiv x$, *if* $i \leq t$.

(c)′ $x'\mathbf{r}_i\mathbf{r}_i^{-1} \equiv' x'$, *if* $i \leq t$.

Proof. (a) We use induction on the length of x. First assume that x is of length 1. Then, since $gd\ x = \langle s,t\rangle$, therefore either $x = \mathbf{q}_s$ and $t = s+1$ or $x = \mathbf{q}_{s-1}^{-1}$ and $t = s-1$. If $x = \mathbf{q}_s$ then, by Lemma 3(a), for any $k < \omega$ there holds

$$\mathbf{q}_{s+k}\mathbf{q}_{s+k}^{-1}x = \mathbf{q}_{s+k}\mathbf{q}_{s+k}^{-1}\mathbf{q}_s \equiv \mathbf{q}_s\mathbf{q}_{s+k+1}\mathbf{q}_{s+k+1}^{-1} = x\mathbf{q}_{t+k}\mathbf{q}_{t+k}^{-1}\ .$$

If $x = \mathbf{q}_{s-1}^{-1}$ then, by Lemma 3(b), for any $k < \omega$ there holds:

$$\mathbf{q}_{s+k}\mathbf{q}_{s+k}^{-1}x = \mathbf{q}_{s+k}\mathbf{q}_{s+k}^{-1}\mathbf{q}_{s-1}^{-1} \equiv \mathbf{q}_{s-1}^{-1}\mathbf{q}_{s+k-1}\mathbf{q}_{s+k-1}^{-1} = x\mathbf{q}_{t+k}\mathbf{q}_{t+k}^{-1}\ .$$

Let $n \geq 1$ and assume as inductive hypothesis that for any y in $F(\Gamma(\tilde{\mathbf{q}}))$ of length $m \leq n$, if $gd\ y = \langle s',t'\rangle$ and $k < \omega$, then $\mathbf{q}_{s'+k}\mathbf{q}_{s'+k}^{-1}y \equiv y\mathbf{q}_{t'+k}\mathbf{q}_{t'+k}^{-1}$. Consider any w in $F(\Gamma(\tilde{\mathbf{q}}))$ of length $n+1$. Then either $w = \mathbf{q}_j x$ or $w = \mathbf{q}_j^{-1}x$ for some $j < \omega$ and some x in $F(\Gamma(\tilde{\mathbf{q}}))$ of length n. Let $gd\ x = \langle s,t\rangle$.

Suppose that $w = \mathbf{q}_j x$ and that $j \leq s-1$. Then gd $w = gd\ \mathbf{q}_j x = \langle j, j+1\rangle \odot \langle s,t\rangle = \langle s-1, t\rangle$. Then, by two uses of the inductive hypothesis, given any $k < \omega$, there hold:

$$\mathbf{q}_{s-1+k}\mathbf{q}_{s-1+k}^{-1}\mathbf{q}_j x \equiv \mathbf{q}_j\mathbf{q}_{s+k}\mathbf{q}_{s+k}^{-1}x \equiv \mathbf{q}_j x\mathbf{q}_{t+k}\mathbf{q}_{t+k}^{-1}\ .$$

Suppose that $w = \mathbf{q}_j x$ and that $j \geq s-1$. Then $gd\,w = gd\ \mathbf{q}_j x = \langle j, t+j+1-s\rangle$. Then by two uses of the inductive hypothesis, given any $k < \omega$, there hold:

$$\begin{aligned}\mathbf{q}_{j+k}\mathbf{q}_{j+k}^{-1}\mathbf{q}_j x &\equiv \mathbf{q}_j\mathbf{q}_{j+k+1}\mathbf{q}_{j+k+1}^{-1}x\\ &= \mathbf{q}_j\mathbf{q}_{s+(j+1-s+k)}\mathbf{q}_{s+(j+1-s+k)}^{-1}x\\ &\equiv \mathbf{q}_j x\mathbf{q}_{t+(j+1-s+k)}\mathbf{q}_{t+(j+1-s+k)}^{-1}\end{aligned}$$

Suppose that $w = \mathbf{q}_j^{-1}x$ and that $j \leq s$. Then $gd\,w = \langle j+1, j\rangle \odot \langle s,t\rangle = \langle s+1, t\rangle$. Then by two uses of the inductive hypothesis, given any $k < \omega$, there hold:

$$\mathbf{q}_{s+1+k}\mathbf{q}_{s+1+k}^{-1}\mathbf{q}_j^{-1}x \equiv \mathbf{q}_j^{-1}\mathbf{q}_{s+k}\mathbf{q}_{s+k}^{-1}x \equiv q_j^{-1}x\mathbf{q}_{t+k}\mathbf{q}_{t+k}^{-1}\ .$$

Suppose that $w = \mathbf{q}_j^{-1}x$ and that $j \geq s$. Then $gd\, w = gd\ \mathbf{q}_j x = \langle j+1, t+j-s\rangle$. Then by two uses of the inductive hypothesis, given any $k < \omega$, there hold:

$$\begin{aligned}
\mathbf{q}_{j+1+k}\mathbf{q}_{j+1+k}^{-1}\mathbf{q}_j^{-1}x &\equiv \mathbf{q}_j^{-1}\mathbf{q}_{j+k}\mathbf{q}_{j+k}^{-1}x \\
&= q_j^{-1}\mathbf{q}_{s+(j-s+k)}\mathbf{q}_{s+(j-s+k)}^{-1}x \\
&\equiv \mathbf{q}_j^{-1}x\mathbf{q}_{t+(j-s+k)}\mathbf{q}_{t+(j-s+k)}^{-1} .
\end{aligned}$$

Since any w in $F(\Gamma(\tilde{\mathbf{q}}))$ of length $n+1$ is either some $\mathbf{q}_j x$ or some $\mathbf{q}_j^{-1}x$ such that x is of length n, there now follows that the inductive hypothesis holds for any y in $F(\Gamma(\tilde{\mathbf{q}}))$ of length $m \leq n+1$. Hence, part (a) follows by induction.

(a)$'$ We use induction on the length of x'. First assume that x' is of length 1. Then, since $gd\ x' = \langle s+1, t+1\rangle$, therefore either $x' = \mathbf{r}_s$, $gd\ x' = \langle s+1, s+2\rangle$, and $t = s+1$ or $x' = \mathbf{r}_{s-1}^{-1}$, $gd\ x' = \langle s+1, s\rangle$ and $t = s-1$. If $x' = \mathbf{r}_s$ then, by Lemma 3(a)$'$, for any $k < \omega$ there holds:

$$\mathbf{r}_{s+k}\mathbf{r}_{s+k}^{-1}x' = \mathbf{r}_{s+k}\mathbf{r}_{s+k}^{-1}\mathbf{r}_s \equiv \mathbf{r}_s\mathbf{r}_{s+k+1}\mathbf{r}_{s+k+1}^{-1} = x'\mathbf{r}_{t+k}\mathbf{r}_{t+k}^{-1} .$$

If $x' = \mathbf{r}_{s-1}^{-1}$ then, by Lemma 3(b)$'$, for any $k < \omega$ there holds:

$$\mathbf{r}_{s+k}\mathbf{r}_{s+k}^{-1}x' = \mathbf{r}_{s+k}\mathbf{r}_{s+k}^{-1}\mathbf{r}_{s-1}^{-1} \equiv \mathbf{r}_{s-1}^{-1}\mathbf{r}_{s+k-1}\mathbf{r}_{s+k-1}^{-1} = x'\mathbf{r}_{t+k}\mathbf{r}_{t+k}^{-1} .$$

The rest of the proof of part (a)$'$ by induction is similar to the corresponding part of the proof of part (a).

(b) We use induction on the length of x. First assume that x is of length 1 and hence either $x = \mathbf{q}_s$ or $x = \mathbf{q}_{s-1}^{-1}$. If $x = \mathbf{q}_s$ then, by Lemma 3(c), $\mathbf{q}_i\mathbf{q}_i^{-1}x \equiv x$ if $i \leq s$. If $x = \mathbf{q}_{s-1}^{-1}$ then, by Lemma 3(d), $\mathbf{q}_i\mathbf{q}_i^{-1}x \equiv x$ if $i \leq (s-1)+1 = s$.

Consider any $n \geq 1$ and any x of length n with $gd\ x = \langle s, t\rangle$ for some t and assume as inductive hypothesis that $\mathbf{q}_i\mathbf{q}_i^{-1}x \equiv x$ for any $i \leq s$.

Let $w = \mathbf{q}_j x$. If $j \geq s-1$ and hence $gd\ w = j$ then, for any $i \leq gd\ w = j$, there follows from Lemma 3(c) that $\mathbf{q}_i\mathbf{q}_i^{-1}\mathbf{q}_j \equiv \mathbf{q}_j$ hence that $\mathbf{q}_i\mathbf{q}_i^{-1}\mathbf{q}_j x \equiv \mathbf{q}_j x$, and hence that $\mathbf{q}_i\mathbf{q}_i^{-1}w \equiv w$. Now assume that $j \leq s-1$ and hence that $gd\ w = s-1$. Then, for any $i \leq gd\ w = s-1$, there follows from part (a) or by the inductive hypothesis, respectively, that

$$\mathbf{q}_i\mathbf{q}_i^{-1}\mathbf{q}_j x \equiv \mathbf{q}_j\mathbf{q}_{i+1}\mathbf{q}_{i+1}^{-1}x \equiv \mathbf{q}_j x ,$$

and hence again that $\mathbf{q}_i\mathbf{q}_i^{-1}w \equiv w$.

Let $w = \mathbf{q}_j^{-1}x$. If $j \geq s$ and hence $gd\ w = j+1$ then, for any $i \leq gd\ w = j+1$, there follows from Lemma 3(d) that $\mathbf{q}_i\mathbf{q}_i^{-1}\mathbf{q}_j \equiv \mathbf{q}_j$, hence that $\mathbf{q}_i\mathbf{q}_i^{-1}\mathbf{q}_j x \equiv \mathbf{q}_j x$, and hence that $\mathbf{q}_i\mathbf{q}_i^{-1}w \equiv w$. Now assume that $j \leq s$ and hence that $gd\ w = s+1$.

Then, for any $i \leq gd\ w = s+1$ there holds by part (a) or the inductive hypothesis, respectively, that

$$\mathbf{q}_i\mathbf{q}_i^{-1}\mathbf{q}_j^{-1}x \equiv \mathbf{q}_j^{-1}\mathbf{q}_{i-1}\mathbf{q}_{i-1}^{-1}x \equiv \mathbf{q}_j^{-1}x \ ,$$

and hence again that $\mathbf{q}_i\mathbf{q}_i^{-1}w \equiv \mathbf{q}_i\mathbf{q}_i^{-1}w$.

Thus the inductive hypothesis holds for any word of length $n+1$. Hence part (b) follows by induction.

(b)′ The proof is similar to that for part (b).

(c) Assume that $gd\ x = \langle s,t\rangle$ and that $i \leq t$. Then $gd\ x^{-1} = \langle t,s\rangle$ Then from part (b) there follows that

$$x\mathbf{q}_i\mathbf{q}_i^{-1} = (\mathbf{q}_i\mathbf{q}_i^{-1}x^{-1})^{-1} \equiv (x^{-1})^{-1} = x \ .$$

(c)′ The proof is similar to that for part (c). □

A word in $F(\Gamma(\tilde{\mathbf{q}})) = F_S(\Gamma(\tilde{\mathbf{q}}))$ or in $F(\Gamma(\tilde{\mathbf{r}})) = F_S(\Gamma(\mathbf{r}))$ shall be ***strictly descending*** if and only if, for some $m \geq 1$, it is a word $\mathbf{q}_{i_0}^{-1}\ldots\mathbf{q}_{i_{m-1}}^{-1}$ or a word $\mathbf{r}_{i_0}^{-1}\ldots\mathbf{r}_{i_{m-1}}^{-1}$, respectively, such that, whenever $k < k' < m$, then $i_k > i_{k'}$. Thus a word x in $F(\Gamma(\tilde{\mathbf{q}}))$ or in $F(\Gamma(\tilde{\mathbf{r}}))$ is strictly descending if and only if x^{-1} is strictly ascending.

A word w shall be **q-*normal*** if and only if it is a word $\mathbf{q}_{s'}\mathbf{q}_{s'}^{-1}xy$ in $F(\Gamma(\tilde{\mathbf{q}}))$ such that x is either empty or strictly descending, y is either empty or strictly ascending and, if xy is nonempty and $gd(xy) = \langle s,t\rangle$, then $s' \geq s$ and hence $gd\ w = \langle s', t+s'-s\rangle$. Note that if m is the length of x, n the length of y, and $gd(xy) = \langle s,t\rangle$, then $t = s-m+n$.

A word w shall be **r-*normal*** if and only if it is a word $\mathbf{r}_{s'}\mathbf{r}_{s'}^{-1}xy$ in $F(\Gamma(\tilde{\mathbf{r}}))$ such that x is either empty or strictly descending, y is either empty or strictly ascending and, if xy is nonempty and $gd(xy) = \langle s+1,t+1\rangle$, then $s' \geq s$ and hence $gd\ w = \langle s', t+s'-s\rangle$.

Henceforth, when $\mathbf{q}_{s'}\mathbf{q}_{s'}^{-1}xy$ and $\mathbf{r}_{s'}\mathbf{r}_{s'}^{-1}xy$ are used to refer to a word that is **q**-normal or **r**-normal, respectively, it will be understood that x is either empty or strictly descending and that y is either empty or strictly ascending.

If w is **q**-normal, then $h(w)$ can be formed in a fairly simple manner. Suppose that $w = q_{s'}q_{s'}^{-1}xy$ where $x \neq \emptyset$ and $y \neq \emptyset$, so that x is a word $\mathbf{q}_{i_0}^{-1}\ldots\mathbf{q}_{i_m}^{-1}$ that is strictly descending and y is a word $\mathbf{q}_{j_0}\ldots\mathbf{q}_{j_n}$ that is strictly ascending. Then $h(x)$ is the $\{i_0,\ldots,i_{m-1}\}$-excision from ${}^{[i_0+1,\omega)}U$ onto ${}^{[i_m+1-m,\omega)}U$, $h(y)$ is the converse of the $\{j_{n-1},\ldots,j_0\}$-excision from ${}^{[j_{n-1}+1,\omega)}U$ onto ${}^{[j_n-1-n,\omega)}U$, $s' \geq \max(i_0+1, j_{n-1}+1-n+m)$, and $h(w) = {}_{[s',\omega)}\overset{\circ}{1} \circ h(x) \circ h(y)$.

If v is $\mathbf{r}$-normal, then $h'(v)$ also can be formed in a fairly simple manner. For example, let $v = \mathbf{r}_{s'}\mathbf{r}_{s'}^{-1}xy$ where $x = \mathbf{r}_4^{-1}\mathbf{r}_2^{-1}\mathbf{r}_1^{-1}\mathbf{r}_0^{-1}$ and $y = \mathbf{r}_2\mathbf{r}_3$, so that $s' \geq \max((4+1)+1, (3+1)+1-\{\overline{\overline{2,3}}\}+\{\overline{\overline{0,1,2,4}}\})$ and hence $s' \geq 7$. Then $h'(\mathbf{r}_{s'}\mathbf{r}_{s'}^{-1}) = {}_{[s'+1,\omega)}\mathring{1}$, $h'(x) = \dot{r}_4 \circ \dot{r}_2 \circ \dot{r}_1 \circ \dot{r}_0 = e_{4,5} \circ \dot{q}_4 \circ e_{2,3} \circ \dot{q}_2 \circ e_{1,2} \circ q_1 \circ e_{0,1} \circ \dot{q}_0$ is the restriction of the $\{4,2,1,0\}$-excision from ${}^{[5,\omega)}U$ onto ${}^{[1,\omega)}U$ to $Do(e_{4,5} \circ e_{2,3} \circ e_{1,2} \circ e_{0,1}) = \{x \in {}^{[6,\omega)}U : x_5 = x_4,\ x_3 = x_2 = x_1 = x_0\}$, and $h'(y) = r_2 \circ r_3 = q_2 \circ e_{2,3} \circ q_3 \circ e_{3,4}$ is the converse of $e_{3,4} \circ \dot{q}_3 \circ e_{2,3} \circ \dot{q}_2$, which is the restriction of the $\{3,2\}$-excision from ${}^{[4,\omega)}U$ onto ${}^{[2,\omega)}U$ to $Do(e_{3,4} \circ e_{2,3}) = \{x \in {}^{[5,\omega)}U : x_4 = x_3 = x_2\}$. (Note that the expressive power of $F(\Gamma(\tilde{\mathbf{r}}))$ is fairly limited. For example, from Lemma 2 and from Lemmas 5 and 6 below there follows that whenever $j \geq i+2$, then there is no v in $F(\Gamma(\tilde{\mathbf{r}}))$ such that $h(v) = e_{i,j} \circ \dot{q}_i$.)

LEMMA 5. *For every word v in $F(\Gamma(\tilde{\mathbf{q}}))$ there is a $\mathbf{q}$-normal word $w = \mathbf{q}_{s'}\mathbf{q}_{s'}^{-1}xy$ such that $v \equiv w$. Also, for every word v in $F(\Gamma(\tilde{\mathbf{r}}))$, there is an $\mathbf{r}$-normal word w such that $v \equiv' w$.*

Proof. We will carry out the proof for v in $F(\Gamma(\tilde{\mathbf{q}}))$. For v in $F(\Gamma(\tilde{\mathbf{r}}))$ the proof is similar. The proof will also show that xy is not longer than v. For any $i < \omega$, an occurrence in v of $\mathbf{q}_i$ shall be ***coupled*** if and only if it is immediately followed by an occurrence of $\mathbf{q}_i^{-1}$ and an occurrence in v of $\mathbf{q}_i^{-1}$ shall be ***coupled*** if and only if it is immediately preceded by an occurrence of $\mathbf{q}_i$. In other words, these occurrences are coupled if and only if they are an occurrence within an occurrence of $\mathbf{q}_i\mathbf{q}_i^{-1}$.

SUBLEMMA 1 *Let $\equiv_2$ be the congruence on $\mathbf{F}(\Gamma(\tilde{\mathbf{q}}))$ that is generated by Q_2. Then, for any v in $F(\Gamma(\tilde{\mathbf{q}})$, there are x and y such that $v \equiv_2 xy$, where x contains no noncoupled occurrence of any $\mathbf{q}_j$, and y contains no noncoupled occurrence of any $\mathbf{q}_i^{-1}$. Moreover, x and y can be so chosen that xy is of the same length as v,* $\max\{i : \mathbf{q}_i^{-1}$ occurs in $xy\} \leq \max\{i : \mathbf{q}_i^{-1}$ occurs in $v\}$, *and* $\max\{j : \mathbf{q}_j$ occurs in $xy\} \leq \max\{j : \mathbf{q}_j$ occurs in $v\}$.

Proof. By induction on the length of v. Consider any v of length 1. If v is $\mathbf{q}_i^{-1}$ let $x = \mathbf{q}_i^{-1} = v$ and $y = \emptyset$ and if v is $\mathbf{q}_j$ let $x = \emptyset$ and $y = \mathbf{q}_j = v$. In either case, the desired conditions are satisfied by v, x, y. Now let $1 \leq n < \omega$ and assume as inductive hypothesis that for any v of length n there are x and y such that v, x, y satisfy the desired conditions. Consider any v' of length $n+1$. Then, for some v of length n, either v' is $v\mathbf{q}_j$ for some $j < \omega$, or v' is $v\mathbf{q}_i^{-1}$, for some $i < \omega$. In either case, there are x and y such that v, x, y satisfy the desired conditions.

First suppose that v' is $v\mathbf{q}_j$. Then $v', x, y\mathbf{q}_j$ satisfy the desired conditions. Now suppose that v' is $v\mathbf{q}_i^{-1}$. Concerning y, one now distinguishes the following cases: Either $y = \emptyset$ or, for some y', either y is $y'\mathbf{q}_i$, or y is $y'\mathbf{q}_j$ for some $j < i$, or y is $y'\mathbf{q}_j$ for some $j > i$.

If $y = \emptyset$, then v', $x\mathbf{q}_i^{-1}$, $\emptyset$ satisfy the desired conditions. If y is $y'\mathbf{q}_i$, then $v', x, y\mathbf{q}_i^{-1} = y'\mathbf{q}_i\mathbf{q}_i^{-1}$ satisfy the desired conditions.

Now suppose that y is $y'\mathbf{q}_j$ for some $j < i$. Then, by Q_2, $y'\mathbf{q}_j\mathbf{q}_i^{-1} \equiv_2 y'\mathbf{q}_{i-1}^{-1}\mathbf{q}_j$. By our inductive hypothesis, there are x'' and y'' such that $xy'\mathbf{q}_{i-1}^{-1}, x'', y''$ satisfy the desired conditions. There follows that $xy'\mathbf{q}_{i-1}^{-1}\mathbf{q}_j, x'', y''\mathbf{q}_j$ satisfy the desired conditions. Hence, so do $v' = v\mathbf{q}_i^{-1} = xy\mathbf{q}_i^{-1} = xy'\mathbf{q}_j\mathbf{q}_i^{-1}, x'', y''\mathbf{q}_j$.

Finally, suppose that y is $y'\mathbf{q}_j$ for some $j > i$. Then, using Q_2 for the second assertion one obtains:

$$y'\mathbf{q}_j\mathbf{q}_i^{-1} = y'(\mathbf{q}_i\mathbf{q}_j^{-1})^{-1} \equiv_2 y'(\mathbf{q}_{j-1}^{-1}\mathbf{q}_i)^{-1} = y'\mathbf{q}_i^{-1}\mathbf{q}_{j-1} .$$

By our inductive hypothesis, there are x'' and y'' such that $xy'\mathbf{q}_i^{-1}, x'', y''$ satisfy the desired conditions. There follows that $xy'\mathbf{q}_i^{-1}\mathbf{q}_{j-1}, x'', y''\mathbf{q}_{j-1}$ satisfy the desired conditions. Hence, so do $v' = v\mathbf{q}_i^{-1} = xy\mathbf{q}_i^{-1} = xy'\mathbf{q}_j\mathbf{q}_i^{-1}, x'', y''\mathbf{q}_{j-1}$. □

SUBLEMMA 2 *Consider any xy in $F(\Gamma(\mathbf{q}))$ such that x contains no noncoupled occurrence of any $\mathbf{q}_j$ and y contains no noncoupled occurrence of any $\mathbf{q}_i^{-1}$. Let x' consist of the noncoupled occurrences of $\mathbf{q}_i^{-1} < \omega$, if any, in the order in which they occur in x, let y' consist of the noncoupled occurrences of $\mathbf{q}_j$, $j < \omega$, if any, in the order in which they occur in y, and let $gd(xy) = \langle s, t\rangle$. Then $xy \equiv \mathbf{q}_s\mathbf{q}_s^{-1}x'y'$.*

Proof. By Lemma 4(b), $xy \equiv \mathbf{q}_s\mathbf{q}_s^{-1}xy$. Suppose that, for some k, y contains an occurrence of $\mathbf{q}_k\mathbf{q}_k^{-1}$, so that $y = \dot{y}\mathbf{q}_k\mathbf{q}_k^{-1}\ddot{y}$ for some $\dot{y}$ and $\ddot{y}$. Let $gd(xy) = \langle s', t'\rangle$. In the case where $k \leq t'$, there holds, by Lemma 4(c), $x\dot{y}\mathbf{q}_k\mathbf{q}_k^{-1} \equiv x\dot{y}$, so that $\mathbf{q}_s\mathbf{q}_s^{-1}xy \equiv \mathbf{q}_s\mathbf{q}_s^{-1}x\dot{y}\mathbf{q}_k\mathbf{q}_k^{-1}\ddot{y} \equiv \mathbf{q}_s\mathbf{q}_s^{-1}x\dot{y}\ddot{y}$.

Now consider the case where $k > t'$. Let $r = k - t'$. Then $gd(x\dot{y}\mathbf{q}_k\mathbf{q}_k^{-1}) = gd(x\dot{y}\mathbf{q}_{t'+r}\mathbf{q}_{t'+r}^{-1}) = \langle s'+r, t'+r\rangle$. Also, by Lemma 4(a), $x\dot{y}\mathbf{q}_k\mathbf{q}_k^{-1} = x\dot{y}\mathbf{q}_{t'+r}\mathbf{q}_{t'+r}^{-1} \equiv \mathbf{q}_{s'+r}\mathbf{q}_{s'+r}x\dot{y}$. Since $x\dot{y}''\mathbf{q}_k\mathbf{q}_k^{-1}$ is an initial segment of xy, $gd(x\dot{y}\mathbf{q}_k\mathbf{q}_k^{-1}) = \langle s' + r, t' + r\rangle$, and $gd(xy) = \langle s, t\rangle$, therefore $s' + r \leq s$. Hence, by Lemma 3(g), $\mathbf{q}_s\mathbf{q}_s^{-1}\mathbf{q}_{s'+r}\mathbf{q}_{s'+r}^{-1} \equiv \mathbf{q}_s\mathbf{q}_s^{-1}$. There now follows

$$\mathbf{q}_s\mathbf{q}_s^{-1}xy = \mathbf{q}_s\mathbf{q}_s^{-1}x\dot{y}\mathbf{q}_k\mathbf{q}_k^{-1}\ddot{y} \equiv \mathbf{q}_s\mathbf{q}_s^{-1}\mathbf{q}_{s'+r}\mathbf{q}_{s'+r}^{-1}x\dot{y}\ddot{y} \equiv \mathbf{q}_s\mathbf{q}_s^{-1}x\dot{y}\ddot{y} .$$

Thus, in either case, in $\mathbf{q}_s\mathbf{q}_s^{-1}xy$ one can replace $y = \dot{y}\mathbf{q}_k\mathbf{q}_k^{-1}\ddot{y}$ by $\dot{y}\ddot{y}$.

If $\dot{y}\ddot{y}$ contains occurrences of $\mathbf{q}_{k'}\mathbf{q}_{k'}^{-1}$, $k' < \omega$, then, by a similar argument, one can omit all of these also, thereby obtaining the desired y'. If x contains occurrences of $\mathbf{q}_{k'}\mathbf{q}_{k'}^{-1}$, $k' < \omega$, then, by a similar argument, one can omit all of these, thereby obtaining the desired x'. □

Let $\equiv_1$ be the congruence on $\mathbf{F}(\Gamma(\tilde{\mathbf{q}}))$ that is generated by Q_1. As one can see from the proof of Theorem 1, if y' is a word formed from $\{\mathbf{q}_j : j < \omega\}$, then there is some y'' that is strictly ascending and, moreover, of the same length as y', such that $y' \equiv_1 y''$. There follows that if x' is a word formed from $\{\mathbf{q}_i : i < \omega\}$, then there is some x'' that is strictly descending and, moreover, of the same length as x', such that $x' \equiv_1 x''$.

From this observation, from Sublemmas 1 and 2, and from the definition of $\mathbf{q}$-normal, there now follows Lemma 5 for $\mathbf{F}(\Gamma(\tilde{\mathbf{q}}))$. For $\mathbf{F}(\Gamma(\tilde{\mathbf{r}}))$ the proof is similar. □

LEMMA 6. *If either v and w are $\mathbf{q}$-normal and $h(v) = h(w)$ or v and w are $\mathbf{r}$-normal and $h'(v) = h'(w)$, then $v = w$.*

Proof. Assume that $v = \mathbf{r}_s\mathbf{r}_s^{-1}xy$ and $w = \mathbf{r}_{s'}\mathbf{r}_{s'}^{-1}x'y'$ are $\mathbf{r}$-normal and that $h'(v) = h'(w)$. Since $h'(v) = h'(w)$, therefore $gd\ v = gd\ w$ and hence $s = s'$. Since $h'(v) = h'(w)$ and $\overline{\overline{U}} \geq 2$, therefore either $x = x' = \emptyset$ or x and x' are the same strictly descending word in $\mathbf{F}(\Gamma(\tilde{\mathbf{r}}))$. Likewise, since $h'(v) = h'(w)$ and $\overline{\overline{U}} \geq 2$, therefore either $y = y' = \emptyset$ or y and y' are the same strictly ascending word in $\mathbf{F}(\Gamma(\tilde{\mathbf{r}}))$. There follows that $v = w$. For the case where v and w are $\mathbf{q}$-normal and $h(v) = h(w)$ the argument is similar. □

We repeat the definitions of $\equiv$, $\equiv'$, h and h' that we have been using: $\equiv$ is the congruence on $\mathbf{F}(\Gamma(\tilde{\mathbf{q}})) = \mathbf{F}_{SI}(\Gamma(\tilde{\mathbf{q}}))$ that is generated by $Q_1 \cup Q_2 \cup Q_3 \cup Q_4$ and $\equiv'$ is the congruence on $\mathbf{F}(\Gamma(\tilde{\mathbf{r}})) = \mathbf{F}_{SI}(\Gamma(\tilde{\mathbf{r}}))$ that is generated by $R_1 \cup R_2 \cup R_3 \cup R_4$; also h is the homomorphism of $\mathbf{F}(\Gamma(\tilde{\mathbf{q}}))$ onto $\mathbf{S}_q$ such that $h(\mathbf{q}_i) = q_i$ for each $i < \omega$ and h' is the homomorphism of $\mathbf{F}(\Gamma(\tilde{\mathbf{r}}))$ onto $\mathbf{S}_r$ such that $h'(\mathbf{r}_i) = r_i$ for each $i < \omega$. Our presentation of $\mathbf{S}_q$ and that of $\mathbf{S}_r$ can now be stated as follows.

THEOREM 7. (a) *$\equiv$ coincides with $h \circ h^{\smile}$ and hence $\mathbf{F}(\Gamma(\tilde{\mathbf{q}}))/\!\equiv$ is isomorphic to $\mathbf{S}_q$.*

(b) *$\equiv'$ coincides with $h' \circ h'^{\smile}$ and hence $\mathbf{F}(\Gamma(\tilde{\mathbf{r}}))/\!\equiv'$ is isomorphic to $\mathbf{S}_r$.*

(c) *The mapping h'' such that $h''(q_i) = r_i$ for each $i < \omega$ can be extended to an isomorphism of $\mathbf{S}_q$ onto $\mathbf{S}_r$.*

Proof. Parts (a) and (b) follow from Lemmas 2, 5 and 6. Part (c) follows from parts (a) and (b). The isomorphism h''' of $\mathbf{S}_q$ onto $\mathbf{S}_r$ that extends h'' can also be described as follows: If $X = {}_s[\![(f_J^{\smile} \circ f_K) \cap (s \times t)]\!]_t$, then $h'''(X) = {}_{s+1}[\![\widehat{(f_J^{\smile} \circ f_K)^+} \cap ((s+1) \times (t+1))]\!]_{t+1}$. □

Recall that near the end of Chapter I we introduced for any $\lambda < \kappa = \overline{\overline{U}}$ and any $j < \omega$ the following one-one mapping of ${}_{[j,\omega)}U$ into ${}_{[j+1,\omega)}U$:

$$s_{\lambda,j} = \{\langle xz, x^\frown\langle u_\lambda\rangle^\frown z\rangle : |x|=j,\ |z| < \omega\}\ .$$

Now let any $\lambda < \kappa$ be given. We let $S_{s,\lambda}$ be the closure under $\circ$ of $\{s_{\lambda,j} : j < \omega\} \cup \{s_{\lambda,j}^{\smile} : j < \omega\}$. Also, we let $\mathbf{S}_{s,\lambda} = \langle S_{s,\lambda}, \circ, {}^{\smile}\rangle$. Further, letting $\Gamma(\tilde{\mathbf{s}}_\lambda) = \{\mathbf{s}_{\lambda,j} : j < \omega\}$ be disjoint from $\Gamma(\tilde{\mathbf{q}}) \cup \Gamma(\tilde{\mathbf{r}})$, we let $S_{\lambda,1}, S_{\lambda,2}, S_{\lambda,3}, S_{\lambda,4}$ be the relation on $F(\Gamma(\tilde{\mathbf{s}}_\lambda))$ that results from Q_1, Q_2, Q_3, Q_4 respectively when each $\mathbf{q}_j$ is replaced by $\mathbf{s}_{\lambda,j}$, and we let $\equiv_\lambda$ be the congruence on $\mathbf{F}(\Gamma(\tilde{\mathbf{s}}_\lambda))$ that is generated by $S_{\lambda,1} \cup S_{\lambda,2} \cup S_{\lambda,3} \cup S_{\lambda,4}$. Finally, we let h_λ be the homomorphism of $\mathbf{F}(\Gamma(\tilde{\mathbf{s}}_\lambda))$ onto $\mathbf{S}_{s,\lambda}$ such that $h_\lambda(\mathbf{s}_{\lambda,j}) = s_{\lambda,j}$ for each $j < \omega$.

One can verify that each of $S_{\lambda,1}, S_{\lambda,2}, S_{\lambda,3}, S_{\lambda,4}$ is included in $h_\lambda \circ h_\lambda{}^{\smile}$. Hence $\equiv_\lambda$ is included in $h_\lambda \circ h_\lambda{}^{\smile}$. Also, for $\equiv_\lambda$ and h_λ there holds an analogue of Lemma 5 or Lemma 6, respectively. Hence Theorem 7 above can be augmented as follows.

THEOREM 7. (d) *$\equiv_\lambda$ coincides with $h_\lambda \circ h_\lambda{}^{\smile}$ and hence $\mathbf{F}(\Gamma(\tilde{\mathbf{s}}_\lambda))/\equiv$ is isomorphic to $\mathbf{S}_{s,\lambda}$.*

(e) *$\mathbf{S}_{s,\lambda}$ is isomorphic to $\mathbf{S}_q$ and to $\mathbf{S}_r$.* □

We now turn to the task of giving a presentation of $\langle \mathbf{S}_q, \subseteq\rangle = \langle S_q, \circ, {}^{\smile}, \subseteq\rangle$ and one of $\langle \mathbf{S}_r, \subseteq\rangle = \langle S_r, \circ, {}^{\smile}, \subseteq\rangle$. We let Q_5, Q_6, R_5 and R_6 be the following relations on $F(\Gamma(\tilde{\mathbf{q}}))$ or on on $F(\Gamma(\tilde{\mathbf{r}}))$, respectively:

$$\begin{aligned}
Q_5 &= \{\langle \mathbf{q}_{i+1}\mathbf{q}_{i+1}^{-1}, \mathbf{q}_i\mathbf{q}_i^{-1}\rangle : i < \omega\} \\
Q_6 &= \{\langle \mathbf{q}_{i+1}\mathbf{q}_{i+1}^{-1}, \mathbf{q}_i^{-1}\mathbf{q}_i\rangle : i < \omega\} \\
R_5 &= \{\langle \mathbf{r}_{i+1}\mathbf{r}_{i+1}^{-1}, \mathbf{r}_i\mathbf{r}_i^{-1}\rangle : i < \omega\} \\
R_6 &= \{\langle \mathbf{r}_{i+1}\mathbf{r}_{i+1}^{-1}, \mathbf{r}_i^{-1}\mathbf{r}_i\rangle : i < \omega\}
\end{aligned}$$

For $\equiv$ as before we let $\lesssim$ be the least quasiordering of $F(\Gamma(\tilde{\mathbf{q}}))$ that includes $\equiv \cup Q_5 \cup Q_6$ and is compatible with $\mathbf{F}(\Gamma(\tilde{\mathbf{q}}))$ in the sense that $v \lesssim w$ implies $v^{-1} \lesssim w^{-1}$ and that $v \lesssim w$ and $v' \lesssim w'$ together imply that $v^\frown v' \lesssim w^\frown w'$. Evidently in this definition of $\preceq$ one may replace the clause that $\preceq$ includes $\equiv \cup Q_5 \cup Q_6$ by the clause that $\preceq$ includes $Q_1 \cup \cdots \cup Q_4 \cup Q_1{}^{\smile} \cdots \cup Q_4{}^{\smile} \cup Q_5 \cup Q_6$.

For $\equiv'$ as before we let $\preceq'$ be the least quasiordering of $F(\Gamma(\tilde{\mathbf{r}}))$ that includes $\equiv' \cup R_5 \cup R_6{}^{\smile}$ and is compatible with $\mathbf{F}(\Gamma(\tilde{\mathbf{r}}))$. Thus $\preceq$ is the least quasiordering of $F(\Gamma(\tilde{\mathbf{r}}))$ that is compatible with $\mathbf{F}(\Gamma(\tilde{\mathbf{r}}))$ and that includes $R_1 \cup \cdots \cup R_4 \cup R_1{}^{\smile} \cdots \cup R_4{}^{\smile} \cup R_5 \cup R_6{}^{\smile}$.

LEMMA 8. *If $v \preceq w$ then $h(v) \subseteq h(w)$. Also, if $v \preceq' w$ then $h'(v) \subseteq h'(w)$.*

Proof. By Lemma 2, $v \equiv w$ implies that $h(v) \equiv h(w)$ and hence that $h(v) \subseteq h(w)$. Also by Lemma 2, $v \equiv' w$ implies that $h'(v) \equiv h'(w)$ and hence that $h'(v) \subseteq h'(w)$.

Since $q_j \circ q_j{}^{\smile} = {}_{[j,\omega)}\overset{\circ}{1}$, therefore $q_{i+1} \circ q_{i+1}^{\smile} \subseteq q_i \circ q_i{}^{\smile}$. Therefore $\langle v, w\rangle \in Q_5$ implies that $h(v) \subseteq h(w)$.

As one can verify, and as we saw in Chapter I,

$$\begin{aligned} q_i{}^{\smile} \circ q_i &= c_i \\ &= \{\langle x, y\rangle \in (\bigcup_{n<\omega} {}^nU)^2 : i < |x| = |y| < \omega,\ x_n = y_n \text{ if } n \in |x| \cap -\{i\}\}\ . \end{aligned}$$

Since $q_{i+1} \circ q_{i+1}^{\smile} = {}_{[i+1,\omega)}\overset{\circ}{1}$, there follows that $q_{i+1} \circ q_{i+1}^{\smile} \subseteq q_i{}^{\smile} \circ q_i$. Therefore $\langle v, w\rangle \in Q_6$ implies that $h(v) \subseteq h(w)$.

Since $r_j \circ r_j{}^{\smile} = {}_{[j+1,\omega)}\overset{\circ}{1}$, therefore $r_{i+1} \circ r_{i+1}^{\smile} \subseteq r_i \circ r_i{}^{\smile}$. Therefore $\langle v, w\rangle \in R_5$ implies that $h'(v) \subseteq h'(w)$.

As one can verify, and as we saw in Chapter I, $r_i{}^{\smile} \circ r_i = e_{i,i+1}$. There follows that $r_i{}^{\smile} \circ r_i \subseteq r_{i+1} \circ r_{i+1}^{\smile} = {}_{[i+2,\omega)}\overset{\circ}{1}$. Therefore, $\langle v, w\rangle \in R_6{}^{\smile}$ implies that $h'(v) \subseteq h'(w)$. □

LEMMA 9. *Assume that N' and N are one-one and bisectable and that $N' \subseteq N \neq N'$. Then for some n, $0 < n < \omega$, there are exactly n pairs $\langle j_0, k_0\rangle, \ldots, \langle j_{n-1}, k_{n-1}\rangle$ in N that are not in N'. Moreover, if $N_0 = N$ and, for each $m < n$, $N_{m+1} = N_m \cap -\{\langle j_m, k_m\rangle\}$, then $N' = N_n$ and, for each m, $N_{m+1} = f_{\{j_m\}}^{\smile} \circ f_{\{j_m\}} \circ N_m$.*

Proof. Since N' and N both are bisectable and sine $N' \subseteq N$, therefore some $\langle s, t\rangle$ bisects them both and $N \cap -N'$ is a finite set $\{\langle j_0, k_0\rangle, \ldots, \langle j_{n-1}, k_{n-1}\rangle\}$. Then from the definition of $N_0, \ldots, N_n$ there follows that $N_n = N'$.

Let $m < n$. Since $f_{\{j_m\}}^{\smile} \circ f_{\{j_m\}} = \{\langle i, i\rangle \in \omega \times \omega : i \neq j_m\}$, therefore $f_{\{j_m\}}^{\smile} \circ f_{\{j_m\}} \circ N_m$ is the set of those pairs $\langle j, k\rangle$ in N_m such that $j \neq j_m$. Since N is one-one, therefore the only pair in N_m that is not in $f^{\smile}_{\{j_m\}} \circ f_{\{j_m\}} \circ N_m$ is the pair $\langle j_m, k_m \rangle$. Hence $f^{\smile}_{\{j_m\}} \circ f_{\{j_m\}} \circ N_m = N_m \cap -\{\langle j_m, k_m\rangle\} = N_{m+1}$. □

LEMMA 10. (a) *Consider any w and w' in $F(\Gamma(\mathbf{q}))$ such that $h(w) \subseteq h(w')$ and gd $w =$ gd w'. Let $\leq_6$ be the least quasiordering of $F(\Gamma(\mathbf{q}))$ that is compatible with $\mathbf{F}(\Gamma(\tilde{\mathbf{q}}))$ and includes $\equiv \cup\, Q_6$. Then $w \lesssim_6 w'$.*

(b) *Consider any v' and v in $F(\Gamma(\mathbf{r}))$ such that $h'(v') \subseteq h'(v)$ and gd $v' =$ gd v. Let ${}_6\leq'$ be the least quasiordering of $F(\Gamma(\mathbf{r}))$ that is compatible with $\mathbf{F}(\Gamma(\tilde{\mathbf{r}}))$ and includes $\equiv' \cup\, R^{\smile}_6$. Then $v' \;{}_6\lesssim'\, v$.*

Proof. (a) Let gd $w =$ gd $w' = \langle s, t\rangle$. By Theorems IV.11 and III.14, $h(w) = {}_s[\![N \cap (s \times t)]\!]_t$ and $h(w') = {}_s[\![N' \cap (s \times t)]\!]_t$ for some N and N' such that N and N' are one-one and bisected by $\langle s, t\rangle$. (Although N, $N^{\smile}$, N' and $N'^{\smile}$ are also order-preserving, this fact will not be used in our proof.) Since ${}_s[\![N \cap (s \times t)]\!]_t \subseteq {}_s[\![N' \cap (s \times t)]\!]_t$ and $\overset{=}{U} \geq 2$, therefore by Lemma III.5(c), $N' \cap (s \times t) \subseteq N \cap (s \times t)$. Since $\langle s, t\rangle$ bisects N and N', there follows that $N' \subseteq N$. If $N' = N$ then $h(w) = h(w')$, hence $w \equiv w'$, by Theorem 7(a), and therefore $w \lesssim w'$. Now assume that $N' \neq N$. Then, by Lemma 9, there is some $n > 0$ and there are exactly n pairs $\langle j_0, k_0\rangle, \ldots, \langle j_{n-1}, k_{n-1}\rangle$ in $N \cap -N'$ such that, if $N_0 = N$ and, for each $m < n$, $N_{m+1} = N_m \cap -\{\langle j_m, k_m\rangle\}$, then $N' = N_n$ and, for each $m < n$, $N_{m+1} = f^{\smile}_{\{j_m\}} \circ f_{\{j_m\}} \circ N_m$, hence $N_{m+1} \cap (s \times t) = (f^{\smile}_{\{j_m\}} \circ f_{\{j_m\}} \circ N_m) \cap (s \times t)$, and hence, since $\langle s, t\rangle$ bisects N and therefore $j_m < s$,

$$ {}_s[\![N_{m+1} \cap (s \times t)]\!]_t = {}_s[\![(f^{\smile}_{\{j_m\}} \circ f_{\{j_m\}} \circ N_m) \cap (s \times t)]\!]_t \;. $$

Let $w_0 = w$ and, for $m < n$, let $w_{m+1} = \mathbf{q}^{-1}_{j_m}\mathbf{q}_{j_m} w_m$. Since $j_m < s$ for each $m < n$, therefore, by induction, gd $w_m = \langle s, t\rangle$ for each $m \leq n$. Now let $m < n$ and assume as inductive hypothesis that $h(w_m) = {}_s[\![N_m \cap (s \times t)]\!]_t$. Then

$$
\begin{aligned}
h(w_{m+1}) &= q^{\smile}_{j_m} \circ q_{j_m} \circ h(w_m) = q^{\smile}_{j_m} \circ q_{j_m} \circ {}_s[\![N_m \cap (s \times t)]\!]_t \\
&= ({}_{[s,\omega)}\overset{\circ}{1} \circ q^{\smile}_{j_m} \circ q_{j_m}) \circ {}_s[\![N_m \cap (s \times t)]\!]_t \\
&= {}_s[\![(f^{\smile}_{\{j_m\}} \circ f_{\{j_m\}}) \cap (s \times s)]\!]_s \circ {}_s[\![N_m \cap (s \times t)]\!]_t \\
&= {}_s[\![(f^{\smile}_{\{j_m\}} \circ f_{\{j_m\}}) \cap (s \times s)) \circ (N_m \cap (s \times t))]\!]_t \\
&= {}_s[\![(f^{\smile}_{\{j_m\}} \circ f_{\{j_m\}} \circ N_m) \cap (s \times t)]\!]_t
\end{aligned}
$$

Together with the equality given earlier this implies that $h(w_{m+1}) = {}_s[\![N_{m+1} \cap (s \times t)]\!]_t$. Hence, by induction, $h(w_m) = {}_s[\![N_m \cap (s \times t)]\!]_t$ for any $m \leq n$. In particular, $h(w_n) = {}_s[\![N' \cap (s \times t)]\!]_t = h(w')$.

Let $m<n$. Since $gd\ w_m=\langle s,t\rangle$ and $j_m<s$, therefore $h(w_m)=h(\mathbf{q}_{j_m+1}\mathbf{q}_{j_m+1}^{-1}w_m)$. Hence by Lemma 4(b), $w_m \equiv \mathbf{q}_{j_m+1}\mathbf{q}_{j_m+1}^{-1}w_m$. By Q_6, $\mathbf{q}_{j_m+1}\mathbf{q}_{j_m+1}^{-1} \dot{\le}_6\ \mathbf{q}_{j_m}^{-1}\mathbf{q}_{j_m}$. There follows that $w_m \dot{\le}_6\ \mathbf{q}_{j_m}^{-1}\mathbf{q}_{j_m}w_m = w_{m+1}$. By induction $w = w_0 \dot{\le}_6\ w_n$. Since $h(w_n) = h(w')$ therefore, by Theorem 7, $w_n \equiv w'$. There now follows that $w \dot{\le}_6\ w'$.

(b) Let $gd\ v' = gd\ v = \langle s+1,t+1\rangle$. By Lemma 5 and Theorem 7(b) we may assume without loss of generality that v' and v are $\mathbf{r}$-normal. Let w' and w result from v' or v, respectively, by replacing every $\mathbf{r}_i$ by $\mathbf{q}_i$ and every $\mathbf{r}_i^{-1}$ by $\mathbf{q}_i^{-1}$. Then w' and w are $\mathbf{q}$-normal and $gd\ w' = gd\ w = \langle s,t\rangle$. Since $h(w)$ and $h(w')$ have the same grade, therefore $h(w) \subseteq h(w')$ if and only if $\overset{\omega}{\rightarrow}(h(w)) \subseteq \overset{\omega}{\rightarrow}(h(w'))$. Similarly, since $h'(v')$ and $h'(v)$ have the same grade, therefore $h'(v') \subseteq h'(v)$ if and only if $\overset{\omega}{\rightarrow}(h'(v')) \subseteq \overset{\omega}{\rightarrow}(h'(v))$. For some N and N' that are one-one and bisected by $\langle s,t\rangle$, there hold $\overset{\omega}{\rightarrow}(h(w)) = {}_\omega[N]$, $\overset{\omega}{\rightarrow}(h(w')) = {}_\omega[N']$, $\overset{\omega}{\rightarrow}(h'(v)) = {}_\omega[\overset{\frown}{N^+}]$, and $\overset{\omega}{\rightarrow}(h'(v')) = {}_\omega[\overset{\frown}{N'^+}]$. Since $h'(v') \subseteq h'(v)$, hence $\overset{\omega}{\rightarrow}(h'(v')) \subseteq \overset{\omega}{\rightarrow}(h'(v))$, and hence ${}_\omega[\overset{\frown}{N'^+}] \subseteq {}_\omega[\overset{\frown}{N^+}]$, therefore $\overset{\frown}{N^+} \subseteq \overset{\frown}{N'^+}$. There follows that $N' \subseteq N$. Since N is bisected by $\langle s,t\rangle$ and $\overset{\omega}{\rightarrow}(h(w)) = {}_\omega[N]$, therefore $h(w) = {}_s[\![N\cap(s\times t)]\!]_t$. Similarly $h(w') = {}_s[\![N'\cap(s\times t)]\!]_t$.

By the proof of part (a), there are $w_0,\dots,w_n$ such that $w_0 = w$, $w_n \equiv w'$ and, for each $m<n$, $w_{m+1} = \mathbf{q}_{j_m}^{-1}\mathbf{q}_{j_m}w_m$ for some $j_m < s$. For each $m \le n$, let v_m result from w_m by replacing each $\mathbf{q}_i$ by $\mathbf{r}_i$ and each $\mathbf{q}_i^{-1}$ by $\mathbf{r}_i^{-1}$. Thus, for each $m<n$, $v_{m+1} = \mathbf{r}_{j_m}^{-1}\mathbf{r}_{j_m}v_m$.

Let $m<n$. Since $gd\ v_m = \langle s+1,t+1\rangle$ and $j_m < s$, therefore $h'(v_m) = h'(\mathbf{r}_{j_m+1}\mathbf{r}_{j_m+1}^{-1}v_m)$. Hence, by Theorem 7(b), $v_m \equiv' \mathbf{r}_{j_m+1}\mathbf{r}_{j_m+1}^{-1}v_m$. By $\breve{R_6}$, $\mathbf{r}_{j_m}^{-1}\mathbf{r}_{j_m}\ {}_6\dot{\le}'\ \mathbf{r}_{j_m+1}\mathbf{r}_{j_m+1}^{-1}$. There follows that $v_{m+1} = \mathbf{r}_{j_m}^{-1}\mathbf{r}_{j_m}v_m\ {}_6\dot{\le}'\ v_m$. By induction, $v_n\ {}_6\dot{\le}' v_0$. Hence $v'\ {}_6\dot{\le}'\ v$. □

LEMMA 11. (a) *If* $h(w) \subseteq h(w')$, *then* $w \dot{\le} w'$.
(b) *If* $h'(w) \subseteq h'(w')$, *then* $w \dot{\le}' w'$.

Proof. (a) Assume that $h(w) \subseteq h(w')$. Let $gd\ w = \langle s,t\rangle$ and $gd\ w' = \langle s',t'\rangle$. Then, by Lemma II.5(e), $\langle s,t\rangle = \langle s'+n, t'+n\rangle$ for some n. From Theorem II.7(f) there follows that $h(w) \subseteq h(\mathbf{q}_s\mathbf{q}_s^{-1}w')$. Then, since $gd\ w = gd(\mathbf{q}_s\mathbf{q}_s^{-1}w')$, therefore, by Lemma 10(a), $w \dot{\le} \mathbf{q}_s\mathbf{q}_s^{-1}w'$. Since $s \ge s'$ therefore, by Q_5, $\mathbf{q}_s\mathbf{q}_s^{-1}w \dot{\le} \mathbf{q}_{s'}\mathbf{q}_{s'}^{-1}w'$. Since $gd\ w' = \langle s',t'\rangle$ therefore, by Lemma 4(b), $\mathbf{q}_{s'}\mathbf{q}_{s'}^{-1}w' \equiv w'$. There now follows that $w \dot{\le} w'$.

(b) One uses Lemma 10(b) and R_5 for a similar argument. □

We now give a presentation of $\langle \mathbf{S}_q, \subseteq\rangle$ and one of $\langle \mathbf{S}_r, \subseteq\rangle$.

THEOREM 12. (a) *For any v and w in $F(\Gamma(\tilde{\mathbf{q}}))$, $h(v) \subseteq h(w)$ if and only if $v \lessdot w$. There follows that $\langle \mathbf{F}(\Gamma(\tilde{\mathbf{q}})), \lessdot \rangle / \equiv$ is isomorphic to $\langle \mathbf{S}_q, \subseteq \rangle$.*

(b) *For any v and w in $F(\Gamma(\tilde{\mathbf{r}}))$, $h'(v) \subseteq h'(w)$ if and only if $v \lessdot' w$. There follows that $\langle \mathbf{F}(\Gamma(\tilde{\mathbf{r}})), \lessdot' \rangle / \equiv$ is isomorphic to $\langle \mathbf{S}_r, \subseteq \rangle$.*

Proof. (a) By Lemma 8, Lemma 11(a), and Theorem 7(a).

(b) By Lemma 8, Lemma 11(b), and Theorem 7(b). □

Since $\lessdot$ includes Q_5 and Q_6 but not $Q_5{}^{\smile}$, while $\lessdot\,'$ includes $R_5{}^{\smile}$ and R_6 but not R_5, therefore, in contrast to Theorem 7(c), $\langle \mathbf{S}_q, \subseteq \rangle$ is neither isomorphic to $\langle \mathbf{S}_r, \subseteq \rangle$ nor isomorphic to $\langle \mathbf{S}_r, \supseteq \rangle$.

There follows that Q_5 is independent from $\{Q_1, Q_2, Q_3, Q_4, Q_6\}$ in the following sense: If $\equiv$ is the congruence on $\mathbf{F}(\Gamma(\tilde{\mathbf{q}}))$ that is generated by $Q_1 \cup Q_2 \cup Q_3 \cup Q_4$ and $\lessdot_6$ is the least quasiordering of $F(\Gamma(\tilde{\mathbf{q}}))$ that is compatible with $\mathbf{F}(\Gamma(\tilde{\mathbf{q}}))$ and includes $\equiv \cup Q_6$ then $\lessdot_6$ does not include Q_5. There also follows that R_5 is independent in a similar sense from $\{R_1, R_2, R_3, R_4, R_6\}$.

Given any $\lambda < \kappa$, we let $S_{\lambda,5}$ and $S_{\lambda,6}$ be the relation on $F(\Gamma(\tilde{\mathbf{s}}_\lambda))$ that results from Q_5 or Q_6, respectively, when each $\mathbf{q}_j$ is replaced by $\mathbf{s}_{\lambda,j}$, and we let $\lessdot_\lambda$ be the least quasiordering of $F(\Gamma(\tilde{\mathbf{s}}_\lambda))$ that is compatible with $\mathbf{F}(\Gamma(\tilde{\mathbf{s}}_\lambda))$ and that includes $\equiv_\lambda \cup S_{\lambda,5} \cup S^{\smile}_{\lambda,6}$. An argument analogous to the proof of Theorem 12(b) shows that to Theorem 12 one can add the following part (c). From it and from Theorem 12(b) there follows part (d).

THEOREM 12. (c) *For any v and w in $F(\Gamma(\tilde{\mathbf{s}}_\lambda))$, $h_\lambda(v) \subseteq h_\lambda(w)$ if and only if $v \lessdot_\lambda w$. There follows that $(\mathbf{F}(\Gamma(\tilde{\mathbf{s}}_\lambda), \lessdot_\lambda) / \equiv_\lambda$ is isomorphic to $\langle \mathbf{S}_{s,\lambda}, \subseteq \rangle$.*

(d) *$\langle \mathbf{S}_{s,\lambda}, \subseteq \rangle$ is isomorphic to $\langle \mathbf{S}_r, \subseteq \rangle$.* □

Recall that ${}_{[\omega]}S_q$ and ${}_{[\omega]}S_r$ are the closure under $\circ$ and ${}^{\smile}$ of $\{\overset{\omega}{\to} q_i : i < \omega\}$ or of $\{\overset{\omega}{\to} r_i : i < \omega\}$, respectively. By Theorem IV.7(a), ${}_{[\omega]}\mathbf{S}_q = \langle {}_{[\omega]}S_q, \circ, {}^{\smile} \rangle$ and ${}_{[\omega]}\mathbf{S}_r = \langle {}_{[\omega]}S_r, \circ, {}^{\smile} \rangle$ are isomorphic. (This also follows from Theorem 7(c), since they are the homomorphic image under $\overset{\omega}{\to}$ of $\mathbf{S}_q$ or of $\mathbf{S}_r$, respectively.) Then, by Theorems III.12(a), III.12(b), and III.17, $\langle {}_{[\omega]}\mathbf{S}_q, \subseteq \rangle$ and $\langle {}_{[\omega]}\mathbf{S}_r, \supseteq \rangle$ are isomorphic. This contrasts with what we remarked a moment ago about $\langle \mathbf{S}_q, \subseteq \rangle$ and $\langle \mathbf{S}_r, \supseteq \rangle$.

Recall that $f_{\{i\}}$ and $(\widehat{f^{\smile}_{\{i\}}})^+$ are the successor or predecessor function, respectively, that delays until i. Let $S' = \{N \in Bsl : N \text{ and } N^{\smile} \text{ are order-preserving}\}$, so that S' is the closure of $\{f_{\{i\}} : i < \omega\}$ under $\circ$ and ${}^{\smile}$, and let $S'' = \{\widehat{N^+} : N \in S'\}$ so that S'' is the closure of $\{(\widehat{f^{\smile}_{\{i\}}})^+ : i < \omega\}$ under $\bullet$ and ${}^{\smile}$. By Theorem III.17, $\mathbf{S}' = \langle S', \circ, {}^{\smile} \rangle$ and $\mathbf{S}'' = \langle S'', \bullet, {}^{\smile} \rangle$ are isomorphic, and also $\langle \mathbf{S}', \subseteq \rangle$ and $\langle \mathbf{S}'', \supseteq \rangle$ are isomorphic. Moreover, by Theorem IV.3(d), $\mathbf{S}'$ is isomorphic to ${}_{[\omega]}\mathbf{S}_q$ and $\langle \mathbf{S}', \supseteq \rangle$ to $\langle {}_{[\omega]}\mathbf{S}_q, \subseteq \rangle$.

Since in each case it is clear what set Γ and what homomorphism to use, there follows that giving a presentation of ${}_{[\omega]}\mathbf{S}_q$ amounts to also giving a presentation of ${}_{[\omega]}\mathbf{S}_r$, $\mathbf{S}'$, and $\mathbf{S}''$, and that giving a presentation of $\langle {}_{[\omega]}\mathbf{S}_q, \subseteq \rangle$ amounts to also giving a presentation of $\langle {}_{[\omega]}\mathbf{S}_r, \supseteq \rangle$, $\langle \mathbf{S}', \supseteq \rangle$, and $\langle \mathbf{S}'', \subseteq \rangle$. In Theorem 13 below we will give a presentation of ${}_{[\omega]}\mathbf{S}_q$ and one of $\langle {}_{[\omega]}\mathbf{S}_q, \subseteq \rangle$.

In Theorem 13 below we let $\equiv^+$ be the least congruence on $\mathbf{F}(\Gamma(\tilde{\mathbf{q}}))$ that includes $\equiv \cup Q_5$ and $\lessdot^+$ the least quasiordering of $F(\Gamma(\tilde{\mathbf{q}}))$ compatible with $\mathbf{F}(\Gamma(\tilde{\mathbf{q}}))$ that includes $\lessdot \cup Q_5^{\smile}$. Since $\lessdot$ includes Q_5, therefore $\lessdot^+$ is also the least quasiordering of $F(\Gamma(\tilde{\mathbf{q}}))$ compatible with $\mathbf{F}(\Gamma(\tilde{\mathbf{q}}))$ that includes $\equiv^+ \cup Q_6$. As is easily shown, Q_4 is included in the least congruence on $\mathbf{F}(\Gamma(\tilde{\mathbf{q}}))$ that includes $Q_3 \cup Q_5$. There follows that $\equiv^+$ coincides with the least congruence on $\mathbf{F}(\Gamma(\tilde{\mathbf{q}}))$ that includes $Q_1 \cup Q_2 \cup Q_3 \cup Q_5$.

THEOREM 13. *Let h be the homomorphism of $\mathbf{F}(\Gamma(\tilde{\mathbf{q}}))$ onto ${}_{[\omega]}\mathbf{S}_q$ such that $h(\mathbf{q}_i) = \overset{\omega}{\rightarrow} q_i$ for each $i < \omega$.*

(a) *$\equiv^+$ coincides with $h \circ h^{\smile}$ and hence $\mathbf{F}(\Gamma(\tilde{\mathbf{q}}))/{\equiv^+}$ is isomorphic to ${}_{[\omega]}\mathbf{S}_q$.*

(b) *$h(v) \subseteq h(w)$ if and only if $v \lessdot^+ w$. Hence $\langle \mathbf{F}(\Gamma(\tilde{\mathbf{q}})), \lessdot^+ \rangle/{\equiv^+}$ is isomorphic to $\langle {}_{[\omega]}\mathbf{S}_q, \subseteq \rangle$.*

Proof. (a) Since ${}_{[\omega]}\mathbf{S}_q$ is a homomorphic image of $\mathbf{S}_q$, there follows from Lemma 2 that $Q_1 \cup Q_2 \cup Q_3 \cup Q_4 \subseteq h \circ h^{\smile}$. Since, for every $i < \omega$,

$$\overset{\omega}{\rightarrow} q_i \circ (\overset{\omega}{\rightarrow} q_i)^{\smile} = \{\langle x, x \rangle : x \in {}^{\omega}U\} = {}_{[\omega]}\overset{\circ}{1} ,$$

therefore $Q_5 \subseteq h \circ h^{\smile}$. There now follows that $\equiv^+$ is included in $h \circ h^{\smile}$.

Since $\equiv$ is included in $\equiv^+$ therefore, by Lemma 5, for any v in $\mathbf{F}(\Gamma(\tilde{\mathbf{q}}))$ there is some w that is $\mathbf{q}$-normal such that $v \equiv^+ w$.

Assume that $h(v) = h(v')$. Let $v \equiv^+ w$ and $v' \equiv^+ w'$, where w and w' are $\mathbf{q}$-normal. Since $\equiv^+$ is included in $h \circ h^{\smile}$, therefore $h(w) = h(w')$. Let ${}_{\omega}\langle N \rangle$ and ${}_{\omega}\langle N' \rangle$ be the $\langle \omega, \omega \rangle$ triple associated with w or with w', respectively. Then $h(w) = {}_{\omega}[N]$ and $h'(w) = {}_{\omega}[N']$. Since $h(w) = h(w')$, therefore ${}_{\omega}[N] = {}_{\omega}[N']$. Since ${}_{\omega}\langle N \rangle$ and ${}_{\omega}\langle N' \rangle$ are canonical, there follows that $N = N'$. Since $N = f_J^{\smile} \circ f_K$ for some J and K, therefore for some s and s', w and w' are the $\mathbf{q}$-normal word $\mathbf{q}_s \mathbf{q}_s^{-1} xy$ or $\mathbf{q}_{s'} \mathbf{q}_{s'}^{-1} xy$, respectively, such that $\{j : \mathbf{q}_j^{-1} \text{ occurs in } x\} = J$ and $\{k : \mathbf{q}_k^{-1} \text{ occurs in } y\} = K$. If $s \geq s'$ then, by Q_5 and induction, $w = \mathbf{q}_s \mathbf{q}_s^{-1} xy \equiv^+ \mathbf{q}_{s'} \mathbf{q}_{s'}^{-1} xy = w$. If $s \leq s'$ then, again by Q_5 and induction, $w' = \mathbf{q}_{s'} \mathbf{q}_{s'}^{-1} xy \equiv^+ \mathbf{q}_s \mathbf{q}_s^{-1} xy = w$. Hence, in either case, $w \equiv^+ w'$ and hence $v \equiv^+ v'$. This concludes our proof that

$h \circ h^{\smile}$ is included in $\equiv^+$ and thus our proof that they coincide. There follows that $\mathbf{F}(\Gamma(\tilde{\mathbf{q}}))/\equiv^+$ is isomorphic to ${}_{[\omega]}\mathbf{S}_q$.

(b) Since $\langle {}_{[\omega]}\mathbf{S}_q, \subseteq \rangle$ is the homomorphic image under $\overset{\omega}{\rightarrow}$ of $\langle \mathbf{S}_q, \subseteq \rangle$, there follows from Lemma 8 that $\langle v, v' \rangle \in Q_6$ implies that $h(v) \subseteq h(v')$. There now follows that if $v \lessdot^+ v'$ then $h(v) \subseteq h(v')$.

Assume that $h(v) \subseteq h(v')$. Then there are w and w' that are $\mathbf{q}$-normal such that $v \equiv^+ w$, $v' \equiv^+ w'$, and $h(w) \subseteq h(w')$. Then a proof similar to that of Lemma 10(a), but somewhat simpler, shows that $w \lessdot^+ w'$. Hence $v \lessdot^+ v'$. This concludes our proof that $h(v) \subseteq h(v')$ implies $v \lessdot^+ v'$ and thus our proof that $h(v) \subseteq h(v')$ if and only if $v \lessdot^+ v'$. There follows that $\langle \mathbf{F}(\Gamma(\tilde{\mathbf{q}})), \lessdot^+ \rangle/\equiv^+$ and $\langle {}_{[\omega]}\mathbf{S}_q, \subseteq \rangle$ are isomorphic. □

Recall from Chapter I that for any $i < \omega$, $p_i = p_{(i,i+1)}$ is the permutation of order 2 of ${}_{[i+2,\omega]}U$ such that, for any $n \geq i+2$ and any $x = \langle x_0, \ldots, x_{i-1}, x_i, x_{i+1}, \ldots, x_{n-1} \rangle$ in ${}^nU$,

$$\begin{aligned} &p_i(\langle x_0, \ldots, x_{i-1}, x_i, x_{i+1}, x_{i+2}, \ldots, x_{n-1} \rangle) \\ &= \langle x_0, \ldots, x_{i-1}, x_{i+1}, x_i, x_{i+2}, \ldots, x_{n-1} \rangle \ . \end{aligned}$$

Also recall that S_p is the closure of $\{p_i : i < \omega\}$ under $\circ$ and $^{\smile}$ and hence, since $p_i^{\smile} = p_i$, also the closure of $\{p_i : i < \omega\}$ under $\circ$, that $\mathbf{S}_p = \langle S_p, \circ, {}^{\smile} \rangle$ and that $\langle \mathbf{S}_p, \subseteq \rangle = \langle S_p, \circ, {}^{\smile}, \subseteq \rangle$. Further, recall that ${}_{[\omega]}S_p$ is the closure under $\circ$ and $^{\smile}$ of $\{\overset{\omega}{\rightarrow} p_i : i < \omega\}$, that ${}_{[\omega]}\mathbf{S}_p = \langle {}_{[\omega]}S_p, \circ, {}^{\smile} \rangle$, and that ${}_{[\omega]}\mathbf{S}_p$ is the homomorphic image under $\overset{\omega}{\rightarrow}$ of $\mathbf{S}_p$.

We now turn to the second major task of this chapter, that of giving a presentation of $\mathbf{S}_p$ and one of $\langle \mathbf{S}_p, \subseteq \rangle$. This will be done by modifying a presentation of ${}_{[\omega]}\mathbf{S}_p$. This presentation, in turn, will be obtained by making use, for each n such that $2 \leq n < \omega$, of a well-known presentation of the group of all permutations of $n = \{0, \ldots, n-1\}$.

Consider any δ such that $2 \leq \delta \leq \omega$. We let $S_{p,\delta} = S_\delta$ be the closure under $\circ$ and $^{\smile}$ of $\{(i, i+1) \cap (\delta \times \delta) : i+1 < \delta\}$. Also we let $\mathbf{S}_{p,\delta} = \mathbf{S}_\delta = \langle S_\delta, \circ, {}^{\smile} \rangle$. Thus, if $2 \leq n < \omega$, then $\mathbf{S}_{p,n} = \mathbf{S}_n$ is the symmetric group whose universe S_n is the set of all permutations of $n = \{0, \ldots, n-1\}$. Also, $\mathbf{S}_{p,\omega} = \mathbf{S}_\omega$ is the group whose universe $S_{p,\omega} = S_\omega$ is the set of those permutations π of ω such that π is bisectable or, equivalently, such that $\{i : \pi(i) \neq i\}$ is finite. From Theorem IV.3(b) there follows that ${}_{[\omega]}\mathbf{S}_p$ is isomorphic under ${}_\omega[\ \ ]$ to $\mathbf{S}_{p,\omega} = \mathbf{S}_\omega$.

For $\mathbf{S}_n$, $2 \leq n < \omega$, there is a beautiful presentation, as a group, that was given by E. H. Moore in 1897 and uses $\{(i, i+1) \cap (n \times n) : i+1 < n\}$ as a set of generators. (See [Moo], or [CM], sec.6.2, or [Jo], pp.46–49.) We will adapt this presentation to obtain, for $2 \leq n < \omega$, a presentation, with the same set of generators, of $\mathbf{S}_n$ as a semigroup with involution. These presentations of $\mathbf{S}_n$, $2 \leq n < \omega$, yield in a straightforward way a presentation, also as a semigroup with involution, of $\mathbf{S}_\omega = \mathbf{S}_{p,\omega}$ and thus one of its isomorph ${}_{[\omega]}\mathbf{S}_p$. Suitable changes in this presentation of ${}_{[\omega]}\mathbf{S}_p$ will then yield a presentation, again as a semigroup with involution, of its pre-homomorphic image $\mathbf{S}_p$.

(Our main purpose in presenting $\mathbf{S}_p$ and ${}_{[\omega]}\mathbf{S}_p$ is to make use of the presentation later on, in presenting $\mathbf{S}_{pqe}$ or ${}_{[\omega]}\mathbf{S}_{pqe}$, respectively. Thus, even though $\mathbf{S}_p$ is an inverse semigroup and ${}_{[\omega]}\mathbf{S}_p$ is a group, a presentation as such would not suit this purpose, since, as can be seen from a remark after Theorem III.11, neither $\mathbf{S}_{pqe}$ nor ${}_{[\omega]}\mathbf{S}_{pqe}$ is an inverse semigroup.)

Until further notice, we let ${}_\omega\Gamma(\tilde{\mathbf{p}}) = {}_\omega\Gamma$ be a set $\{\mathbf{p}_i : i < \omega\}$ disjoint from $\Gamma(\tilde{\mathbf{q}})$ and from $\Gamma(\tilde{\mathbf{r}})$, such that, if $i \neq j$, then $\mathbf{p}_i \neq \mathbf{p}_j$. Also, for any n such that $2 \leq n < \omega$ we let ${}_n\Gamma(\tilde{\mathbf{p}}) = \{\mathbf{p}_i \in \Gamma : i + 1 < n\}$. Further, for any δ such that $2 \leq \delta \leq \omega$, we let ${}_\delta P_1, \ldots, {}_\delta P_6$ be the relation on $F_{SI}({}_\delta\Gamma) = F({}_\delta\Gamma)$ that is given below, where, for $i + 1 < \delta$ and $0 \leq m < \omega$, $\mathbf{p}_i^1 = \mathbf{p}_i$ and $\mathbf{p}_i^{m+1} = \mathbf{p}_i\mathbf{p}_i^m$. For brevity, we often let $\Gamma = {}_\omega\Gamma$ and $P_1 = {}_\omega P_1, \ldots, P_6 = {}_\omega P_6$.

$$\begin{aligned}
{}_\delta P_1 &= \{\langle \mathbf{p}_i^{-1}, \mathbf{p}_i\rangle : i + 1 < \delta\} \\
{}_\delta P_2 &= \{\langle \mathbf{p}_i^2\mathbf{p}_i, \mathbf{p}_i\rangle : i + 1 < \delta\} \\
{}_\delta P_3 &= \{\langle \mathbf{p}_i^2\mathbf{p}_{i+1}, \mathbf{p}_{i+1}\rangle : i + 2 < \delta\} \\
{}_\delta P_4 &= \{\langle \mathbf{p}_{i+1}^2\mathbf{p}_i, \mathbf{p}_i\rangle : i + 2 < \delta\} \\
{}_\delta P_5 &= \{\langle \mathbf{p}_i\mathbf{p}_{i+1}\mathbf{p}_i, \mathbf{p}_{i+1}\mathbf{p}_i\mathbf{p}_{i+1}\rangle : i + 2 < \delta\} \\
{}_\delta P_6 &= \{\langle \mathbf{p}_i\mathbf{p}_j, \mathbf{p}_j\mathbf{p}_i\rangle : i \leq j - 2,\ j + 1 < \delta\}
\end{aligned}$$

LEMMA 14. *Let $2 \leq \delta \leq \omega$.*

(a) *The congruence on $\mathbf{F}({}_\delta\Gamma) = \mathbf{F}_{SI}({}_\delta\Gamma(\tilde{\mathbf{p}}))$ that is generated by ${}_\delta P_2 \cup {}_\delta P_3 \cup {}_\delta P_4$ coincides with the congruence on $\mathbf{F}({}_\delta\Gamma)$ that is generated by*

$${}_\delta P_{2,3,4} = \{\langle \mathbf{p}_i^2\mathbf{p}_j, \mathbf{p}_j\rangle : i+1 < \delta,\ j+1 < \delta\}$$

(b) *Let ${}_\delta P_1'$, ${}_\delta P_2'$, and ${}_\delta P_3'$ be the following relations on $F({}_\delta\Gamma)$:*

$$\begin{aligned}
{}_\delta P_1' &= \{\langle \mathbf{p}_i^2, \mathbf{p}_0^2\rangle : i + 1 < \delta\}\ , \\
{}_\delta P_2' &= \{\langle \mathbf{p}_0^2 x, x\rangle : x \in F({}_\delta\Gamma)\}\ , \\
{}_\delta P_3' &= \{\langle x^{-1}x, \mathbf{p}_0^2\rangle : x \in F({}_\delta\Gamma)\}\ ,
\end{aligned}$$

Then the congruence relations on $\mathbf{F}({}_\delta\Gamma)$ *that are generated by* ${}_\delta P_1 \cup {}_\delta P_2 \cup {}_\delta P_3 \cup {}_\delta P_4$, *by* ${}_\delta P'_1 \cup {}_\delta P'_2 \cup {}_\delta P'_3$, *or by* ${}_\delta P_1 \cup {}_\delta P'_2 \cup {}_\delta P'_3$, *respectively, coincide.*

Proof. (a) ${}_\delta P_{2,3,4}$ includes ${}_\delta P_2 \cup {}_\delta P_3 \cup {}_\delta P_4$. A proof by induction on $|i-j|$ shows that ${}_\delta P_{2,3,4}$ is included in the congruence on $\mathbf{F}({}_\delta\Gamma)$ that is generated by ${}_\delta P_2 \cup {}_\delta P_3 \cup {}_\delta P_4$.

(b) Let $\equiv$ be the congruence on $\mathbf{F}({}_\delta\Gamma)$ that is generated by ${}_\delta P_1 \cup {}_\delta P_2 \cup {}_\delta P_3 \cup {}_\delta P_4$. By part (a) and induction on the length of x, ${}_\delta P'_2$ is included in $\equiv$. Let $i+1<\delta$ and $j+1<\delta$. Then, by the definition of ${}^{-1}$, by ${}_\delta P_1$, or by ${}_\delta P_{2,3,4}$, respectively,

$$\mathbf{p}_j\mathbf{p}_i^2 = (\mathbf{p}_i^{-1})^2\mathbf{p}_j^{-1} \equiv \mathbf{p}_i^2\mathbf{p}_j \equiv \mathbf{p}_j \ .$$

There follows that $\mathbf{p}_j^2 \equiv \mathbf{p}_j^2\mathbf{p}_0^2$. Since, by ${}_\delta P_{2,3,4}$, $\mathbf{p}_j^2\mathbf{p}_0 \equiv \mathbf{p}_0$, there now follows that $\mathbf{p}_j^2 \equiv \mathbf{p}_0^2$. Hence ${}_\delta P'_1$ is included in $\equiv$. Consider any $x = x_0 \ldots x_{n-1}$ in $F({}_\delta\Gamma)$. Then $x^{-1}x = x_{n-1}^{-1} \ldots x_0^{-1}x_0 \ldots x_n$. By ${}_\delta P_1$ and ${}_\delta P'_1$, $x_m^{-1}x_m \equiv \mathbf{p}_0^2$ for each $m<n$. There follows that $x^{-1}x \equiv \mathbf{p}_0^2$. Hence ${}_\delta P'_3$ is included in $\equiv$. There now follows that the congruence on $\mathbf{F}({}_\delta\Gamma)$ that is generated by $P'_1 \cup P'_2 \cup P'_3$ is included in $\equiv$.

Now let $\equiv$ be the congruence on $\mathbf{F}({}_\delta\Gamma)$ that is generated by ${}_\delta P'_1 \cup {}_\delta P'_2 \cup {}_\delta P'_3$. By ${}_\delta P'_1$ and ${}_\delta P'_2$, ${}_\delta P_2 \cup {}_\delta P_3 \cup {}_\delta P_4$ is included in $\equiv$. By Theorem 1.12 in [Ro], since ${}_\delta P'_2$ and ${}_\delta P'_3$ are included in $\equiv$, therefore, for any x in $F({}_\delta\Gamma)$, there hold $xx^{-1} \equiv \mathbf{p}_0^2$ and $x\mathbf{p}_0^2 \equiv x$. Using these for the third or fourth step, respectively, and ${}_\delta P'_2$ and ${}_\delta P'_1$ for the first or second step, respectively, one obtains:

$$\mathbf{p}_i^{-1} \equiv \mathbf{p}_0^2\mathbf{p}_i^{-1} \equiv \mathbf{p}_i\mathbf{p}_i\mathbf{p}_i^{-1} \equiv \mathbf{p}_i\mathbf{p}_0^2 \equiv \mathbf{p}_i \ .$$

Hence ${}_\delta P_1$ is included in $\equiv$ and therefore so is ${}_\delta P_1 \cup {}_\delta P_2 \cup {}_\delta P_3 \cup {}_\delta P_4$.

Let $\equiv$ be the congruence on $\mathbf{F}({}_\delta\Gamma)$ that is generated by ${}_\delta P_1 \cup {}_\delta P'_3$ and let $i+1<\delta$. Then $\mathbf{p}_i^2 \equiv \mathbf{p}_i^{-1}\mathbf{p}_i \equiv \mathbf{p}_0^2$. Hence ${}_\delta P'_1$ is included in $\equiv$. □

THEOREM 15. *Assume that* $2 \le \delta \le \omega$, *that* ${}_\delta\!\equiv$ *is the congruence on* $\mathbf{F}({}_\delta\Gamma) = \mathbf{F}_{SI}({}_\delta\Gamma)$ *that is generated by* ${}_\delta P_1 \cup {}_\delta P_2 \cup {}_\delta P_3 \cup {}_\delta P_4 \cup {}_\delta P_5 \cup {}_\delta P_6$, *that* $\mathbf{F}_{Gp}({}_\delta\Gamma)$ *is the free group having* ${}_\delta\Gamma$ *as free generating set, and that* ${}_\delta\dot{\equiv}$ *is the congruence on* $\mathbf{F}_{Gp}({}_\delta\Gamma)$ *that is generated by* ${}_\delta P'_1 \cup {}_\delta P_5 \cup {}_\delta P_6$.

(a) $\mathbf{F}({}_\delta\Gamma)/{}_\delta\!\equiv$ *is isomorphic to* $\mathbf{F}_{Gp}({}_\delta\Gamma)/{}_\delta\dot{\equiv}$.

(b) *If* $\delta = n < \omega$, *then* $\mathbf{F}({}_\delta\Gamma)/{}_\delta\!\equiv$ *is isomorphic to the symmetric group* $\mathbf{S}_n$.

(c) $\mathbf{F}({}_\omega\Gamma)/{}_\omega\!\equiv$ *is isomorphic to* $\mathbf{S}_\omega = \mathbf{S}_{p,\omega}$ *and hence also to its isomorph* ${}_{[\omega]}\mathbf{S}_p$. *Moreover, for any* n *such that* $2 \le n < \omega$, $({}_\omega\!\equiv) \cap (F({}_n\Gamma))^2$ *coincides with* ${}_n\!\equiv$.

Proof. (a) By Lemma 14(b), ${}_\delta\!\equiv$ coincides with the congruence on $\mathbf{F}({}_\delta\Gamma)$ that is generated by $({}_\delta P_2' \cup {}_\delta P_3') \cup ({}_\delta P_1' \cup {}_\delta P_5 \cup {}_\delta P_6)$. Let ${}_\delta\!\equiv_{2,3}$ be the congruence on $\mathbf{F}({}_\delta\Gamma)$ that is generated by ${}_\delta P_2' \cup {}_\delta P_3'$. By the above-mentioned Theorem 1.12 in [Ro], $\mathbf{F}(\Gamma_\delta)/{}_\delta\!\equiv_{2,3}$ is isomorphic to $\mathbf{F}_{Gp}({}_\delta\Gamma)$. From the Second Isomorphism Theorem (cf. [BS], II.sec.6), there now follows that $\mathbf{F}(\Gamma_\delta)/{}_\delta\!\equiv$ is isomorphic to $\mathbf{F}_{Gp}({}_\delta\Gamma)/{}_\delta\dot{\equiv}$.

(b) According to the presentation of $\mathbf{S}_n$ given by E. H. Moore (cf. [Moo], or [CM], sec. 6.2, or [Jo], pp.46–49), $\mathbf{S}_n$ is isomorphic to $\mathbf{F}_{Gp}({}_n\Gamma)/{}_n\dot{\equiv}$. Hence, by part (a), $\mathbf{F}({}_n\Gamma)/{}_n\!\equiv$ is isomorphic to $\mathbf{S}_n$.

(c) Since $\mathbf{S}_\omega$ is generated by $\{(i, i+1) : i < \omega\}$, there is a unique homomorphism h of $\mathbf{F}({}_\omega\Gamma)$ onto $\mathbf{S}_\omega$ such that $h(\mathbf{p}_i) = (i, i+1)$ for each $i < \omega$. Consider any n such that $2 \leq n < \omega$ and let ${}_nh$ be the restriction of h to $F({}_n\Gamma)$. Then ${}_nh$ is a homomorphism of $\mathbf{F}({}_n\Gamma)$ onto the subgroup $\mathbf{S}_n'$ of $\mathbf{S}_\omega$ whose universe S_n' consists of those permutations π in S_ω such that $\pi(m) = m$ for each $m \geq n$. The mapping ${}_nh'$ of S_n' onto S_n such that ${}_nh'(\pi) = \pi \cap (n \times n)$ for every π in S_n' is an isomorphism of $\mathbf{S}_n'$ onto the symmetric group $\mathbf{S}_n$. Now consider any x and y such that ${}_nh(x) = {}_nh(y)$ and hence such that ${}_nh'({}_nh(x)) = {}_nh'({}_nh(y))$. From part (b) there follows that $\langle x, y\rangle$ is in ${}_n\!\equiv$. Thus ${}_nh \circ {}_nh^{\smile}$ is included in ${}_n\!\equiv$. As is easy to verify, each of the relations ${}_\omega P_1, \ldots, {}_\omega P_6$ is included in $h \circ h^{\smile}$. Therefore ${}_\omega\!\equiv$ is included in $h \circ h^{\smile}$ and hence $({}_\omega\!\equiv) \cap (F({}_n\Gamma))^2$ is included in ${}_nh \circ {}_nh^{\smile}$. Since ${}_n\!\equiv$ is included in $({}_\omega\!\equiv) \cap (F({}_n\Gamma))^2$, there now follows that ${}_nh \circ {}_nh^{\smile}$, ${}_n\!\equiv$, and $({}_\omega\!\equiv) \cap (F_{SI}({}_n\Gamma))^2$ coincide. There also follows that $h \circ h^{\smile}$ and ${}_\omega\!\equiv$ coincide and hence that $\mathbf{F}({}_\omega\Gamma)/{}_\omega\!\equiv$ is isomorphic to $\mathbf{S}_\omega$ and also to its isomorph ${}_{[\omega]}\mathbf{S}_p$. □

As mentioned earlier, one may think of ${}_\delta\!\equiv$ as the positive atomic diagram of $\mathbf{F}({}_\delta\Gamma)/{}_\delta\!\equiv$, or of any structure to which it is isomorphic, and of ${}_\delta P_1 \cup \cdots \cup {}_\delta P_6$ as an axiomatization of this diagram. Thus one may think of the second part of Theorem 15(c) as asserting that ${}_\omega P_1 \cup \cdots \cup {}_\omega P_6$ and ${}_\omega\!\equiv$ conservatively extend ${}_nP_1 \cup \cdots \cup {}_nP_6$ or ${}_n\!\equiv$, respectively.

After these preparations, we now turn to the task of presenting $\mathbf{S}_p = \langle S_p, \circ, {}^{\smile}\rangle$ and $\langle \mathbf{S}_p, \subseteq\rangle = \langle S_p, \circ, {}^{\smile}, \subseteq\rangle$. For the rest of this chapter we let $h(\mathbf{p}_i) = p_i = p_{(i,i+1)}$ for any $i < \omega$. Also, for any word w in $F(\Gamma) = F(\Gamma(\tilde{\mathbf{p}}))$, we again let ***the grade associated with*** w be the grade $gd(h(w))$ of $h(w)$ and denote it by $gd\ w$. We may sometimes make tacit use of the following simple fact:

$$\text{If } \ s = \max\{i : \mathbf{p}_i \text{ occurs in } w\}, \ \text{ then } \ gd\ w = \langle s+2, s+2\rangle.$$

The following Lemma will allow us to make use of the presentation of ${}_{[\omega]}\mathbf{S}_p$ given by Theorem 15(c) to obtain a presentation of $\mathbf{S}_p$.

LEMMA 16. (a) *$h(v) = h(w)$ if and only if $\overset{\omega}{\rightarrow}(h(v)) = \overset{\omega}{\rightarrow}(h(w))$ and $gd\ v = gd\ w$.*

(b) *If $gd\ v = \langle s+2, s+2\rangle$, then $h(\mathbf{p}_s^2 v) = h(v)$.*

Proof. (a) Assume that $\overset{\omega}{\rightarrow}(h(v)) = \overset{\omega}{\rightarrow}(h(w))$. Then either $\overset{n}{\rightarrow}(h(v)) = (h(w))$ for some $n \geq 0$ or $\overset{n}{\rightarrow}(h(w)) = (h(v))$ for some $n \geq 0$. Hence, if also $gd\ v = gd\ w$, then $h(v) = h(w)$. (Cf. Theorem II.12(b).) That the converse holds is obvious.

(b) This follows from $h(\mathbf{p}_s^2) = {}_{[s+2,\omega]}\overset{\circ}{1}$. □

As one can see from the proof of Lemma 16(a), analogues hold, for example, for $F_{SI}(\Gamma(\tilde{\mathbf{q}}))$, or for $F_{SI}(\Gamma(\tilde{\mathbf{q}}) \cup \Gamma(\tilde{\mathbf{p}}))$ and the appropriate h.

Consider any $\langle v, w\rangle$ belonging to one of the relations P_1, P_2, P_3, P_5, P_6. As one can verify, $gd\ v = gd\ w$. Also, as we saw in the proof of Theorem 15(c), $\overset{\omega}{\rightarrow}(h(v)) = \overset{\omega}{\rightarrow}(h(w))$. Hence, by Lemma 16(a), $h(v) = h(w)$. There follows that each of the relations P_1, P_2, P_3, P_5, P_6 is included in $h \circ h^{\smile}$.

In contrast, if $\langle v, w\rangle = \langle \mathbf{p}_i^2, \mathbf{p}_0^2\rangle$ and $i \neq 0$, then $gd\ v = \langle i+2, i+2\rangle \neq gd\ w$ and hence $h(v) \neq h(w)$. This is why ${}_\omega P_1'$ is less suitable than ${}_\omega P_1$ for presentations of ${}_{[\omega]}\mathbf{S}_p$ when one wants to compare them with presentations of $\mathbf{S}_p$. Also, if $\langle v, w\rangle = \langle \mathbf{p}_{i+1}^2\mathbf{p}_i, \mathbf{p}_i\rangle$, then $gd\ v = \langle i+3, i+3\rangle \neq \langle i+2, i+2\rangle = gd\ w$, and hence $h(v) \neq h(w)$. Hence P_4 is not included in $h \circ h^{\smile}$. This suggests that, for any δ such that $2 \leq \delta \leq \omega$, we consider the following relation ${}_\delta P_4^{\#}$ on $F_{SI}({}_\delta\Gamma)$:

$$ {}_\delta P_4^{\#} = \{\langle \mathbf{p}_{i+1}^2\mathbf{p}_i, \mathbf{p}_i\mathbf{p}_{i+1}^2\rangle : i+2 < \delta\} \ . $$

Consider any $\langle v, w\rangle = \langle \mathbf{p}_{i+1}^2\mathbf{p}_i, \mathbf{p}_i\mathbf{p}_{i+1}^2\rangle$ in $P_4^{\#} = {}_\omega P_4^{\#}$. Then $gd\ v = \langle i+3, i+3\rangle = gd\ w$. Also, $\langle v, w\rangle$ is included in the congruence relation $\equiv_{1,4}$ on $\mathbf{F}(\Gamma)$ that is generated by $P_1 \cup P_4$. Hence, by Theorem 15(c), $\overset{\omega}{\rightarrow}(h(v)) = \overset{\omega}{\rightarrow}(h(w))$ and therefore, by Lemma 16(a), $h(v) = h(w)$. There now follows that the congruence relation on $\mathbf{F}(\Gamma)$ that is generated by $P_1 \cup P_2 \cup P_3 \cup P_4^{\#} \cup P_5 \cup P_6$ is included in $h \circ h^{\smile}$.

LEMMA 17. *Let $\equiv$ be the congruence on $\mathbf{F}(\Gamma) = \mathbf{F}_{SI}({}_\omega\Gamma(\tilde{\mathbf{p}}))$ that is generated by $P_1 \cup P_2 \cup P_3 \cup P_4^{\#} \cup P_6$.*

(a) *$\mathbf{p}_i^2 x \equiv x \equiv x\mathbf{p}_i^2$, if $gd\ x \geq \langle i+2, i+2\rangle$.*

(b) *$\mathbf{p}_i^2 x \equiv x\mathbf{p}_i^2$, if $gd\ x \leq \langle i+2, i+2\rangle$.*

Proof. (a) By P_3 and induction, $\mathbf{p}_i^2\mathbf{p}_j \equiv \mathbf{p}_j$, if $i < j$. Hence, by P_2, $\mathbf{p}_i^2\mathbf{p}_j \equiv \mathbf{p}_j$, if $i \leq j$. Since, by P_1 and the definition of ${}^{-1}$, $\mathbf{p}_i^2\mathbf{p}_j \equiv \mathbf{p}_j\mathbf{p}_i^2$, there follows that $\mathbf{p}_j \equiv \mathbf{p}_j\mathbf{p}_i^2$ if $i \leq j$. Part (a) now follows by induction on the length of x.

(b) By P_6, $\mathbf{p}_i^2\mathbf{p}_j \equiv \mathbf{p}_j\mathbf{p}_i^2$ if $i \geq j+2$. By $P_4^{\#}$, $\mathbf{p}_i^2\mathbf{p}_j \equiv \mathbf{p}_j\mathbf{p}_i^2$ if $i \geq j+1$. Since $\mathbf{p}_i^2\mathbf{p}_i = \mathbf{p}_i^3 = \mathbf{p}_i\mathbf{p}_i^2$, there now follows that $\mathbf{p}_i^2\mathbf{p}_j \equiv \mathbf{p}_j\mathbf{p}_i^2$ if $i \geq j$. Now assume that *gd* $x \leq \langle i+2, i+2\rangle$. Then $i \geq j$ for any $\mathbf{p}_j$ that occurs in x. Hence by induction on the length of x there follows that $\mathbf{p}_i^2 x \equiv x\mathbf{p}_i^2$. □

THEOREM 18. *Let* $\equiv$ *be the congruence on* $\mathbf{F}(\Gamma) = \mathbf{F}_{SI}({}_\omega\Gamma(\tilde{\mathbf{p}}))$ *that is generated by* $P_1 \cup P_2 \cup P_3 \cup P_4^{\#} \cup P_5 \cup P_6$ *and let* h *be the homomorphism of* $\mathbf{F}(\Gamma)$ *onto* $\mathbf{S}_p$ *such that* $h(\mathbf{p}_i) = p_i = p_{(i,i+1)}$ *for each* $i < \omega$.

(a) $\equiv$ *coincides with* $h \circ h^{\smile}$ *and hence* $\mathbf{F}(\Gamma)/\equiv$ *is isomorphic to* $\mathbf{S}_p$.

(b) *Let* $\lessdot$ *be the least quasiordering of* $F(\Gamma)$ *that is compatible with* $\mathbf{F}(\Gamma)$ *and includes* $\equiv \cup P_4$. *Then for any* v *and* w *in* $F(\Gamma)$, $h(v) \subseteq h(w)$ *if and only if* $v \lessdot w$. *There follows that* $\langle \mathbf{F}(\Gamma), \lessdot\rangle/\equiv$ *is isomorphic to* $\langle \mathbf{S}_p, \subseteq\rangle$.

Proof. (a) As we saw a while earlier, $\equiv$ is included in $h \circ h^{\smile}$. Now assume that $h(v) = h(w)$. Then, by Lemma 16(a), *gd* $v =$ *gd* w and $\overset{\omega}{\rightarrow}(h(v)) = \overset{\omega}{\rightarrow}(h(w))$. Let *gd* $v =$ *gd* $w = \langle s+2, s+2\rangle$. Then v and w are in $F({}_{s+2}\Gamma)$. Let $\doteqdot$ be the congruence on $\mathbf{F}({}_{s+2}\Gamma)$ that is generated by ${}_{s+2}P_1 \cup {}_{s+2}P_2 \cup {}_{s+2}P_3 \cup {}_{s+2}P_4 \cup {}_{s+2}P_5 \cup {}_{s+2}P_6$. Then, by Theorem 15(c), $v \doteqdot w$. We now want to show that this implies that $p_s^2 v \equiv p_s^2 w$, from which, by Lemma 17(a), there follows that $v \equiv w$.

Assume that $v \doteqdot w$. Since $\doteqdot$ is the congruence on $\mathbf{F}_{SI}({}_{s+2}\Gamma)$ that is generated by $\equiv \cup\ {}_{s+2}P_4$, therefore there is a sequence $x_0, \ldots, x_n$ such that $x_0 = v$, $x_n = w$, and, for each $m < n$, either $x_m \equiv x_{m+1}$ or, for some i such that $i+2 < s+2$ and hence $i < s$, there are y_m and z_m, possibly empty, such that either

(i) $x_m = y_m \mathbf{p}_{i+1}^2 \mathbf{p}_i z_m$ and $x_{m+1} = y_m \mathbf{p}_i z_m$ or

(ii) $x_m = y_m \mathbf{p}_i z_m$ and $x_{m+1} = y_m \mathbf{p}_{i+1}^2 \mathbf{p}_i z_m$.

If $x_{m'} \equiv x_{m'+1}$ for each $m' < n$, then $v \equiv w$. Now suppose that there are $m' < n$ such that not $x_{m'} \equiv x_{m'+1}$. Let m be the least of these. Then $x_0 \equiv x_m$ and hence also $\mathbf{p}_s^2 x_0 \equiv \mathbf{p}_s^2 x_m$. Also, for some $i < s$ and some y_m and z_m there holds either (i) or (ii).

Now, of the following three assertions, the first, second, and third hold by Lemma 17(b), 17(a), or 17(b), respectively:

$$\mathbf{p}_s^2 y_m \mathbf{p}_{i+1}^2 \mathbf{p}_i z_m \equiv y_m \mathbf{p}_s^2 \mathbf{p}_{i+1}^2 \mathbf{p}_i z_m \equiv y_m \mathbf{p}_s^2 \mathbf{p}_i z_m \equiv \mathbf{p}_s^2 y_m \mathbf{p}_i z_m \ .$$

Thus $\mathbf{p}_s^2 x_m \equiv \mathbf{p}_s^2 x_{m+1}$, if either (i) or (ii). There follows that $\mathbf{p}_s^2 x_0 \equiv \mathbf{p}_s^2 x_{m+1}$. By induction there follows that $\mathbf{p}_s^2 x_0 \equiv \mathbf{p}_s^2 x_n$ and hence that $\mathbf{p}_s^2 v \equiv \mathbf{p}_s^2 w$. By Lemma 17(a) there follows that $v \equiv w$. This concludes our proof that $h \circ h^{\smile}$ is included in $\equiv$ and thus our proof that they coincide. There follows that $\mathbf{F}_{SI}(\Gamma)/\equiv$ is isomorphic to $\mathbf{S}_p$.

(b) Since $h(\mathbf{p}_{i+1}^2 \mathbf{p}_i) \subseteq h(\mathbf{p}_i)$ for any $\langle \mathbf{p}_{i+1}^2 \mathbf{p}_i, \mathbf{p}_i\rangle$ in P_4 and since $\equiv$ is included in $h \circ h^{\smile}$, therefore $v \lessdot w$ implies that $h(v) \subseteq h(w)$. Now assume that $h(v) \subseteq h(w)$.

Let $gd\ v = \langle s+2, s+2\rangle$ and $gd\ w = \langle t+2, t+2\rangle$. Then $s \geq t$ and $h(\mathbf{p}_s^2 w) = h(v)$. By part (a), $v \equiv \mathbf{p}_s^2 v$ and $\mathbf{p}_s^2 w \equiv v$. Also, since P_4 and P_2 are included in $\lesssim$, therefore $\mathbf{p}_s^2 \lesssim \mathbf{p}_t^2$. There follows that $v \lesssim w$. □

According to Theorems 18(a) and 15(a), our presentation of $\mathbf{S}_p$ can be obtained from our presentation of ${}_{[\omega]}\mathbf{S}_p$ by replacing $P_4 = {}_\omega P_4$ by $P_4^{\#}$. Hence also a presentation of ${}_{[\omega]}\mathbf{S}_p$ can be obtained from our presentation of $\mathbf{S}_p$ by adjoining P_4.

From the proof of Theorem 18(a) one can see that if $2 \leq n < \omega$ and ${}_n\equiv$ is the congruence on $\mathbf{F}_{SI}({}_n\Gamma)$ that is generated by ${}_nP_1 \cup {}_nP_2 \cup {}_nP_3 \cup {}_nP_4^{\#} \cup {}_nP_5 \cup {}_nP_6$, then ${}_\omega\equiv$ extends ${}_n\equiv$ conservatively. A similar remark applies to Theorem 18(b).

CHAPTER VI

Presentation of $\mathbf{S}_{pq}$, $\mathbf{S}_{pe}$ and Related Structures

The main purpose of this chapter is to give a presentation of $\mathbf{S}_{pq} = \langle S_{pq}, \circ, ^{\smile} \rangle$ and $\langle \mathbf{S}_{pq}, \subseteq \rangle$ and also of $\mathbf{S}_{pe} = \langle S_{pe}, \circ, ^{\smile} \rangle$ and $\langle \mathbf{S}_{pe} \subseteq \rangle$. We will also give a presentation of the homomorphic images of these under $\xrightarrow{\omega}$.

(Let $T_{p\dot{q}}$ be the closure under $\circ$ of the $\{p_i : i < \omega\} \cup \{\dot{q}_i : i < \omega\}$. Presentations of $\mathbf{T}_{p\dot{q}} = \langle T_{p\dot{q}}, \circ \rangle$ and of related structures may also be of intrest, but we have not investigated this topic.)

We first take up $\mathbf{S}_{pq}$. Until further notice we let $\mathbf{q}_0 = \mathbf{q}$ and we let $\Gamma = \Gamma(\tilde{\mathbf{p}}, \mathbf{q}) = \{\mathbf{p}_i : i < \omega\} \cup \{\mathbf{q}\}$, where $\mathbf{p}_i \neq \mathbf{p}_j$ if $i \neq j$ and where $\mathbf{q} = \mathbf{q}_0 \notin \{\mathbf{p}_i, i < \omega\}$. Also, we let ${}_1Q_3$, PQ_1, PQ_2, PQ_3, PQ_4 be the relations on $F(\Gamma) = F_{SI}(\Gamma(\tilde{\mathbf{p}}, \mathbf{q}))$ that are given below. The first four of these are singletons and do not involve any $\mathbf{p}_i$ other than $\mathbf{p}_0$.

$$\begin{aligned}
{}_1Q_3 &= \{\langle \mathbf{q}\mathbf{q}^{-1}\mathbf{q}, \mathbf{q} \rangle\} \\
PQ_1 &= \{\langle \mathbf{q}\mathbf{q}^{-1}\mathbf{p}_0, \mathbf{p}_0 \rangle\} \\
PQ_2 &= \{\langle \mathbf{q}\mathbf{p}_0\mathbf{q}^{-1}, \mathbf{q}^{-1}\mathbf{q} \rangle\} \\
PQ_3 &= \{\langle \mathbf{q}^2\mathbf{p}_0, \mathbf{q}^2 \rangle\} \\
PQ_4 &= \{\langle \mathbf{p}_i\mathbf{q}, \mathbf{q}\mathbf{p}_{i+1} \rangle : i < \omega\}
\end{aligned}$$

Until the end of the proof of Theorem 12 we let $\equiv_P$ be the congruence on $\mathbf{F}(\Gamma) = \mathbf{F}_{SI}(\Gamma(\tilde{\mathbf{p}}, \mathbf{q}))$ that is generated by the union $P_1 \cup P_2 \cup P_3 \cup P_4^{\#} \cup P_5 \cup P_6$ of the six relations we considered in Chapter V and we let $\equiv$ be the congruence on $\mathbf{F}_{SI}(\Gamma(\tilde{\mathbf{p}}, \mathbf{q}))$ that is generated by $\equiv_P \cup {}_1Q_3 \cup PQ_1 \cup PQ_2 \cup PQ_3 \cup PQ_4$. As was shown in the proof of Theorem I.1(b), S_{pq} is the closure under $\circ$ and $^{\smile}$ of $\{p_i : i < \omega\} \cup \{q_0\}$. Until the end of the proof of Theorem 12, we let h be the homomorphism of $\mathbf{F}(\Gamma) = \mathbf{F}_{SI}(\Gamma(\tilde{\mathbf{p}}, \mathbf{q}))$ onto $\mathbf{S}_{pq}$ such that $h(\mathbf{q}) = q_0 = q$ and $h(\mathbf{p}_i) = p_i = p_{(i,i+1)}$ for each $i < \omega$.

LEMMA 1. (a) *If* $\mathbf{q}$ *does not occur in* v *or in* w*, then* $v \equiv_P w$ *if and only if* $h(v) = h(w)$.

(b) $\equiv$ *is included in* $h \circ h^{\smile}$.

Proof. (a) By Theorem V.18(a).

(b) From a remark just before Lemma V.17, there follows that $\equiv_P$ is included in $h \circ h^{\smile}$. Since $h(\mathbf{q}\mathbf{q}^{-1}) = q_0 \circ q_0^{\smile} = {}_{[0,\omega)}\overset{\circ}{1}$, therefore ${}_1Q_3$ and PQ_1 are included in $h \circ h^{\smile}$.

From $p_0 = p_{(0,1)} = {}_{[2,\omega)}\overset{\circ}{1} \circ p_0$ and $q_0 \circ p_0 = q_0 \circ {}_{[2,\omega)}\overset{\circ}{1} \circ p_0$ there follow the first three equalities below. The fourth follows from the definition of q_0.

$$\begin{aligned} q_0 \circ {}_{[2,\omega)}\overset{\circ}{1} &= \{\langle xz, yxz\rangle : |x|=|y|=1,\ 0 \leq |z| < \omega\} \\ q_0 \circ p_0 &= \{\langle xz, xyz\rangle : |x|=|y|=1,\ 0 \leq |z| < \omega\} \\ q_0 \circ p_0 \circ q_0^{\smile} &= \{\langle xz, yz\rangle : |x|=|y|=1,\ 0 \leq |z| < \omega\} \\ q_0^{\smile} \circ q_0 &= \{\langle xz, yz\rangle : |x|=|y|=1,\ 0 \leq |z| < \omega\} \end{aligned}$$

From the last two equalities there follows that PQ_2 is included in $h \circ h^{\smile}$.

One sees from the following that PQ_3 is included in $h \circ h^{\smile}$.

$$q_0^2 p_0 = \{\langle z, xyz\rangle : |x|=|y|=1,\ |z| < \omega\} = q_0^2 .$$

The first and third equality below follow from the definitions involved. The second and fourth equality follow from the first or third, respectively.

$$\begin{aligned} {}_{[i+2,\omega)}\overset{\circ}{1} \circ q_0 &= \{\langle vwxz, yvwxz\rangle : |v|=i,\ |w|=|x|=|y|=1,\ |z| < \omega\} \\ {}_{[i+2,\omega)}\overset{\circ}{1} \circ q_0 \circ p_{i+1} &= \{\langle vwxz, yvxwz\rangle : |v|=i,\ |w|=|x|=|y|=1,\ |z| < \omega\} \\ p_i &= \{\langle vwxz, vxwz\rangle : |v|=i,\ |w|=|x|=1,\ |z| < \omega\} \\ p_i \circ q_0 &= \{\langle vwxz, yvxwz\rangle : |v|=i,\ |w|=|x|=|y|=1,\ |z| < \omega\} \end{aligned}$$

Since $q_0 \circ p_{i+1} = q_0 \circ {}_{[i+3,\omega)}\overset{\circ}{1} \circ p_{i+1} = {}_{[i+2,\omega)}\overset{\circ}{1} \circ q_0 \circ p_{i+1}$, there follows that $q_0 \circ p_{i+1} = p_i \circ q_0$. Hence $(PQ_4)^{\smile}$ is included in $h \circ h^{\smile}$ and therefore so is PQ_4. There now follows that $\equiv$ is included in $h \circ h^{\smile}$. □

For proving that, conversely, $h \circ h^{\smile}$ is included in $\equiv$ it will be useful to introduce, for certain elements w of $F(\Gamma) = F_{SI}(\Gamma(\tilde{\mathbf{p}}, \mathbf{q}))$, a name that is suggestive and whose length is generally much shorter than that of w. Recalling that $\mathbf{q}_0 = \mathbf{q}$, we let

$$\mathbf{q}_{i+1} = \mathbf{q}_i \mathbf{p}_i\ , \quad \text{for any } i < \omega\ .$$

Three consequences are:

$$\begin{aligned} \mathbf{q}_{i+1} &= \mathbf{q}\mathbf{p}_0 \dots \mathbf{p}_i\ , && \text{for any } i < \omega\ , \\ \mathbf{q}_{i+1}^{-1} &= \mathbf{p}_i^{-1} \dots \mathbf{p}_0^{-1}\mathbf{q}^{-1}\ , && \text{for any } i < \omega \\ \mathbf{q}_k &= \mathbf{q}_i\mathbf{p}_i \dots \mathbf{p}_{k-1}\ , && \text{for any } i < k\ . \end{aligned}$$

Among other consequences are the following

LEMMA 2. (a) $\mathbf{q}_{i+1}^{-1} \equiv_P \mathbf{p}_i \dots \mathbf{p}_0\mathbf{q}^{-1}$, *for any* $i < \omega$.

(b) $\mathbf{q}_{i+1}\mathbf{q}_{i+1}^{-1} \equiv_P \mathbf{q}\mathbf{p}_i^2\mathbf{q}^{-1}$, *for any* $i < \omega$.

(c) $h(\mathbf{q}_i) = q_i$, *for any* $i < \omega$.

(d) *If* $\mathbf{F}(\{\mathbf{q}_i : i < \omega\})$ *is the subalgebra of* $\mathbf{F}(\Gamma(\tilde{\mathbf{p}}, \mathbf{q}))$ *that is generated by* $\{\mathbf{q}_i : i < \omega\}$, *then the restriction of* h *to its universe is a homomorphsim of this subalgebra onto* $\mathbf{S}_q = \langle S_q, \circ, {}^{\smile} \rangle$.

Proof. (a) By P_1.

(b) Since $j \le i$ implies that $\mathbf{p}_j\mathbf{p}_i^2 p_j \equiv_P \mathbf{p}_j^2\mathbf{p}_i^2 \equiv_P p_i^2$ this follows from part (a).

(c)Since $\mathbf{q}_0 = \mathbf{q}$, therefore $h(\mathbf{q}_0) = h(\mathbf{q}) = q_0$. Now consider any $i < \omega$. Then the following three equalities imply the fourth:

$$\begin{aligned} q_i \circ p_i &= q_i \circ {}_{[i+2,\omega)}\mathring{1} \circ p_i = {}_{[i+1,\omega)}\mathring{1} \circ q_i \circ p_i \\ {}_{[i+1,\omega)}\mathring{1} \circ q_i &= \{\langle vwz, vxwz \rangle : |v| = i,\ |w| = |x| = 1,\ |z| < \omega\} \\ {}_{[i+1,\omega)}\mathring{1} \circ q_i \circ p_i &= \{\langle vxwz, vwxz \rangle : |v| = i,\ |w| = |x| = 1,\ |z| < \omega\} \\ q_i \circ p_i &= \{\langle vwz, vwxz \rangle : |v| = i,\ |w| = |x| = 1,\ |z| < \omega\} \end{aligned}$$

From the definition of q_{i+1} there follows that $q_i \circ p_i = q_{i+1}$. Hence $h(\mathbf{q}_{i+1}) = h(\mathbf{q}_i\mathbf{p}_i) = h(\mathbf{q}_i) \circ h(\mathbf{p}_i) = q_i \circ p_i = q_{i+1}$. By induction, $h(\mathbf{q}_i) = q_i$ for any $i < \omega$.

(d) By part (c). □

Each of the relations ${}_1Q_3$, PQ_1, PQ_2, PQ_3, PQ_4 is included in a certain more comprehensive relation. From parts (c),...,(g) of the next lemma there follows that each of these is also included in $\equiv$.

LEMMA 3. (a) *Let* $\equiv_{1,4}$ *be the congruence on* $\mathbf{F}_{SI}(\Gamma(\tilde{\mathbf{p}}, \mathbf{q}))$ *that is generated by* $\equiv_P \cup PQ_1 \cup PQ_4$. *Then*

$$\mathbf{q}_{i+2}\mathbf{q}_{i+2}^{-1} \equiv_{1,4} \mathbf{p}_i^2\ , \quad \textit{for any } i < \omega\ .$$

(b) *Let* $\equiv_4$ *be the congruence on* $\mathbf{F}_{SI}(\Gamma(\tilde{\mathbf{p}}, \mathbf{q}))$ *that is generated by* $\equiv_P \cup PQ_4$. *Then*

$$\mathbf{p}_i^2\mathbf{q}_{i+2} \equiv_4 \mathbf{q}_{i+2} , \quad \textit{for any} \ \ i < \omega .$$

(c) *Let* $\dot{\equiv}$ *be the congruence on* $\mathbf{F}_{SI}(\Gamma(\tilde{\mathbf{p}}, \mathbf{q}))$ *that is generated by* $\equiv_P \cup_1 Q_3 \cup PQ_1 \cup PQ_2 \cup PQ_4$. *Then*

$$\mathbf{q}_i\mathbf{q}_i^{-1}\mathbf{q}_i \dot{\equiv} \mathbf{q}_i , \quad \textit{for any} \ \ i < \omega .$$

(d) *Let* $\equiv_{1,4}$ *be the congruence on* $\mathbf{F}_{SI}(\Gamma(\tilde{\mathbf{p}}, \mathbf{q}))$ *that is generated by* $\equiv_P \cup PQ_1 \cup PQ_4$. *Then*

$$\mathbf{q}_i\mathbf{q}_i^{-1}\mathbf{p}_i \equiv_{1,4} \mathbf{p}_i , \quad \textit{for any} \ \ i < \omega .$$

(e) *Let* $\equiv_{2,4}$ *be the congruence on* $\mathbf{F}_{SI}(\Gamma(\tilde{\mathbf{p}}, \mathbf{q}))$ *that is generated by* $\equiv_P \cup PQ_2 \cup PQ_4$. *Then*

$$\mathbf{q}_i\mathbf{p}_i\mathbf{q}_i^{-1} \equiv_{2,4} \mathbf{q}_i^{-1}\mathbf{q}_i , \quad \textit{for any} \ \ i < \omega .$$

(f) *Let* $\equiv_{3,4}$ *be the congruence on* $\mathbf{F}_{SI}(\Gamma(\tilde{\mathbf{p}}, \mathbf{q}))$ *that is generated by* $\equiv_P \cup PQ_3 \cup PQ_4$. *Then*

$$\mathbf{q}_i^2\mathbf{p}_i \equiv_{3,4} \mathbf{q}_i^2 , \quad \textit{for any} \ \ i < \omega .$$

(g) *Let* $\equiv_4$ *be the congruence on* $\mathbf{F}_{SI}(\Gamma(\tilde{\mathbf{p}}, \mathbf{q}))$ *that is generated by* $\equiv_P \cup PQ_4$. *Then*

$$\mathbf{p}_i\mathbf{q}_j \equiv_4 \mathbf{q}_j\mathbf{p}_{i+1} \quad \textit{and} \quad \mathbf{p}_{i+1}\mathbf{q}_j^{-1} \equiv_4 \mathbf{q}_j^{-1}\mathbf{p}_i , \qquad \textit{if} \ \ i \geq j .$$

Proof. (a) Of the following assertions the first holds by Lemma 2(b), the second by $(PQ_4)^{\smile}$, the third and sixth by P_1, and the fifth by PQ_1:

$$\begin{aligned} \mathbf{q}_2\mathbf{q}_2^{-1} &\equiv_P \ \mathbf{q}\mathbf{p}_1^2\mathbf{q}^{-1} \equiv_{1,4} \mathbf{p}_0^2\mathbf{q}\mathbf{q}^{-1} \equiv_P \mathbf{p}_0\mathbf{p}_0^{-1}\mathbf{q}\mathbf{q}^{-1} \\ &= \ \mathbf{p}_0(\mathbf{q}\mathbf{q}^{-1}\mathbf{p}_0)^{-1} \equiv_{1,4} \mathbf{p}_0\mathbf{p}_0^{-1} \equiv_P \mathbf{p}_0^2 . \end{aligned}$$

Assume as inductive hypothesis that $\mathbf{q}_{i+2}\mathbf{q}_{i+2}^{-1} \equiv_{1,4} \mathbf{p}_i^2$. Then of the following assertions the first holds by Lemma 2(b), the second by $(PQ_4)^{\smile}$, and the sixth by PQ_1. The remaining assertions, which involve $\equiv_P$, make use of Theorem V.18(a). (From now on, this use will not always be mentioned.)

$$\begin{aligned} \mathbf{q}_{i+3}\mathbf{q}_{i+3}^{-1} &\equiv_P \ \mathbf{q}\mathbf{p}_{i+2}^2\mathbf{q}^{-1} \equiv_{1,4} \mathbf{p}_{i+1}^2\mathbf{q}\mathbf{q}^{-1} \equiv_P \mathbf{p}_{i+1}^2\mathbf{p}_0\mathbf{p}_0\mathbf{q}\mathbf{q}^{-1} \\ &= \ \mathbf{p}_{i+1}^2\mathbf{p}_0(\mathbf{q}\mathbf{q}^{-1}\mathbf{p}_0^{-1})^{-1} \equiv_P \mathbf{p}_{i+1}^2\mathbf{p}_0(\mathbf{q}\mathbf{q}_0^{-1}\mathbf{p}_0)^{-1} \\ &\equiv_{1,4} \ \mathbf{p}_{i+1}^2\mathbf{p}_0\mathbf{p}_0^{-1} \equiv_P \mathbf{p}_{i+1}^2 . \end{aligned}$$

There follows by induction that $\mathbf{q}_{i+2}\mathbf{q}_{i+2}^{-1} \equiv_{1,4} \mathbf{p}_i^2$, for any $i < \omega$.

(b) The second of the following assertions holds by PQ_4:

$$\begin{aligned}\mathbf{p}_i^2\mathbf{q}_{i+2} &= \mathbf{p}_i^2\mathbf{q}_i\mathbf{p}_0\cdots\mathbf{p}_{i+1} \equiv_4 \mathbf{q}\mathbf{p}_{i+1}^2\mathbf{p}_0\cdots\mathbf{p}_{i+1}\\ &\equiv_P \mathbf{q}\mathbf{p}_0\cdots\mathbf{p}_{i+1}\mathbf{p}_{i+1}^2 \equiv_P \mathbf{q}\mathbf{p}_0\cdots\mathbf{p}_{i+1} = \mathbf{q}_{i+2}\ .\end{aligned}$$

(c) Since $\mathbf{q}_0 = \mathbf{q}$ therefore, by ${}_1Q_3$, $\mathbf{q}_0\mathbf{q}_0^{-1}\mathbf{q}_0\doteq\mathbf{q}_0$. Of the following nine assertions the first holds by Lemma 2(b), the second by $(PQ_2)^{\smile}$, the third by PQ_4, the fifth by PQ_4 and properties of ${}^{-1}$, the seventh by PQ_2, the next-to-last by ${}_3Q_1$:

$$\begin{aligned}\mathbf{q}_1\mathbf{q}_1^{-1}\mathbf{q}_1 &\equiv_P \mathbf{q}\mathbf{p}_0^2\mathbf{q}^{-1}\mathbf{q}\mathbf{p}_0 \doteq \mathbf{q}\mathbf{p}_0^2\mathbf{q}\mathbf{p}_0\mathbf{q}^{-1}\mathbf{p}_0 \doteq \mathbf{q}\mathbf{q}\mathbf{p}_1^2\mathbf{p}_0\mathbf{q}^{-1}\mathbf{p}_0\\ &\equiv_P \mathbf{q}\mathbf{q}\mathbf{p}_0\mathbf{p}_1^2\mathbf{q}^{-1}\mathbf{p}_0\doteq\mathbf{q}\mathbf{q}\mathbf{p}_0\mathbf{q}^{-1}\mathbf{p}_0^2\mathbf{p}_0 \equiv_P \mathbf{q}\mathbf{q}\mathbf{p}_0\mathbf{q}^{-1}\mathbf{p}_0\\ &\doteq \mathbf{q}\mathbf{q}^{-1}\mathbf{q}\mathbf{p}_0 \doteq \mathbf{q}\mathbf{p}_0 = \mathbf{q}_1\ .\end{aligned}$$

By part (a) or (b), respectively, there hold, for any $i < \omega$,

$$\mathbf{q}_{i+2}\mathbf{q}_{i+2}^{-1}\mathbf{q}_{i+2} \doteq \mathbf{p}_i^2\mathbf{q}_{i+2} \doteq \mathbf{q}_{i+2}\ .$$

(d) The first assertion below holds by Lemma 2(b), the second and fourth by PQ_4 and properties of ${}^{-1}$, and the sixth by PQ_1.

$$\begin{aligned}\mathbf{q}_{i+1}\mathbf{q}_{i+1}^{-1}\mathbf{p}_{i+1} &\equiv_P \mathbf{q}\mathbf{p}_i^2\mathbf{q}^{-1}\mathbf{p}_{i+1} \equiv_{1,4} \mathbf{q}\mathbf{p}_i^2\mathbf{p}_{i+2}\mathbf{q}^{-1}\\ &\equiv_P \mathbf{q}\mathbf{p}_{i+2}\mathbf{q}^{-1} \equiv_{1,4} \mathbf{q}\mathbf{q}^{-1}\mathbf{p}_{i+1} \equiv_P \mathbf{q}\mathbf{q}^{-1}\mathbf{p}_0\mathbf{p}_0\mathbf{p}_{i+1}\\ &\equiv_{1,4} \mathbf{p}_0\mathbf{p}_0\mathbf{p}_{i+1} \equiv_P \mathbf{p}_{i+1}\ .\end{aligned}$$

(e) Assume as inductive hypothesis that $\mathbf{q}_i\mathbf{p}_i\mathbf{q}_i^{-1} \equiv_{2,4} \mathbf{q}_i^{-1}\mathbf{q}_i$. Then the second, sixth, and tenth assertion below hold by P_1, the third by P_5, the fourth and seventh by $(PQ_4)^{\smile}$, and the ninth by the inductive hypothesis.

$$\begin{aligned}\mathbf{q}_{i+1}\mathbf{p}_{i+1}\mathbf{q}_{i+1}^{-1} &= \mathbf{q}_i\mathbf{p}_i\mathbf{p}_{i+1}\mathbf{p}_i^{-1}\mathbf{q}_i^{-1} \equiv_P \mathbf{q}_i\mathbf{p}_i\mathbf{p}_{i+1}\mathbf{p}_i\mathbf{q}_i^{-1}\\ &\equiv_P \mathbf{q}_i\mathbf{p}_{i+1}\mathbf{p}_i\mathbf{p}_{i+1}\mathbf{q}_i^{-1} \equiv_{2,4} \mathbf{p}_i\mathbf{q}_i\mathbf{p}_i\mathbf{p}_{i+1}\mathbf{q}_i^{-1}\\ &= \mathbf{p}_i\mathbf{q}_i\mathbf{p}_i(\mathbf{q}_i\mathbf{p}_{i+1}^{-1})^{-1} \equiv_P \mathbf{p}_i\mathbf{q}_i\mathbf{p}_i(\mathbf{q}_i\mathbf{p}_{i+1})^{-1}\\ &\equiv_{2,4} \mathbf{p}_i\mathbf{q}_i\mathbf{p}_i(\mathbf{p}_i\mathbf{q}_i)^{-1} = \mathbf{p}_i\mathbf{q}_i\mathbf{p}_i\mathbf{q}_i^{-1}\mathbf{p}_i^{-1}\\ &\equiv_{2,4} \mathbf{p}_i\mathbf{q}_i^{-1}\mathbf{q}_i\mathbf{p}_i^{-1} \equiv_P \mathbf{p}_i^{-1}\mathbf{q}_i^{-1}\mathbf{q}_i\mathbf{p}_i = \mathbf{q}_{i+1}^{-1}\mathbf{q}_{i+1}\ .\end{aligned}$$

From PQ_2, therefore, there follows by induction that, for any $i < \omega$, $\mathbf{q}_i\mathbf{p}_i\mathbf{q}_i^{-1} \equiv_{2,4} \mathbf{q}_i^{-1}\mathbf{q}_i$.

(f) Assume as inductive hypothesis that $\mathbf{q}_i^2\mathbf{p}_i \equiv_{3,4} \mathbf{q}_i^2$. Then the second assertion below holds by PQ_4, the third by P_5, the fourth by the inductive hypothesis, and the fifth by $(PQ_4)^{\smile}$.

$$\begin{aligned}\mathbf{q}_{i+1}^2\mathbf{p}_{i+1} &= \mathbf{q}_i\mathbf{p}_i\mathbf{q}_i\mathbf{p}_i\mathbf{p}_{i+1} \equiv_{3,4} \mathbf{q}_i\mathbf{q}_i\mathbf{p}_{i+1}\mathbf{p}_i\mathbf{p}_{i+1} \\ &\equiv_P \mathbf{q}_i^2\mathbf{p}_i\mathbf{p}_{i+1}\mathbf{p}_i \equiv_{3,4} \mathbf{q}_i^2\mathbf{p}_{i+1}\mathbf{p}_i \\ &\equiv_{3,4} \mathbf{q}_i\mathbf{p}_i\mathbf{q}_i\mathbf{p}_i = \mathbf{q}_{i+1}^2 .\end{aligned}$$

From PQ_3, therefore, there follows by induction that, for any $i < \omega$, $\mathbf{q}_i^2\mathbf{p} \equiv_{3,4} \mathbf{q}_i^2$.

(g) Given any $j \geq 0$, assume as inductive hypothesis that, if $i \geq j$, then $\mathbf{p}_i\mathbf{q}_j \equiv_4 \mathbf{q}_j\mathbf{p}_{i+1}$. Let $i+1 \geq j+1$. Then the second assertion below holds by the inductive hypothesis and the third by P_6.

$$\mathbf{p}_{i+1}\mathbf{q}_{j+1} = \mathbf{p}_{i+1}\mathbf{q}_j\mathbf{p}_j \equiv_4 \mathbf{q}_j\mathbf{p}_{i+2}\mathbf{p}_j \equiv_P \mathbf{q}_j\mathbf{p}_j\mathbf{p}_{i+2} = \mathbf{q}_{j+1}\mathbf{p}_{i+2} .$$

From PQ_4, therefore, there follows by induction that, for any j, if $i \geq j$, then $\mathbf{p}_i\mathbf{q}_j \equiv_4 \mathbf{q}_j\mathbf{p}_{i+1}$. There follows that, if $i \geq j$, then $\mathbf{p}_{i+1}\mathbf{q}_j^{-1} = (\mathbf{q}_j\mathbf{p}_{i+1}^{-1})^{-1} \equiv_P (\mathbf{q}_j\mathbf{p}_{i+1})^{-1} \equiv_4 (\mathbf{p}_i\mathbf{q}_j)^{-1} = \mathbf{q}_j^{-1}\mathbf{p}_i$. □

Lemma 3(c) concerns an analogue on $F(\Gamma) = F_{SI}(\Gamma(\tilde{\mathbf{p}}, \mathbf{q}))$ of the relation Q_3 on $F_{SI}(\Gamma(\mathbf{q}))$ that was used in the presentation of $\mathbf{S}_q$ that was given in Theorem V.7. Parts (a),(b),(c) of the next lemma concern a relation on $F_{SI}(\Gamma(\tilde{\mathbf{p}}, \mathbf{q}))$ that is in a similar sense an analogue of the relation Q_1, Q_2, or Q_4, respectively, that was used there. Thus, Lemma 3(c) and Lemma 4 together show that the congruence $\equiv$ on $\mathbf{F}_{SI}(\Gamma(\mathbf{q})) = \mathbf{F}(\Gamma)$ that was used there has an interpretation within the congruence on $\mathbf{F}_{SI}(\Gamma(\tilde{\mathbf{p}}, \mathbf{q}))$ that we are using now.

LEMMA 4. (a) *Let $\equiv_{3,4}$ be the congruence on $\mathbf{F}_{SI}(\Gamma(\tilde{\mathbf{p}}, \mathbf{q}))$ that is generated by $\equiv_P \cup PQ_3 \cup PQ_4$. Then $\mathbf{q}_i\mathbf{q}_j \equiv_{3,4} \mathbf{q}_{j-1}\mathbf{q}_i$, if $i < j$.*

(b) *Let $\equiv_{2,4}$ be the congruence on $\mathbf{F}_{SI}(\Gamma(\tilde{\mathbf{p}}, \mathbf{q}))$ that is generated by $\equiv_P \cup PQ_2 \cup PQ_4$. Then $\mathbf{q}_i\mathbf{q}_j^{-1} \equiv_{2,4} \mathbf{q}_{j-1}^{-1}\mathbf{q}_i$, if $i < j$.*

(c) *Let $\doteq$ be the congruence on $\mathbf{F}(\Gamma(\tilde{\mathbf{p}}, \mathbf{q}))$ that is generated by $\equiv_P \cup_1 Q_3 \cup P_1 \cup P_4$. Then $\mathbf{q}_i\mathbf{q}_i^{-1}\mathbf{q}_{i+1} \doteq \mathbf{q}_{i+1}$, if $i < \omega$.*

Proof. (a) Consider any $i < \omega$. By Lemma 3(f),

$$\mathbf{q}_i\mathbf{q}_i \equiv_{3,4} \mathbf{q}_i\mathbf{q}_i\mathbf{p}_i = \mathbf{q}_i\mathbf{q}_{i+1} .$$

Hence $\mathbf{q}_{j-1}\mathbf{q}_i \equiv_{3,4} \mathbf{q}_i\mathbf{q}_j$, if $j = i+1$. Now let $j > i+1$. Then, by Lemma 3(g),

$$\mathbf{q}_{j-1}\mathbf{q}_i = \mathbf{q}_i\mathbf{p}_i \cdots \mathbf{p}_{j-2}\mathbf{q}_i \equiv_4 \mathbf{q}_i\mathbf{q}_i\mathbf{p}_{i+1} \cdots \mathbf{p}_{j-1} = \mathbf{q}_i\mathbf{q}_j .$$

(b) Consider any $i < \omega$. By Lemma 3(e),

$$\mathbf{q}_i\mathbf{q}_{i+1}^{-1} = \mathbf{q}_i\mathbf{p}_i\mathbf{q}_i^{-1} \equiv_{2,4} \mathbf{q}_i^{-1}\mathbf{q}_i .$$

Hence $\mathbf{q}_i\mathbf{q}_j^{-1} \equiv_{2,4} \mathbf{q}_{j-1}^{-1}\mathbf{q}_i$, if $j = i+1$. Now let $j \geq i+1$ and assume as inductive hypothesis that $\mathbf{q}_i\mathbf{q}_j^{-1} \equiv_{2,4} \mathbf{q}_{j-1}^{-1}\mathbf{q}_i$. Then of the following assertions the second holds by Lemma 3(g) and the fourth by the inductive hypothesis:

$$\begin{aligned} \mathbf{q}_i\mathbf{q}_{j+1}^{-1} &\equiv_P \mathbf{q}_i\mathbf{p}_j \cdots \mathbf{p}_i\mathbf{q}_i^{-1} \equiv_4 \mathbf{p}_{j-1}\mathbf{q}_i\mathbf{p}_{j-1} \cdots \mathbf{p}_i\mathbf{q}_i^{-1} \\ &\equiv_P \mathbf{p}_{j-1}\mathbf{q}_i\mathbf{q}_j^{-1} \equiv_{2,4} \mathbf{p}_{j-1}\mathbf{q}_{j-1}^{-1}\mathbf{q}_i \equiv_P \mathbf{q}_j^{-1}\mathbf{q}_i \,. \end{aligned}$$

The desired conclusion therefore follows by induction on $j - i$.

(c) By ${}_1Q_3$ one has $\mathbf{q}_0\mathbf{q}_0^{-1}\mathbf{q}_1 = \mathbf{q}\mathbf{q}^{-1}\mathbf{q}\mathbf{p}_0 \doteq \mathbf{q}\mathbf{p}_0 = \mathbf{q}_1$.

Of the following assertions the second holds by $(PQ_2)^{\smile}$, the third by PQ_4, the fifth by PQ_3, the sixth by $(PQ_4)^{\smile}$ and properties of ${}^{-1}$, and the eighth by PQ_1:

$$\begin{aligned} \mathbf{q}_1\mathbf{q}_1^{-1}\mathbf{q}_2 &\equiv_P \mathbf{q}\mathbf{p}_0^2\mathbf{q}^{-1}\mathbf{q}\mathbf{p}_0\mathbf{p}_1 \doteq \mathbf{q}\mathbf{p}_0^2\mathbf{q}\mathbf{p}_0\mathbf{q}^{-1}\mathbf{p}_0\mathbf{p}_1 \\ &\doteq \mathbf{q}\mathbf{q}\mathbf{p}_1^2\mathbf{p}_0\mathbf{q}^{-1}\mathbf{p}_0\mathbf{p}_1 \equiv_P \mathbf{q}^2\mathbf{p}_0\mathbf{p}_1^2\mathbf{q}^{-1}\mathbf{p}_0\mathbf{p}_1 \doteq \mathbf{q}^2\mathbf{p}_1^2\mathbf{q}^{-1}\mathbf{p}_0\mathbf{p}_1 \\ &\doteq \mathbf{q}^2\mathbf{q}^{-1}\mathbf{p}_0^2\mathbf{p}_0\mathbf{p}_1 \equiv_P \mathbf{q}\mathbf{q}\mathbf{q}^{-1}\mathbf{p}_0\mathbf{p}_1 \equiv \mathbf{q}\mathbf{p}_0\mathbf{p}_1 = \mathbf{q}_2 \end{aligned}$$

Given any $i \geq 0$, of the following assertions the first holds by Lemma 3(a) and the third by PQ_4:

$$\begin{aligned} \mathbf{q}_{i+2}\mathbf{q}_{i+2}^{-1}\mathbf{q}_{i+3} &\doteq \mathbf{p}_i^2\mathbf{q}_{i+3} = \mathbf{p}_i^2\mathbf{q}\mathbf{p}_0 \cdots \mathbf{p}_{i+2} \doteq \mathbf{q}\mathbf{p}_{i+1}^2\mathbf{p}_0 \cdots \mathbf{p}_{i+2} \\ &\equiv_P \mathbf{q}\mathbf{p}_0 \cdots \mathbf{p}_{i+2}\mathbf{p}_{i+1}^2 \equiv_P \mathbf{q}\mathbf{p}_0 \cdots \mathbf{p}_{i+2} = \mathbf{q}_{i+3} \end{aligned}$$

□

Parts (a),(b),(c) of Lemma V.4 concerned a congruence relation on $\mathbf{F}_{SI}(\Gamma(\tilde{\mathbf{q}}))$. Parts (c),(d),(e) of the following lemma assert a respective analogue concerning our congruence $\equiv$ on $\mathbf{F}_{SI}(\Gamma(\tilde{\mathbf{p}}, \mathbf{q}))$.

LEMMA 5. (a) $\mathbf{q}_i\mathbf{q}_i^{-1}\mathbf{p}_j \equiv \mathbf{p}_j\mathbf{q}_i\mathbf{q}_i^{-1}$, *if* $i \geq j+2$.
(b) $\mathbf{q}_i\mathbf{q}_i^{-1}\mathbf{p}_j \equiv \mathbf{p}_j \equiv \mathbf{p}_j\mathbf{q}_i\mathbf{q}_i^{-1}$, *if* $0 \leq i \leq j+2$.
(c) $\mathbf{q}_{s+r}\mathbf{q}_{s+r}^{-1}x \equiv x\mathbf{q}_{t+r}\mathbf{q}_{t+r}^{-1}$, *if* $gd\ x = \langle s,t\rangle$.
(d) $\mathbf{q}_i\mathbf{q}_i^{-1}x \equiv x$, *if* $gd\ x = \langle s,t\rangle$ *and* $i \leq s$.
(e) $x\mathbf{q}_i\mathbf{q}_i^{-1} \equiv x$, *if* $gd\ x = \langle s,t\rangle$ *and* $i \leq t$.
(f) $x\mathbf{q}_i\mathbf{q}_i^{-1}y \equiv \mathbf{q}_s\mathbf{q}_s^{-1}xy \equiv xy\mathbf{q}_t\mathbf{q}_t^{-1}$, *if* $gd(x\mathbf{q}_i\mathbf{q}_i^{-1}y) = \langle s,t\rangle$.

Proof. (a) Assume that $i \geq j+2$. Then from Lemma 3(a) there follows

$$\mathbf{q}_i\mathbf{q}_i^{-1}\mathbf{p}_j \equiv \mathbf{p}_{i-2}^2\mathbf{p}_j \equiv_P \mathbf{p}_j\mathbf{p}_{i-2}^2 \equiv \mathbf{p}_j\mathbf{q}_i\mathbf{q}_i^{-1} \,.$$

(b) First, let $i = 0$ and $j \geq i = 0$. Then, by PQ_1, there hold:

$$\mathbf{q}_0\mathbf{q}_0^{-1}\mathbf{p}_j \equiv_P \mathbf{q}_0\mathbf{q}_0^{-1}\mathbf{p}_0\mathbf{p}_0\mathbf{p}_j \equiv \mathbf{p}_0\mathbf{p}_0\mathbf{p}_j \equiv_P \mathbf{p}_j$$

Next, let $i = 1$ and $j = 0$. Then of the following assertions the third holds by PQ_4, the sixth by $(PQ_4)^{\smile}$, and the last by PQ_1:

$$\begin{array}{rcl}\mathbf{q}_i\mathbf{q}_i^{-1}\mathbf{p}_0 & = & (\mathbf{p}_0^{-1}\mathbf{q}_1\mathbf{q}_1^{-1})^{-1} \equiv_P (\mathbf{p}_0\mathbf{q}\mathbf{p}_0^2\mathbf{q}^{-1})^{-1} \equiv (\mathbf{q}\mathbf{p}_1\mathbf{p}_0^2\mathbf{q}_1^{-1})^{-1} \\ & \equiv_P & (\mathbf{q}\mathbf{p}_1\mathbf{q}^{-1})^{-1} = \mathbf{q}(\mathbf{q}\mathbf{p}_1)^{-1} \equiv \mathbf{q}(\mathbf{p}_0\mathbf{q})^{-1} \\ & = & \mathbf{q}\mathbf{q}^{-1}\mathbf{p}_0^{-1} \equiv_P \mathbf{q}\mathbf{q}^{-1}\mathbf{p}_0 \equiv \mathbf{p}_0\end{array}$$

Now let $i = 1$ and $j \geq 1$. Then from Lemma 3(d) there follows:

$$\mathbf{q}_1\mathbf{q}_1^{-1}\mathbf{p}_j \equiv_P \mathbf{q}_1\mathbf{q}_1^{-1}\mathbf{p}_1\mathbf{p}_1\mathbf{p}_j \equiv \mathbf{p}_1\mathbf{p}_1\mathbf{p}_j \equiv_P \mathbf{p}_j \ .$$

Finally, let $i \geq 2$ and $j \geq i - 2$. Then from Lemma 3(a) there follows:

$$\mathbf{q}_i\mathbf{q}_i^{-1}\mathbf{p}_j \equiv \mathbf{p}_{i-2}^2\mathbf{p}_j \equiv_P \mathbf{p}_j \ .$$

There now follows that, if $0 \leq i \leq j + 2$, then $\mathbf{q}_i\mathbf{q}_i^{-1}\mathbf{p}_j \equiv \mathbf{p}_j$. Hence also, if $0 \leq i \leq j + 2$, then

$$\mathbf{p}_j \equiv_P \mathbf{p}_j^{-1} \equiv (\mathbf{q}_i\mathbf{q}_i^{-1}\mathbf{p}_j)^{-1} = \mathbf{p}_j^{-1}\mathbf{q}_i\mathbf{q}_i^{-1} \equiv_P \mathbf{p}_j\mathbf{q}_i\mathbf{q}_i^{-1} \ .$$

(c),(d),(e). A proof similar to that of the Lemmas V.3(a),...,V.3(g) shows that from Lemma 3(c) and Lemma 4 there follows that an analogue of each of these lemmas holds for our present $\equiv$. From these analogues and from parts (a) and (b) there follow parts (c), (d), and (e) by a proof similar to that of Lemma V.4.

(f) Let $gd(x\mathbf{q}_i\mathbf{q}_i^{-1}y) = \langle s, t\rangle$ and $gd\ x = \langle s', t'\rangle$. First assume that $i \leq t'$. Then $x\mathbf{q}_i\mathbf{q}_i^{-1} \equiv x$, by part (e). Since, by part (d), $x\mathbf{q}_i\mathbf{q}_i^{-1}y \equiv \mathbf{q}_s\mathbf{q}_s^{-1}x\mathbf{q}_i\mathbf{q}_i^{-1}y$, there now follows that $x\mathbf{q}_i\mathbf{q}_i^{-1}y \equiv \mathbf{q}_s\mathbf{q}_s^{-1}xy$. Now assume that $i \geq t'$. Let $r = i - t'$. By part (c),

$$x\mathbf{q}_i\mathbf{q}_i^{-1} = x\mathbf{q}_{t'+r}\mathbf{q}_{t'+r}^{-1} \equiv \mathbf{q}_{s'+r}\mathbf{q}_{s'+r}^{-1}x \ .$$

Thus, $gd(x\mathbf{q}_i\mathbf{q}_i^{-1}) = \langle s'+r, t'+r\rangle$. Since $\langle s, t\rangle = gd(x\mathbf{q}_i\mathbf{q}_i^{-1}y) = gd(x\mathbf{q}_i\mathbf{q}_i^{-1}) \odot gd\ y$, there follows that $s \geq s' + r$. Hence, by part (e),

$$\mathbf{q}_s\mathbf{q}_s^{-1}\mathbf{q}_{s'+r}\mathbf{q}_{s'+r}^{-1} \equiv \mathbf{q}_s\mathbf{q}_s^{-1}$$

Using also part (d), one now has:

$$x\mathbf{q}_i\mathbf{q}_i^{-1}y \equiv \mathbf{q}_s\mathbf{q}_s^{-1}x\mathbf{q}_i\mathbf{q}_i^{-1}y \equiv \mathbf{q}_s\mathbf{q}_s^{-1}\mathbf{q}_{s'+r}\mathbf{q}_{s'+r}^{-1}xy \equiv \mathbf{q}_s\mathbf{q}_s^{-1}xy \ .$$

The proof that $x\mathbf{q}_i\mathbf{q}_i^{-1}y \equiv xy\mathbf{q}_t\mathbf{q}_t^{-1}$ is similar. □

Mostly to prepare for Lemma 7 below, we now prove the following.

LEMMA 6. (a) $\mathbf{q}_i\mathbf{p}_j \equiv \mathbf{p}_{j-1}\mathbf{q}_i$ *and* $\mathbf{p}_j\mathbf{q}_i^{-1} \equiv \mathbf{q}_i^{-1}\mathbf{p}_{j-1}$, *if* $i < j$.
(a)′ $\mathbf{p}_i\mathbf{q}_j \equiv \mathbf{q}_j\mathbf{p}_{i+1}$ *and* $\mathbf{q}_j^{-1}\mathbf{p}_i \equiv \mathbf{p}_{i+1}\mathbf{q}_j^{-1}$ *if* $i \geq j$.
(b) $\mathbf{q}_i\mathbf{p}_i = \mathbf{q}_{i+1}$ *and* $\mathbf{q}_{i+1}^{-1} \equiv_P \mathbf{p}_i\mathbf{q}_i^{-1}$.
(c) $\mathbf{q}_i\mathbf{p}_{i-1} \equiv \mathbf{q}_i\mathbf{q}_i^{-1}\mathbf{q}_{i-1}$ *and* $\mathbf{p}_{i-1}\mathbf{q}_i^{-1} \equiv \mathbf{q}_{i-1}^{-1}\mathbf{q}_i\mathbf{q}_i^{-1} \equiv \mathbf{q}_{i+1}\mathbf{q}_{i+1}^{-1}\mathbf{q}_{i-1}^{-1}$.
(d) $\mathbf{q}_i\mathbf{p}_j \equiv \mathbf{p}_j\mathbf{q}_i$ *and* $\mathbf{p}_j\mathbf{q}_i^{-1} \equiv \mathbf{q}_i^{-1}\mathbf{p}_j$ *if* $i \geq j+2$.
(d)′ $\mathbf{p}_i\mathbf{q}_j \equiv \mathbf{q}_j\mathbf{p}_i$ *and* $\mathbf{q}_j^{-1}\mathbf{p}_i \equiv \mathbf{p}_i\mathbf{q}_j^{-1}$ *if* $i+1 < j$.
(e) $\mathbf{p}_i\mathbf{q}_{i+1} \equiv \mathbf{q}_{i+1}\mathbf{p}_i\mathbf{p}_{i+1}\mathbf{p}_i$ *and* $\mathbf{q}_{i+1}^{-1}\mathbf{p}_i \equiv \mathbf{p}_i\mathbf{p}_{i+1}\mathbf{p}_i\mathbf{q}_{i+1}^{-1}$.

Proof. (a) by Lemma 3(g).

(b) By the definition of $\mathbf{q}_{i+1}$ and by $\mathbf{p}_i^{-1} \equiv_P \mathbf{p}_i$.

(c) Using, respectively, the definition of $\mathbf{q}_i$, Lemma 3(a), and Lemma 5(c) one obtains:

$$\mathbf{q}_i\mathbf{p}_{i-1} = \mathbf{q}_{i-1}\mathbf{p}_{i-1}\mathbf{p}_{i-1} \equiv \mathbf{q}_{i-1}\mathbf{q}_{i+1}\mathbf{q}_{i+1}^{-1} \equiv \mathbf{q}_i\mathbf{q}_i^{-1}\mathbf{q}_{i-1}$$

The second assertion of part (c) is equivalent to the first.

(d) Assume that $i \geq 2$. Then the third assertion below holds by Lemma 3(g).

$$\begin{aligned}\mathbf{q}_i\mathbf{p}_{i-2} &= \mathbf{q}_{i-2}\mathbf{p}_{i-2}\mathbf{p}_{i-1}\mathbf{p}_{i-2} \equiv_P \mathbf{q}_{i-2}\mathbf{p}_{i-1}\mathbf{p}_{i-2}\mathbf{p}_{i-1} \\ &\equiv \mathbf{p}_{i-2}\mathbf{q}_{i-2}\mathbf{p}_{i-2}\mathbf{p}_{i-12} = \mathbf{p}_{i-2}\mathbf{q}_i \ .\end{aligned}$$

Now let $j \geq 0$ and $i \geq j+2$ and assume as inductive hypothesis that $\mathbf{q}_i\mathbf{p}_j \equiv \mathbf{p}_j\mathbf{q}_i$. Then the second assertion below holds by $i \geq j+2$ and the third by the inductive hypothesis

$$\mathbf{q}_{i+1}\mathbf{p}_j = \mathbf{q}_i\mathbf{p}_i\mathbf{p}_j \equiv_P \mathbf{q}_i\mathbf{p}_j\mathbf{p}_i \equiv \mathbf{p}_j\mathbf{q}_i\mathbf{p}_i = \mathbf{p}_j\mathbf{q}_{i+1} \ .$$

The second assertion of part (d) is equivalent to the first.

(a)′ This is equivalent to part (a).

(e) The second assertion below holds by part (a)′.

$$\mathbf{p}_i\mathbf{q}_{i+1} = \mathbf{p}_i\mathbf{q}_i\mathbf{p}_i \equiv \mathbf{q}_i\mathbf{p}_{i+1}\mathbf{p}_i \equiv_P \mathbf{q}_i\mathbf{p}_i\mathbf{p}_i\mathbf{p}_{i+1}\mathbf{p}_i = \mathbf{q}_{i+1}\mathbf{p}_i\mathbf{p}_{i+1}\mathbf{p}_i \ .$$

The second assertion of part (e) is equivalent to the first.

(d)′ This is equivalent to part (d). □

A $\langle \mathbf{q}^{-1}, \mathbf{p}, \mathbf{q}\rangle$ ***word*** shall be any $w^\frown x^\frown y = wxy$ such that the following three conditions are satisfied:

(1) Either $w = \emptyset$ or $w = \mathbf{q}_{i_0}^{-1} \dots \mathbf{q}_{i_{n-1}}^{-1}$ for some $n \geq 1$ and some $\mathbf{q}_{i_0}^{-1}, \dots, \mathbf{q}_{i_{n-1}}^{-1}$.

(2) Either $x = \emptyset$ or $x = \mathbf{p}_{i_0} \dots \mathbf{p}_{i_{n-1}}$ for some $n \geq 1$ and some $\mathbf{p}_{i_0}, \dots, \mathbf{p}_{i_{n-1}}$.

(3) Either $y = \emptyset$ or $y = \mathbf{q}_{i_0} \cdots \mathbf{q}_{i_{n-1}}$ for some $n \geq 1$ and some $\mathbf{q}_{i_0}, \ldots, \mathbf{q}_{i_{n-1}}$.

Also, for example, a $\langle \mathbf{q}^{-1} \rangle$ ***word*** shall be any w for which (1) holds and a $\langle \mathbf{q}^{-1}, \mathbf{q} \rangle$ ***word*** any wy for which (1) and (3) hold. Thus, for example, the set of nonempty $\langle \mathbf{p} \rangle$ words coincides with $F(\Gamma(\tilde{\mathbf{p}}))$, while the nonempty $\langle \mathbf{q}^{-1}, \mathbf{q} \rangle$ words are elements of $F(\Gamma\{\mathbf{q}_i : i < \omega\})$ of a special form.

Henceforth in using wxy to refer to a $\langle \mathbf{q}^{-1}, \mathbf{p}, \mathbf{q} \rangle$ word we will assume that w, x, y satisfy the conditions (1),(2),(3), respectively. Similar conventions apply, for example, to $\langle \mathbf{q}, \mathbf{p} \rangle$ words.

LEMMA 7. *Let* $gd\, v = \langle s, t \rangle$.
(a) *If* $\mathbf{q}^{-1}$ *does not occur in* v, *then* $v \equiv \mathbf{q}_s \mathbf{q}_s^{-1} xy$ *for some* $\langle \mathbf{p}, \mathbf{q} \rangle$ *word* xy.
(b) *If* $\mathbf{q}$ *does not occur in* v, *then* $v \equiv \mathbf{q}_s \mathbf{q}_s^{-1} wx$ *for some* $\langle \mathbf{q}^{-1}, \mathbf{p} \rangle$ *word* wx.
(c) $v \equiv \mathbf{q}_s \mathbf{q}_s^{-1} wxy$ *for some* $\langle \mathbf{q}^{-1}, \mathbf{p}, \mathbf{q} \rangle$ *word* wxy.

Proof. (a) and (b). By Lemmas 6, 4(c), 5, and induction on the length of v.
(c) By Lemmas 6, 4(a), 4(b), 4(c), 5, and induction on the length of v. □

One can see from the proofs that one can choose xy, wx, or wxy, respectively, so that its length does not exceed that of v.

An element of $F_{SI}(\Gamma(\tilde{\mathbf{p}}, \mathbf{q}))$ shall be ***strictly ascending*** if and only if it is a word $\mathbf{q}_{i_0} \cdots \mathbf{q}_{i_n}$ such that $n \geq 0$ and, if $m < m' \leq n$ then $i_m < i_{m'}$. An element of $F_{SI}(\Gamma(\tilde{\mathbf{p}}, \mathbf{q}))$ shall be ***strictly descending*** if and only if it is a word $\mathbf{q}_{i_0}^{-1} \cdots \mathbf{q}_{i_n}^{-1}$ such that $n \geq 0$ and, if $m < m' \leq n$ then $i_m > i_{m'}$. Thus the elements of $F_{SI}(\Gamma(\tilde{\mathbf{p}}, \mathbf{q}))$ that are strictly ascending or strictly descending, respectively, are nonempty $\langle \mathbf{q} \rangle$ words or nonempty $\langle \mathbf{q}^{-1} \rangle$ words, respectively, of a certain special form.

A word v in $F_{SI}(\Gamma(\tilde{\mathbf{p}}, \mathbf{q}))$ shall be $\{\tilde{\mathbf{p}}, \mathbf{q}\}$-***normal*** if and only if $v = \mathbf{q}_r \mathbf{q}_r^{-1} wxy$ for some $\langle \mathbf{q}^{-1}, \mathbf{p}, \mathbf{q} \rangle$ word wxy such that w is either empty or strictly descending, such that y is either empty or strictly ascending, and such that, if wxy is nonempty and $gd(wxy) = \langle s, t \rangle$, then $r \geq s$ and hence $gd\, v = \langle r, t + r - s \rangle$.

LEMMA 8. *Consider any* $\mathbf{q}_s \mathbf{q}_s^{-1} wxy$ *such that* wxy *is a* $\langle \mathbf{q}^{-1}, \mathbf{p}, \mathbf{q} \rangle$ *word. Then for some* w', y', *and* $r \geq s$, $\mathbf{q}_r \mathbf{q}_r^{-1} w'xy'$ *is* $\{\tilde{\mathbf{p}}, \mathbf{q}\}$-*normal and* $\mathbf{q}_r \mathbf{q}_r^{-1} w'xy' \equiv \mathbf{q}_s \mathbf{q}_s^{-1} wxy$.

Proof. If $wxy \neq \emptyset$, $gd(wxy) = \langle r, r' \rangle$, and $r \geq s$ then, by Lemma 4(c) and Lemma 5, $\mathbf{q}_r \mathbf{q}_r^{-1} wxy \equiv \mathbf{q}_s \mathbf{q}_s^{-1} wxy$. If $w \neq \emptyset$ then, by Lemma 4(a) there is some

w' that is strictly descending such that $w' \equiv w$. If $y \neq \emptyset$ then, again by Lemma 4(a), there is some y' that is strictly ascending such that $y' \equiv y$. $\square$

Consider any $\mathbf{q}_r\mathbf{q}_r^{-1}wxy$ that is $\{\tilde{\mathbf{p}}, \mathbf{q}\}$-normal. With w, x and y we now associate a unique special triple ${}_{s_1}\langle M_1\rangle_{t_1}$, ${}_{s_2}\langle M_2\rangle$, or ${}_{s_3}\langle M_3\rangle_{t_3}$, respectively, as follows. If $w = \emptyset$ then ${}_{s_1}\langle M_1\rangle_{t_1}$ shall be ${}_0\langle\emptyset\rangle_0 = {}_0\langle\emptyset\rangle$. Likewise, if $x = \emptyset$ then ${}_{s_2}\langle M_2\rangle$ shall be ${}_0\langle\emptyset\rangle$ and if $y = \emptyset$ then ${}_{s_3}\langle M_3\rangle_{t_3}$ shall be ${}_0\langle\emptyset\rangle$. Now suppose that x is nonempty, hence $gd\, x = \langle s_2, s_2\rangle$ for some $s_2 \geq 2$, and hence $h(x) = {}_{s_2}[\![\pi]\!]$ for a unique permutation π of $s_2 = \{0, \dots s_2 - 1\}$. Then we let ${}_{s_2}\langle M_2\rangle = {}_{s_2}\langle\pi\rangle$. Next suppose that w is nonempty and hence is a word $\mathbf{q}_{j_0}^{-1} \dots \mathbf{q}_{j_m}^{-1}$ that is strictly descending. Then we let $s_1 = j_0 + 1 = \{0, \dots, j_0\}$, $t_1 = j_0 - m$, $J = \{j_0, \dots, j_m\}$ and $M_1 = f_J{}^\smile \cap (s_1 \times t_1)$. Also if y is nonempty and hence is a word $\mathbf{q}_{k_0} \dots \mathbf{q}_{k_n}$ that is strictly ascending, then we let $t_3 = k_n + 1$, $s_3 = k_n - n$, $K = \{k_0, \dots, k_n\}$, and $N_3 = f_K \cap (s_3 \times t_3)$.

We shall say of the ${}_{s_1}\langle M_1\rangle_{t_1}$, ${}_{s_2}\langle M_2\rangle$, ${}_{s_3}\langle M_3\rangle_{t_3}$ just described that it is ***the special triple associated with*** w, x, or y, respectively. There follows that if $w \neq \emptyset$ then $h(x) = {}_{s_1}[\![M_1]\!]_{t_1}$ and if $w = \emptyset$ then ${}_{s_1}[\![M_1]\!]_{t_1} = {}_{[0,\omega)}\overset{\circ}{1}$. Also, if $x \neq \emptyset$ then, as we already saw, $h(x) = {}_{s_2}[\![M_2]\!]$ and if $x = \emptyset$ then ${}_{s_2}[\![M_2]\!] = {}_{[0,\omega)}\overset{\circ}{1}$. Further, if $y \neq \emptyset$ then $h(y) = {}_{s_3}[\![M_3]\!]_{t_3}$ and if $y = \emptyset$ then ${}_{s_3}[\![M_3]\!]_{t_3} = {}_{[0,\omega)}\overset{\circ}{1}$. There follows that, whether w, x or y is empty or not, there holds:

$$h(\mathbf{q}_r\mathbf{q}_r^{-1}wxy) = {}_{[r,\omega)}\overset{\circ}{1} \circ {}_{s_1}[\![M_1]\!]_{t_1} \circ {}_{s_2}[\![M_2]\!] \circ {}_{s_3}[\![M_3]\!]_{t_3} \ .$$

LEMMA 9. *Assume that* $\mathbf{q}_r\mathbf{q}_r^{-1}wxy$ *and* $\mathbf{q}_{r'}\mathbf{q}_{r'}^{-1}w'x'y'$ *are* $\{\tilde{\mathbf{p}}, \mathbf{q}\}$*-normal and that* $h(\mathbf{q}_r\mathbf{q}_r^{-1}wxy) = h(\mathbf{q}_{r'}\mathbf{q}_{r'}^{-1}w'x'y')$. *Then* $r = r'$, $w = w'$, $y = y'$, *and, if* ${}_{s_2}\langle M_2\rangle$ *is associated with* x, ${}_{s'}\langle M_2'\rangle$ *is associated with* x', *and* $s_2'' = \max(s_2, s_2')$, *then* $h(\mathbf{q}_{s_2''}\mathbf{q}_{s_2''}^{-1}x) = h(\mathbf{q}_{s_2''}\mathbf{q}_{s_2''}^{-1}x')$. *There follows that* $\mathbf{q}_r\mathbf{q}_r^{-1}wxy \equiv \mathbf{q}_{r'}\mathbf{q}_{r'}^{-1}w'x'y'$.

Proof. Since $h(\mathbf{q}_r\mathbf{q}_r^{-1}wxy) = h(\mathbf{q}_{r'}\mathbf{q}_{r'}^{-1}w'x'y')$ therefore $\mathbf{q}_r\mathbf{q}_r^{-1}wxy$ and $\mathbf{q}_{r'}\mathbf{q}_{r'}^{-1}w'x'y'$ have the same grade. Since they are $\{\tilde{\mathbf{p}}, \mathbf{q}\}$-normal, there follows that $r = r'$.

Let ${}_{s_1}\langle M_1\rangle_{t_1}$ and ${}_{s_1'}\langle M_1'\rangle_{t_1'}$ be the special triple that is associated with w or with w', respectively. First assume that $w \neq \emptyset$ and $w' \neq \emptyset$. Then, for some J and J', $M_1 = f_J{}^\smile \cap (s_1 \times t_1)$ and $M_1' = f_{J'}{}^\smile \cap (s_1' \times t_1')$. Now suppose that $w = w'$ did not hold. Then, since s_1 and t_1 are determined by J and s_1' and t_1' are determined by J', then there would hold $J \neq J'$. Suppose that j is in J but not in J'. Let u and u' be distinct elements of U and let v be the sequence in ${}^rU$ such that $v_j = u'$ and $v_m = u$ for any $m < r$ such that $m \neq j$. Then the only sequence v' such that $\langle v, v'\rangle$ is in $h(wx)$ is the sequence v' of length $r - \overline{\overline{J}}$ such that $v_n' = u$ for every $n < r - \overline{\overline{J}}$. In contrast, the only sequence v'' such that $\langle v, v''\rangle$ is in $h(w'x')$ contains an occurrence of u'. There follows that the set $\{v' : \langle v, v'\rangle \in h(wxy)\}$ has the property that if $v'_m = u'$ then any v'' such that $|v''| = |v'|$ and $v''_n = v'_n$ for any $n < |v'|$ such that $n \neq m$ also belongs to the set. In contrast, the set $\{v' : \langle v, v'\rangle \in h(w'x'y')\}$ has the property that there is some m such that $v'_m = u'$ for every v' in the set. There follows that $\{v' : \langle v, v'\rangle \in h(wxy)\} \neq \{v' : \langle v, v'\rangle \in h(w'x'y')\}$ and hence that $h(\mathbf{q}_r\mathbf{q}_r^{-1}wxy) \neq h(\mathbf{q}_r\mathbf{q}_r^{-1}w'x'y') = h(\mathbf{q}_{r'}\mathbf{q}_{r'}^{-1}w'x'y')$, contrary to the hypothesis of the lemma. There follows that $J \subseteq J'$. Similarly $J' \subseteq J$. Hence if $w \neq \emptyset$ and $w' \neq \emptyset$, then $w = w'$. A similar argument shows that the supposition that $w = \emptyset$ and $w' \neq \emptyset$ or that $w \neq \emptyset$ and $w' = \emptyset$ again is incompatible with the hypothesis of the lemma. There now follows that $w = w'$.

Using an argument that is dual but similar one can show that the supposition that $y \neq y'$ is incompatible with the hypothesis of the lemma. Thus $y = y'$.

There now follows that if both $x = \emptyset$ and $x' = \emptyset$, then $\mathbf{q}_r\mathbf{q}_r^{-1}wxy = \mathbf{q}_{r'}\mathbf{q}_{r'}^{-1}w'x'y'$ and hence $\mathbf{q}_r\mathbf{q}_r^{-1}wxy \equiv \mathbf{q}_{r'}\mathbf{q}_{r'}^{-1}w'x'y'$. For the rest of the proof assume that $x = \emptyset$ and $x' = \emptyset$ do not both hold. Let ${}_{s_2}\langle M_2\rangle$ and ${}_{s'_2}\langle M'_2\rangle$ be the special triple associated with x or with x', respectively and let $s''_2 = \max(s_2, s'_2)$. Then $s''_2 \geq 2$.

First assume that $x \neq \emptyset$ and $x' \neq \emptyset$. Then M_2 is a permutation π_2 of $s_2 = \{0, \ldots, s_{2-1}\}$ and M'_2 is a permutation π'_2 of s'_2, Suppose that $h(\mathbf{q}_{s''_2}\mathbf{q}_{s''_2}^{-1}x) \neq h(\mathbf{q}_{s''_2}\mathbf{q}_{s''_2}^{-1}x')$. Then for some $k < \min(s_2, s'_2)$, $\pi(k) \neq \pi'(k)$. Let u and u' be distinct elements of U and let v be the sequence in ${}^{s''_2}U$ such that $v_k = u'$ and $v_m = u$ for any $m < s''_2$ such that $m \neq k$. Then the only sequence v' such that $\langle v, v'\rangle$ is in ${}_{s_2}[\![\pi]\!]$ satisfies $v'_{\pi(k)} = u'$ and $v'_n = u$ if $n < s''_2$ and $n \neq \pi(k)$, whereas the only sequence v' such that $\langle v, v'\rangle$ is in ${}_{s'_2}[\![\pi']\!]$ satisfies $v'_{\pi'(k)} = u'$ and $v'_n = u$ if $n < s''_2$ and $n \neq \pi'(k)$. There follows that there is a pair in $h(\mathbf{q}_r\mathbf{q}_r^{-1}wxy)$ which is not in $h(\mathbf{q}_r\mathbf{q}_r^{-1}wx'y) = h(\mathbf{q}_{r'}\mathbf{q}_{r'}^{-1}w'x'y')$, contrary to the hypothesis of the lemma. This shows that if $x \neq \emptyset$ and $x' \neq \emptyset$ then $h(\mathbf{q}_{s''_2}\mathbf{q}_{s''_2}^{-1}x) = h(\mathbf{q}_{s''_2}\mathbf{q}_{s''_2}^{-1}x')$.

The supposition that $h(\mathbf{q}_{s''_2}\mathbf{q}_{s''_2}^{-1}x) \neq {}_{[s''_2,\omega)}\overset{\circ}{1}$ and that $x' = \emptyset$ and hence $h(\mathbf{q}_{s''_2}\mathbf{q}_{s''_2}^{-1}x') = {}_{[s''_2,\omega)}\overset{\circ}{1}$ likewise can be shown to be incompatible with the hypothesis of the lemma. The same can be shown for the supposition that $h(\mathbf{q}_{s''_2}\mathbf{q}_{s''_2}^{-1}x') \neq {}_{[s''_2,\omega)}\overset{\circ}{1}$ and that $x = \emptyset$ and hence $h(\mathbf{q}_{s''_2}\mathbf{q}_{s''_2}^{-1}x) = {}_{[s''_2,\omega)}\overset{\circ}{1}$. There now follows that, in any case, $h(\mathbf{q}_{s''_2}\mathbf{q}_{s''_2}^{-1}x) = h(\mathbf{q}_{s''_2}\mathbf{q}_{s''_2}^{-1}x')$.

Since $s_2'' \geq 2$ therefore $q_{s_2''}q_{s_2''}{}^{\smile} = p^2_{s_2''-2}$ and hence $h(\mathbf{p}^2_{s_2''-2}x) = h(\mathbf{p}^2_{s''-2}x')$. Also, by Lemma 3(a), $\mathbf{q}_{s_2''}\mathbf{q}_{s_2''}^{-1} \equiv \mathbf{p}^2_{s_2''-2}$. There follows

$$\mathbf{q}_{s_2''}\mathbf{q}_{s_2''}^{-1}x \equiv \mathbf{p}^2_{s_2''-2}x \equiv_P \mathbf{p}^2_{s_2''-2}x' \equiv \mathbf{q}_{s_2''}\mathbf{q}_{s_2''}^{-1}x' \ .$$

By Lemmas 5(c) and 4(c) and by the choice of r and s_2'' there holds $\mathbf{q}_r\mathbf{q}_r^{-1}w \equiv \mathbf{q}_r\mathbf{q}_r^{-1}w\mathbf{q}_{s_2''}\mathbf{q}_{s_2''}^{-1}$. There now follows

$$\begin{aligned} \mathbf{q}_r\mathbf{q}_r^{-1}wxy &\equiv \mathbf{q}_r\mathbf{q}_r^{-1}w\mathbf{q}_{s_2''}\mathbf{q}_{s_2''}^{-1}xy \equiv \mathbf{q}_r\mathbf{q}_r^{-1}w\mathbf{q}_{s_2''}\mathbf{q}_{s_2''}^{-1}x'y \\ &\equiv \mathbf{q}_r\mathbf{q}_r^{-1}wx'y = \mathbf{q}_{r'}\mathbf{q}_{r'}^{-1}w'x'y' \ . \end{aligned}$$

□

We now give a presentation of $\mathbf{S}_{pq}$.

THEOREM 10. *Let $\equiv_P$ be the congruence on $\mathbf{F}_{SI}(\Gamma(\tilde{\mathbf{p}}, \mathbf{q}))$ that is generated by $P_1 \cup P_2 \cup P_3 \cup P_4^{\#} \cup P_5 \cup P_6$ and let $\equiv$ be the congruence on $\mathbf{F}_{SI}(\Gamma(\tilde{\mathbf{p}}, \mathbf{q}))$ that is generated by $\equiv_P \cup_1 Q_3 \cup PQ_1 \cup PQ_2 \cup PQ_3 \cup PQ_4$. Let h be the homomorphism of $\mathbf{F}_{SI}(\Gamma(\tilde{\mathbf{p}}, \mathbf{q}))$ onto $\mathbf{S}_{pq}$ such that $h(\mathbf{q}) = q_0$ and $h(\mathbf{p}_i) = p_i$ for each $i < \omega$. Then $\equiv$ coincides with $h \circ h^{\smile}$ and $\mathbf{F}_{SI}(\Gamma(\tilde{\mathbf{p}}, \mathbf{q}))/\!\equiv$ is isomorphic to $\mathbf{S}_{pq}$.*

Proof. By Lemmas 1(b), 7(c), 8, and 9. □

In presenting $\langle \mathbf{S}_{pq}, \subseteq \rangle = \langle \mathbf{S}_{pq}, \circ, {}^{\smile}, \subseteq \rangle$, we will also make use of the following two relations on $F_{SI}(\Gamma(\tilde{\mathbf{p}}, \mathbf{q}))$, each consisting of a single pair. Also, neither involves any $\mathbf{p}_i$ other than $\mathbf{p}_0 = \mathbf{p}_{(0,1)}$.

$$\begin{aligned} PQ_5 &= \{\langle \mathbf{q}\mathbf{p}_0^2\mathbf{q}^{-1}, \mathbf{q}\mathbf{q}^{-1} \rangle\} \\ PQ_6 &= \{\langle \mathbf{q}\mathbf{p}_0^2\mathbf{q}^{-1}, \mathbf{q}^{-1}\mathbf{q} \rangle\} \end{aligned}$$

THEOREM 11. *Let $\equiv$ be as in Theorem 10. Let $\preceq$ be the least quasiordering of $F_{SI}(\Gamma(\tilde{\mathbf{p}}, \mathbf{q}))$ that is compatible with $\mathbf{F}_{SI}(\Gamma(\tilde{\mathbf{p}}, \mathbf{q}))$ and includes $\equiv \cup P_4 \cup PQ_5 \cup PQ_6$. Then for any w and w' in $F_{SI}(\Gamma(\tilde{\mathbf{p}}, \mathbf{q}))$, $h(w) \subseteq h(w')$ if and only if $w \preceq w'$. There follows that $\langle \mathbf{F}_{SI}(\Gamma(\tilde{\mathbf{p}}, \mathbf{q}), \preceq \rangle/\!\equiv$ is isomorphic to $\langle \mathbf{S}_{pq}, \subseteq \rangle$.*

Proof. Let $\equiv_P$ be as in Theorem 10. Whereas Q_5 and Q_6 were defined in Chapter V to be relations on $F_{SI}(\Gamma(\tilde{\mathbf{q}}))$, when dealing with $F_{SI}(\Gamma(\tilde{\mathbf{p}}, \mathbf{q}))$, as we are now, we let them be the relation on $F_{SI}(\Gamma(\tilde{\mathbf{p}}, \tilde{\mathbf{q}}))$ that was defined earlier in the present chapter. A singleton included in the present Q_5 or Q_6, respectively, are ${}_2Q_5 = \{\langle \mathbf{q}_1\mathbf{q}_1^{-1}, \mathbf{q}_0\mathbf{q}_0^{-1} \rangle\}$ or ${}_2Q_6 = \{\langle \mathbf{q}_1\mathbf{q}_1^{-1}, \mathbf{q}_0^{-1}\mathbf{q}_0 \rangle\}$, respectively. Since $\mathbf{q}_1\mathbf{q}_1^{-1} = \mathbf{q}\mathbf{p}_0(\mathbf{q}\mathbf{p}_0)^{-1} \equiv_P \mathbf{q}\mathbf{p}_0^2\mathbf{q}^{-1}$ and $\mathbf{q}_0 = \mathbf{q}$, there follows that one can also define $\preceq$ to be the least quasiordering of $F_{SI}(\Gamma(\tilde{\mathbf{p}}, \mathbf{q}))$ that is compatible with $\mathbf{F}_{SI}(\Gamma(\tilde{\mathbf{p}}, \mathbf{q}))$ and includes $\equiv \cup P_4 \cup {}_2Q_5 \cup {}_2Q_6$.

Given any $i \geq 0$, there hold by Lemma 3(a), Theorem V.18(a), P_4, or Lemma 3(a), respectively:

$$\mathbf{q}_{i+2}\mathbf{q}_{i+2}^{-1} \equiv \mathbf{q}\mathbf{p}_{i+1}^{2}\mathbf{q}^{-1} \equiv_P \mathbf{q}\mathbf{p}_{i+1}^{2}\mathbf{p}_i\mathbf{p}_i\mathbf{q}^{-1} \;\leq\; \mathbf{q}\mathbf{p}_i\mathbf{p}_i\mathbf{q}^{-1} \equiv \mathbf{q}_{i+1}\mathbf{q}_{i+1}^{-1} \;.$$

Since ${}_2Q_5$ is included in $\leq$, there follows that Q_5 is included in $\leq$.

Assume as inductive hypothesis that $\mathbf{q}_{i+1}\mathbf{q}_{i+1}^{-1} \leq \mathbf{q}_i^{-1}\mathbf{q}_i$, so that, by Lemma 3(a), $\mathbf{q}\mathbf{p}_i^2\mathbf{q}^{-1} \leq \mathbf{q}_i^{-1}\mathbf{q}_i$. Then

$$\mathbf{p}_i\mathbf{q}\mathbf{p}_i^{2}\mathbf{q}^{-1}\mathbf{p}_i \;\leq\; \mathbf{p}_i\mathbf{q}_i^{-1}\mathbf{q}_i\mathbf{p}_i \;.$$

Also, using PQ_4 for the second assertion one obtains

$$\begin{aligned} \mathbf{p}_i\mathbf{q}\mathbf{p}_i^{2}\mathbf{q}^{-1}\mathbf{p}_i \;&\equiv_P\; \mathbf{p}_i\mathbf{q}\mathbf{p}_i^{2}(\mathbf{p}_i\mathbf{q})^{-1} \equiv \mathbf{q}\mathbf{p}_{i+1}\mathbf{p}_i^{2}(\mathbf{q}\mathbf{p}_{i+1})^{-1} \\ &\equiv_P\; \mathbf{q}\mathbf{p}_{i+1}\mathbf{p}_i^{2}\mathbf{p}_{i+1}\mathbf{q}^{-1} \equiv_P \mathbf{q}\mathbf{p}_{i+1}\mathbf{p}_{i+1}\mathbf{q}^{-1} = \mathbf{q}\mathbf{p}_{i+1}^{2}\mathbf{q}^{-1} \;. \end{aligned}$$

There follows that $\mathbf{q}\mathbf{p}_{i+1}^2\mathbf{q}^{-1} \leq \mathbf{p}_i\mathbf{q}_i^{-1}\mathbf{q}_i\mathbf{p}_i$. Hence, by Lemma 3(a) and the definition of $\mathbf{q}_{i+1}$, $\mathbf{q}_{i+2}\mathbf{q}_{i+2}^{-1} \leq \mathbf{q}_{i+1}^{-1}\mathbf{q}_{i+1}$. Since ${}_2Q_6$ is included in $\leq$, there now follows by induction that Q_6 is included in $\leq$.

There now follows that $\leq$ can also be defined to be the least quasiordering of $F_{SI}(\Gamma(\tilde{\mathbf{p}}, \mathbf{q}))$ that is compatible with $\mathbf{F}_{SI}(\Gamma(\tilde{\mathbf{p}}, \mathbf{q}))$ and includes $\equiv \cup P_4 \cup Q_5 \cup Q_6$.

As we have seen earlier, if $\langle w, w'\rangle$ is in P_4, Q_5, or Q_6, then $h(w) \subseteq h(w')$. Hence if $w \leq w'$, then $h(w) \subseteq h(w')$.

Now assume that $h(w) \subseteq h(w')$. First, also assume that $gd\ w = gd\ w'$. Then, by a proof similar to that of Lemma V.10, except that for the last paragraph one uses Theorem 10, instead of Theorem V.7, and except that we now let $\leq_6$ be the least quasiordering of $F_{SI}(\Gamma(\tilde{\mathbf{p}}, \mathbf{q}))$ that is compatible with $\mathbf{F}_{SI}(\Gamma(\tilde{\mathbf{p}}, \mathbf{q}))$ and includes P_4 and the present $\equiv$ and Q_6, we can conclude that $w \leq_6 w'$.

Now consider any w and w' in $F_{SI}(\Gamma(\tilde{\mathbf{p}}, \mathbf{q}))$ such that $h(w) \subseteq h(w')$. Let $gd\ w = \langle s, t\rangle$ and $gd\ w' = \langle s', t'\rangle$. Then $s \geq s'$, $gd\ w = gd(\mathbf{q}_s\mathbf{q}_s^{-1}w')$ and, by Lemma II.5(e) and Theorem II.7(f), $h(w) \subseteq h(\mathbf{q}_s\mathbf{q}_s^{-1}w')$. Hence, as we just saw, $w \leq_6 \mathbf{q}_s\mathbf{q}_s^{-1}w'$. Since $gd\ w' = \langle s', t'\rangle$ therefore, by Q_5, $\mathbf{q}_s\mathbf{q}_s^{-1}w' \leq \mathbf{q}_{s'}\mathbf{q}_{s'}^{-1}w'$. Since $gd\ w' = \langle s', t'\rangle$ therefore, by Lemma 5(d), $\mathbf{q}_{s'}\mathbf{q}_{s'}^{-1}w' \equiv w'$. There now follows that $w \leq w'$. This concludes our proof that $h(w) \subseteq h(w')$ if and only if $w \leq w'$. There follows that $\langle \mathbf{F}_{SI}(\Gamma(\tilde{\mathbf{p}}, \mathbf{q})), \leq\rangle/\!\equiv$ is isomorphic to $\langle \mathbf{S}_{pq}, \subseteq\rangle$. □

We conclude with a presentation of ${}_{[\omega]}\mathbf{S}_{pq} = \langle {}_{[\omega]}S_{pq}, \circ, {}^{\smile}\rangle$ and of its isomorph $\langle\{N \in Bsl : N \text{ is one-one}\}, \circ, {}^{\smile}\rangle$ and a presentation of $\langle {}_{[\omega]}\mathbf{S}_{pq}, \subseteq\rangle$ and of its isomorph $\langle\{N \in Bsl : N \text{ is one-one}\}, \circ, {}^{\smile}, \supseteq\rangle$.

THEOREM 12. (a) *Let* $\equiv$ *be as in Theorem 10 and let* $\equiv^+$ *be the congruence on* $\mathbf{F}_{SI}(\Gamma(\tilde{\mathbf{p}}, \mathbf{q}))$ *that is generated by* $\equiv \cup P_4 \cup PQ_5$.

Let h' *be the homomorphism of* $\mathbf{F}_{SI}(\Gamma(\tilde{\mathbf{p}}, \mathbf{q}))$ *onto* ${}_{[\omega]}\mathbf{S}_{pq}$ *such that* $h'(\mathbf{q}) = \overset{\omega}{\rightarrow} q_0$ *and* $h'(\mathbf{p}_i) = \overset{\omega}{\rightarrow} p_i$ *for each* $i < \omega$ *and let* h'' *be the homomorphism of* $\mathbf{F}_{SI}(\Gamma(\tilde{\mathbf{p}}, \mathbf{q}))$ *onto* $\langle\{N \in Bsl : N \text{ is one-one}\}, \circ, {}^{\smile}\rangle$ *such that* $h''(\mathbf{q}) = f_{\{0\}}$ *and* $h''(\mathbf{p}_i) = (i, i+1)$ *for each* $i < \omega$. *Then* $\equiv^+$, $h' \circ h'^{\smile}$, *and* $h'' \circ h''^{\smile}$ *coincide. There follows that* $\mathbf{F}_{SI}(\Gamma(\tilde{\mathbf{p}}, \mathbf{q}))/\equiv^+$ *is isomorphic to* ${}_{[\omega]}\mathbf{S}_{pq} = \langle S_{pq}, \circ, {}^{\smile}\rangle$ *and also to* $\langle\{N \in Bsl : N \text{ is one-one}\}, \circ, {}^{\smile}\rangle$.

(b) *Let* $\equiv^+$, h' *and* h'' *be as in part* (a) *and let* $\leq^+$ *be the least quasiordering of* $F_{SI}(\Gamma(\tilde{\mathbf{p}}, \mathbf{q}))$ *that is compatible with* $\mathbf{F}_{SI}(\Gamma(\tilde{\mathbf{p}}, \mathbf{q}))$ *and includes* $\equiv^+ \cup PQ_6$. *Then* $h(v) \subseteq h(w)$ *if and only if* $v \leq^+ w$. *There follows that* $\langle \mathbf{F}_{SI}(\Gamma(\tilde{\mathbf{p}}, \mathbf{q})), \leq^+\rangle/\equiv^+$ *is isomorphic to* $\langle {}_{[\omega]}\mathbf{S}_{pq}, \subseteq\rangle$ *and also to* $\langle\{N \in Bsl : N \text{ is one-one}\}, \circ, {}^{\smile}, \supseteq\rangle$.

Proof. (a) We continue to let h be the homomorphism of $\mathbf{F}_{SI}(\Gamma(\tilde{\mathbf{p}}, \mathbf{q}))$ onto $\mathbf{S}_{pq}$ such that $h(\mathbf{q}) = q_0$ and $h(\mathbf{p}_i) = p_i$ for each $i < \omega$. Then, for any x in $F_{SI}(\Gamma(\tilde{\mathbf{p}}, \mathbf{q}))$, $h'(x) = \overset{\omega}{\rightarrow}(h(x))$. By Lemma 1(b), $\equiv$ is included in $h \circ h^{\smile}$ and hence also in $h' \circ h'^{\smile}$. One can verify that P_4 and PQ_5 also are included in $h' \circ h'^{\smile}$. Hence $\equiv^+$ is included in $h' \circ h'^{\smile}$. Since ${}_{[\omega]}[\ ]$ is an isomorphism between $\langle\{N \in Bsl : N \text{ is one-one}\}, \circ, {}^{\smile}\rangle$ and ${}_{[\omega]}\mathbf{S}_{pq}$, therefore $\equiv^+$ is included in $h'' \circ h''^{\smile}$.

Since P_4 and PQ_5 are included in $\equiv^+$, the first two paragraphs of the proof of Theorem 11 can be adapted to show that Q_5 is included in $\equiv^+$. Hence, by induction, $\mathbf{q}_i\mathbf{q}_i^{-1} \equiv^+ \mathbf{q}_0\mathbf{q}_0^{-1}$ for any $i < \omega$. Then, by Lemma 5(d), for any $i < \omega$ and any x in $F_{SI}(\Gamma(\tilde{\mathbf{p}}, \mathbf{q}))$ there holds $\mathbf{q}_i\mathbf{q}_i^{-1}x \equiv^+ x$.

Assume that $h'(v) = h'(w)$. Then $\overset{\omega}{\rightarrow}(h(v)) = \overset{\omega}{\rightarrow}(h(w))$. Then, for some $n \geq 0$, either $\overset{n}{\rightarrow}(h(v)) = h(w)$ or $\overset{n}{\rightarrow}(h(w)) = h(v)$. Suppose that $\overset{n}{\rightarrow}(h(v)) = h(w)$. Let $gd(h(w)) = \langle s, t\rangle$. Then, by Theorem II.7(f),

$$h(\mathbf{q}_s\mathbf{q}_s^{-1}v) = q_s \circ q_s^{-1} \circ h(v) = {}_{[s,\omega)}\overset{\circ}{1} \circ h(v) = h(w)\ .$$

By Theorem 10, $\langle \mathbf{q}_s\mathbf{q}_s^{-1}v, w\rangle$ belongs to $\equiv$ and hence to $\equiv^+$. Since $\mathbf{q}_s\mathbf{q}_s^{-1}v \equiv^+ v$, there follows that $v \equiv^+ w$. If $\overset{n}{\rightarrow}(h(w)) = h(v)$, then a similar argument shows that $v \equiv^+ w$.

This concludes our proof that $\equiv^+$ coincides with $h' \circ h'^{\smile}$. Since $h' \circ h'^{\smile}$ coincides with $h'' \circ h''^{\smile}$, therefore $\equiv^+$ also coincides with $h'' \circ h''^{\smile}$. There follows that $\mathbf{F}_{SI}(\Gamma(\tilde{\mathbf{p}},\mathbf{q}))/\equiv^+$ is isomorphic to ${}_{[\omega]}\mathbf{S}_{pq}$ and also to $\langle\{N \in Bsl : N \text{ is one-one}\}, \circ, {}^{\smile}\rangle$.

(b) As we saw in the proof of Theorem 11, $h(\mathbf{q}\mathbf{p}_0^2\mathbf{q}^{-1}) \subseteq h(\mathbf{q}^{-1}\mathbf{q})$. From the definition of $\leq^+$ in terms of $\equiv^+$ and PQ_6, and from part (a), there therefore follows that if $v \leq^+ w$ then $h'(v) \subseteq h'(w)$.

Assume that $h'(v) \subseteq h'(w)$. Then $\stackrel{\omega}{\rightarrow}(h(v)) \subseteq \stackrel{\omega}{\rightarrow}(h(w))$. Let $gd(h(w)) = \langle s, t\rangle$. Then, by Lemma II.5 and Theorem II.7(f), $gd(h(\mathbf{q}_s\mathbf{q}_s^{-1}v)) = \langle s, t\rangle$ and $h(\mathbf{q}_s\mathbf{q}_s^{-1}v) \subseteq h(w)$. Then, for the relation $\leq$ of Theorem 11 there holds $\mathbf{q}_s\mathbf{q}_s^{-1}v \leq w$. Since this $\leq$ is included in $\leq^+$ and since $\mathbf{q}_s\mathbf{q}_s^{-1}v \equiv^+ v$, there follows that $v \leq^+ w$.

This concludes our proof that $h'(v) \subseteq h'(w)$ if and only if $v \leq^+ w$. There follows that $h''(v) \subseteq h''(w)$ if and only if $v \leq^+ w$. There also follows that $\langle\mathbf{F}_{SI}(\Gamma(\tilde{\mathbf{p}},\mathbf{q})), \leq^+\rangle/\equiv^+$ is isomorphic to $\langle{}_{[\omega]}\mathbf{S}_{pq}, \subseteq\rangle$ and to $\langle\{N \in Bsl : N \text{ is one-one}\}, \circ, {}^{\smile}, \supseteq\rangle$. □

The presentation given in Theorem 12(b) of the structure $\langle\{N \in Bsl : N$ is one-to-one$\}, \circ, {}^{\smile}, \supseteq\rangle$ may turn out to be a useful ingredient of a presentation of the larger, and probably more important, structure $\langle Bsl, \circ, {}^{\smile}, \supseteq\rangle$.

We now turn to the task of presenting $\mathbf{S}_{pe} = \langle S_{pe}, \circ, {}^{\smile}\rangle$. Our presentation will suggest, but will not give explicitly, a presentation of the semilattice $\langle S_e, \circ\rangle$ and also one of its homomorphic image $\langle{}_{[\omega]}S_e, \circ\rangle$ which, by Theorems IV.3(e) and IV.6(e), is isomorphic to $\langle\{N \in Bsl : N = eq_\omega N\}, \bullet\rangle$.

We let $\mathbf{e} = \mathbf{e}_{0,1}$ be an object that is not in $\{\mathbf{p}_i : i < \omega\} \cup \{\mathbf{q}\}$. We let $\Gamma(\tilde{\mathbf{p}},\mathbf{e}) = \{\mathbf{p}_i : i < \omega\} \cup \{\mathbf{e}\}$ and we let $\mathbf{F}(\Gamma(\tilde{\mathbf{p}},\mathbf{e})) = \mathbf{F}_{SI}(\Gamma(\tilde{\mathbf{p}},\mathbf{e}))$ be the free semigroup with involution that has $\Gamma(\tilde{\mathbf{p}},\mathbf{e})$ as free generating set. We let $E_1, \ldots, PE_4$ be the following relations on $F(\Gamma(\tilde{\mathbf{p}},\mathbf{e}))$. Note that both PE_3 and PE_4 consist of a pair $\langle\mathbf{e}w, w\mathbf{e}\rangle$ such that the word w, when read from right to left, is again w.

$$\begin{aligned}
E_1 &= \{\langle\mathbf{e}^{-1}, \mathbf{e}\rangle\}\\
E_2 &= \{\langle\mathbf{e}^2, \mathbf{e}\rangle\}\\
PE_1 &= \{\langle\mathbf{p}_0\mathbf{e}, \mathbf{e}\rangle\}\\
PE_2 &= \{\langle\mathbf{p}_i\mathbf{e}, \mathbf{e}\mathbf{p}_i\rangle : 2 \leq i < \omega\}\\
PE_3 &= \{\langle\mathbf{e}\mathbf{p}_1\mathbf{e}\mathbf{p}_1, \mathbf{p}_1\mathbf{e}\mathbf{p}_1\mathbf{e}\rangle\}\\
PE_4 &= \{\langle\mathbf{e}\mathbf{p}_1\mathbf{p}_0\mathbf{p}_2\mathbf{p}_1\mathbf{e}\mathbf{p}_1\mathbf{p}_2\mathbf{p}_0\mathbf{p}_1, \mathbf{p}_1\mathbf{p}_0\mathbf{p}_2\mathbf{p}_1\mathbf{e}\mathbf{p}_1\mathbf{p}_2\mathbf{p}_0\mathbf{p}_1\mathbf{e}\rangle\}
\end{aligned}$$

For the rest of this chapter we let $\equiv_P$ be the congruence on $\mathbf{F}(\Gamma(\tilde{\mathbf{p}},\mathbf{e}))$ that is generated by the union $P_1 \cup P_2 \cup P_3 \cup P_4^{\#} \cup P_5 \cup P_6$ of the six relations on $\mathbf{F}(\Gamma(\tilde{\mathbf{p}}))$ that we considered in Chapter V and we let $\equiv$ be the congruence on $\mathbf{F}(\Gamma(\tilde{\mathbf{p}},\mathbf{e}))$ that is generated by $\equiv_P \cup E_1 \cup E_2 \cup PE_1 \cup PE_2 \cup PE_3 \cup PE_4$. Since $e_{i+1,i+1} = p_i^2 = p_{(i,i+1)}^2$ for each $i < \omega$, therefore, as one can see from the proof of Theorem I.1(b), S_{pe} is the closure under $\circ$ and $\smile$ of $\{p_i : i < \omega\} \cup \{e_{0,1}\}$. For the rest of this chapter we let h be the homomorphism of $\mathbf{F}(\Gamma) = \mathbf{F}_{SI}(\Gamma(\tilde{\mathbf{p}}, \mathbf{e}))$ onto $\mathbf{S}_{pe}$ such that $h(\mathbf{e}) = e = e_{0,1}$ and $h(\mathbf{p}_i) = p_i = p_{(i,i+1)}$ for each $i < \omega$.

LEMMA 13. (a) *If* $\mathbf{e}$ *does not occur in* v *or in* w, *then* $v \equiv_P w$ *if and only if* $h(v) = h(w)$.

(b) $\equiv$ *is included in* $h \circ h^{\smile}$.

Proof. (a) By Theorem V.18(a).

(b) From a remark before V.17 there follows that $\equiv_P$ is included in $h \circ h^{\smile}$. Since $e_{0,1} \subseteq {}_{[0,\omega)}\overset{\circ}{1}$, therefore E_1 and E_2 are included in $h \circ h^{\smile}$. Now consider any $x = \langle x_0, x_1, x_2, \ldots, x_{n-1} \rangle$ in $Do\ e_{0,1}$. Then $x_0 = x_1$. There follows that $e_{0,1} \circ p_{(0,1)} = e_{0,1}$ and, since $e_{0,1}^{\smile} = e_{0,1}$ and $p_{(0,1)}^{\smile} = p_{0,1}$, that also $p_{(0,1)} \circ e_{0,1} = e_{0,1}$. Hence PE_1 is included in $h \circ h^{\smile}$. Now consider any $i \geq 2$ and any $x = \langle x_0, \ldots, x_{i-1}, x_i, x_{i+1}, x_{i+2}, \ldots x_{n-1} \rangle$ such that $n \geq i + 2$. Then

$$p_i(x) = p_{(i,i+1)}(x) = \langle x_0, \ldots, x_{i-1}, x_{i+1}, x_i, x_{i+2}, \ldots x_{n-1} \rangle .$$

Then the two conditions $x \in Do(e_{0,1} \circ p_i)$ and $x \in Do(p_i \circ e_{0,1})$ are equivalent to each other and, if x satisfies them, then $(p_i \circ e_{0,1})(x) = (e_{0,1} \circ p_i)(x)$. There follows that PE_2 is included in $h \circ h^{\smile}$. Since $h(\mathbf{p}_1\mathbf{e}\mathbf{p}_1) = p_{(1,2)} \circ e_{0,1} \circ p_{(1,2)} = e_{0,2}$ and since $e_{0,1} \circ e_{0,2} = e_{0,2} \circ e_{0,1}$, therefore PE_3 is included in $h \circ h^{\smile}$. Since $h(\mathbf{p}_1\mathbf{e}\mathbf{p}_1) = e_{0,2}$, therefore $h(\mathbf{p}_2\mathbf{p}_1\mathbf{e}\mathbf{p}_1\mathbf{p}_2) = p_{(2,3)} \circ e_{0,2} \circ p_{(2,3)} = e_{0,3}$, hence $h(\mathbf{p}_0\mathbf{p}_2\mathbf{p}_1\mathbf{e}\mathbf{p}_1\mathbf{p}_2\mathbf{p}_0) = p_{(0,1)} \circ e_{0,3} \circ p_{(0,1)} = e_{1,3}$, and hence $h(\mathbf{p}_1\mathbf{p}_0\mathbf{p}_2\mathbf{p}_1\mathbf{e}\mathbf{p}_1\mathbf{p}_2\mathbf{p}_0\mathbf{p}_1) = p_{(1,2)} \circ e_{1,3} \circ p_{(1,2)} = e_{2,3}$. Since $e_{0,1} \circ e_{2,3} = e_{2,3} \circ e_{0,1}$, there follows that PE_4 is included in $h \circ h^{\smile}$. There now follows that $\equiv$ is included in $h \circ h^{\smile}$. □

For certain elements of $F(\Gamma(\tilde{\mathbf{p}}, \mathbf{e}))$, we now introduce names that mostly are briefer or more suggestive. We let:

$$\begin{aligned}
\mathbf{e}_{0,k+1} &= \mathbf{p}_k\mathbf{e}_{0,k}\mathbf{p}_k && \text{if } 1 \le k < \omega \\
\mathbf{e}_{j+1,k} &= \mathbf{p}_j\mathbf{e}_{j,k}\mathbf{p}_j && \text{if } 0 \le j < k-1 < \omega \\
\mathbf{e}_{j,k} &= \mathbf{e}_{k,j} && \text{if } j > k \ge 0 \\
\mathbf{e}_{j,j} &= \mathbf{p}_{j-1}^2 && \text{if } 1 \le j < \omega \\
\mathbf{p}_{(j,j+1)} &= \mathbf{p}_j && \text{if } 0 \le j < \omega \\
\mathbf{p}_{(j,k+1)} &= \mathbf{p}_k\mathbf{p}_{(j,k)}\mathbf{p}_k && \text{if } 0 \le j < k < \omega \\
\mathbf{p}_{(j,k)} &= \mathbf{p}_{(k,j)} && \text{if } j > k \ge 0 \\
\mathbf{p}_{(j,j)} &= \mathbf{p}_{j-1}^2 && \text{if } 1 \le j < \omega \ .
\end{aligned}$$

There follows

$$\begin{aligned}
\mathbf{e}_{0,k} &= \mathbf{p}_{k-1}\cdots\mathbf{p}_1\mathbf{e}\mathbf{p}_1\cdots\mathbf{p}_{k-1}\ , \quad \text{if } 2 \le k < \omega\ , \\
\mathbf{e}_{j,k} &= \mathbf{p}_{j-1}\cdots\mathbf{p}_0\mathbf{e}_{0,k}\mathbf{p}_0\cdots\mathbf{p}_{j-1}\ , \quad \text{if } 0 < j < k < \omega\ , \\
\mathbf{p}_{(j,k)} &= \mathbf{p}_{k-1}\cdots\mathbf{p}_{j+1}\mathbf{p}_j\mathbf{p}_{j+1}\cdots\mathbf{p}_{k-1}\ , \quad \text{if } 0 \le j \le k-2 < \omega\ .
\end{aligned}$$

Henceforth, in referring to $\mathbf{e}_{j,j}$ or to $\mathbf{p}_{(j,j)}$ we shall tacitly assume that $j \ge 1$.

To bring out the effect or power of PE_2, in Lemma 14(c) below and also in some later places, we make use of the subcongruence of $\equiv$ that is generated by $\equiv_P \cup PE_2$. To bring out the effects of other subsets of $\{E_1, E_2, \dots PE_4\}$ we will also, when appropriate, make use of other subcongruences of $\equiv$.

LEMMA 14. (a) $\mathbf{e}_{j,k+1} \equiv_P \mathbf{p}_k\mathbf{e}_{j,k}\mathbf{p}_k$, *if* $0 \le j < k$.

(b) $\mathbf{p}_i\mathbf{e}_{j,j} \equiv_P \mathbf{p}_i \equiv_P \mathbf{e}_{j,j}\mathbf{p}_i$, *if* $0 \le i \le j-1$.

$\mathbf{p}_i\mathbf{e}_{j,j} \equiv_P \mathbf{e}_{j,j}\mathbf{p}_i$, *if* $i \ge j-1$.

(c) *Let* $\equiv_2$ *be the congruence on* $\mathbf{F}(\Gamma(\tilde{\mathbf{p}}, \mathbf{e}))$ *that is generated by* $\equiv_P \cup PE_2$. *Then*

$$\mathbf{p}_{(i,i')}\mathbf{e}_{j,k} \equiv_2 \mathbf{e}_{j,k}\mathbf{p}_{(i,i')}\ , \quad \textit{if} \quad j < k < i < i'\ .$$

Proof. Henceforth in using $=$ we will often make tacit use of definitions involved. Also, in using $\equiv_P$ we will often make tacit use of Theorem V.18(b), which allows us to interpret $\equiv_P$ as making assertions about $\mathbf{S}_p$ whose truth, if not evident, can be tested by letting $\overset{=}{U} = 2$.

(a) Assume that $0 \le j < k$. Then, the second assertion holds because $k \ge j'+2$ for any j' in $\{0, \dots, j-1\}$, while for the first assertion one uses the definition of

$e_{j,k+1}$ and that of $e_{0,k+1}$.

$$\begin{aligned}\mathbf{e}_{j,k+1} &= \mathbf{p}_{j-1}\cdots\mathbf{p}_0\mathbf{p}_k\mathbf{p}_{k-1}\cdots\mathbf{p}_1\mathbf{e}\mathbf{p}_1\cdots\mathbf{p}_{k-1}\mathbf{p}_k\mathbf{p}_0\cdots\mathbf{p}_{j-1}\\ &\equiv_P \mathbf{p}_k\mathbf{p}_{j-1}\cdots\mathbf{p}_0\mathbf{p}_{k-1}\cdots\mathbf{p}_1\mathbf{e}\mathbf{p}_1\cdots\mathbf{p}_{k-1}\mathbf{p}_0\cdots\mathbf{p}_{j-1}\mathbf{p}_k\\ &= \mathbf{p}_k\mathbf{e}_{j,k}\mathbf{p}_k\ .\end{aligned}$$

(b) This follows from $\mathbf{e}_{j,j} = \mathbf{p}_{j-1}^2$ and, respectively:

$$\begin{aligned}\mathbf{p}_i\mathbf{p}_{j-1}^2 &\equiv_P \mathbf{p}_i \equiv_P \mathbf{p}_{j-1}^2\mathbf{p}_i\ , \quad \text{if } 0\leq i\leq j-1\ .\\ \mathbf{p}_i\mathbf{p}_{j-1}^2 &\equiv_P \mathbf{p}_{j-1}^2\mathbf{p}_i \quad \text{if } i\geq j-1\ .\end{aligned}$$

(c) We will first consider the case where $i' = i+1$ and hence $\mathbf{p}_{(i,i')} = \mathbf{p}_i$. First, suppose that $j \neq 0$. Then the second and fourth assertion below follow from $i \geq m+2$ for any m in $\{0,\ldots,j-1\}\cup\{0,\ldots,k-1\}$. The third assertion holds by PE_2.

$$\begin{aligned}\mathbf{p}_i\mathbf{e}_{j,k} &= \mathbf{p}_i\mathbf{p}_{j-1}\cdots\mathbf{p}_0\mathbf{p}_{k-1}\cdots\mathbf{p}_1\mathbf{e}\mathbf{p}_1\cdots\mathbf{p}_{k-1}\mathbf{p}_0\cdots\mathbf{p}_{j-1}\\ &\equiv_P \mathbf{p}_{j-1}\cdots\mathbf{p}_0\mathbf{p}_{k-1}\cdots\mathbf{p}_1\mathbf{p}_i\mathbf{e}\mathbf{p}_1\cdots\mathbf{p}_{k-1}\mathbf{p}_0\cdots\mathbf{p}_{j-1}\\ &\equiv_2 \mathbf{p}_{j-1}\cdots\mathbf{p}_0\mathbf{p}_{k-1}\cdots\mathbf{p}_1\mathbf{e}\mathbf{p}_i\mathbf{p}_1\cdots\mathbf{p}_{k-1}\mathbf{p}_0\cdots\mathbf{p}_{j-1}\\ &\equiv_P \mathbf{p}_{j-1}\cdots\mathbf{p}_0\mathbf{p}_{k-1}\cdots\mathbf{p}_1\mathbf{e}\mathbf{p}_1\cdots\mathbf{p}_{k-1}\mathbf{p}_0\cdots\mathbf{p}_{j-1}\mathbf{p}_i = \mathbf{e}_{j,k}\mathbf{p}_i\ .\end{aligned}$$

Now, suppose that $j = 0$. Then, by an argument that is similar but somewhat simpler, again $\mathbf{p}_i\mathbf{e}_{j,k} \equiv_2 \mathbf{e}_{j,k}\mathbf{p}_i$. This concludes our proof that whenever $i > j > k$ then $\mathbf{p}_i\mathbf{e}_{j,k} \equiv_2 \mathbf{e}_{j,k}\mathbf{p}_i$.

Now let $i < i'$. Then $\mathbf{p}_{(i,i')}$ is a word w formed from letters $\mathbf{p}_{i''}$ such that $i \leq i'' < i'$. Hence, by what has already been shown and by induction on the length of w there follows that $\mathbf{p}_{(i,i')}\mathbf{e}_{j,k} \equiv_2 \mathbf{e}_{j,k}\mathbf{p}_{(i,i')}$. ☐

LEMMA 15. *Let* $\equiv_2$ *be the congruence on* $\mathbf{F}(\Gamma(\tilde{\mathbf{p}},\mathbf{e}))$ *that is generated by* $\equiv_P \cup PE_2$. *Consider the following four cases:*

(a) $j<k<k'$; (b) $k<k'<j$; (c) $k'<k<j$; (d) $j<k'<k$.

$$\begin{aligned}\text{Then }\ \mathbf{e}_{j,k'} &\equiv_2 \mathbf{p}_{(k,k')}\,\mathbf{e}_{j,k}\,\mathbf{p}_{(k,k')}\ , \quad \text{if (a), (b), or (c),}\\ \mathbf{p}_{k-1}^2\mathbf{e}_{j,k'} &\equiv_2 \mathbf{p}_{(k,k')}\,\mathbf{e}_{j,k}\,\mathbf{p}_{(k,k')}\ , \quad \text{if (d).}\end{aligned}$$

Proof. From now on we will freely change from $\mathbf{p}_i$ to $\mathbf{p}_{(i,i+1)}$ and vice versa.

(a) Assume that $j < k < k'$. For proof by induction on $k'-k$, we first assume that $k'-k = 1$. Then from Lemma 14(a) and from $\mathbf{p}_k = \mathbf{p}_{(k,k+1)} = \mathbf{p}_{(k,k')}$ there follows that $\mathbf{e}_{j,k'} \equiv_P \mathbf{p}_{(k,k')}\mathbf{e}_{j,k}\mathbf{p}_{(k,k')}$. Now let $k'-k \geq 1$ and assume as inductive hypothesis that $\mathbf{e}_{j,k'} \equiv_2 \mathbf{p}_{(k,k')}\,\mathbf{e}_{j,k}\,\mathbf{p}_{(k,k')}$. Then the first assertion below holds by

Lemma 14(a), the second by the inductive hypothesis, and the fourth by Lemma 14(c).

$$\begin{aligned}
\mathbf{e}_{j,k'+1} &\equiv_P \mathbf{p}_{(k',k'+1)}\,\mathbf{e}_{j,k'}\,\mathbf{p}_{(k',k'+1)} \\
&\equiv_2 \mathbf{p}_{(k',k'+1)}\mathbf{p}_{(k,k')}\,\mathbf{e}_{j,k}\,\mathbf{p}_{(k,k')}\mathbf{p}_{(k',k'+1)} \\
&\equiv_P \mathbf{p}_{(k,k'+1)}\mathbf{p}_{(k',k'+1)}\,\mathbf{e}_{j,k}\,\mathbf{p}_{(k',k'+1)}\mathbf{p}_{(k,k'+1)} \\
&\equiv_2 \mathbf{p}_{(k,k'+1)}\,\mathbf{e}_{j,k}\,\mathbf{p}_{k'}\mathbf{p}_{k'}\mathbf{p}_{(k,k'+1)} \equiv_P \mathbf{p}_{(k,k'+1)}\,\mathbf{e}_{j,k}\,\mathbf{p}_{(k,k'+1)}
\end{aligned}$$

(b) Assume that $k < k' < j$. We again use induction on $k' - k$. First, let $k' - k = 1$. Then, by the definitions involved,

$$\mathbf{e}_{j,k'} = \mathbf{e}_{j,k+1} = \mathbf{e}_{k+1,j} = \mathbf{p}_k\,\mathbf{e}_{k,j}\,\mathbf{p}_k = \mathbf{p}_{k,k'}\,\mathbf{e}_{j,k}\,\mathbf{p}_{k,k'}\ .$$

Now let $k' - k \geq 1$ and $k' < j - 1$, and assume as inductive hypothesis that $\mathbf{e}_{k',j} = \mathbf{e}_{j,k'} \equiv_2 \mathbf{p}_{k,k'}\,\mathbf{e}_{j,k}\,\mathbf{p}_{k,k'}$. Then, of the following assertions the second follows from the inductive hypothesis and, if for $k = 0$ we let $\mathbf{p}_{k-1} \dots \mathbf{p}_0$ and $\mathbf{p}_0 \dots \mathbf{p}_{k-1}$ be the empty sequence, the fourth from the definition of $\mathbf{e}_{j,k} = \mathbf{e}_{k,j}$, the fifth from $k' \geq k'' + 2$ for any k'' in $\{0, \dots, k-1\}$, and the sixth by part (a) of Lemma 15.

$$\begin{aligned}
\mathbf{e}_{k'+1,j} &= \mathbf{p}_{k'}\,\mathbf{e}_{k',j}\,\mathbf{p}_{k'} \equiv_2 \mathbf{p}_{k'}\mathbf{p}_{(k,k')}\,\mathbf{e}_{j,k}\,\mathbf{p}_{(k,k')}\mathbf{p}_{k'} \\
&\equiv_P \mathbf{p}_{(k,k'+1)}\mathbf{p}_{(k',k'+1)}\,\mathbf{e}_{j,k}\mathbf{p}_{(k',k'+1)}\mathbf{p}_{(k,k'+1)} \\
&= \mathbf{p}_{(k,k'+1)}\mathbf{p}_{(k',k'+1)}\mathbf{p}_{k-1}\cdots\mathbf{p}_0\,\mathbf{e}_{j,0}\,\mathbf{p}_0\cdots\mathbf{p}_{k-1}\mathbf{p}_{(k',k'+1)}\mathbf{p}_{(k,k'+1)} \\
&\equiv_P \mathbf{p}_{(k,k'+1)}\mathbf{p}_{k-1}\cdots\mathbf{p}_0\mathbf{p}_{(k',k'+1)}\,\mathbf{e}_{j,0}\,\mathbf{p}_{(k',k'+1)}\mathbf{p}_0\cdots\mathbf{p}_{k-1}\mathbf{p}_{(k,k'+1)} \\
&\equiv_2 \mathbf{p}_{(k,k'+1)}\mathbf{p}_{k-1}\cdots\mathbf{p}_0\mathbf{p}_{(k',k'+1)}\mathbf{p}_{(1,j)}\,\mathbf{e}\,\mathbf{p}_{(1,j)}\mathbf{p}_{(k',k'+1)}\mathbf{p}_0\cdots\mathbf{p}_{k-1}\mathbf{p}_{(k,k'+1)}
\end{aligned}$$

First, suppose that $k' = 1$ and hence $k = 0$. Then the third assertion below holds by $\mathbf{e} = \mathbf{e}_{0,1}$ and Lemma 14(c).

$$\begin{aligned}
&\mathbf{p}_{(k',k'+1)}\mathbf{p}_{(1,j)}\,\mathbf{e}\,\mathbf{p}_{(1,j)}\mathbf{p}_{(k',k'+1)} = \mathbf{p}_{(1,2)}\mathbf{p}_{(1,j)}\,\mathbf{e}\,\mathbf{p}_{(1,j)}\mathbf{p}_{(1,2)} \\
&\equiv_P \mathbf{p}_{(1,j)}\mathbf{p}_{(2,j)}\,\mathbf{e}\,\mathbf{p}_{(2,j)}\mathbf{p}_{(1,j)} \equiv_2 \mathbf{p}_{(1,j)}\,\mathbf{e}\,\mathbf{p}_{(2,j)}\mathbf{p}_{(2,j)}\mathbf{p}_{(1,j)} \equiv_P \mathbf{p}_{(1,j)}\,\mathbf{e}\,\mathbf{p}_{(1,j)}
\end{aligned}$$

Now suppose that $k' \geq 2$. Then the first assertion below follows from $\{1, j\} \cap \{k', k+1\} = \emptyset$, the second holds by PE_2, and the third follows from $j > k' + 1$.

$$\begin{aligned}
&\mathbf{p}_{(k',k'+1)}\mathbf{p}_{(1,j)}\,\mathbf{e}\,\mathbf{p}_{(1,j)}\mathbf{p}_{(k',k'+1)} \equiv_P \mathbf{p}_{(1,j)}\mathbf{p}_{(k',k'+1)}\,\mathbf{e}\,\mathbf{p}_{(k',k'+1)}\mathbf{p}_{(1,j)} \\
&\equiv_2 \mathbf{p}_{(1,j)}\,\mathbf{e}\,\mathbf{p}_{(k',k'+1)}\mathbf{p}_{(k',k'+1)}\mathbf{p}_{(1,j)} \equiv_P \mathbf{p}_{(1,j)}\,\mathbf{e}\,\mathbf{p}_{(1,j)}
\end{aligned}$$

From the last three paragraphs there follows that, whether $k' = 1$ or $k' \geq 2$, there holds the first assertion below. The second assertion follows from part (a).

$$\begin{aligned}
\mathbf{e}_{k'+1,j} &\equiv_2 \mathbf{p}_{(k,k'+1)}\mathbf{p}_{k-1}\cdots\mathbf{p}_0\mathbf{p}_{(k',k'+1)}\mathbf{p}_{(1,j)}\,\mathbf{e}\,\mathbf{p}_{(k',k'+1)}\mathbf{p}_{(1,j)}\mathbf{p}_0\cdots\mathbf{p}_{k-1}\mathbf{p}_{(k,k'+1)}\\
&\equiv_2 \mathbf{p}_{(k,k'+1)}\mathbf{p}_{k-1}\cdots\mathbf{p}_0\mathbf{p}_{(1,j)}\,\mathbf{e}\,\mathbf{p}_{(1,j)}\mathbf{p}_0\cdots\mathbf{p}_{k-1}\mathbf{p}_{(k,k'+1)}\\
&= \mathbf{p}_{(k,k'+1)}\,\mathbf{e}_{j,k}\,\mathbf{p}_{(k,k'+1)}\ .
\end{aligned}$$

By induction on $k' - k$ there now follows that $\mathbf{e}_{j,k'} \equiv_2 \mathbf{p}_{(k,k')}\,\mathbf{e}_{j,k}\,\mathbf{p}_{(k,k')}$.

(c) Assume that $k' < k < j$. Then, by part (b) of Lemma 15,

$$\mathbf{e}_{j,k} \equiv_2 \mathbf{p}_{(k,k')}\,\mathbf{e}_{j,k'}\,\mathbf{p}_{(k,k')}$$

Since $\max(k, k') < j$, therefore

$$\mathbf{p}^2_{(k,k')}\,\mathbf{e}_{j,k'} \equiv_P \mathbf{e}_{j,k'} \equiv_P \mathbf{e}_{j,k'}\,\mathbf{p}^2_{(k,k')}$$

There now follows that

$$\mathbf{p}_{(k,k')}\,\mathbf{e}_{j,k}\,\mathbf{p}_{(k,k')} \equiv_2 \mathbf{p}_{(k,k')}\mathbf{p}_{(k,k')}\,\mathbf{e}_{j,k'}\,\mathbf{p}_{(k,k')}\mathbf{p}_{(k,k')} \equiv_P \mathbf{e}_{j,k'}\ .$$

(d) Assume that $j < k' < k$. Then, by part (a) of Lemma 15,

$$\begin{aligned}
\mathbf{e}_{j,k} &\equiv_2 \mathbf{p}_{(k',k)}\,\mathbf{e}_{j,k'}\,\mathbf{p}_{(k,k')}\ ,\\
\mathbf{p}_{(k',k)}\,\mathbf{e}_{j,k}\,\mathbf{p}_{(k',k)} &\equiv_2 \mathbf{p}^2_{(k',k)}\,\mathbf{e}_{j,k'}\,\mathbf{p}^2_{(k',k)}\ .
\end{aligned}$$

Since $\mathbf{p}_{k-1} = \mathbf{p}_{(k-1,k)}$, and since $k' < k$, therefore

$$\mathbf{p}^2_{(k',k)} \equiv_P \mathbf{p}^2_{k-1}\ .$$

Since $k > \max\{j, k'\}$ therefore, by Lemma 14(c),

$$\mathbf{e}_{j,k'}\,\mathbf{p}^2_{k-1} \equiv_2 \mathbf{p}^2_{k-1}\,\mathbf{e}_{j,k'}$$

There now follows:

$$\begin{aligned}
\mathbf{p}_{(k',k)}\,\mathbf{e}_{j,k}\,\mathbf{p}_{(k',k)} &\equiv_2 \mathbf{p}^2_{(k',k)}\,\mathbf{e}_{j,k'}\,\mathbf{p}^2_{(k',k)} \equiv_P \mathbf{p}^2_{k-1}\,\mathbf{e}_{j,k'}\,\mathbf{p}^2_{k-1}\\
&\equiv_2 \mathbf{p}^2_{k-1}\mathbf{p}^2_{k-1}\mathbf{e}_{j,k'} \equiv_P \mathbf{p}^2_{k-1}\,\mathbf{e}_{j,k'}\ .
\end{aligned}$$

□

From Lemma 15(a) or, respectively, Lemmas 15(a) and 15(b) there follows:

$$\begin{aligned}
\mathbf{e}_{0,k} &\equiv_2 \mathbf{p}_{(1,k)}\,\mathbf{e}\,\mathbf{p}_{(1,k)} && \text{if } 1 < k\ ,\\
\mathbf{e}_{j,k} &\equiv_2 \mathbf{p}_{(0,j)}\mathbf{p}_{(1,k)}\,\mathbf{e}\,\mathbf{p}_{(1,k)}\mathbf{p}_{(0,j)} && \text{if } 0 < j < k
\end{aligned}$$

For any congruence on $\mathbf{F}(\Gamma(\tilde{\mathbf{p}}, \mathbf{e}))$ that includes $\equiv_2$, one may therefore choose as definition of $e_{j,k}$ where $e_{j,k} \neq e_{0,1}$ and $j < k$, the one that the above two assertions suggest, in place of the definition that we chose earlier.

Each of the relations E_1, E_2, PE_1, PE_2, PE_3, PE_4 can be generalized in certain ways. As will be shown by some of the lemmas that now follow, these generalizations also are included in $\equiv$.

LEMMA 16. *Let $\equiv_{E_1}$ be the congruence on $\mathbf{F}(\Gamma(\tilde{\mathbf{p}}, \mathbf{e}))$ that is generated by $\equiv_P \cup E_1$. Then,*

$$\mathbf{e}_{j,k}^{-1} \equiv_{E_1} \mathbf{e}_{j,k} \ , \quad \textit{if} \quad \{j,k\} \neq \{0\} \ .$$

Proof. If $j = k \neq 0$, then $\mathbf{e}_{j,k}^{-1} = (\mathbf{p}_{j-1}^2)^{-1} \equiv_P \mathbf{p}_{j-1}^2 = \mathbf{e}_{j,j}$. From $\mathbf{e}_{0,1}^{-1} \equiv_{E_1} \mathbf{e}_{0,1}$, $\mathbf{e}_{0,k+1} = \mathbf{p}_k\, \mathbf{e}_{0,k}\, \mathbf{p}_k$ if $k \geq 2$, and $\mathbf{p}_k^{-1} \equiv_P \mathbf{p}_k$ there follows by induction that $\mathbf{e}_{0,k}^{-1} \equiv_{E_1} \mathbf{e}_{0,k}$ for any $k \geq 1$. From this, $\mathbf{e}_{j+1,k} = \mathbf{p}_j\, \mathbf{e}_{j,k}\, \mathbf{p}_j$ if $0 \leq j < k-1$, and $\mathbf{p}_j^{-1} \equiv_P \mathbf{p}_j$, there then follows by induction that $\mathbf{e}_{j,k}^{-1} \equiv_{E_1} \mathbf{e}_{j,k}$ whenever $0 \leq j < k$. Finally, if $j > k \geq 0$ then $\mathbf{e}_{j,k}^{-1} = \mathbf{e}_{k,j}^{-1} \equiv_{E_1} \mathbf{e}_{k,j} = \mathbf{e}_{j,k}$. □

LEMMA 17. *Let $\equiv_{E_2}$ be the congruence on $\mathbf{F}(\Gamma(\tilde{\mathbf{p}}, \mathbf{e}))$ that is generated by $\equiv_P \cup E_2$. Then,*

$$\mathbf{e}_{j,k}\mathbf{e}_{j,k} \equiv_{E_2} \mathbf{e}_{j,k} \ , \quad \textit{if} \quad \{j,k\} \neq \{0\} \ .$$

Proof. One uses E_2 and a proof by induction similar to that of Lemma 16. □

LEMMA 18. *Let $\equiv_1$ be the congruence on $\mathbf{F}(\Gamma(\tilde{\mathbf{p}}, \mathbf{e}))$ that is generated by $\equiv_P \cup PE_1$. Then for any j and k such that $e_{j,k}$ is defined and hence such that $\{j,k\} \neq \{0\}$,*

$$\mathbf{p}_{(j,k)}\mathbf{e}_{j,k} \equiv_1 \mathbf{e}_{j,k} \ .$$

Proof. If $j = 0$, then by PE_1 there holds $\mathbf{p}_j\mathbf{e}_{j,j+1} = \mathbf{p}_0\mathbf{e}_{0,1} = \mathbf{p}_0\mathbf{e} \equiv_1 \mathbf{e} = \mathbf{e}_{j,j+1}$.

Now let $j \geq 0$ and assume as inductive hypothesis that $\mathbf{p}_j\mathbf{e}_{j,j+1} \equiv_1 \mathbf{e}_{j,j+1}$. Then the third of the following assertions holds by the inductive hypothesis:

$$\begin{aligned}
\mathbf{p}_{j+1}\mathbf{e}_{j+1,j+2} &= \mathbf{p}_{j+1}\mathbf{p}_j\mathbf{p}_{j+1}\, \mathbf{e}_{j,j+1}\, \mathbf{p}_{j+1}\mathbf{p}_j \\
&\equiv_P \mathbf{p}_j\mathbf{p}_{j+1}\mathbf{p}_j\, \mathbf{e}_{j,j+1}\, \mathbf{p}_{j+1}\mathbf{p}_j \\
&\equiv_1 \mathbf{p}_j\mathbf{p}_{j+1}\, \mathbf{e}_{j,j+1}\, \mathbf{p}_{j+1}\mathbf{p}_j = \mathbf{e}_{j+1,j+2}
\end{aligned}$$

There follows by induction that, for any $j \geq 0$, $\mathbf{p}_{(j,j+1)}\mathbf{e}_{j,j+1} = \mathbf{p}_j\mathbf{e}_{j,j+1} \equiv_1 \mathbf{e}_{j,j+1}$.

Now let $j < k$ and assume as inductive hypothesis that $\mathbf{p}_{(j,k)}\mathbf{e}_{j,k} \equiv_1 \mathbf{e}_{j,k}$. Then of the following assertions the first and last follows from Lemma 14(a) and the fourth from the inductive hypothesis:

$$\begin{aligned}\mathbf{p}_{(j,k+1)}\mathbf{e}_{j,k+1} &\equiv_P \mathbf{p}_{(j,k+1)}\mathbf{p}_k\,\mathbf{e}_{j,k}\,\mathbf{p}_k \equiv_P \mathbf{p}_k\mathbf{p}_{(j,k)}\mathbf{p}_k\mathbf{p}_k\,\mathbf{e}_{j,k}\,\mathbf{p}_k \\ &\equiv_P \mathbf{p}_k\mathbf{p}_{(j,k)}\,\mathbf{e}_{j,k}\,\mathbf{p}_k \equiv_1 \mathbf{p}_k\,\mathbf{e}_{j,k}\,\mathbf{p}_k \equiv_P \mathbf{e}_{j,k+1}\,.\end{aligned}$$

There follows by induction that, if $j < k$, then $\mathbf{p}_{(j,k)}\mathbf{e}_{j,k} \equiv_1 \mathbf{e}_{j,k}$. Hence also, if $j > k$, then $\mathbf{p}_{(j,k)}\mathbf{e}_{j,k} = \mathbf{p}_{(k,j)}\mathbf{e}_{k,j} \equiv_1 \mathbf{e}_{k,j} = \mathbf{e}_{j,k}$.

Finally, if $j = k \neq 0$, then $\mathbf{p}_{(j,k)}\mathbf{e}_{j,k} = \mathbf{p}_{(j,j)}\mathbf{e}_{j,j} = \mathbf{p}_{j-1}^2\mathbf{p}_{j-1}^2 \equiv_P \mathbf{p}_{j-1}^2 = \mathbf{e}_{j,j}$. □

By the lemma just proved, if $i = j = k-1$, then $\mathbf{p}_i\mathbf{e}_{j,k} = \mathbf{p}_j\mathbf{e}_{j,j+1} \equiv_1 \mathbf{e}_{j,j+1} = \mathbf{e}_{j,k}$. Since $\mathbf{e}_{j,k} = \mathbf{e}_{k,j}$, the next lemma deals with $\mathbf{p}_i\mathbf{e}_{j,k}$ in every other case where $j \neq k$. (Lemma 14(b) deals with $\mathbf{p}_i\mathbf{e}_{j,k}$ in the case where $j = k$.)

LEMMA 19. *Let $\equiv_2$ be the congruence on $\mathbf{F}(\Gamma(\tilde{\mathbf{p}}, \mathbf{e}))$ that is generated by $\equiv_P \cup PE_2$. Also, let $j < k$.*

(a) $\mathbf{p}_i\mathbf{e}_{j,k} \equiv_2 \mathbf{e}_{j,k}\mathbf{p}_i$, *if* $i < j-1$.
(b) $\mathbf{p}_i\mathbf{e}_{j,k} \equiv_2 \mathbf{e}_{j-1,k}\mathbf{p}_i$, *if* $i = j-1$.
(c) $\mathbf{p}_i\mathbf{e}_{j,k} \equiv_2 \mathbf{e}_{j+1,k}\mathbf{p}_i$, *if* $i = j < k-1$.
(d) $\mathbf{p}_i\mathbf{e}_{j,k} \equiv_2 \mathbf{e}_{j,k}\mathbf{p}_i$, *if* $j < i < k-1$.
(e) $\mathbf{p}_i\mathbf{e}_{j,k} \equiv_2 \mathbf{e}_{j,k-1}\mathbf{p}_i$, *if* $j < i = k-1$.
(f) $\mathbf{p}_i\mathbf{e}_{j,k} \equiv_2 \mathbf{e}_{j,k+1}\mathbf{p}_i$, *if* $i = k$.
(g) $\mathbf{p}_i\mathbf{e}_{j,k} \equiv_2 \mathbf{e}_{j,k}\mathbf{p}_i$, *if* $i < k$.

Proof. (a) First assume that $i \geq 2$. Then the first and the last assertion below hold by Lemmas 15(a) and 15(b), the second and fourth by $i < j-1 < k-1$, and the third by PE_2.

$$\begin{aligned}\mathbf{p}_i\mathbf{e}_{j,k} &\equiv_2 \mathbf{p}_i\mathbf{p}_{(0,j)}\mathbf{p}_{(1,k)}\,\mathbf{e}\,\mathbf{p}_{(1,k)}\mathbf{p}_{(0,j)} \equiv_P \mathbf{p}_{(0,j)}\mathbf{p}_{(1,k)}\mathbf{p}_i\,\mathbf{e}\,\mathbf{p}_{(1,k)}\mathbf{p}_{(0,j)} \\ &\equiv_2 \mathbf{p}_{(0,j)}\mathbf{p}_{(1,k)}\,\mathbf{e}\,\mathbf{p}_i\mathbf{p}_{(1,k)}\mathbf{p}_{(0,j)} \\ &\equiv_P \mathbf{p}_{(0,j)}\mathbf{p}_{(1,k)}\,\mathbf{e}\,\mathbf{p}_{(1,k)}\mathbf{p}_{(0,j)}\mathbf{p}_i \equiv_2 \mathbf{e}_{j,k}\mathbf{p}_i\,.\end{aligned}$$

Now assume that $i = 1$. Then the first and last assertion below hold by Lemma 15(a) and 15(b), the second and sixth by $i = 1 < j - 1$, and the fourth by PE_2.

$$\begin{aligned}
\mathbf{p}_1\mathbf{e}_{j,k} &\equiv_2 \mathbf{p}_{(1,2)}\mathbf{p}_{(0,j)}\mathbf{p}_{(1,k)}\,\mathbf{e}\,\mathbf{p}_{(1,k)}\mathbf{p}_{(0,j)} \\
&\equiv_P \mathbf{p}_{(0,j)}\mathbf{p}_{(1,2)}\mathbf{p}_{(1,k)}\,\mathbf{e}\,\mathbf{p}_{(1,k)}\mathbf{p}_{(0,j)} \\
&\equiv_P \mathbf{p}_{(0,j)}\mathbf{p}_{(1,k)}\mathbf{p}_{(2,k)}\,\mathbf{e}\,\mathbf{p}_{(1,k)}\mathbf{p}_{(0,j)} \\
&\equiv_2 \mathbf{p}_{(0,j)}\mathbf{p}_{(1,k)}\,\mathbf{e}\,\mathbf{p}_{(2,k)}\mathbf{p}_{(1,k)}\mathbf{p}_{(0,j)} \\
&\equiv_P \mathbf{p}_{(0,j)}\mathbf{p}_{(1,k)}\,\mathbf{e}\,\mathbf{p}_{(1,k)}\mathbf{p}_{(1,2)}\mathbf{p}_{(0,j)} \\
&\equiv_P \mathbf{p}_{(0,j)}\mathbf{p}_{(1,k)}\,\mathbf{e}\,\mathbf{p}_{(1,k)}\mathbf{p}_{(0,j)}\mathbf{p}_{(1,2)} \\
&\equiv_2 \mathbf{e}_{j,k}\mathbf{p}_1 \ .
\end{aligned}$$

Finally assume that $i = 0$. Using first Lemmas 15(a) and 15(b), and then $1 < j < k$ we obtain the following, respectively:

$$\begin{aligned}
\mathbf{p}_0\mathbf{e}_{j,k} &\equiv_2 \mathbf{p}_0\mathbf{p}_{(2,j)}\mathbf{p}_{(3,k)}\,\mathbf{e}_{2,3}\,\mathbf{p}_{(3,k)}\mathbf{p}_{(2,j)} \equiv_P \mathbf{p}_{(2,j)}\mathbf{p}_{(3,k)}\mathbf{p}_0\,\mathbf{e}_{2,3}\mathbf{p}_{(3,k)}\mathbf{p}_{(2,j)} \\
\mathbf{e}_{j,k}\mathbf{p}_0 &\equiv_2 \mathbf{p}_{(2,j)}\mathbf{p}_{(3,k)}\,\mathbf{e}_{2,3}\,\mathbf{p}_{(3,k)}\mathbf{p}_{(0,j)}\mathbf{p}_0 \equiv_P \mathbf{p}_{(2,j)}\mathbf{p}_{(3,k)}\,\mathbf{e}_{2,3}\,\mathbf{p}_0\mathbf{p}_{(3,k)}\mathbf{p}_{(0,j)} \ .
\end{aligned}$$

Hence, to show that $\mathbf{p}_0\mathbf{e}_{j,k} \equiv_2 \mathbf{e}_{j,k}\mathbf{p}_0$, it suffices to show that $\mathbf{p}_0\mathbf{e}_{2,3} \equiv_2 \mathbf{e}_{2,3}\mathbf{p}_0$. Since $\mathbf{p}_0\mathbf{e}_{2,3} \equiv_P \mathbf{p}_0\mathbf{e}_{2,3}\mathbf{p}_0\mathbf{p}_0$, this can be shown by showing that $\mathbf{p}_0\mathbf{e}_{2,3} \equiv_2 \mathbf{e}_{2,3}$. This can be shown as follows, where the fifth assertion holds by PE_2 and the sixth by $\mathbf{p}_2\mathbf{p}_2\mathbf{p}_1\mathbf{p}_2 \equiv_P \mathbf{p}_1\mathbf{p}_2$:

$$\begin{aligned}
\mathbf{p}_0\mathbf{e}_{2,3}\mathbf{p}_0 &= \mathbf{p}_0\mathbf{p}_1\mathbf{p}_0\,\mathbf{e}_{0,3}\mathbf{p}_0\mathbf{p}_1\mathbf{p}_0 = \mathbf{p}_0\mathbf{p}_1\mathbf{p}_0\mathbf{p}_2\mathbf{p}_1\,\mathbf{e}\,\mathbf{p}_1\mathbf{p}_2\mathbf{p}_0\mathbf{p}_1\mathbf{p}_0 \\
&\equiv_P \mathbf{p}_1\mathbf{p}_0\mathbf{p}_1\mathbf{p}_2\mathbf{p}_1\,\mathbf{e}\,\mathbf{p}_1\mathbf{p}_2\mathbf{p}_1\mathbf{p}_0\mathbf{p}_1 \\
&\equiv_P \mathbf{p}_1\mathbf{p}_0\mathbf{p}_2\mathbf{p}_1\mathbf{p}_2\,\mathbf{e}\,\mathbf{p}_2\mathbf{p}_1\mathbf{p}_2\mathbf{p}_0\mathbf{p}_1 \\
&\equiv_2 \mathbf{p}_1\mathbf{p}_0\mathbf{p}_2\mathbf{p}_1\,\mathbf{e}\,\mathbf{p}_2\mathbf{p}_2\mathbf{p}_1\mathbf{p}_2\mathbf{p}_0\mathbf{p}_1 \\
&\equiv_P \mathbf{p}_1\mathbf{p}_0\mathbf{p}_2\mathbf{p}_1\,\mathbf{e}\,\mathbf{p}_1\mathbf{p}_2\mathbf{p}_0\mathbf{p}_1 = \mathbf{e}_{2,3}
\end{aligned}$$

(b) First assume that $j \geq 3$. Then of the following two assertions the first holds by Lemma 15(b).

$$\mathbf{p}_{j-1}\mathbf{e}_{j,k} \equiv_2 \mathbf{p}_{(j-1,j)}\mathbf{p}_{(j-1,j)}\,\mathbf{e}_{j-1,k}\,\mathbf{p}_{(j-1,j)} \equiv_P \mathbf{e}_{j-1,k}\mathbf{p}_{j-1} \ .$$

Now assume that $j = 2$. Then the third assertion below holds because $\mathbf{e}_{0,k} \equiv_2 \mathbf{p}_{(1,k)}\,\mathbf{e}\,\mathbf{p}_{(1,k)}$ and $\mathbf{p}_1\mathbf{p}_1\mathbf{p}_0\mathbf{p}_{(1,k)} \equiv_P \mathbf{p}_0\mathbf{p}_{(1,k)}$.

$$\begin{aligned}
\mathbf{p}_{j-1}\mathbf{e}_{j,k} &= \mathbf{p}_1\mathbf{e}_{2,k} = \mathbf{p}_1\mathbf{p}_1\mathbf{p}_0\mathbf{e}_{(0,k)}\mathbf{p}_0\mathbf{p}_1 \equiv_2 \mathbf{p}_0\mathbf{e}_{0,k}\mathbf{p}_0\mathbf{p}_1 \\
&= \mathbf{e}_{1,k}\mathbf{p}_1 = \mathbf{e}_{j-1,k}\mathbf{p}_{j-1} \ .
\end{aligned}$$

Finally, assume that $j = 1$. Then the second and fourth assertion below holds by Lemma 15(a) and 15(b) or by Lemma 15(a) respectively.

$$\begin{aligned} \mathbf{p}_{j-1}\mathbf{e}_{j,k} &= \mathbf{p}_0\mathbf{e}_{1,k} \equiv_2 \mathbf{p}_0\mathbf{p}_0\mathbf{p}_{(1,k)}\,\mathbf{e}\,\mathbf{p}_{(1,k)}\mathbf{p}_0 \equiv_P \mathbf{p}_{(1,k)}\,\mathbf{e}\,\mathbf{p}_{(1,k)}\mathbf{p}_0 \\ &\equiv_2 \mathbf{e}_{0,k}\mathbf{p}_0 = \mathbf{e}_{j-1,k}\mathbf{p}_{j-1}\ . \end{aligned}$$

(c) By Lemma 15(b), $\mathbf{e}_{j+1,k} \equiv_2 \mathbf{p}_j\mathbf{e}_{j,k}\mathbf{p}_j$. Hence, if $i = j$, then

$$\mathbf{p}_i\mathbf{e}_{j,k} = \mathbf{p}_j\mathbf{e}_{j,k} \equiv_P \mathbf{p}_j\mathbf{e}_{j,k}\mathbf{p}_j\mathbf{p}_j \equiv_2 \mathbf{e}_{j+1,k}\mathbf{p}_j = \mathbf{e}_{j+1,k}\mathbf{p}_i\ .$$

(d) First assume that $j \geq 1$. Then of the assertions below the first and fifth hold by Lemmas 15(a) and 15(b), the second and fourth because $2 \leq i < k-1$, and the third by PE_2.

$$\begin{aligned} \mathbf{p}_i\mathbf{e}_{j,k} &\equiv_2 \mathbf{p}_i\mathbf{p}_{(0,j)}\mathbf{p}_{(1,k)}\,\mathbf{e}\,\mathbf{p}_{(1,k)}\mathbf{p}_{(0,j)} \equiv_P \mathbf{p}_{(0,j)}\mathbf{p}_{(1,k)}\mathbf{p}_i\,\mathbf{e}\,\mathbf{p}_{(1,k)}\mathbf{p}_{(0,j)} \\ &\equiv_2 \mathbf{p}_{(0,j)}\mathbf{p}_{(1,k)}\,\mathbf{e}\,\mathbf{p}_i\mathbf{p}_{(1,k)}\mathbf{p}_{(0,j)} \equiv_P \mathbf{p}_{(0,j)}\mathbf{p}_{(1,k)}\,\mathbf{e}\,\mathbf{p}_{(1,k)}\mathbf{p}_{(0,j)}\mathbf{p}_i \\ &\equiv_2 \mathbf{e}_{j,k}\mathbf{p}_i \end{aligned}$$

Now assume that $j = 0$ and $i \geq 2$. Then of the assertions below, the second and sixth hold by Lemma 15(a), the third and fifth because $2 \leq i < k-1$, and the fourth by PE_2.

$$\begin{aligned} \mathbf{p}_i\mathbf{e}_{j,k} &= \mathbf{p}_i\mathbf{e}_{0,k} \equiv_2 \mathbf{p}_i\mathbf{p}_{(1,k)}\,\mathbf{e}\,\mathbf{p}_{(1,k)} \equiv_P \mathbf{p}_{(1,k)}\mathbf{p}_i\,\mathbf{e}\,\mathbf{p}_{(1,k)} \\ &\equiv_2 \mathbf{p}_{(1,k)}\,\mathbf{e}\,\mathbf{p}_i\mathbf{p}_{(1,k)} \equiv_P \mathbf{p}_{(1,k)}\,\mathbf{e}\,\mathbf{p}_{(1,k)}\mathbf{p}_i \equiv_2 \mathbf{e}_{0,k}\mathbf{p}_i = \mathbf{e}_{j,k}\mathbf{p}_i\ . \end{aligned}$$

Finally assume that $j = 0$ and $i = 1$. Then of the following assertions the second and sixth hold by Lemma 15(a) and the fourth by Lemma 14(c).

$$\begin{aligned} \mathbf{p}_i\mathbf{e}_{j,k} &= \mathbf{p}_1\mathbf{e}_{0,k} \equiv_2 \mathbf{p}_{(1,2)}\mathbf{p}_{(1,k)}\,\mathbf{e}\,\mathbf{p}_{(1,k)} \\ &\equiv_P \mathbf{p}_{(1,k)}\mathbf{p}_{(2,k)}\,\mathbf{e}\,\mathbf{p}_{(1,k)} \equiv_2 \mathbf{p}_{(1,k)}\,\mathbf{e}\,\mathbf{p}_{(2,k)}\mathbf{p}_{(1,k)} \\ &\equiv_P \mathbf{p}_{(1,k)}\,\mathbf{e}\,\mathbf{p}_{(1,k)}\mathbf{p}_{(1,2)} \equiv_2 \mathbf{e}_{0,k}\mathbf{p}_1 = \mathbf{e}_{j,k}\mathbf{p}_1\ . \end{aligned}$$

(e) The second assertion below holds by Lemma 14(a).

$$\mathbf{p}_i\mathbf{e}_{j,k} = \mathbf{p}_{k-1}\mathbf{e}_{j,k} \equiv_P \mathbf{p}_{k-1}\mathbf{p}_{k-1}\mathbf{e}_{j,k-1}\mathbf{p}_{k-1} \equiv_P \mathbf{e}_{j,k-1}\mathbf{p}_{k-1} = \mathbf{e}_{j,k-1}\mathbf{p}_i\ .$$

(f) The third assertion below holds by Lemma 14(a).

$$\mathbf{p}_i\mathbf{e}_{j,k} = \mathbf{p}_k\mathbf{e}_{j,k} \equiv_P \mathbf{p}_k\mathbf{e}_{j,k}\mathbf{p}_k\mathbf{p}_k \equiv_P \mathbf{e}_{j,k+1}\mathbf{p}_k = \mathbf{e}_{j,k+1}\mathbf{p}_i\ .$$

(g) This follows from Lemma 14(c). □

Until the end of this chapter let $\equiv'$ be the congruence on $\mathbf{F}(\Gamma(\tilde{\mathbf{p}}, \mathbf{e}))$ that is generated by $\equiv_P \cup E_1 \cup PE_1 \cup PE_2$. A consequence of Lemmas 16, 18 and 19 is the following (cf. Theorem III.16(a)).

THEOREM 20. *Consider any sequence* $v = \langle v_0, \dots, v_{n-1}\rangle$ *such that, for each* $n' < n$, *either* (i) $v_{n'} = \mathbf{p}_i$ *for some* i *or* (ii) $v_{n'} = \mathbf{e}_{j,k}$ *for some* $j \neq k$. *Let* m *be the number of* $n' < n$ *such that* (i) *holds and assume that both* $m \neq 0$ *and* $n - m \neq 0$. *Let* x *be the word* $v_0{}^\frown \dots {}^\frown v_{n-1}$ *in* $F(\Gamma(\tilde{\mathbf{p}}, \mathbf{e}))$ *that results from* v *by concatenating, in their given order, its* n *terms and let* $\mathbf{p}_{i_0} \cdots \mathbf{p}_{i_{m-1}}$ *be the word in* $F(\Gamma(\tilde{\mathbf{p}}))$ *that results by concatenating, in the order in which they occur in* v, *the* m *terms of* v *that are in* $\Gamma(\tilde{\mathbf{p}})$. *Then there are sequences* $w = \langle w_0, \dots, w_{n-m-1}\rangle$ *and* $w' = \langle w'_0, \dots, w'_{n-m-1}\rangle$ *of length* $n - m$ *such that each* w_k *and each* w'_k *is some* $\mathbf{e}_{j,k}$, $j \neq k$, *and such that, if* y *is the word* $w_0{}^\frown \dots {}^\frown w_{n-m-1}$ *that results from* w *by concatenating, in their given order, its* $n - m$ *terms, and if* y' *is the word* $w'_0{}^\frown \dots {}^\frown w'_{n-m-1}$ *that similarly results from* w', *then both*

$$x \equiv' y\mathbf{p}_{i_0} \cdots \mathbf{p}_{i_{m-1}} \qquad \text{and} \qquad x \equiv' \mathbf{p}_{i_0} \cdots \mathbf{p}_{i_{m-1}} y' \ .$$

Proof. Consider any j. By Lemma 18, $\mathbf{p}_j \mathbf{e}_{j,j+1} \equiv_1 \mathbf{e}_{j,j+1}$. Since $\mathbf{p}_j = \mathbf{p}_j^{-1}$ and since, by Lemma 16, $\mathbf{e}_{j,j+1}^{-1} \equiv_{E_1} \mathbf{e}_{j,j+1}$, therefore $\mathbf{e}_{j,j+1} \equiv' \mathbf{e}_{j,j+1}\mathbf{p}_j$. There now follows

$$\mathbf{p}_j \mathbf{e}_{j,j+1}^{-1} \equiv' \mathbf{e}_{j,j+1}^{-1} \mathbf{p}_j \ , \quad \text{for any } \ j < \omega \ .$$

From this result, from Lemma 19, and from $\mathbf{e}_{j,k} = \mathbf{e}_{k,j}$, there follows that for any i, j and k such that $j \neq k$ there is some j' and k' such that $j' \neq k'$ and such that $\mathbf{p}_i \mathbf{e}_{j,k} \equiv' \mathbf{e}_{j',k'} \mathbf{p}_i$ and hence also

$$\mathbf{e}_{j,k}\mathbf{p}_i = (\mathbf{p}_i^{-1}\mathbf{e}_{j,k}^{-1})^{-1} \equiv' (\mathbf{p}_i \mathbf{e}_{j,k})^{-1} \equiv' (\mathbf{e}_{j',k'}\mathbf{p}_i)^{-1} = \mathbf{p}_i^{-1}\mathbf{e}_{j',k'}^{-1} \equiv' \mathbf{p}_i \mathbf{e}_{j',k'} \ .$$

The desired conclusion now follows by induction on n. □

The next lemma supplements Lemma 15.

LEMMA 21. *Let* $\equiv'$ *be the congruence on* $\mathbf{F}(\Gamma(\tilde{\mathbf{p}}, \mathbf{e}))$ *that is generated by* $\equiv_P \cup E_1 \cup PE_1 \cup PE_2$. *Consider the following two cases:* (a) $k < j < k'$; (b) $k' < j < k$.

$$\begin{aligned} \text{Then } \ \mathbf{e}_{j,k'} \ &\equiv' \ \mathbf{p}_{(k,k')}\, \mathbf{e}_{j,k}\, \mathbf{p}_{(k,k')} \ , \quad \text{if (a)} \ , \\ \mathbf{p}_{k-1}^2 \mathbf{e}_{j,k'} \ &\equiv' \ \mathbf{p}_{(k,k')}\, \mathbf{e}_{j,k}\, \mathbf{p}_{(k,k')} \ , \quad \text{if (b)} \ . \end{aligned}$$

Proof. (a) First, let $j = k+1$ and $k' = j+1 = k+2$. Then the second assertion below holds by Lemma 14(a) and the third by Lemmas 18 and 16.

$$\begin{aligned}
\mathbf{e}_{j,k'} &= \mathbf{e}_{k+1,k+2} \equiv_P \mathbf{p}_k\mathbf{p}_{k+1}\,\mathbf{e}_{k,k+1}\,\mathbf{p}_{k+1}\mathbf{p}_k \\
&\equiv' \mathbf{p}_k\mathbf{p}_{k+1}\mathbf{p}_k\,\mathbf{e}_{k,k+1}\,\mathbf{p}_k\mathbf{p}_{k+1}\mathbf{p}_k \\
&\equiv_P \mathbf{p}_{k+1}\mathbf{p}_k\mathbf{p}_{k+1}\,\mathbf{e}_{k,k+1}\,\mathbf{p}_{k+1}\mathbf{p}_k\mathbf{p}_{k+1} \\
&= \mathbf{p}_{(k,k+2)}\,\mathbf{e}_{k,k+1}\,\mathbf{p}_{(k,k+2)} = \mathbf{p}_{(k,k')}\mathbf{e}_{k,j}\mathbf{p}_{(k,k')} \\
&= \mathbf{p}_{(k,k')}\,\mathbf{e}_{j,k}\,\mathbf{p}_{(k,k')}\ .
\end{aligned}$$

Now let $j = k+1$ and $k' \geq j+1 = k+2$ and assume as inductive hypothesis that $\mathbf{e}_{k+1,k'} \equiv' \mathbf{p}_{(k,k')}\mathbf{e}_{k+1,k'}\mathbf{p}_{(k,k')}$. Then of the following assertions, the second and fifth hold by Lemma 14(a) and the third by the inductive hypothesis:

$$\begin{aligned}
\mathbf{e}_{j,k'+1} &= \mathbf{e}_{k+1,k'+1} \equiv_P \mathbf{p}_{k'}\,\mathbf{e}_{k+1,k'}\,\mathbf{p}_{k'} \\
&\equiv' \mathbf{p}_{k'}\mathbf{p}_{(k,k')}\,\mathbf{e}_{k+1,k'}\,\mathbf{p}_{(k,k')}\mathbf{p}_{k'} \\
&\equiv_P \mathbf{p}_{(k,k'+1)}\mathbf{p}_{(k',k'+1)}\,\mathbf{e}_{k+1,k'}\,\mathbf{p}_{(k',k'+1)}\mathbf{p}_{(k',k'+1)} \\
&\equiv_P \mathbf{p}_{(k,k'+1)}\,\mathbf{e}_{k+1,k'+1}\,\mathbf{p}_{(k,k'+1)} \\
&= \mathbf{p}_{(k,k'+1)}\,\mathbf{e}_{j,k'+1}\mathbf{p}_{(k,k'+1)}\ .
\end{aligned}$$

There follows by induction that, if $j = k+1$ and $k' > j$, then $\mathbf{e}_{j,k'} \equiv' \mathbf{p}_{(k,k')}\,\mathbf{e}_{j,k}\,\mathbf{p}_{(k,k')}$.

Now let $k+1 \leq j < k'$ and $k \neq 0$, and assume as inductive hypo- thesis that $\mathbf{e}_{j,k'} \equiv' \mathbf{p}_{(k,k')}\,\mathbf{e}_{j,k}\,\mathbf{p}_{(k,k')}$ and hence

$$\mathbf{p}_{(k,k')}\,\mathbf{e}_{j,k'}\,\mathbf{p}_{(k,k')} \equiv' \mathbf{p}^2_{(k,k')}\,\mathbf{e}_{j,k}\,\mathbf{p}^2_{(k,k')} \equiv_P \mathbf{p}^2_{k'-1}\,\mathbf{e}_{j,k}\ .$$

Since $k-1 < k < j$, therefore by Lemma 15(c),

$$\mathbf{e}_{j,k-1} \equiv_2 \mathbf{p}_{(k,k-1)}\,\mathbf{e}_{j,k}\,\mathbf{p}_{(k,k-1)} = \mathbf{p}_{k-1}\,\mathbf{e}_{j,k}\,\mathbf{p}_{k-1}\ .$$

Since $k' \geq k+2$ there follows that

$$\mathbf{p}^2_{k'-1}\,\mathbf{e}_{j,k-1} \equiv_2 \mathbf{p}^2_{k'-1}\mathbf{p}_{k-1}\,\mathbf{e}_{j,k}\mathbf{p}_{k-1} \equiv_P \mathbf{p}_{k-1}\mathbf{p}^2_{k'-1}\,\mathbf{e}_{j,k}\,\mathbf{p}_{k-1}\ .$$

Of the following assertions, the first follows from the inductive hypo- thesis and the third from Lemma 19(a).

$$\begin{aligned}
\mathbf{p}_{k-1}\mathbf{p}^2_{k'-1}\,\mathbf{e}_{j,k}\,\mathbf{p}_{k-1} &\equiv' \mathbf{p}_{k-1}\mathbf{p}_{(k,k')}\,\mathbf{e}_{j,k'}\,\mathbf{p}_{(k,k')}\mathbf{p}_{k-1} \\
&\equiv_P \mathbf{p}_{(k-1,k')}\mathbf{p}_{(k-1,k)}\,\mathbf{e}_{j,k'}\,\mathbf{p}_{(k-1,k)}\mathbf{p}_{(k-1,k')} \\
&\equiv_2 \mathbf{p}_{(k-1,k')}\mathbf{e}_{j,k'}\mathbf{p}_{(k-1,k)}\mathbf{p}_{(k-1,k)}\mathbf{p}_{(k-1,k')} \\
&\equiv_P \mathbf{p}_{(k-1,k')}\,\mathbf{e}_{j,k'}\,\mathbf{p}_{(k-1,k')}\ .
\end{aligned}$$

There now follows that $\mathbf{p}^2_{k'-1}\,\mathbf{e}_{j,k-1} \equiv' \mathbf{p}_{(k-1,k')}\,\mathbf{e}_{j,k'}\mathbf{p}_{(k-1,k')}$. Hence, by induction, if $k<j< k'$, then $\mathbf{p}^2_{k'-1}\,\mathbf{e}_{j,k} \equiv' \mathbf{p}_{(k,k')}\,\mathbf{e}_{j,k'}\mathbf{p}_{(k,k')}$. There follows that

$$\begin{aligned}\mathbf{e}_{j,k'} &\equiv_P \mathbf{p}_{(k,k')}\mathbf{p}_{(k,k')}\,\mathbf{e}_{j,k'}\,\mathbf{p}_{(k,k')}\mathbf{p}_{(k,k')} \\ &\equiv' \mathbf{p}_{(k,k')}\mathbf{p}^2_{k'-1}\,\mathbf{e}_{j,k}\,\mathbf{p}_{(k,k')} \\ &\equiv_P \mathbf{p}_{(k,k')}\,\mathbf{e}_{j,k}\,\mathbf{p}_{(k,k')}\ .\end{aligned}$$

(b) Assume that $k' < j < k$. Then the second assertion below follows from part (a), with k and k' interchanged, the third from $k > k'$, and the fourth from Lemma 14(c).

$$\begin{aligned}\mathbf{p}_{(k,k')}\,\mathbf{e}_{j,k}\,\mathbf{p}_{(k,k')} &= \mathbf{p}_{(k',k)}\,\mathbf{e}_{j,k}\,\mathbf{p}_{(k',k)} \equiv' \mathbf{p}_{(k',k)}\mathbf{p}_{(k',k)}\,\mathbf{e}_{j,k'}\,\mathbf{p}_{(k',k)}\mathbf{p}_{(k',k)} \\ &\equiv_P \mathbf{p}^2_{k-1}\,\mathbf{e}_{j,k'}\,\mathbf{p}^2_{k-1} \equiv_2 \mathbf{p}^2_{k-1}\mathbf{p}^2_{k-1}\mathbf{e}_{j,k'} \equiv_P \mathbf{p}^2_{k-1}\,\mathbf{e}_{j,k'}\end{aligned}$$

□

According to the next lemma, parts (a), (d), (g) of Lemma 19 can be generalized if one replaces $\equiv_2$ by a congruence on $\mathbf{F}(\Gamma(\tilde{\mathbf{p}},\mathbf{e}))$ that includes it as well as E_1 and PE_1.

LEMMA 22. *Let* $\equiv'$ *be the congruence on* $\mathbf{F}(\Gamma(\tilde{\mathbf{p}},\mathbf{e}))$ *that is generated by* $\equiv_P \cup E_1 \cup PE_1 \cup PE_2$. *Assume that either* $i = i'$, $j = k$, *or* $\{i,i'\}\cap\{j,k\} = \emptyset$. *Then*

$$\mathbf{p}_{(i,i')}\mathbf{e}_{j,k} \equiv' \mathbf{e}_{j,k}\mathbf{p}_{(i,i')}$$

Proof. Assume that $j = k$. Since $\mathbf{p}_{(i,i')}$ is a word w formed from letters $\mathbf{p}_{i''}$, therefore by Lemma 14(b) and induction on the length of $\mathbf{p}_{(i,i')} = w$,

$$\mathbf{p}_{(i,i')}\mathbf{e}_{j,k} = \mathbf{p}_{(i,i')}\mathbf{e}_{j,j} \equiv_P \mathbf{e}_{j,j}\mathbf{p}_{(i,i')} = \mathbf{e}_{j,k}\mathbf{p}_{(i,i')}\ .$$

Assume that $i = i'$ and hence $\mathbf{p}_{(i,i')} = \mathbf{p}_{(i,i)} = \mathbf{p}^2_{(i-1)}$. Also assume that $j \neq k$. Since $\mathbf{e}_{j,k} = \mathbf{e}_{k,j}$, we will assume without loss of generality, that $j < k$. First suppose that $i \leq k$. Then $\mathbf{p}_{(i,i)}\mathbf{p}_{(j,k)} \equiv_P \mathbf{p}_{(j,k)} \equiv_P \mathbf{p}_{(j,k)}\mathbf{p}_{(i,i)}$. By Lemma 18, $\mathbf{e}_{j,k} \equiv_1 \mathbf{p}_{(j,k)}\mathbf{e}_{j,k}$ and hence, by Lemma 16, $\mathbf{e}_{j,k} \equiv' \mathbf{e}_{j,k}\mathbf{p}_{(j,k)}$. There follows that $\mathbf{p}_{(i,i)}\mathbf{e}_{j,k} \equiv_1 \mathbf{e}_{j,k} \equiv' \mathbf{e}_{j,k}\mathbf{p}_{(i,i)}$. Next suppose that $i = k+1$. Since $\mathbf{p}_{(k+1,k+1)} = \mathbf{p}^2_k$ and since by Lemmas 19(f) or 19(e), respectively, $\mathbf{p}_k\mathbf{e}_{j,k} \equiv_2 \mathbf{e}_{j,k+1}\mathbf{p}_k$ and $\mathbf{p}_k\mathbf{e}_{j,k+1} \equiv_2 \mathbf{e}_{j,k}\mathbf{p}_k$, there follows that $\mathbf{p}_{(i,i)}\mathbf{e}_{j,k} = \mathbf{p}_{(k+1,k+1)}\mathbf{e}_{j,k} \equiv' \mathbf{e}_{j,k}\mathbf{p}_{(i,i)}$. Finally suppose that $i > k+1$. Then, by Lemma 14(c), $\mathbf{p}_{(i,i)}\mathbf{e}_{j,k} = \mathbf{p}^2_{i-1}\mathbf{e}_{j,k} \equiv_2 \mathbf{e}_{j,k}\mathbf{p}^2_{i-1} = \mathbf{e}_{j,k}\mathbf{p}_{(i,i)}$.

Now assume that $i \neq i'$, $j \neq k$ and $\{i, i'\} \cap \{j, k\} = \emptyset$. Since $\mathbf{p}_{(i',i)} = \mathbf{p}_{(i,i')}$ and $\mathbf{e}_{k,j} = \mathbf{e}_{j,k}$, we will assume, without loss of generality, that $i < i'$ and $j < k$. Then there are the following six cases:

$$\begin{array}{lll} (1)\ i < i' < j < k\ ; & (2)\ i < j < i' < k\ ; & (3)\ i < j < k < i'\ ; \\ (4)\ j < i < i' < k\ ; & (5)\ j < i < k < i'\ ; & (6)\ j < k < i < i'\ . \end{array}$$

CASE 1: $i < i' < j < k$. Then $\mathbf{p}_{(i,i')}$ is a word w formed from letters $\mathbf{p}_{i''}$ such that $i'' < j - 1$ and hence such that, by Lemma 19(a), $\mathbf{p}_{i''}\mathbf{e}_{j,k} \equiv_2 \mathbf{e}_{j,k}\mathbf{p}_{i''}$. By induction on the length of $\mathbf{p}_{(i,i')} = w$ there follows that

$$\mathbf{p}_{(i,i')}\mathbf{e}_{j,k} = w\mathbf{e}_{j,k} \equiv_2 \mathbf{e}_{j,k}w = \mathbf{e}_{j,k}\mathbf{p}_{(i,i')}\ .$$

CASE 4: $j < i < i' < k$. Then $\mathbf{p}_{(i,i')}$ is a word w formed from letters $\mathbf{p}_{i''}$ such that $j < i'' < k - 1$ and hence such that, by Lemma 19(d), $\mathbf{p}_{i''}\mathbf{e}_{j,k} \equiv_2 \mathbf{e}_{j,k}\mathbf{p}_{i''}$. By induction on the length of $\mathbf{p}_{(i,i')} = w$ there now again follows that

$$\mathbf{p}_{(i,i')}\mathbf{e}_{j,k} \equiv_2 \mathbf{e}_{j,k}\mathbf{p}_{(i,i')}\ .$$

CASE 6: $j < k < i < i'$. Then $\mathbf{p}_{(i,i')}\mathbf{e}_{j,k} \equiv_2 \mathbf{e}_{j,k}\mathbf{p}_{(i,i')}$ holds by Lemma 14(c).

CASE 2: $i < j < i' < k$. There are several subcases.

Subcase 2a: $i \geq 2$. Then of the following assertions, the first and last hold by Lemmas 15(a) and 15(b), the second and fourth hold because $i \notin \{0, 1, j, k\}$, and the third holds by Lemma 14(c).

$$\begin{aligned} \mathbf{p}_{(i,i')}\mathbf{e}_{j,k} &\equiv_2 \mathbf{p}_{(i,i')}\mathbf{p}_{(0,j)}\mathbf{p}_{(1,k)}\,\mathbf{e}\,\mathbf{p}_{(1,k)}\mathbf{p}_{(0,j)} \\ &\equiv_P \mathbf{p}_{(0,j)}\mathbf{p}_{(1,k)}\mathbf{p}_{(i,i')}\,\mathbf{e}\,\mathbf{p}_{(1,k)}\mathbf{p}_{(0,j)} \\ &\equiv_2 \mathbf{p}_{(0,j)}\mathbf{p}_{(1,k)}\mathbf{e}\,\mathbf{p}_{(i,i')}\mathbf{p}_{(1,k)}\mathbf{p}_{(0,j)} \\ &\equiv_P \mathbf{p}_{(0,j)}\mathbf{p}_{(1,k)}\,\mathbf{e}\,\mathbf{p}_{(1,k)}\mathbf{p}_{(0,j)}\mathbf{p}_{(i,i')} \\ &\equiv_2 \mathbf{e}_{j,k}\mathbf{p}_{(i,i')} \end{aligned}$$

Subcase 2b: $i = 1$. Then of the following assertions the first and last hold by Lemmas 15(a) and 15(b), the second and sixth hold because $\{1, i'\} \cap \{0, j\} = \emptyset$, the third and fifth hold because $\mathbf{p}_{(i,i')}\mathbf{p}_{(1,k)} \equiv_P \mathbf{p}_{(1,k)}\mathbf{p}_{(i',k)}$ or $\mathbf{p}_{(i',k)}\mathbf{p}_{(1,k)} \equiv_P$

$\mathbf{p}_{(1,k)}\mathbf{p}_{(i',k)}$, respectively, and the fourth holds by Lemma 14(c).

$$\begin{aligned}
\mathbf{p}_{(i,i')}\mathbf{e}_{j,k} &\equiv_2 \mathbf{p}_{(1,i')}\mathbf{p}_{(0,j)}\mathbf{p}_{(1,k)}\,\mathbf{e}\,\mathbf{p}_{(1,k)}\mathbf{p}_{(0,j)} \\
&\equiv_P \mathbf{p}_{(0,j)}\mathbf{p}_{(1,i')}\mathbf{p}_{(1,k)}\,\mathbf{e}\,\mathbf{p}_{(1,k)}\mathbf{p}_{(0,j)} \\
&\equiv_P \mathbf{p}_{(0,j)}\mathbf{p}_{(1,k)}\mathbf{p}_{(i',k)}\,\mathbf{e}\,\mathbf{p}_{(1,k)}\mathbf{p}_{(0,j)} \\
&\equiv_2 \mathbf{p}_{(0,j)}\mathbf{p}_{(1,k)}\,\mathbf{e}\,\mathbf{p}_{(i',k)}\mathbf{p}_{(1,k)}\mathbf{p}_{(0,j)} \\
&\equiv_P \mathbf{p}_{(0,j)}\mathbf{p}_{(1,k)}\,\mathbf{e}\,\mathbf{p}_{(1,k)}\mathbf{p}_{(1,i')}\mathbf{p}_{(0,j)} \\
&\equiv_P \mathbf{p}_{(0,j)}\mathbf{p}_{(1,k)}\,\mathbf{e}\,\mathbf{p}_{(1,k)}\mathbf{p}_{(0,j)}\mathbf{p}_{(1,i')} \\
&\equiv_2 \mathbf{e}_{j,k}\mathbf{p}_{(1,i')}
\end{aligned}$$

Subcase 2c: $i = 0$ and $j \geq 2$. Then of the following assertions the first and last hold by Lemmas 15(a) and 15(b), the second and sixth hold because $\{1, k\}$ is disjoint from $\{0, j\}$ and also from $\{0, i'\}$, the third and fifth hold because $\mathbf{p}_{(0,i')}\mathbf{p}_{(0,j)} \equiv_P \mathbf{p}_{(0,j)}\mathbf{p}_{(i',j)}$ and $\mathbf{p}_{(i',j)}\mathbf{p}_{(0,j)} \equiv_P \mathbf{p}_{(0,j)}\mathbf{p}_{(0,i')}$ respectively, and the fourth holds by Lemma 14(c).

$$\begin{aligned}
\mathbf{p}_{(0,i')}\mathbf{e}_{j,k} &\equiv_2 \mathbf{p}_{(0,i')}\mathbf{p}_{(0,j)}\mathbf{p}_{(1,k)}\,\mathbf{e}\,\mathbf{p}_{(1,k)}\mathbf{p}_{(0,j)} \\
&\equiv_P \mathbf{p}_{(1,k)}\mathbf{p}_{(0,i')}\mathbf{p}_{(0,j)}\,\mathbf{e}\,\mathbf{p}_{(0,j)}\mathbf{p}_{(1,k)} \\
&\equiv_P \mathbf{p}_{(1,k)}\mathbf{p}_{(0,j)}\mathbf{p}_{(i',j)}\,\mathbf{e}\,\mathbf{p}_{(0,j)}\mathbf{p}_{(1,k)} \\
&\equiv_2 \mathbf{p}_{(1,k)}\mathbf{p}_{(0,j)}\,\mathbf{e}\,\mathbf{p}_{(i',j)}\mathbf{p}_{(0,j)}\mathbf{p}_{(1,k)} \\
&\equiv_P \mathbf{p}_{(1,k)}\mathbf{p}_{(0,j)}\,\mathbf{e}\,\mathbf{p}_{(0,j)}\mathbf{p}_{(0,i')}\mathbf{p}_{(1,k)} \\
&\equiv_P \mathbf{p}_{(0,j)}\mathbf{p}_{(1,k)}\,\mathbf{e}\,\mathbf{p}_{(1,k)}\mathbf{p}_{(0,j)}\mathbf{p}_{(0,i')} \\
&\equiv_2 \mathbf{e}_{j,k}\mathbf{p}_{(0,i')}
\end{aligned}$$

Subcase 2d: $i = 0$ and $j = 1$. Then of the following assertions, the second and next-to-last hold by $\mathbf{e} = \mathbf{e}_0$ and Lemma 21(a) and the fourth by Lemma 14(c).

$$\begin{aligned}
\mathbf{p}_{(i,i')}\mathbf{e}_{j,k} &\equiv_2 \mathbf{p}_{(0,i')}\mathbf{p}_{(1,k)} \equiv' \mathbf{p}_{(0,i')}\mathbf{p}_{(0,k)}\,\mathbf{e}\,\mathbf{p}_{(0,k)} \\
&\equiv_P \mathbf{p}_{(0,k)}\mathbf{p}_{(i',k)}\,\mathbf{e}\,\mathbf{p}_{(0,k)} \equiv_2 \mathbf{p}_{(0,k)}\,\mathbf{e}\,\mathbf{p}_{(i',k)}\mathbf{p}_{(0,k)} \\
&\equiv_P \mathbf{p}_{(0,k)}\,\mathbf{e}\,\mathbf{p}_{(0,k)}\mathbf{p}_{(0,i')} \\
&\equiv' \mathbf{e}_{1,k}\mathbf{p}_{(0,i')} = \mathbf{e}_{j,k}\mathbf{p}_{(i,i')}
\end{aligned}$$

CASE 3: $i < j < k < i'$. This can be broken down into the following four subcases, similar to those for case 2. Subcase 3a: $i \geq 2$; subcase 3b: $i = 1$; subcase 3c: $i = 0$ and $j \geq 2$; subcase 3d: $i = 0$ and $j = 1$. For each of these subcases, the argument used for the corresponding subcase of case 2 shows again that $\mathbf{p}_{(i,i')}\mathbf{e}_{j,k} \equiv' \mathbf{e}_{j,k}\mathbf{p}_{(i,i')}$.

CASE 5: $j < i < k < i'$. There are three subcases.
Subcase 5a: $j \geq 1$. Then for reasons that by now should be familiar there hold

$$\begin{aligned}
\mathbf{p}_{(i,i')}\mathbf{e}_{j,k} &\equiv_2 \mathbf{p}_{(i,i')}\mathbf{p}_{(0,j)}\mathbf{p}_{(1,k)}\,\mathbf{e}\,\mathbf{p}_{(1,k)}\mathbf{p}_{(0,j)} \\
&\equiv_P \mathbf{p}_{(0,j)}\mathbf{p}_{(1,k)}\mathbf{p}_{(i,i')}\,\mathbf{e}\,\mathbf{p}_{(1,k)}\mathbf{p}_{(0,j)} \\
&\equiv_2 \mathbf{p}_{(0,j)}\mathbf{p}_{(1,k)}\,\mathbf{e}\,\mathbf{p}_{(i,i')}\mathbf{p}_{(1,k)}\mathbf{p}_{(0,j)} \\
&\equiv_P \mathbf{p}_{(0,j)}\mathbf{p}_{(1,k)}\,\mathbf{e}\,\mathbf{p}_{(1,k)}\mathbf{p}_{(0,j)}\mathbf{p}_{(i,i')} \\
&\equiv_2 \mathbf{e}_{j,k}\mathbf{p}_{(i,i')} \ .
\end{aligned}$$

Subcase 5b: $j = 0$ and $i \geq 2$. Then there hold

$$\begin{aligned}
\mathbf{p}_{(i,i')}\mathbf{e}_{0,k} &\equiv_2 \mathbf{p}_{(i,i')}\mathbf{p}_{(1,k)}\,\mathbf{e}\,\mathbf{p}_{(1,k)} \equiv_P \mathbf{p}_{(1,k)}\mathbf{p}_{(i,i')}\,\mathbf{e}\,\mathbf{p}_{(1,k)} \\
&\equiv_2 \mathbf{p}_{(1,k)}\,\mathbf{e}\,\mathbf{p}_{(i,i')}\mathbf{p}_{(1,k)} \equiv_2 \mathbf{p}_{(1,k)}\,\mathbf{e}\,\mathbf{p}_{(1,k)}\mathbf{p}_{(i,i')} \\
&\equiv_2 \mathbf{e}_{0,k}\mathbf{p}_{(i,i')} \ .
\end{aligned}$$

Subcase 5c: $j = 0$ and $i = 1$. Since $k \geq 2$ therefore, by Lemma 15(c), $\mathbf{e}_{k,0} \equiv_2 \mathbf{p}_{(1,0)}\,\mathbf{e}_{k,1}\,\mathbf{p}_{(1,0)}$ and hence $\mathbf{e}_{0,k} \equiv_2 \mathbf{p}_{(0,1)}\,\mathbf{e}_{1,k}\,\mathbf{p}_{(0,1)}$. From this there follows the second and next-to-last assertion below. The fourth assertion follows from what was shown for the subcase 3(d).

$$\begin{aligned}
\mathbf{p}_{(i,i')}\mathbf{e}_{j,k} &= \mathbf{p}_{(1,i')}\mathbf{e}_{0,k} \equiv_2 \mathbf{p}_{(1,i')}\mathbf{p}_{(0,1)}\,\mathbf{e}_{1,k}\,\mathbf{p}_{(0,1)} \\
&\equiv_P \mathbf{p}_{(0,1)}\mathbf{p}_{(0,i')}\,\mathbf{e}_{1,k}\,\mathbf{p}_{(0,1)} \equiv' \mathbf{p}_{(0,1)}\,\mathbf{e}_{1,k}\,\mathbf{p}_{(0,i')}\mathbf{p}_{(0,1)} \\
&\equiv_P \mathbf{p}_{(0,1)}\,\mathbf{e}_{1,k}\,\mathbf{p}_{(0,1)}\mathbf{p}_{(1,i')} \equiv_2 \mathbf{e}_{0,k}\mathbf{p}_{(1,i')} = \mathbf{e}_{j,k}\mathbf{p}_{(i,i')}.
\end{aligned}$$

□

Until the end of the proof of Theorem 24 we let $\equiv''$ be the congruence on $\mathbf{F}(\Gamma(\tilde{\mathbf{p}}, \mathbf{e}))$ that is generated by $\equiv_P \cup E_1 \cup PE_1 \cup PE_2 \cup PE_3 \cup PE_4$ and hence by $\equiv' \cup PE_3 \cup PE_4$. Note that according to the definitions of $\mathbf{e}_{0,1}$, $\mathbf{e}_{0,2}$, or $\mathbf{e}_{2,3}$, respectively, PE_3 and PE_4 are, respectively, the following relation on $F(\Gamma(\tilde{\mathbf{p}}, \mathbf{e}))$:

$$\begin{aligned}
PE_3 &= \{\langle \mathbf{e}_{0,1}\mathbf{e}_{0,2},\ \mathbf{e}_{0,2}\mathbf{e}_{0,1}\rangle\} \ , \\
PE_4 &= \{\langle \mathbf{e}_{0,1}\mathbf{e}_{2,3},\ \mathbf{e}_{2,3}\mathbf{e}_{0,1}\rangle\} \ .
\end{aligned}$$

They postulate that $e_{0,1}$ commutes with $e_{0,2}$ or with $e_{2,3}$, respectively. The next lemma states that, in conjunction with $\equiv'$, they imply that any $e_{j,k}$ commutes with any $e_{j',k'}$.

LEMMA 23. $\mathbf{e}_{j,k}\mathbf{e}_{j',k'} \equiv'' \mathbf{e}_{j',k'}\mathbf{e}_{j,k}$ *for any* j, k, j', k'.

Proof. Henceforth we will freely change from $\mathbf{e}_{i,i'}$ to $\mathbf{e}_{i',i}$. First, we consider the case where either $j = k$ or $j' = k'$. Suppose that $j = k$. Then $j \geq 1$ and $\mathbf{e}_{j,k} = \mathbf{e}_{j,j} = \mathbf{p}_{j-1}^2$. Also, by Lemmas 18 and 16, $\mathbf{e}_{j',k'} \equiv' \mathbf{p}_{(j',k')}\mathbf{e}_{j',k'} \equiv' \mathbf{e}_{j',k'}\mathbf{p}_{(j',k')}$. If $j - 1 \leq \max\{j', k'\}$, then $\mathbf{p}_{j-1}^2\mathbf{e}_{j',k'} \equiv' \mathbf{p}_{j-1}^2\mathbf{p}_{(j',k')}\mathbf{e}_{j',k'} \equiv_P \mathbf{p}_{(j',k')}\mathbf{e}_{j',k'} \equiv' \mathbf{e}_{j',k'}\mathbf{p}_{(j',k')} \equiv' \mathbf{e}_{j',k'}\mathbf{p}_{j-1}^2$. If $j - 1 > \max\{j', k'\}$, then, by Lemma 14(c), $\mathbf{p}_{j-1}^2\mathbf{e}_{j',k'} \equiv_2 \mathbf{e}_{j',k'}\mathbf{p}_{j-1}^2$. Thus, in either case, $\mathbf{e}_{jj}\mathbf{e}_{j',k'} \equiv' \mathbf{e}_{j',k'}\mathbf{e}_{j,j}$. Since this holds for any j, j', k' there follows that $\mathbf{e}_{j,k}\mathbf{e}_{j',j'} \equiv' \mathbf{e}_{j',j'}\mathbf{e}_{j,k}$.

For the rest of the proof we assume that $j \neq k$ and $j' \neq k'$. We first consider the case where $\{j,k\} \cap \{j',k'\} \neq \emptyset$. Since $\{j,k\} = \{j',k'\}$ implies that $e_{j,k} = e_{j',k'}$, and hence that $e_{j,k}e_{j',k'} = e_{j',k'}e_{j,k}$, we also assume that $\{j,k\} \neq \{j',k'\}$. Since $e_{i,i'} = e_{i',i}$ for any i and i', we may then assume without loss of generality that $j' = j$ and that j, k, k' are distinct. Using PE_3, we want to show that in this case $\mathbf{e}_{j,k}\mathbf{e}_{j,k'} \equiv'' \mathbf{e}_{j,k'}\mathbf{e}_{j,k}$.

Let $k' \geq 3$. Then, of the following assertions, the first and fifth hold by Lemma 15(a), the second and fourth by Lemma 14(c), and the third by PE_3.

$$\begin{aligned} \mathbf{e}_{0,1}\mathbf{e}_{0,k'} &\equiv_2 \mathbf{e}_{0,1}\mathbf{p}_{(2,k')}\mathbf{e}_{0,2}\mathbf{p}_{(2,k')} \equiv_2 \mathbf{p}_{(2,k')}\mathbf{e}_{0,1}\mathbf{e}_{0,2}\mathbf{p}_{(2,k')} \\ &\equiv'' \mathbf{p}_{(2,k')}\mathbf{e}_{0,2}\mathbf{e}_{0,1}\mathbf{p}_{(2,k')} \equiv_2 \mathbf{p}_{(2,k')}\mathbf{e}_{0,2}\mathbf{p}_{(2,k')}\mathbf{e}_{0,1} \equiv \mathbf{e}_{0,k'}\mathbf{e}_{0,1} \ . \end{aligned}$$

From this and from PE_3 there follows that

$$\mathbf{e}_{0,1}\mathbf{e}_{0,k'} \equiv'' \mathbf{e}_{0,k'}\mathbf{e}_{0,1} \ , \quad \text{if } 2 \leq k'.$$

Let $2 \leq k < k'$. Then, of the following assertions, the first and fifth hold by Lemma 15(a), the second and fourth by Lemma 22, and the third by what has just been shown.

$$\begin{aligned} \mathbf{e}_{0,k}\mathbf{e}_{0,k'} &\equiv_2 \mathbf{p}_{(1,k)}\mathbf{e}_{0,1}\mathbf{p}_{(1,k)}\mathbf{e}_{0,k'} \equiv' \mathbf{p}_{(1,k)}\mathbf{e}_{0,1}\mathbf{e}_{0,k'}\mathbf{p}_{(1,k)} \\ &\equiv'' \mathbf{p}_{(1,k)}\mathbf{e}_{0,k'}\mathbf{e}_{0,1}\mathbf{p}_{(1,k)} \equiv' \mathbf{e}_{0,k'}\mathbf{p}_{(1,k)}\mathbf{e}_{0,1}\mathbf{p}_{(1,k)} \\ &\equiv_2 \mathbf{e}_{0,k'}\mathbf{e}_{0,k} \ . \end{aligned}$$

From this and from the preceding paragraph there follows that

$$\mathbf{e}_{0,k}\mathbf{e}_{0,k'} \equiv'' \mathbf{e}_{0,k'}\mathbf{e}_{0,k} \ , \quad \text{if } 1 \leq k < k'.$$

Let $1 \leq j < k < k'$. Then of the following assertions, the first and seventh each follows by two uses of Lemma 15(b), the third and fifth hold by Lemma 19(d), and the fourth holds by what has just been shown.

$$\begin{aligned} \mathbf{e}_{j,k}\mathbf{e}_{j,k'} &\equiv_2 \mathbf{p}_{(0,j)}\mathbf{e}_{0,k}\mathbf{p}_{(0,j)}\mathbf{p}_{(0,j)}\mathbf{e}_{0,k'}\mathbf{p}_{(0,j)} \equiv_P \mathbf{p}_{(0,j)}\mathbf{e}_{0,k}\mathbf{p}_{j-1}^2\mathbf{e}_{0,k'}\mathbf{p}_{(0,j)} \\ &\equiv_2 \mathbf{p}_{(0,j)}\mathbf{p}_{j-1}^2\mathbf{e}_{0,k}\mathbf{e}_{0,k'}\mathbf{p}_{(0,j)} \equiv'' \mathbf{p}_{(0,j)}\mathbf{p}_{j-1}^2\mathbf{e}_{0,k'}\mathbf{e}_{0,k}\mathbf{p}_{(0,j)} \\ &\equiv_2 \mathbf{p}_{(0,j)}\mathbf{e}_{0,k'}\mathbf{p}_{j-1}^2\mathbf{e}_{0,k}\mathbf{p}_{(0,j)} \equiv_P \mathbf{p}_{(0,j)}\mathbf{e}_{0,k'}\mathbf{p}_{(0,j)}\mathbf{p}_{(0,j)}\mathbf{e}_{0,k}\mathbf{p}_{(0,j)} \\ &\equiv_2 \mathbf{e}_{j,k'}\mathbf{e}_{j,k} \end{aligned}$$

From this and from the preceding paragraph there follows that

$$\mathbf{e}_{j,k}\mathbf{e}_{j,k'} \equiv'' \mathbf{e}_{j,k'}\mathbf{e}_{j,k}\ , \quad \text{if } 0 \leq j < k < k'.$$

Let $0 \leq k < j < k'$. Choose some $k'' > k'$. Then, of the following assertions, the first and fifth hold by Lemma 21(a), the second and fourth by Lemma 22, and the third by $j < k' < k''$ and by what has just been shown.

$$\begin{aligned} \mathbf{e}_{j,k}\mathbf{e}_{j,k'} &\equiv' \mathbf{p}_{(k,k'')}\mathbf{e}_{j,k''}\mathbf{p}_{(k,k'')}\mathbf{e}_{j,k'} \equiv' \mathbf{p}_{(k,k'')}\mathbf{e}_{j,k''}\mathbf{e}_{j,k'}\mathbf{p}_{(k,k'')} \\ &\equiv'' \mathbf{p}_{(k,k'')}\mathbf{e}_{j,k'}\mathbf{e}_{j,k''}\mathbf{p}_{(k,k'')} \equiv' \mathbf{e}_{j,k'}\mathbf{p}_{(k,k'')}\mathbf{e}_{j,k''}\mathbf{p}_{(k,k'')} \\ &\equiv' \mathbf{e}_{j,k'}\mathbf{e}_{j,k} \end{aligned}$$

Let $0 \leq k < k' < j$. Then, by what was shown in the next-to-last paragraph:

$$\mathbf{e}_{k,k'}\mathbf{e}_{k,j} \equiv'' \mathbf{e}_{k,j}\mathbf{e}_{k,k'}\ .$$

There follows the third assertion below. The first and fifth assertion hold by Lemma 21(a), and the second and fourth by Lemmas 16 and 18.

$$\begin{aligned} \mathbf{e}_{j,k}\mathbf{e}_{j,k'} &\equiv' \mathbf{e}_{k,j}\mathbf{p}_{(k,j)}\mathbf{e}_{k,k'}\mathbf{p}_{(k,j)} \equiv' \mathbf{p}_{(k,j)}\mathbf{e}_{k,j}\mathbf{e}_{k,k'}\mathbf{p}_{(k,j)} \\ &\equiv'' \mathbf{p}_{(k,j)}\mathbf{e}_{k,k'}\mathbf{e}_{k,j}\mathbf{p}_{(k,j)} \\ &\equiv' \mathbf{p}_{(k,j)}\mathbf{e}_{k,k'}\mathbf{p}_{(k,j)}\mathbf{e}_{k,j} \equiv' \mathbf{e}_{j,k'}\mathbf{e}_{k,j} \end{aligned}$$

This concludes our proof that $\mathbf{e}_{j,k}\mathbf{e}_{j,k'} \equiv'' \mathbf{e}_{j,k'}\mathbf{e}_{j,k}$ whenever j, k, k' are distinct and hence either $j < k < k'$ or $k < j < k'$ or $k < k' < j$.

We now turn to the remaining case where j, k, j', k' are distinct. Since $\mathbf{e}_{i,i'} = \mathbf{e}_{i',i}$ for any i and i', we may assume without loss of generality that $j < k$ and that $j' < k'$. Using PE_4, we want to show that in this case $\mathbf{e}_{j,k}\mathbf{e}_{j',k'} \equiv'' \mathbf{e}_{j',k'}\mathbf{e}_{j,k}$.

First, let $0 \leq j < k < j' < k'$. From $\mathbf{e}_{0,1}\mathbf{e}_{2,3} \equiv'' \mathbf{e}_{2,3}\mathbf{e}_{0,1}$ we obtain, in succession, $\mathbf{e}_{0,1}\mathbf{e}_{2,k'} \equiv'' \mathbf{e}_{2,k'}\mathbf{e}_{0,1}$, $\mathbf{e}_{0,1}\mathbf{e}_{j',k'} \equiv'' \mathbf{e}_{j',k'}\mathbf{e}_{0,1}$, $\mathbf{e}_{0,k}\mathbf{e}_{j',k'} \equiv'' \mathbf{e}_{j',k'}\mathbf{e}_{0,k}$, and $\mathbf{e}_{j,k}\mathbf{e}_{j',k'} \equiv'' \mathbf{e}_{j',k'}\mathbf{e}_{j,k}$ in the same manner in which from $\mathbf{e}_{0,1}\mathbf{e}_{0,2} \equiv \mathbf{e}_{0,2}\mathbf{e}_{0,1}$ we earlier obtained in succession, for $1 \le k < k'$, $\mathbf{e}_{0,1}\mathbf{e}_{0,k'} \equiv'' \mathbf{e}_{0,k'}\mathbf{e}_{0,1}$ and $\mathbf{e}_{0,k}\mathbf{e}_{0,k'} \equiv'' \mathbf{e}_{0,k'}\mathbf{e}_{0,k}$.

Now let $0 \le j < j' < k < k'$. Then from $\mathbf{p}^2_{(j',k)}\mathbf{p}_{(k,k')} \equiv_P \mathbf{p}_{(k,k')} \equiv_P \mathbf{p}_{(k,k')}\mathbf{p}^2_{(j',k)}$ and from $\mathbf{p}_{(k,k')}\mathbf{e}_{k,k'} \equiv' \mathbf{e}_{k,k'} \equiv' \mathbf{e}_{k,k'}\mathbf{p}_{(k,k')}$ there follows

$$\mathbf{p}_{(j',k)}\mathbf{p}_{(j',k)}\mathbf{e}_{k,k'} \equiv' \mathbf{e}_{k,k'} \equiv' \mathbf{e}_{k,k'}\mathbf{p}_{(j',k)}\mathbf{p}_{(j',k)} \ .$$

From this there follows the second and the fourth assertion below. For the first and the fifth assertion below one uses Lemmas 15(a) and 15(c). The third assertion follows from $j < j' < k < k'$ and from what was shown in the last paragraph.

$$\begin{aligned}
\mathbf{e}_{j,k}\mathbf{e}_{j,k'} &\equiv_2 \mathbf{p}_{(j',k)}\mathbf{e}_{j,j'}\mathbf{p}_{(j',k)}\mathbf{p}_{(j',k)}\mathbf{e}_{k,k'}\mathbf{p}_{(j',k)} \\
&\equiv' \mathbf{p}_{(j',k)}\mathbf{e}_{j,j'}\mathbf{e}_{k,k'}\mathbf{p}_{(j',k)} \\
&\equiv'' \mathbf{p}_{(j',k)}\mathbf{e}_{k,k'}\mathbf{e}_{j,j'}\mathbf{p}_{(j',k)} \\
&\equiv' \mathbf{p}_{(j',k)}\mathbf{e}_{k,k'}\mathbf{p}_{(j',k)}\mathbf{p}_{(j',k)}\mathbf{e}_{j,j'}\mathbf{p}_{(j',k)} \\
&\equiv_2 \mathbf{e}_{j',k'}\mathbf{e}_{j,k}
\end{aligned}$$

Let $0 \le j < j' < k' < k$. Then from $\mathbf{p}^2_{(k',k)}\mathbf{p}_{(j',k)} \equiv_P \mathbf{p}_{(j',k)} \equiv_P \mathbf{p}_{(j',k)}\mathbf{p}^2_{(k',k)}$ and $\mathbf{p}_{(j',k)}\mathbf{e}_{j',k} \equiv' \mathbf{e}_{j',k}\mathbf{p}_{(j',k)}$ there follows:

$$\mathbf{p}^2_{(k',k)}\mathbf{e}_{j',k} \equiv' \mathbf{e}_{j',k} \equiv' \mathbf{e}_{j',k}\mathbf{p}^2_{(k',k)} \ .$$

From this there follows the fourth and the sixth assertion below. The first and the last assertion hold by Lemma 15(a), while the third and the seventh assertion hold by Lemma 15(d). The eighth assertion holds by Lemma 14(c). The fifth assertion follows from $j < j' < k' < k$ and from what was shown in the last paragraph.

$$\begin{aligned}
\mathbf{e}_{j,k}\mathbf{e}_{j,k'} &\equiv_2 \mathbf{p}_{(k',k)}\mathbf{e}_{j,k'}\mathbf{p}_{(k',k)}\mathbf{e}_{j',k'} \equiv_P \mathbf{p}_{(k',k)}\mathbf{e}_{j,k'}\mathbf{p}_{(k',k)}\mathbf{p}^2_{k-1}\mathbf{e}_{j',k'} \\
&\equiv_2 \mathbf{p}_{(k',k)}\mathbf{e}_{j,k'}\mathbf{p}_{(k',k)}\mathbf{p}_{(k',k)}\mathbf{e}_{j',k}\mathbf{p}_{(k',k)} \equiv' \mathbf{p}_{(k',k)}\mathbf{e}_{j',k}\mathbf{e}_{j',k}\mathbf{p}_{(k',k)} \\
&\equiv'' \mathbf{p}_{(k',k)}\mathbf{e}_{j',k}\mathbf{e}_{j',k}\mathbf{p}_{(k,k')} \equiv' \mathbf{p}_{(k',k)}\mathbf{e}_{j',k}\mathbf{p}_{(k',k)}\mathbf{p}_{(k',k)}\mathbf{e}_{j,k'}\mathbf{p}_{(k,k')} \\
&\equiv_2 \mathbf{p}^2_{k-1}\mathbf{e}_{j',k'}\mathbf{p}_{(k',k)}\mathbf{e}_{j,k'}\mathbf{p}_{(k,k')} \equiv_2 \mathbf{e}_{j',k'}\mathbf{p}^2_{k-1}\mathbf{p}_{(k',k)}\mathbf{e}_{j,k'}\mathbf{p}_{(k,k')} \\
&\equiv_P \mathbf{e}_{j',k'}\mathbf{p}_{(k',k)}\mathbf{e}_{j',k}\mathbf{p}_{(k,k')} \equiv_2 \mathbf{e}_{j',k'}\mathbf{e}_{j,k}
\end{aligned}$$

This concludes our proof that $\mathbf{e}_{j,k}\mathbf{e}_{j',k'} \equiv'' \mathbf{e}_{j',k'}\mathbf{e}_{j,k}$ whenever either $j < k < j' = k'$ or $j < j' < k < k'$ or $j < j' < k' < k$. Since $\mathbf{e}_{i,i'} = \mathbf{e}_{i',i}$ for any i and i', there follows that $\mathbf{e}_{j,k}\mathbf{e}_{j',k'} \equiv'' \mathbf{e}_{j',k'}\mathbf{e}_{j,k}$ whenever j, k, j', k' are distinct. □

Henceforth we let $\mathbf{F}(\Gamma(\tilde{\mathbf{e}}))$ be the subalgebra of $\mathbf{F}(\Gamma(\tilde{\mathbf{p}}, \mathbf{e}))$ that is generated by $\{\mathbf{e}_{j,k} : \{j,k\} \ne \{0\}\}$.

THEOREM 24. *Consider any v and w in $\mathbf{F}(\Gamma(\tilde{\mathbf{e}}))$.*
(a) $vw \equiv'' wv$.
(b) $v^{-1} \equiv'' v$.

Proof. Let $v = v_0 \dots v_{m-1}$ and $w = w_0 \dots w_{n-1}$ where each $v_{m'}$ and $w_{n'}$ is in $\mathbf{F}(\Gamma(\tilde{\mathbf{e}}))$.
(a) This follows from Lemma 23 by induction on $m + n$.
(b) By Lemma 16, Lemma 23, and induction on m. □

LEMMA 25. (a) $h(\mathbf{e}_{j,k}) = e_{j,k}$, *if* $\{j, k\} \neq \{0\}$.
(b) *The restriction of h to $\mathbf{F}(\Gamma(\tilde{\mathbf{e}}))$ is a homomorphism of $\mathbf{F}(\Gamma(\tilde{\mathbf{e}}))$ onto $\mathbf{S}_e = \langle S_e, \circ, ^{\smile} \rangle$.*

Proof. (a) By the definition of h and of $\mathbf{e}_{0,1}$, $h(\mathbf{e}) = h(\mathbf{e}_{0,1}) = e_{0,1}$. Now let $k \geq 1$ and assume as inductive hypothesis that $h(\mathbf{e}_{0,k}) = e_{0,k}$. Then, since $\mathbf{e}_{0,k+1} = \mathbf{p}_k \mathbf{e}_{0,k} \mathbf{p}_k$, there follows that

$$h(\mathbf{e}_{0,k+1}) = h(\mathbf{p}_k) \circ h(\mathbf{e}_{0,k}) \circ h(\mathbf{p}_k) = p_{(k,k+1)} \circ e_{0,k} \circ p_{(k,k+1)} = e_{0,k+1} \ .$$

Now let $0 \leq j < k$ and assume as inductive hypothesis that $h(\mathbf{e}_{j,k}) = e_{j,k}$. Then, since $\mathbf{e}_{j+1,k} = \mathbf{p}_j \mathbf{e}_{j,k} \mathbf{p}_j$, there follows that

$$h(\mathbf{e}_{j+1,k}) = h(\mathbf{p}_j) \circ h(\mathbf{e}_{j,k}) \circ h(\mathbf{p}_j) = p_{(j,j+1)} \circ e_{j,k} \circ p_{(j,j+1)} = e_{j+1,k} \ .$$

There now follows by induction that whenever $j < k$ then $h(\mathbf{e}_{j,k}) = e_{j,k}$ and hence, since $\mathbf{e}_{k,j} = \mathbf{e}_{j,k}$ and $e_{j,k} = e_{k,j}$, also $h(\mathbf{e}_{k,j}) = e_{k,j}$. Now let $j = k \neq 0$. Then $\mathbf{e}_{j,k} = \mathbf{e}_{j,j} = \mathbf{p}_{j-1}^2$ and hence $h(\mathbf{e}_{j,k}) = h(\mathbf{p}_{j-1}^2) = {}_{[j+1,\omega)}\overset{\circ}{1} = e_{j,j} = e_{j,k}$.
(b) This follows from part (a). □

We finally turn to the congruence $\equiv$ on $\mathbf{F}(\Gamma(\tilde{\mathbf{p}}, \mathbf{e}))$ that is generated by $\equiv_P \cup E_1 \cup E_2 \cup PE_1 \cup \dots \cup PE_4$. The first assertion of the next lemma is similar to Theorem 1.3.7 in [HMT I].

LEMMA 26. $\mathbf{e}_{i,j}\mathbf{e}_{j,k} \equiv \mathbf{e}_{i,j}\mathbf{e}_{i,k} \equiv \mathbf{e}_{i,j}\mathbf{e}_{j,k}\mathbf{e}_{i,k}$ *for any i, j, k.*

Proof. To prove the first assertion, first assume that $j = i$. Then the assertion holds trivially. Secondly, assume that $i = k$ and hence $i \neq 0$. Then, the second assertion below holds by Lemma 17 and the third follows from $\mathbf{p}_{(i,j)} \equiv_P \mathbf{p}_{(i,j)}\mathbf{p}_{i-1}^2$ and from Lemmas 18 and 16.

$$\mathbf{e}_{i,j}\mathbf{e}_{j,k} = \mathbf{e}_{i,j}\mathbf{e}_{j,i} \equiv \mathbf{e}_{i,j} \equiv \mathbf{e}_{i,j}\mathbf{p}_{i-1}^2 = \mathbf{e}_{i,j}\mathbf{e}_{i,i} = \mathbf{e}_{i,j}\mathbf{e}_{i,k} \ .$$

The last line also shows that $\mathbf{e}_{j,i}\mathbf{e}_{i,j} \equiv \mathbf{e}_{j,i}\mathbf{e}_{j,j}$ and hence $\mathbf{e}_{j,i}\mathbf{e}_{j,j} \equiv \mathbf{e}_{j,i}\mathbf{e}_{i,j}$. Now thirdly assume that $j = k$. There then follows that, again,

$$\mathbf{e}_{i,j}\mathbf{e}_{j,k} = \mathbf{e}_{i,j}\mathbf{e}_{j,j} \equiv \mathbf{e}_{j,i}\mathbf{e}_{i,j}\ \mathbf{e}_{i,j}\mathbf{e}_{i,j} = \mathbf{e}_{i,j}\mathbf{e}_{i,k}\ .$$

Finally assume that i, j, k are distinct. First, also assume that $j > i$ and $j > k$. Then of the following assertions the first and third hold by Lemma 23, the second and fourth by Lemmas 18 and 16, and the fifth by Lemma 15(d).

$$\begin{aligned} \mathbf{e}_{i,j}\mathbf{e}_{j,k} &\equiv \mathbf{e}_{j,k}\mathbf{e}_{i,j} \equiv \mathbf{e}_{j,k}\mathbf{e}_{i,j}\mathbf{p}_{(i,j)} \equiv \mathbf{e}_{i,j}\mathbf{e}_{j,k}\mathbf{p}_{(i,j)} \\ &\equiv \mathbf{e}_{i,j}\mathbf{p}_{(i,j)}\mathbf{e}_{j,k}\mathbf{p}_{(i,j)} \equiv \mathbf{e}_{i,j}\mathbf{p}_{j-1}^{2}\mathbf{e}_{i,k} \equiv \mathbf{e}_{i,j}\mathbf{e}_{i,k} \end{aligned}$$

If $j \leq i$ or $j \leq k$, then by Lemmas 15(a), 15(b), or 21(a), $\mathbf{p}_{(i,j)}\mathbf{e}_{j,k}\mathbf{p}_{(i,j)} \equiv \mathbf{e}_{i,k}$, and $\mathbf{e}_{i,j}\mathbf{e}_{j,k} \equiv \mathbf{e}_{i,j}\mathbf{e}_{i,k}$ by a similar, but somewhat simpler, argument. This concludes the proof of our first assertion. From it there follows the third assertion below. The first assertion below holds by Lemma 17, and the second and fourth by Lemma 23.

$$\mathbf{e}_{i,j}\mathbf{e}_{j,k} \equiv \mathbf{e}_{i,j}\mathbf{e}_{j,k}\mathbf{e}_{j,k} \equiv \mathbf{e}_{j,k}\mathbf{e}_{i,j}\mathbf{e}_{j,k} \equiv \mathbf{e}_{j,k}\mathbf{e}_{i,j}\mathbf{e}_{i,k} \equiv \mathbf{e}_{i,j}\mathbf{e}_{j,k}\mathbf{e}_{i,k}\ .$$

□

Consider any words v and w in $F(\Gamma(\tilde{\mathbf{e}}))$. Then for some $m \geq 1$ and some $n \geq 1$, v is a sequence $v_0 \dots v_{m-1}$ and w is a sequence $w_0 \dots w_{n-1}$ such that each term $v_{m'}$ of v and each term $w_{n'}$ of w is a word $\mathbf{e}_{j,k}$ in $\Gamma(\tilde{\mathbf{e}})$. We shall say that w ***at least matches*** v if and only if the following two conditions are satisfied.

(i) For every $j \neq k$, if $\mathbf{e}_{j,k}$ is one or more of the terms v_m of v then it is also one or more of the terms w_n of w.

(ii) For every j, if $\mathbf{e}_{j,j}$ is one or more of the terms $v_{m'}$ of v then, for some $k \geq j$, $\mathbf{e}_{k,k}$ is one or more of the terms $w_{n'}$ of w.

Evidently, if w at least matches v then $h(w) \subseteq h(v)$. If w at least matches v and v at least matches w then they shall ***match each other.*** Evidently, if v and w match each other, then $h(v) = h(w)$.

A word $w = w_0 \dots w_{n-1}$ in $F(\Gamma(\tilde{\mathbf{e}}))$ shall be ***closed with respect to transitivity*** if and only if, for any i, j, and k, and for any $n_0 < n$ and $n_1 < n$, whenever $w_{n_0} = \mathbf{e}_{i,j} = \mathbf{e}_{j,i}$ and $w_{n_1} = \mathbf{e}_{j,k} = \mathbf{e}_{k,j}$, then there is some $n_2 < n$ such that $w_{n_2} = \mathbf{e}_{i,k} = \mathbf{e}_{k,i}$.

LEMMA 27. *For any v in $F(\Gamma(\tilde{\mathbf{e}}))$ there is some w in $F(\Gamma(\tilde{\mathbf{e}}))$ that is closed with respect to transitivity such that $v \equiv w$. Moreover w can be so chosen that, if w' is any word in $F(\Gamma(\tilde{\mathbf{e}}))$ that is closed with respect to transitivity and at least matches v, then it also at least matches w.*

Proof. Suppose that $v = v_0 \ldots v_{n-1}$ is a word in $F(\Gamma(\tilde{\mathbf{e}}))$ for which there are i, j, k, m_0, m_1 such that $i \neq j$, $j \neq k$, $m_0 < m$, and $m_1 < m$ such that $v_{m_0} = \mathbf{e}_{i,j}$ and $v_{m_1} = \mathbf{e}_{j,k}$, but for which there is no $m_2 < m$ such that $v_{m_2} = \mathbf{e}_{i,k}$. Then, by Lemmas 17 and 23 or, respectively, by Lemma 26 there hold:

$$v \equiv v\mathbf{e}_{i,j}\mathbf{e}_{j,k} \equiv v\mathbf{e}_{i,j}\mathbf{e}_{j,k}\mathbf{e}_{i,k} \; .$$

By adding to v in this manner sufficiently often one obtains from v a word w in $F(\Gamma(\tilde{\mathbf{e}}))$ that is closed with respect to transitivity such that $v \equiv w$. Now consider any word w' in $F(\Gamma(\tilde{\mathbf{e}}))$ that at least matches v and that is closed with respect to transitivity. Then any $\mathbf{e}_{i,k}$ that is a term of w but not of v is also a term of w'. There follows that if $i \neq k$ and $\mathbf{e}_{i,k}$ is a term of w then it is also a term of w'. Now consider any term $\mathbf{e}_{j,j}$ of w. If $\mathbf{e}_{j,j}$ is not a term of v then, as we just saw, it is also a term of w'. If $\mathbf{e}_{j,j}$ is a term of v then, since w' at least matches v, there is some $k \geq j$ such that $\mathbf{e}_{k,k}$ is a term of w'. There now follows that w' at least matches w. □

LEMMA 28. *Consider any v and w in $F(\Gamma(\tilde{\mathbf{e}}))$ such that v is closed with respect to transitivity. Then $h(v) \subseteq h(w)$ implies that v at least matches w.*

Proof. We will argue by contraposition. Assume that v is closed with respect to transitivity but does not at least match w. First, suppose that, for some $j \neq k$, $\mathbf{e}_{j,k}$ is a term of w but not a term of v. Since v is closed with respect to transitivity therefore the sets $I = \{i : \mathbf{e}_{i,j} \text{ is a term of } v\}$ and $I' = \{i' : \mathbf{e}_{i',k} \text{ is a term of } v\}$ are disjoint. Since $\overline{\overline{U}} \geq 2$, therefore in U there are elements u and u' such that $u \neq u'$. For some $n = \{0, \ldots n-1\}$ such that $i < n$ and $i' < n$ for any $\mathbf{e}_{i,i'}$ in v, let $x = \langle x_0, \ldots, x_{n-1} \rangle$ be the sequence in ${}^nU$ such that, if $m \in I$ then $x_m = u$ and if $m \in n \cap -I$ then $x_m = u'$. Then $x \in h(\mathbf{e}_{i,i'})$ for every term $\mathbf{e}_{i,i'}$ of v and hence $x \in h(v)$. In contrast, $x \notin h(\mathbf{e}_{j,k})$ and hence $x \notin h(w)$. Hence $h(v) \not\subseteq h(w)$.

Now suppose that there is a term $\mathbf{e}_{j,j}$ of w such that, for no $k \geq j$, $\mathbf{e}_{k,k}$ is a term of v. Since v is closed with respect to transitivity, there follows that $i < j$ and $i' < j$ for any term $\mathbf{e}_{i,i'} = \mathbf{e}_{i',i}$ in v. For some u in U, let $x = \langle x_0, \ldots, x_{j-1} \rangle$ be the sequence of length j such that $x_m = u$ for each $m < j$. Then $x \in h(\mathbf{e}_{i,i'})$ for every term $\mathbf{e}_{i,i'}$ of v and hence $x \in h(v)$. In contrast, $x \notin h(\mathbf{e}_{j,j})$ and hence $x \notin h(w)$. Hence again $h(v) \not\subseteq h(w)$. □

THEOREM 29. *For any v and w in $F(\Gamma(\tilde{\mathbf{e}}))$,*

$$h(v) = h(w) \quad \textit{if and only if} \quad v \equiv w.$$

Proof. By Lemma 12(b) or by Lemma 25(b), if $v \equiv w$ then $h(v) = h(w)$. Now assume that $h(v) = h(w)$. By Lemma 27, there are v' and w' in $F(\Gamma(\tilde{\mathbf{e}}))$ that are closed with respect to transitivity such that $v \equiv v'$ and $w \equiv w'$. Since $v \equiv v'$ therefore $h(v) = h(v')$. Since $w \equiv w'$, therefore $h(w) = h(w')$. Hence $h(v') = h(w')$. By Lemma 28, v' and w' match each other. From Lemmas 17 and 23 and since $\mathbf{e}_{j,j}\mathbf{e}_{k,k} \equiv_P \mathbf{e}_{k,k}$ whenever $j \leq k$, there follows that $v' \equiv w'$. There now follows that $v \equiv w$. □

Given any $n \geq 2$, we now introduce the following abbreviation. Consider any $i_0, \ldots, i_n$ in ω such that $i_m \neq i_{m'}$, if $0 \leq m < m' \leq n$. Then we let

$$\mathbf{p}_{(i_0,\ldots,i_{n-1})} = \mathbf{p}_{(i_{n-1},i_n)}\mathbf{p}_{(i_0,\ldots,i_{n-1})} \ .$$

When using this abbreviation we will tacitly assume that $i_0, \ldots, i_n$ are distinct.

Consider any $w = w_0 \ldots w_{s-1}$ in $F(\Gamma(\tilde{\mathbf{e}}))$ where $s \geq 1$ and, for each $r < s$, $w_r = \mathbf{e}_{j_r,k_r} = \mathbf{e}_{k_r,j_r}$ for some j_r and k_r such that $\{j_r, k_r\} \neq \{0\}$. Also, consider any $x = x_0 \ldots x_{t-1}$ in $F(\Gamma(\tilde{\mathbf{p}}))$ where $t \geq 1$ and where each $x_{t'}$, $t' < t$, is either some $\mathbf{p}_i^2$ or some $\mathbf{p}_{(i_0,\ldots,i_n)}$. Then we shall say that w ***covers*** x if and only if the following hold, for every $t' < t$:

(i) If $x_{t'} = \mathbf{p}_i^2$ then, for some $r < s$, $w_r = \mathbf{e}_{j_r,k_r}$ where either $j_r > i$ or $k_r > i$.
(ii) If $x_{t'} = \mathbf{p}_{(i_0,\ldots,i_n)}$, then, for each $m < n$, there is some $r < s$ such that $\mathbf{e}_{j_r,k_r} = \mathbf{e}_{i_m,i_{m+1}}$.

LEMMA 30. *Consider any W in S_e and any X in S_p.*
(a) *The following five conditions are equivalent to each other:*

$$W \subseteq X, \quad W \subseteq W \circ X, \quad W = W \circ X, \quad W \subseteq X \circ W, \quad W = X \circ W \ .$$

(b) *Let $w = w_0 \ldots w_{s-1}$ be any word in $F(\Gamma(\tilde{\mathbf{e}}))$ as described above such that $W = h(w)$ and let $x = x_0 \ldots x_{t-1}$ be any word in $F(\Gamma(\tilde{\mathbf{p}}))$ as described above such that $X = h(x)$. Also, let w be closed with respect to transitivity. Then the following four conditions are equivalent to each other:*

$$W \subseteq X, \quad w \text{ covers } x, \quad wx \equiv w, \quad xw \equiv w \ .$$

Proof. (a) Assume that $W \subseteq W \circ X$. Since $W \subseteq {}_{[0,\omega)}\overset{\circ}{1}$, therefore $W \circ X \subseteq X$. There follows that $W \subseteq X$.

Assume that $W \subseteq X$. Suppose that $W \neq W \circ X$. Then, since $Rg\, W = Do\, W \subseteq Do\, X$ and hence $Do(W \circ X) = Do\, W$, there would be some $\langle y, y\rangle$ in W and $\langle y, z\rangle$ in X such that $\langle y, z\rangle$ is not in W and hence such that $z \neq y$. Since X is one-one, therefore $\langle y, y\rangle$ would not be in X and hence W would not be included in X. There follows that $W = W \circ X$.

Evidently, $W = W \circ X$ implies that $W \subseteq W \circ X$. There now follows that the conditions $W \subseteq X$, $W \circ X \subseteq X$, $W \circ X = X$ are equivalent to each other. By a similar argument, the conditions $W \subseteq X$, $X \circ W \subseteq X$, $X \circ W = X$ are equivalent to each other.

(b) Assume that w does not cover x because for some $t' < t$, $x_{t'} = \mathbf{p}_i^2 = \mathbf{p}_{(i,i+1)}^2$ but there is no $r < s$ such that $w_r = \mathbf{e}_{j,h_r}$ with either $j_r > i$ or $k_r > i$. Then $X = h(x) \subseteq {}_{[i+2,\omega)}\overset{\circ}{1}$ and hence $X \cap {}_{[i+1]}\overset{\circ}{1} = \emptyset$. In contrast, $e_{0,1} \circ \cdots \circ e_{i-1,i} \subseteq h(w) = W$ and hence $W \cap {}_{[i+1]}\overset{\circ}{1} \neq \emptyset$. Hence $W \not\subseteq X$.

Now assume that w does not cover x because for some $t' < t$, $x_{t'} = \mathbf{p}_{(i_0,\ldots,i_n)}$ but there is some $m < n$ such that no $w_r = \mathbf{e}_{j_r,k_r} = \mathbf{e}_{k_r,j_r}$ is $\mathbf{e}_{i_m,i_{m+1}}$. Also assume that w is closed with respect to transitivity. Then there are no $r < s$ and $r' < s$ such that $j_r = i_m, j_{r'} = i_{m+1}$, and $k_r = k_{r'}$. Since $\overline{\overline{U}} \geq 2$ there is a sequence y in ${}^{[0,\omega)}U$ such that, for distinct elements u and u' of U, the following holds for any k such that, for some $r < s$ and some j, $w_r = \mathbf{e}_{j,k} = \mathbf{e}_{k,j}$:

If there is some $r' < s$ such that $w_{r'} = \mathbf{e}_{i_m,k}$,
then $y_k = u$; otherwise, $y_k = u'$.

Thus, for every $r < s$, if $w_r = \mathbf{e}_{j_r,k_r}$ then either both $y_{j_r} = u$ and $y_{k_r} = u$ or else both $y_{j_r} = u'$ and $y_{k_r} = u'$. There follows that $\langle y, y\rangle \in h(w_r)$ for each $r < s$ and hence that $\langle y, y\rangle \in h(w) = W$. Now from the definition of y there follows that, in particular, $y_{i_m} = u$ and $y_{i_{m+1}} = u'$. Let $\langle y, z\rangle \in h(\mathbf{p}_{(i_0,\ldots,i_n)})$. Then $z_{i_{m+1}} = y_m = u \neq u' = y_{i_{m+1}}$. Hence $z \neq y$. Since $h(\mathbf{p}_{(i_0,\ldots,i_n)})$ is one-one, therefore $\langle y, y\rangle \notin h(\mathbf{p}_{(i_0,\ldots,i_n)})$. Hence $\langle y, y\rangle \notin h(x) = X$. There follows that again $W \not\subseteq X$. This concludes our proof that $W \subseteq X$ implies that w covers x.

Assume that w covers x. Consider any $x_{t'}$ such that $t' < t$. First suppose that $x_{t'} = \mathbf{p}_i^2$. Since w covers x there is some $r < s$ such that $w_r = \mathbf{e}_{j_r,k_r}$ and either $j_r > i$ or $k_r > i$. Now $j_r > i$ implies that

$$\mathbf{e}_{j_r,k_r}\mathbf{p}_i^2 \equiv \mathbf{e}_{j_r,k_r}\mathbf{p}_{j_r-1}^2\mathbf{p}_i^2 \equiv_P \mathbf{e}_{j_r,k_r}\mathbf{p}_{j_r-1}^2 \equiv \mathbf{e}_{j_r,k_r}\ .$$

Similarly, $k_r > i$ implies that $\mathbf{e}_{j_r,k_r}\mathbf{p}_i^2 \equiv \mathbf{e}_{j_r,k_r}$. Also, by Lemmas 17 and 23, $w \equiv w\mathbf{e}_{j_r,k_r}$. There follows that

$$wx_{t'} = w\mathbf{p}_i^2 \equiv w\mathbf{e}_{j_r,k_r}\mathbf{p}_i^2 \equiv w\mathbf{e}_{j_r,k_r} \equiv w\ .$$

Now suppose that $x_{t'} = \mathbf{p}_{(i_0,\ldots,i_n)}$. First, also suppose that $n \geq 2$. Since w covers x therefore, there is some $r < s$ such that $\mathbf{e}_{j_r,k_r} = \mathbf{e}_{i_{n-1},i_n}$. Hence by Lemmas

17 and 23, $w \equiv w\mathbf{e}_{i_{n-1},i_n}$. Also, by Lemmas 16 and 18, $\mathbf{e}_{i_{n-1},i_n}\mathbf{p}_{(i_{n-1},i_n)} \equiv \mathbf{e}_{i_{n-1},i_n}$. Further, $\mathbf{p}_{(i_0,\ldots,i_n)} = \mathbf{p}_{(i_{n-1},i_n)}\mathbf{p}_{(i_0,\ldots,i_{n-1})}$. There now follows:

$$\begin{aligned} w\mathbf{p}_{(i_0,\ldots,i_n)} &\equiv w\mathbf{e}_{i_{n-1},i_n}\mathbf{p}_{(i_{n-1},i_n)}\mathbf{p}_{(i_0,\ldots,i_{n-1})} \\ &= w\mathbf{e}_{i_{n-1},i_n}\mathbf{p}_{(i_0,\ldots,i_{n-1})} \equiv w\mathbf{p}_{(i_0,\ldots,i_{n-1})} \ . \end{aligned}$$

If $n \geq 3$, then a similar argument shows that $w\mathbf{p}_{(i_0,\ldots,i_{n-1})} \equiv w\mathbf{p}_{(i_0,\ldots,i_{n-2})}$. By repeating this argument sufficiently often one shows that $w\mathbf{p}_{(i_0,\ldots,i_n)} \equiv w\mathbf{p}_{(i_0,i_1)}$. A similar, but simpler, argument then shows that $w\mathbf{p}_{(i_0,i_1)} \equiv w$. There follows that $wx_{t'} = w\mathbf{p}_{(i_0,\ldots,i_n)} \equiv w$. If $n = 1$ and hence $\mathbf{p}_{(i_0,\ldots,i_n)} = \mathbf{p}_{(i_0,i_1)}$, then the simpler argument just mentioned shows that $wx_{t'} = w\mathbf{p}_{(i_0,i_1)} \equiv w$. Hence, in all cases, $wx_{t'} \equiv w$.

From the last two paragraphs there follows that $wx_{t'} \equiv w$ for any $t' < t$. By induction there follows that $wx \equiv wx_0 \ldots x_{t-1} \equiv w$. This concludes our proof that if w covers x then $wx \equiv w$. An argument that uses $\mathbf{p}_{(i_0,\ldots,i_n)} \equiv_P \mathbf{p}_{(i_1,\ldots,i_n)}\mathbf{p}_{(i_0,i_1)}$ but otherwise is similar shows that if w covers x then $xw \equiv w$.

Assume that $wx \equiv w$. Then, by Lemma 13, $h(x) \circ h(w) = h(w)$. Hence $X \circ W = W$. Then, by part (a), $W \subseteq X$. There now follows that the four conditions concerned are equivalent to each other. □

We continue to let $\equiv$ be the congruence on $\mathbf{F}(\Gamma(\tilde{\mathbf{p}}, \mathbf{e}))$ that is generated by $\equiv_P \cup E_1 \cup E_2 \cup PE_1 \cup \cdots \cup PE_4$. We can now give a presentation of $\mathbf{S}_{pe}$.

THEOREM 31. *For any v and v' in $\mathbf{F}(\Gamma(\tilde{\mathbf{p}}, \mathbf{e}))$, $h(v) = h(v')$ if and only if $v \equiv v'$. Hence $\mathbf{F}(\Gamma(\tilde{\mathbf{p}}, \mathbf{e}))/\equiv$ is isomorphic to $\mathbf{S}_{pe}$.*

Proof. By Lemma 13, if $v \equiv v'$ then $h(v) = h(v')$. Now assume that $h(v) = h(v')$. By Theorem 20, for some w and w' in $F(\Gamma(\tilde{\mathbf{e}}))$ and some y and z in $F(\Gamma(\tilde{\mathbf{p}}))$, $v \equiv wy$ and $v' \equiv w'z$. Moreover, if $gd\,y = \langle s+1, s+1\rangle$, then $y \equiv_P \mathbf{p}_{s-1}^2 y \equiv \mathbf{e}_{s,s}y$ and $wy \equiv w\mathbf{e}_{s,s}y$. Using $w\mathbf{e}_{s,s}$ in place of w we can therefore assume without loss of generality that $gd\,w \geq gd\,y$. Similarly, we can assume that $gd\,w' \geq gd\,z$. Further, by Lemma 27, we can assume that w and w' are closed with respect to transitivity. From $h(v) = h(v')$, $v \equiv wy$, $v' \equiv w'z$, and Lemma 13 there follows that $h(wy) = h(w'z)$ and hence that $h(w) \circ h(y) = h(w') \circ h(z)$.

Since $gd(h(y)) \leq gd(h(w))$ and $Rg(h(w)) = Do(h(w))$, therefore $Do(h(w)) \subseteq Do(h(w) \circ h(y))$. Now suppose that $h(w) \not\subseteq h(w')$. Since $h(w)$ and $h(w')$ are included in ${}_{[0,\omega)}\overset{\circ}{1}$, there would follow that $Do(h(w)) \not\subseteq Do(h(w'))$, hence that $Do(h(w) \circ h(y)) \not\subseteq Do(h(w') \circ h(z))$, and hence that $h(w) \circ h(y) \not\subseteq h(w') \circ h(z)$. Since

$h(w) \circ h(y) \subseteq h(w') \circ h(z)$, there follows that $h(w) \subseteq h(w')$. A similar argument shows that $h(w') \subseteq h(w)$. Hence $h(w) = h(w')$.

There follows that $h(w) \circ h(y) = h(w) \circ h(z)$. Since $h(w) = h(w')$ and $gd\, w' \geq gd\, z = gd\, z^{-1}$, there follows that

$$h(w) \circ h(y) \circ (h(z))^{\smile} = h(w) \circ h(z) \circ (h(z))^{\smile} = h(w) \ .$$

Let $x = yz^{-1}$. Then $h(x) = h(y) \circ h(z^{-1}) = h(y) \circ (h(z))^{\smile}$. Hence, by what has just been shown, $h(w) \circ h(x) = h(w)$. From Lemma 30 there follows that $w \equiv wx = wyz^{-1}$. Hence $wz \equiv wyz^{-1}z \equiv wy$. Since $h(w) = h(w')$ therefore, by Theorem 29, $w \equiv w'$. From $v \equiv wy$ and $v' \equiv w'z$, there now follows that $v \equiv v'$. □

We now let $\lesssim$ be the least quasiordering of $F(\Gamma(\tilde{\mathbf{p}}, \mathbf{e}))$ that is compatible with $\mathbf{F}(\Gamma(\tilde{\mathbf{p}}, \mathbf{e}))$ and includes $\equiv \cup P_4 \cup PE_5$, where P_4 is the relation on $F(\Gamma(\tilde{\mathbf{p}}))$ that was defined just before Lemma V.14 and where PE_5 is the following relation on $F(\Gamma(\tilde{\mathbf{p}}, \mathbf{e}))$:

$$PE_5 \;=\; \{\langle \mathbf{e}, \mathbf{p}_0^2 \rangle\} \ .$$

Since $h(\mathbf{e}) = e_{0,1}$ and $h(\mathbf{p}_0^2) = p_{0,1}^2 = {}_{[2,\omega)}\overset{\circ}{1}$, therefore $h(\mathbf{e}) \subseteq h(\mathbf{p}_0^2)$. As we saw earlier, $v \equiv w$ implies that $h(v) = h(w)$ and $\langle v, w \rangle \in P_4$ implies that $h(v) \subseteq h(w)$. There follows that $v \lesssim w$ implies that $h(v) \subseteq h(w)$.

LEMMA 32. *Consider any v and w in $F(\Gamma(\mathbf{e}))$. Then $h(v) \subseteq h(w)$ implies that $v \lesssim w$. Hence $h(v) \subseteq h(w)$ if and only if $v \lesssim w$.*

Proof. Let $k \geq 1$ and assume as inductive hypothesis that $\mathbf{e}_{0,k} \lesssim \mathbf{p}_0^2$. Then

$$\mathbf{e}_{0,k+1} \;=\; \mathbf{p}_k \mathbf{e}_{0,k} \mathbf{p}_k \;\lesssim\; \mathbf{p}_k \mathbf{p}_0^2 \mathbf{p}_k \;\equiv_P\; \mathbf{p}_k^2 \;\lesssim\; \mathbf{p}_0^2 \ .$$

Now let $j < k+1$ and assume as inductive hypothesis that $\mathbf{e}_{j,k} \lesssim \mathbf{p}_0^2$. Then

$$\mathbf{e}_{j+1,k} \;=\; \mathbf{p}_j \mathbf{e}_{j,k} \mathbf{p}_j \;\lesssim\; \mathbf{p}_j \mathbf{p}_0^2 \mathbf{p}_j \;\equiv_P\; \mathbf{p}_j^2 \;\lesssim\; \mathbf{p}_0^2 \ .$$

Since $\mathbf{e} = \mathbf{e}_{0,1}$ there follows by induction that, if $0 \leq j < k$, then $\mathbf{e}_{j,k} \lesssim \mathbf{p}_0^2$. Since $\mathbf{e}_{j,k} = \mathbf{e}_{k,j}$ therefore, if $j \neq k$, then $\mathbf{e}_{j,k} \lesssim \mathbf{p}_0^2$. As to $\mathbf{e}_{j,j}$ where $j \geq 1$, there holds $\mathbf{e}_{j,j} = \mathbf{p}_{j-1}^2 \lesssim \mathbf{p}_0^2$. There now follows that

$$\mathbf{e}_{j,k} \;\lesssim\; \mathbf{p}_0^2 \text{ , for any } \mathbf{e}_{j,k} \text{ in } \; F(\Gamma(\mathbf{e})).$$

Since $\mathbf{p}_0^2 \mathbf{p}_0^2 \equiv_P \mathbf{p}_0^2$, induction on the length of w then shows:

$$w \;\lesssim\; \mathbf{p}_0^2 \ , \quad \text{for any } w \text{ in } \; F(\Gamma(\mathbf{e})).$$

Consider any v and w in $F(\Gamma(\mathbf{e}))$ such that $h(v) \subseteq h(w)$. By Lemma 27, there is some v' in $F(\Gamma(\mathbf{e}))$ such that $v \equiv v'$ and such that v' is closed with respect to transitivity. Since $h(v') \subseteq h(w)$ therefore, by Lemma 28, v' at least matches w. Now let $w = w_0 \ldots w_{n-1}$ where $n \geq 1$ and where, for each $n' < n$, $w_{n'}$ is some $\mathbf{e}_{j,k}$. Then, for each $n' < n$, v' also at least matches $w_0 \ldots w_{n'}$. From the definition of this notion and that of $\mathbf{e}_{(j,j)}$, and from $\mathbf{e}_{j,k} \lesssim \mathbf{p}_0^2$ for each $\mathbf{e}_{j,k}$ in $F(\Gamma(\tilde{\mathbf{e}}))$, there follows that $v' \lesssim w_0$ and that, for any $n' < n-1$, if $v' \lesssim w_0 \ldots w_{n'-1}$ then also $v' \lesssim w_0 \ldots w_{n'-1}w_{n'}$. By induction there follows that $v' \lesssim w_0 \ldots w_{n-1} = w$. Since $v \equiv v'$, therefore $v \lesssim w$. □

We now give a presentation of $\langle \mathbf{S}_{pe}, \subseteq \rangle = \langle S_{pe}, \circ, ^{\smile}, \subseteq \rangle$.

THEOREM 33. *For any v and v' in $\mathbf{F}(\Gamma(\tilde{\mathbf{p}}, \mathbf{e}))$, $h(v) \subseteq h(v')$ if and only if $v \lesssim v'$. There follows that $\langle \mathbf{F}(\Gamma(\tilde{\mathbf{p}}, \mathbf{e})), \lesssim \rangle / \equiv$ is isomorphic to $\langle \mathbf{S}_{pe}, \subseteq \rangle$.*

Proof. We already saw that $v \lesssim v'$ implies that $h(v) \subseteq h(v')$. Now assume that $h(v) \subseteq h(v')$. As we saw in the proof of Theorem 31, there are w and w' in $\mathbf{F}(\Gamma(\tilde{\mathbf{e}}))$ and y and z in $\mathbf{F}(\Gamma(\tilde{\mathbf{p}}))$ such that $v \equiv wy$ and $v' \equiv w'z$, such that w and w' are closed with respect to transitivity, and such that $gd\,w \gtrsim gd\,y$ and $gd\,w' \gtrsim gd\,z$. As we also saw there, $h(w) \subseteq h(w')$. Since $h(w)$ and $h(w')$ are included in ${}_{[0,\omega)}\overset{\circ}{1}$, therefore $h(w) \circ h(w') = h(w) \cap h(w') = h(w)$. Since $h(v) \subseteq h(v')$ therefore, by Lemma 13, $h(w) \circ h(y) \subseteq h(w') \circ h(z)$. There now follows:

$$\begin{aligned} h(w) \circ h(y) &= h(w) \circ h(w) \circ h(y) \subseteq h(w) \circ h(w') \circ h(z) \\ &= h(w) \circ h(z)\,. \end{aligned}$$

Since $gd\,w \gtrsim gd\,y = gd\,y^{-1}$, there follows:

$$h(w) = h(w) \circ h(y) \circ (h(y))^{\smile} \subseteq h(w) \circ h(z) \circ (h(y))^{\smile}$$

Let $x = zy^{-1}$. Then $h(x) = h(z) \circ h(y^{-1}) = h(z) \circ (h(y))^{\smile}$. Hence, by what has just been shown, $h(w) \subseteq h(w) \circ h(x)$. From Lemma 30 there follows that $w \equiv wx = wzy^{-1}$. Hence $wy \equiv wzy^{-1}y \equiv wz$. By Lemma 32, from $h(w) \subseteq h(w')$ there follows that $w \lesssim w'$. From $v \equiv wy$, $wy \equiv wz$, $w \lesssim w'$, and $w'z \equiv v'$, there follows that $v \lesssim v'$. □

Let $\equiv^+$ be the congruence on $\mathbf{F}(\Gamma(\tilde{\mathbf{p}}, \mathbf{e}))$ that is generated by $\equiv \cup P_4$. Thus, $\equiv^+$ is also the congruence of $\mathbf{F}(\Gamma(\tilde{\mathbf{p}}, \mathbf{e}))$ that is obtained by replacing, in the definition of $\equiv$, $P_4^{\#}$ by P_4. Let $\lesssim^+$ be the least quasi-ordering of $F(\Gamma(\tilde{\mathbf{p}}, \mathbf{e}))$ that is compatible with $\mathbf{F}(\Gamma(\tilde{\mathbf{p}}, \mathbf{e}))$ and includes $\lesssim \smile P_4^{\smile}$. Thus $\lesssim^+$ is also the least quasi-ordering of $F(\Gamma(\tilde{\mathbf{p}}, \mathbf{e}))$ that is compatible with $\mathbf{F}(\Gamma(\tilde{\mathbf{p}}, \mathbf{e}))$ and includes $\equiv^+ \smile PE_5$. We now give a presentation of ${}_{[\omega]}\mathbf{S}_{pe} = \langle {}_{[\omega]}S_{pe}, \circ, ^{\smile} \rangle$ and one of $\langle {}_{[\omega]}\mathbf{S}_{pe}, \subseteq \rangle = \langle S_{pe}, \circ, ^{\smile}, \subseteq \rangle$.

THEOREM 34. *Let h' be the homomorphism of $\mathbf{F}(\Gamma(\tilde{\mathbf{p}}, \mathbf{e}))$ onto ${}_{[\omega]}\mathbf{S}_{pe}$ such that $h'(\mathbf{e}) = \overset{\omega}{\rightarrow} h(\mathbf{e}) = \overset{\omega}{\rightarrow} e_{0,1}$ and $h'(\mathbf{p}_i) = \overset{\omega}{\rightarrow} h(\mathbf{p}_i) = \overset{\omega}{\rightarrow} p_i$ for each $i < \omega$.*

(a) *$h'(v) = h'(w)$ if and only if $v \equiv^+ w$. There follows that $\mathbf{F}(\Gamma(\tilde{\mathbf{p}}, \mathbf{e}))/\equiv^+$ is isomorphic to ${}_{[\omega]}\mathbf{S}_{pe} = \langle {}_{[\omega]}S_{pe}, \circ, {}^{\smile} \rangle$.*

(b) *$h'(v) \subseteq h'(w)$ if and only if $v \leq^+ w$. There follows that $\langle \mathbf{F}(\Gamma(\tilde{\mathbf{p}}, \mathbf{e})), \leq^+ \rangle/\equiv^+$ is isomorphic to $\langle {}_{[\omega]}\mathbf{S}_{pe}, \subseteq \rangle$.*

Proof. (a) Evidently, $\overset{\omega}{\rightarrow} p_i^2 = \overset{\omega}{\rightarrow} p_0^2 = {}_{[\omega]}\overset{\circ}{1}$ for any $i < \omega$. There follows that $(\overset{\omega}{\rightarrow} p_{i+1})^2 \circ \overset{\omega}{\rightarrow} p_i = \overset{\omega}{\rightarrow} p_i$ for each i. Hence $P_4 \subseteq h' \circ h'^{\smile}$. Since $\equiv \subseteq h^{\smile} \circ h$, as we saw earlier, and since $h \circ h^{\smile} \subseteq h' \circ h'^{\smile}$, there follows that $\equiv^+ \subseteq h' \circ h'^{\smile}$.

Since P_4 is included in $\equiv^+$ therefore, by induction, for each $i \geq 0$ there holds $\mathbf{p}_i^2 \equiv^+ \mathbf{p}_0^2$.

Assume that $h'(v) = h'(w)$. Then for some $n \geq 0$, either $\overset{n}{\rightarrow}(h(v)) = h(w)$ or $\overset{n}{\rightarrow}(h(w)) = h(v)$. Suppose that $\overset{n}{\rightarrow}(h(v)) = h(w)$. Recall that $gd\, w = \langle s+2, s+2 \rangle$ for some $s \geq 0$. Then, by Theorem II.7(f),

$$h(\mathbf{p}_s^2 v) = p_s^2 \circ h(v) = {}_{[s+2,\omega)}\overset{\circ}{1} \circ h(v) = h(w) \ .$$

Hence, by Theorem 31, $\mathbf{p}_s^2 v \equiv w$. Since $\mathbf{p}_0^2 v \equiv v$, there follows that

$$v \equiv \mathbf{p}_0^2 v \equiv^+ \mathbf{p}_s^2 v \equiv w \ .$$

If $\overset{n}{\rightarrow}(h(w)) = h(v)$, then a similar argument shows that $v \equiv^+ w$.

(b) Since $P_4^{\smile}$ and $\leq$ are included in $h' \circ h'^{\smile}$, therefore so is $\leq^+$. Now assume that $h'(v) \subseteq h'(w)$. Let $gd(h(w)) = \langle s+2, s+2 \rangle$. Then, by Lemma II.5 and Theorem II.7(f), $gd(h(\mathbf{p}_s^2 v)) = \langle s+2, s+2 \rangle$ and $h(\mathbf{p}_s^2 v) \subseteq h(w)$. Then, by Theorem 33, $\mathbf{p}_s^2 v \leq w$. Since $\mathbf{p}_s^2 \equiv^+ \mathbf{p}_0^2$ and $\mathbf{p}_0^2 v \equiv v$, there follows that $v \leq^+ w$. □

Let S'' be the closure under $\bullet$ and ${}^{\smile}$ of $\{M \in Bsl : M = eq_\omega M\} \cup \{\pi \in Bsl : \pi$ is a permutation of $\omega\}$. By Theorem IV.3(f), $\langle S'', \bullet, {}^{\smile} \rangle$ is isomorphic under ${}_\omega[\ ]$ to $\langle {}_{[\omega]}S_{pe}, \circ, {}^{\smile} \rangle$ and $\langle S'', \bullet, {}^{\smile}, \supseteq \rangle$ is isomorphic under ${}_\omega[\ ]$ to $\langle {}_{[\omega]}S_{pe}, \circ, {}^{\smile}, \subseteq \rangle$. Hence Theorem 34 also gives a presentation of these two structures.

Given any $n \geq 2$, let ${}_nS$ be the closure under $\bullet$ and ${}^{\smile}$ of $\{M : M = eq_n M\} \cup \{\pi : \pi$ is a permutation of $n\}$. (Thus the symmetric group $\mathbf{S}_n = \langle S_n, \circ, {}^{\smile} \rangle = \langle S_n, \bullet, {}^{\smile} \rangle$ is a subalgebra of $\langle {}_nS, \bullet, {}^{\smile} \rangle$.) It is likely that a conservational result similar to the one described after the proof of Theorem V.18 holds for the presentations of $\langle S'', \bullet, {}^{\smile} \rangle$ and of $\langle S'', \bullet, {}^{\smile}, \supseteq \rangle$ just mentioned.

CHAPTER VII

Presentation of $\mathbf{S}_{pqe}$ and Related Structures

In this chapter we will give a presentation of $\mathbf{S}_{pqe} = \langle S_{pqe}, \circ, {}^{\smile} \rangle$ and of $\langle \mathbf{S}_{pqe}, \subseteq \rangle$. We let $\Gamma = \Gamma(\tilde{\mathbf{p}}, \mathbf{q}, \mathbf{e})$ be the union of the sets $\Gamma(\tilde{\mathbf{p}}, \mathbf{q})$ and $\Gamma(\tilde{\mathbf{p}}, \mathbf{e})$ that were used in presenting $\mathbf{S}_{pq}$ or $\mathbf{S}_{pe}$, respectively, and we let $\mathbf{F}(\Gamma) = \mathbf{F}_{SI}(\Gamma(\tilde{\mathbf{p}}, \mathbf{q}, \mathbf{e}))$ be the free semigroup with involution that has Γ as a free generating set. Again using $\mathbf{F}(\Gamma)$, we will also give a presentation, closely related to that of $\mathbf{S}_{pqe}$ or of $\langle \mathbf{S}_{pqe}, \subseteq \rangle$ respectively, of ${}_{[\omega]}\mathbf{S}_{pqe} = \langle {}_{[\omega]}S_{pqe}, \circ, {}^{\smile} \rangle$ and of $\langle {}_{[\omega]}\mathbf{S}_{pqe}, \subseteq \rangle$.

(Let $T_{p\dot{q}e}$ be the closure under $\circ$ of $\{p_i : i < \omega\} \cup \{\dot{q}_i : i < \omega\} \cup \{e_{ij} : i, j < \omega\}$. Presentations of $\mathbf{T}_{p\dot{q}e} = \langle T_{p\dot{q}e}, \circ \rangle$ and of related structures may also be of interest, but we have not investigated this topic.)

We let QE_1, PQE_1, and PQE_2 be the following relation, each consisting of one pair only, on $F(\Gamma(\mathbf{q}, \mathbf{e}))$ or on $F(\Gamma(\tilde{\mathbf{p}}, \mathbf{q}, \mathbf{e}))$, respectively.

$$\begin{aligned} QE_1 &= \{\langle \mathbf{e}\mathbf{q}^{-1}\mathbf{q}\mathbf{e}, \mathbf{e} \rangle\} \\ PQE_1 &= \{\langle \mathbf{q}\mathbf{e}\mathbf{q}^{-1}, \mathbf{q}\mathbf{p}_0^2\mathbf{q}^{-1} \rangle\} \\ PQE_2 &= \{\langle \mathbf{e}\mathbf{q}\mathbf{p}_0\mathbf{p}_1, \mathbf{q}\mathbf{p}_0\mathbf{p}_1\mathbf{e} \rangle\} \end{aligned}$$

Recall that for presenting $\mathbf{S}_p$ we used the union $P_1 \cup P_2 \cup P_3 \cup P_4^{\#} \cup P_5 \cup P_6$ of six relations on $F(\Gamma(\tilde{\mathbf{p}}))$, that for presenting $\mathbf{S}_{pq}$ we used, in addition to it, the union ${}_1Q_3 \cup PQ_1 \cup PQ_2 \cup PQ_3 \cup PQ_4$ of five relations on $\mathbf{F}(\Gamma(\tilde{\mathbf{p}}, \mathbf{q}))$, and that for presenting $\mathbf{S}_{pe}$ we used, in addition to it, the union $E_1 \cup E_2 \cup PE_1 \cup PE_2 \cup PE_3 \cup PE_4$ of six relations on $F(\Gamma(\tilde{\mathbf{p}}, \mathbf{e}))$. In this chapter we let $\equiv'$ be the congruence on $\mathbf{F}(\Gamma(\tilde{\mathbf{p}}, \mathbf{q}, \mathbf{e}))$ that is generated by the union of the three unions just described. Also, $\equiv$ shall be the congruence on $\mathbf{F}(\Gamma(\tilde{\mathbf{p}}, \mathbf{q}, \mathbf{e}))$ that is generated by $\equiv' \cup QE_1 \cup PQE_1 \cup PQE_2$.

By Theorem I.1(b), S_{pqe} coincides with the closure under $\circ$ and ${}^{\smile}$ of $\{p_i : i < \omega\} \cup \{q\} \cup \{e\}$. Throughout this chapter we let h be the homomorphism of $\mathbf{F}(\Gamma) = \mathbf{F}_{SI}(\Gamma(\tilde{\mathbf{p}}, \mathbf{q}, \mathbf{e}))$ onto $\mathbf{S}_{pqe}$ such that $h(\mathbf{q}) = q = q_0$, $h(\mathbf{e}) = e = e_{0,1}$, and $h(\mathbf{p}_i) = p_i = p_{(i,i+1)}$ for each $i < \omega$.

LEMMA 1. (a) *If* $\mathbf{q}$ *does not occur in either* x *or* y, *or if* $\mathbf{e}$ *does not occur in either* x *or* y, *then* $x \equiv' y$ *if and only if* $h(x) = h(y)$.

(b) $\equiv$ *is included in* $h \circ h^{\smile}$.

Proof. (a) By Theorem VI.31 or Theorem VI.10, respectively.

(b) Since $e_{0,1} \circ q_0^{\smile} \circ q_0 = \{\langle x, y\rangle : |x| = |y| \geq 2,\ x_0 = x_1,\ y_n = x_n \text{ if } n \neq 0\}$, therefore

$$\begin{aligned} e_{0,1} \circ q_0^{\smile} \circ q_0 \circ e_{0,1} &= e_{0,1} \circ c_0 \circ e_{0,1} \\ = \{\langle x,y\rangle : |x| = |y| \geq 2,\ x_0 = x_1,\ y_0 = y_1,\ y_n = x_n &\text{ if } n \neq 0\} = e_{0,1}\ . \end{aligned}$$

Hence QE_1 is included in $h \circ h^{\smile}$.

Since $gd\ q_0 = \langle 0, 1\rangle$ and $gd\ e_{0,1} = \langle 2, 2\rangle$, therefore $gd(q_0 \circ e_{0,1}) = \langle 1, 2\rangle$. Also, $gd\ q_1 = gd(q_0 \circ p_0) = \langle 1, 2\rangle$. Hence $gd(q_0 \circ e_{0,1} \circ q_0^{\smile}) = \langle 1, 1\rangle = gd(q_1 \circ q_1^{\smile}) = gd(q_0 \circ p_0^2 \circ q_0^{\smile})$. There follows that

$$q_0 \circ e_{0,1} \circ q_0^{\smile} = {}_{[1,\omega)}\overset{\circ}{1} = q_1 \circ q_1^{\smile} = q_0 \circ p_0^2 \circ q_0^{\smile}\ .$$

Hence PQE_1 is included in $h \circ h^{\smile}$.

The second assertion below follows from the first and from $q_2 = q_0 \circ p_0 \circ p_1$.

$$\begin{aligned} e_{0,1} \circ q_2 &= \{\langle xxz, xxyz\rangle : |x| = |y| = 1,\ |z| < \omega\} = q_2 \circ e_{0,1}\ , \\ e \circ q_0 \circ p_0 \circ p_1 &= q_0 \circ p_0 \circ p_1 \circ e\ . \end{aligned}$$

There follows that PQE_2 is included in $h \circ h^{\smile}$.

By Lemmas VI.1(b) and VI.13(b), $\equiv'$ is included in $h \circ h^{\smile}$. There now follows that $\equiv$ is included in $h \circ h^{\smile}$. □

LEMMA 2. *Let* $\equiv_1'$ *be the congruence on* $\mathbf{F}(\Gamma(\tilde{\mathbf{p}}, \mathbf{q}, \mathbf{e}))$ *that is generated by* $\equiv' \cup QE_1$. *Then*

$$\mathbf{e}_{j,k}\overset{-1}{\mathbf{q}_k}\mathbf{q}_k\mathbf{e}_{j,k} \equiv_1' \mathbf{e}_{j,k}\ , \quad \text{if } j \neq k\ .$$

Proof. Henceforth, an assertion concerning $\equiv'$ will often be made without justifying it. It can be verified using Lemma 1(a) and either inspection or, if needed, Theorem III.11(a).

First note that, by QE_1,

$$\mathbf{e}_{0,1}\overset{-1}{\mathbf{q}_1}\mathbf{q}_1\mathbf{e}_{0,1} \equiv' \mathbf{e}_{0,1}\mathbf{p}_0\overset{-1}{\mathbf{q}_0}\mathbf{q}_0\mathbf{p}_0\mathbf{e}_{0,1} \equiv' \mathbf{e}_{0,1}\overset{-1}{\mathbf{q}_0}\mathbf{q}_0\mathbf{e}_{0,1} \equiv_1 \mathbf{e}_{0,1}\ .$$

Now let $k \geq 1$ and assume as inductive hypothesis that $\mathbf{e}_{0,k}\overset{-1}{\mathbf{q}_k}\mathbf{q}_k\mathbf{e}_{0,k} \equiv_1' \mathbf{e}_{0,k}$. Then, using this hypothesis for the next-to-last assertion, one has:

$$\begin{aligned} \mathbf{e}_{0,k+1}\overset{-1}{\mathbf{q}_{k+1}}\mathbf{q}_{k+1}\mathbf{e}_{0,k+1} &\equiv' \mathbf{e}_{0,k+1}\mathbf{p}_k\overset{-1}{\mathbf{q}_k}\mathbf{q}_k\mathbf{p}_k\mathbf{e}_{0,k+1} \\ \equiv' \mathbf{p}_k\mathbf{e}_{0,k}\overset{-1}{\mathbf{q}_k}\mathbf{q}_k\mathbf{e}_{0,k}\mathbf{p}_k &\equiv_1' \mathbf{p}_k\mathbf{e}_{0,k}\mathbf{p}_k \equiv' \mathbf{e}_{0,k+1}\ . \end{aligned}$$

There now follows by induction that

$$\mathbf{e}_{0,k}\mathbf{q}_0^{-1}\mathbf{q}_0\mathbf{e}_{0,k} \equiv_1' \mathbf{e}_{0,k} \qquad \text{for any } \ k \geq 1 \ .$$

Next, for $0 \leq j < k-1$ assume as inductive hypothesis that $\mathbf{e}_{j,k}\mathbf{q}_k^{-1}\mathbf{q}_k\mathbf{e}_{j,k} \equiv_1' \mathbf{e}_{j,k}$. Then, using $j < k-1$ for the second assertion and the inductive hypothesis for the fourth one obtains:

$$\begin{aligned}\mathbf{e}_{j,k+1}\mathbf{q}_k^{-1}\mathbf{q}_k\mathbf{e}_{j,k+1} &\equiv' \mathbf{p}_j\mathbf{e}_{j,k}\mathbf{p}_j\mathbf{q}_k^{-1}\mathbf{q}_k\mathbf{p}_j\mathbf{e}_{j,k}\mathbf{p}_j \\ &\equiv' \mathbf{p}_j\mathbf{e}_{j,k}\mathbf{q}_k^{-1}\mathbf{q}_k\mathbf{p}_j\mathbf{p}_j\mathbf{e}_{j,k}\mathbf{p}_j \equiv' \mathbf{p}_j\mathbf{e}_{j,k}\mathbf{q}_k^{-1}\mathbf{q}_k\mathbf{e}_{j,k}\mathbf{p}_j \\ &\equiv_1' \mathbf{p}_j\mathbf{e}_{j,k}\mathbf{p}_j \equiv' \mathbf{e}_{j+1,k} \ .\end{aligned}$$

There now follows by induction that

$$\mathbf{e}_{j,k}\mathbf{q}_k^{-1}\mathbf{q}_k\mathbf{e}_{j,k} \equiv_1' \mathbf{e}_{j,k} \qquad \text{whenever } \ 0 \leq j < k \ .$$

There remains to show that $\mathbf{e}_{j,k}\mathbf{q}_j^{-1}\mathbf{q}_j\mathbf{e}_{j,k} \equiv_1' \mathbf{e}_{j,k}$ whenever $0 \leq j < k$. First, for $k \geq 1$ assume as inductive hypothesis that $\mathbf{e}_{0,k}\mathbf{q}^{-1}\mathbf{q}\mathbf{e}_{0,k} \equiv_1' \mathbf{e}_{0,k}$. Then, using $k \geq 1$ for the second assertion and the inductive hypothesis for the next-to-last assertion one obtains:

$$\begin{aligned}\mathbf{e}_{0,k+1}\mathbf{q}^{-1}\mathbf{q}\mathbf{e}_{0,k+1} &\equiv' \mathbf{p}_k\mathbf{e}_{0,k}\mathbf{p}_k\mathbf{q}^{-1}\mathbf{q}\mathbf{p}_k\mathbf{e}_{0,k}\mathbf{p}_k \\ &\equiv' \mathbf{p}_k\mathbf{e}_{0,k}\mathbf{q}^{-1}\mathbf{p}_{k-1}\mathbf{q}\mathbf{p}_k\mathbf{e}_{0,k}\mathbf{p}_k \equiv' \mathbf{p}_k\mathbf{e}_{0,k}\mathbf{q}^{-1}\mathbf{q}\mathbf{p}_k\mathbf{p}_k\mathbf{e}_{0,k}\mathbf{p}_k \\ &\equiv' \mathbf{p}_k\mathbf{e}_{0,k}\mathbf{q}^{-1}\mathbf{q}\mathbf{p}_k^2\mathbf{p}_k\mathbf{e}_{0,k+1} \equiv' \mathbf{p}_k\mathbf{e}_{0,k}\mathbf{q}^{-1}\mathbf{q}\mathbf{p}_k\mathbf{e}_{0,k+1} \\ &\equiv' \mathbf{p}_k\mathbf{e}_{0,k}\mathbf{q}^{-1}\mathbf{q}\mathbf{e}_{0,k}\mathbf{p}_k \equiv_1' \mathbf{p}_k\mathbf{e}_{0,k}\mathbf{p}_k \equiv' \mathbf{e}_{0,k+1} \ .\end{aligned}$$

Thus, since $\mathbf{e} = \mathbf{e}_{0,1}$ and $\mathbf{q} = \mathbf{q}_0$, therefore, by QE_1 and induction there follows:

$$\mathbf{e}_{0,k}\mathbf{q}_0^{-1}\mathbf{q}_0\mathbf{e}_{0,k} \equiv_1' \mathbf{e}_{0,k} \qquad \text{for any } \ k \geq 1 \ .$$

Finally, for $0 \leq j < k-1$ assume as inductive hypothesis that $\mathbf{e}_{j,k}\mathbf{q}_j^{-1}\mathbf{q}_j\mathbf{e}_{j,k} \equiv_1' \mathbf{e}_{j,k}$. Then, using $j < k-1$ for the second and the inductive hypothesis for the third assertion one obtains:

$$\begin{aligned}\mathbf{e}_{j+1,k}\mathbf{q}_{j+1}^{-1}\mathbf{q}_{j+1}\mathbf{e}_{j+1,k} &\equiv' \mathbf{p}_j\mathbf{e}_{j,k}\mathbf{p}_j\mathbf{p}_j\mathbf{q}_j^{-1}\mathbf{q}_j\mathbf{p}_j\mathbf{p}_j\mathbf{e}_{j,k}\mathbf{p}_j \\ &\equiv' \mathbf{p}_j\mathbf{e}_{j,k}\mathbf{q}_j^{-1}\mathbf{q}_j\mathbf{e}_{j,k}\mathbf{p}_j \equiv_1' \mathbf{p}_j\mathbf{e}_{j,k}\mathbf{p}_j \equiv' \mathbf{e}_{j+1,k} \ .\end{aligned}$$

There now follows by induction that

$$\mathbf{e}_{j,k}\mathbf{q}_j^{-1}\mathbf{q}_j\mathbf{e}_{j,k} \equiv_1' \mathbf{e}_{j,k} \qquad \text{whenever } \ 0 \leq j < k \ .$$

Since $\mathbf{e}_{j,k} = \mathbf{e}_{k,j}$, therefore from the last paragraph and the second paragraph before it there follows that

$$\mathbf{e}_{j,k}\mathbf{q}_k^{-1}\mathbf{q}_k\mathbf{e}_{j,k} \equiv_1' \mathbf{e}_{j,k} \qquad \text{whenever} \quad j \neq k . \qquad \square$$

Henceforth, in connection with $F(\Gamma(\tilde{\mathbf{p}}, \mathbf{q}, \mathbf{e}))$, we will let $\mathbf{e}_{0,0} = \mathbf{q}_1\mathbf{q}_1^{-1}$. This will supplement our use of $\mathbf{e}_{j,k}$ where $\{j, k\} \neq \{0\}$ that was initiated in Chapter VI in connection with $F(\Gamma(\tilde{\mathbf{p}}, \mathbf{e}))$.

LEMMA 3. *Let* $\equiv_1$ *be the congruence on* $\mathbf{F}(\Gamma(\tilde{\mathbf{p}}, \mathbf{q}, \mathbf{e}))$ *that is generated by* $\equiv' \cup PQE_1$.

(a) $\mathbf{q}_k\mathbf{e}_{j,k}\mathbf{q}_k^{-1} \equiv_1 \mathbf{q}_k\mathbf{q}_k^{-1} \equiv_1 \mathbf{q}_j\mathbf{e}_{j,k}\mathbf{q}_j^{-1}$, *if* $j \leq k$.

(b) $\mathbf{q}_{j+1}\mathbf{e}_{j,j+1}\mathbf{q}_j^{-1} \equiv_1 \mathbf{q}_{j+1}\mathbf{q}_{j+1}^{-1} \equiv_1 \mathbf{q}_j\mathbf{e}_{j,j+1}\mathbf{q}_{j+1}^{-1}$.

(c) $\mathbf{q}_{j+n+1}\mathbf{e}_{j,j+n+1}\mathbf{q}_j^{-1} \equiv' \mathbf{q}_{j+n}\mathbf{e}_{j,j+n}\mathbf{q}_j^{-1}\mathbf{p}_{j+n-1}$, *if* $n \geq 1$.

(d) $\mathbf{q}_{j+n}\mathbf{e}_{j,j+n}\mathbf{q}_j^{-1} \equiv_1 \mathbf{p}_j \ldots \mathbf{p}_{j+n-2}$, *if* $n \geq 2$.

(e) $\mathbf{q}_j\mathbf{e}_{j,j+n}\mathbf{q}_{j+n}^{-1} \equiv_1 \mathbf{p}_{j+n-2} \ldots \mathbf{p}_j$, *if* $n \geq 2$.

Proof. (a) First assume that $j = k$. Suppose, first, that $j = k = 0$ and hence $\mathbf{e}_{j,k} = \mathbf{q}_1\mathbf{q}_1^{-1}$. Then there holds

$$\mathbf{q}_k\mathbf{e}_{j,k}\mathbf{q}_k^{-1} = \mathbf{q}_0\mathbf{q}_1\mathbf{q}_1^{-1}\mathbf{q}_0^{-1} \equiv' \mathbf{q}_0\mathbf{q}_0^{-1} = \mathbf{q}_k\mathbf{q}_k^{-1} .$$

Now suppose that $j = k \geq 1$. Then $\mathbf{e}_{j,k} = \mathbf{e}_{j,j} = \mathbf{p}_{j-1}^2 \equiv' \mathbf{q}_{j+1}\mathbf{q}_{j+1}^{-1}$. Then

$$\mathbf{q}_k\mathbf{e}_{j,k}\mathbf{q}_k^{-1} = \mathbf{q}_j\mathbf{q}_{j+1}\mathbf{q}_{j+1}^{-1}\mathbf{q}_j \equiv' \mathbf{q}_j\mathbf{q}_j^{-1} = \mathbf{q}_k\mathbf{q}_k^{-1} .$$

There remains to deal with the cases where $j < k$. First, let $j = 0$ and $k = 1$. Then, using PQE_1 for the third assertion one obtains:

$$\mathbf{q}_0\mathbf{e}_{i,1}\mathbf{q}_0^{-1} = \mathbf{q}\mathbf{e}\mathbf{q}^{-1} \equiv_1 \mathbf{q}\mathbf{p}_0^2\mathbf{q}^{-1} \equiv' \mathbf{q}_1\mathbf{q}_1^{-1} .$$

For $j \geq 0$, assume now as inductive hypothesis that $\mathbf{q}_j\mathbf{e}_{j,j+1}\mathbf{q}_j^{-1} \equiv_1 \mathbf{q}_{j+1}\mathbf{q}_{j+1}^{-1}$. Then, using the inductive hypothesis for the fourth assertion, one obtains:

$$\begin{aligned}
\mathbf{q}_{j+1}\mathbf{e}_{j+1,j+2}\mathbf{q}_{j+1}^{-1} &\equiv' \mathbf{q}_j\mathbf{p}_j\mathbf{p}_j\mathbf{p}_{j+1}\mathbf{e}_{j,j+1}\mathbf{p}_{j+1}\mathbf{p}_j\mathbf{p}_j\mathbf{q}_j^{-1} \\
&\equiv' \mathbf{q}_j\mathbf{p}_{j+1}\mathbf{e}_{j,j+1}\mathbf{p}_{j+1}\mathbf{q}_j^{-1} \\
&\equiv' \mathbf{p}_j\mathbf{q}_j\mathbf{e}_{j,j+1}\mathbf{q}_j^{-1}\mathbf{p}_j \equiv_1 \mathbf{p}_j\mathbf{q}_{j+1}\mathbf{q}_{j+1}^{-1}\mathbf{p}_j \\
&\equiv' \mathbf{q}_{j+1}\mathbf{q}_{j+1}^{-1}\mathbf{p}_j^2 \equiv' \mathbf{q}_{j+1}\mathbf{q}_{j+1}^{-1}\mathbf{q}_{j+2}\mathbf{q}_{j+2}^{-1} \equiv' \mathbf{q}_{j+2}\mathbf{q}_{j+2}^{-1} .
\end{aligned}$$

There now follows by induction that $\mathbf{q}_j\mathbf{e}_{j,j+1}\mathbf{q}_j^{-1} \equiv_1 \mathbf{q}_{j+1}\mathbf{q}_{j+1}^{-1}$ for any $j \geq 0$.

This proves the second assertion of part (a) for the case where $k = j + 1$. To prove this assertion for the general case, let $0 \leq j < k$, and assume as inductive hypothesis that $\mathbf{q}_j\mathbf{e}_{j,k}\mathbf{q}_j^{-1} \equiv_1 \mathbf{q}_k\mathbf{q}_k^{-1}$. Then, using the inductive hypothesis for the third assertion, one obtains:

$$\begin{aligned}\mathbf{q}_j\mathbf{e}_{j,k+1}\mathbf{q}_j^{-1} &\equiv' \mathbf{q}_j\mathbf{p}_k\mathbf{e}_{j,k}\mathbf{p}_k\mathbf{q}_j^{-1} \equiv' \mathbf{p}_{k-1}\mathbf{q}_j\mathbf{e}_{j,k}\mathbf{q}_j^{-1}\mathbf{p}_{k-1} \\ &\equiv_1 \mathbf{p}_{k-1}\mathbf{q}_k\mathbf{q}_k^{-1}\mathbf{p}_{k-1} \equiv' \mathbf{q}_k\mathbf{q}_k^{-1}\mathbf{p}_{k-1}^2 \\ &\equiv' \mathbf{q}_k\mathbf{q}_k^{-1}\mathbf{q}_{k+1}\mathbf{q}_{k+1}^{-1} \equiv' \mathbf{q}_{k+1}\mathbf{q}_{k+1}'\end{aligned}$$

There now follows by induction:

$$\mathbf{q}_j\mathbf{e}_{j,k}\mathbf{q}_j^{-1} \equiv_1 \mathbf{q}_k\mathbf{q}_k^{-1}\ , \qquad \text{if } j < k\ .$$

To prove the first assertion of part (a), first note that, as we showed earlier, that $\mathbf{q}_j\mathbf{e}_{j,j+1}\mathbf{q}_j^{-1} \equiv_1 \mathbf{q}_{j+1}\mathbf{q}_{j+1}^{-1}$. Since $\mathbf{p}_j\mathbf{e}_{j,j+1} \equiv' \mathbf{e}_{j,j+1} \equiv' \mathbf{e}_{j,j+1}\mathbf{p}_j$, there follows that

$$\mathbf{q}_{j+1}\mathbf{e}_{j,j+1}\mathbf{q}_{j+1}^{-1} \equiv' \mathbf{q}_j\mathbf{p}_j\mathbf{e}_{j,j+1}\mathbf{p}_j\mathbf{q}_j^{-1} \equiv' \mathbf{q}_j\mathbf{e}_{j,j+1}\mathbf{q}_j^{-1} \equiv_1 \mathbf{q}_{j+1}\mathbf{q}_{j+1}'\ .$$

For $k > j$ assume now as inductive hypothesis that $\mathbf{q}_k\mathbf{e}_{j,k}\mathbf{q}_k^{-1} \equiv_1 \mathbf{q}_k\mathbf{q}_k^{-1}$. Then, using this hypothesis for the third assertion, one obtains

$$\begin{aligned}\mathbf{q}_{k+1}\mathbf{e}_{j,k+1}\mathbf{q}_{k+1}^{-1} &\equiv' \mathbf{q}_k\mathbf{p}_k\mathbf{p}_k\mathbf{e}_{j,k}\mathbf{p}_k\mathbf{p}_k\mathbf{q}_k^{-1} \\ &\equiv' \mathbf{p}_{k-1}^2\mathbf{q}_k\mathbf{e}_{j,k}\mathbf{q}_k^{-1}\mathbf{p}_{k-1}^2 \equiv_1 \mathbf{p}_{k-1}^2\mathbf{q}_k\mathbf{q}_k^{-1}\mathbf{p}_{k-1}^2 \\ &\equiv' \mathbf{q}_{k+1}\mathbf{q}_{k+1}^{-1}\mathbf{q}_k\mathbf{q}_k^{-1}\mathbf{q}_{k+1}\mathbf{q}_{k+1}^{-1} \equiv' \mathbf{q}_{k+1}\mathbf{q}_{k+1}^{-1}\ .\end{aligned}$$

There now follows by induction that

$$\mathbf{q}_k\mathbf{e}_{j,k}\mathbf{q}_k^{-1} \equiv_1 \mathbf{q}_k\mathbf{q}_k^{-1}\ , \qquad \text{if } j < k\ .$$

(b) Using $\mathbf{e}_{j,j+1}\mathbf{p}_j \equiv' \mathbf{e}_{j,j+1}$ for the second and part (a) for the third assertion one obtains:

$$\mathbf{q}_j\mathbf{e}_{j,j+1}\mathbf{q}_{j+1}^{-1} \equiv' \mathbf{q}_j\mathbf{e}_{j,j+1}\mathbf{p}_j\mathbf{q}_j^{-1} \equiv' \mathbf{q}_j\mathbf{e}_{j,j+1}\mathbf{q}_j^{-1} \equiv_1 \mathbf{q}_{j+1}\mathbf{q}_{j+1}^{-1}\ .$$

Since $\mathbf{q}_{j+1}\mathbf{e}_{j,j+1}\mathbf{q}_j^{-1} = (\mathbf{q}_j\mathbf{e}_{j,j+1}\mathbf{q}_{j+1}^{-1})^{-1}$, there follows that
$\mathbf{q}_{j+1}\mathbf{e}_{j,j+1}\mathbf{q}_j^{-1} \equiv_1 \mathbf{q}_{j+1}\mathbf{q}_{j+1}^{-1}$.

(c) Let $n \geq 1$. Then

$$\begin{aligned}\mathbf{q}_{j+n+1}\mathbf{e}_{j,j+n+1}\mathbf{q}_j^{-1} &\equiv' \mathbf{q}_{j+n}\mathbf{p}_{j+n}\mathbf{p}_{j+n}\mathbf{e}_{j,j+n}\mathbf{p}_{j+n}\mathbf{q}_j^{-1} \\ &\equiv' \mathbf{q}_{j+n}\mathbf{e}_{j,j+n}\mathbf{p}_{j+n}^2\mathbf{p}_{j+n}\mathbf{q}_j^{-1} \\ &\equiv' \mathbf{q}_{j+n}\mathbf{e}_{j,j+n}\mathbf{p}_{j+n}\mathbf{q}_j^{-1} \equiv' \mathbf{q}_{j+n}\mathbf{e}_{j,j+n}\mathbf{q}_j^{-1}\mathbf{p}_{j+n-1}\ .\end{aligned}$$

(d) Using part (c) for the first and part (b) for the second assertion one obtains:

$$\mathbf{q}_{j+2}\mathbf{e}_{j,j+2}\mathbf{q}_j^{-1} \equiv' \mathbf{q}_{j+1}\mathbf{e}_{j,j+1}\mathbf{q}_j^{-1}\mathbf{p}_j \equiv_1 \mathbf{q}_{j+1}\mathbf{q}_{j+1}^{-1}\mathbf{p}_j \equiv' \mathbf{p}_j$$

Now for $n \geq 2$ assume as inductive hypothesis that $\mathbf{q}_{j+n}\mathbf{e}_{j,j+n}\mathbf{q}_j^{-1} \equiv_1 \mathbf{p}_j \cdots \mathbf{p}_{j+n-2}$. Then, by part (c) or the inductive hypothesis, respectively, there holds:

$$\mathbf{q}_{j+n+1}\mathbf{e}_{j,j+n+1}\mathbf{q}_j^{-1} \equiv' \mathbf{q}_{j+n}\mathbf{e}_{j,j+n}\mathbf{q}_j^{-1}\mathbf{p}_{j+n-1} \equiv_1 \mathbf{p}_j \cdots \mathbf{p}_{j+n-2}\mathbf{p}_{j+n-1} \ .$$

Hence, by induction,

$$\mathbf{q}_{j+n}\mathbf{e}_{j,j+n}\mathbf{q}_j^{-1} \equiv_1 \mathbf{p}_j \cdots \mathbf{p}_{j+n-2} \ , \qquad \text{if } \ n \geq 2 \ .$$

(e) Since $(\mathbf{q}_j\mathbf{e}_{j,j+n}\mathbf{q}_{j+n}^{-1})^{-1} \equiv \mathbf{q}_{j+n}\mathbf{e}_{j,j+n}\mathbf{q}_j^{-1}$, this follows from part (d). □

LEMMA 4. *Let* $\equiv_2$ *be the congruence on* $\mathbf{F}(\Gamma(\tilde{\mathbf{p}}, \mathbf{q}, \mathbf{e}))$ *that is generated by* $\equiv' \cup PQE_2$.

(a) $\mathbf{e}_{j,k}\mathbf{q}_i \equiv_2 \mathbf{q}_i\mathbf{e}_{j,k}$ *and* (a)′ $\mathbf{q}_i^{-1}\mathbf{e}_{j,k} \equiv_2 \mathbf{e}_{j,k}\mathbf{q}_i^{-1}$, *if* $j<k<i$

(b) $\mathbf{e}_{j,k}\mathbf{q}_i \equiv_2 \mathbf{q}_i\mathbf{e}_{j,k+1}$ *and* (b)′ $\mathbf{q}_i^{-1}\mathbf{e}_{j,k} \equiv_2 \mathbf{e}_{j,k+1}\mathbf{q}_i^{-1}$, *if* $j<i\leq k$.

(c) $\mathbf{e}_{j,k}\mathbf{q}_i \equiv_2 \mathbf{q}_i\mathbf{e}_{j+1,k+1}$ *and* (c)′ $\mathbf{q}_i^{-1}\mathbf{e}_{j,k} \equiv_2 \mathbf{e}_{j+1,k+1}\mathbf{q}_i^{-1}$, *if* $i\leq j<k$.

(d) $\mathbf{q}_i\mathbf{e}_{j,k} \equiv_2 \mathbf{e}_{j,k}\mathbf{q}_i$ *and* (d)′ $\mathbf{e}_{j,k}\mathbf{q}_i^{-1} \equiv_2 \mathbf{q}_i^{-1}\mathbf{e}_{j,k}$, *if* $j<k<i$.

(e) $\mathbf{q}_i\mathbf{e}_{j,k} \equiv_2 \mathbf{e}_{j,k-1}\mathbf{q}_i$ *and* (e)′ $\mathbf{e}_{j,k}\mathbf{q}_i^{-1} \equiv_2 \mathbf{q}_i^{-1}\mathbf{e}_{j,k-1}$, *if* $j<i<k$.

(f) $\mathbf{q}_i\mathbf{e}_{j,k} \equiv_2 \mathbf{e}_{j-1,k-1}\mathbf{q}_i$ *and* (f)′ $\mathbf{e}_{j,k}\mathbf{q}_i^{-1} \equiv_2 \mathbf{q}_i^{-1}\mathbf{e}_{j-1,k-1}$, *if* $i<j<k$.

Proof. (a) First, for $i \geq 2$ assume as inductive hypothesis that $\mathbf{e}\mathbf{q}_i \equiv_2 \mathbf{q}_i\mathbf{e}$. Then, using this hypothesis for the second assertion, one obtains:

$$\mathbf{e}\mathbf{q}_{i+1} = \mathbf{e}\mathbf{q}_i\mathbf{p}_i \equiv_2 \mathbf{q}_i\mathbf{e}\mathbf{p}_i \equiv' \mathbf{q}_i\mathbf{p}_i\mathbf{e} = \mathbf{q}_{i+1}\mathbf{e} \ .$$

By PQE_2 and $\mathbf{q}_2 = \mathbf{q}\mathbf{p}_0\mathbf{p}_1$, the inductive hypothesis holds for $i = 2$. Therefore, by induction there follows:

$$\mathbf{e}\mathbf{q}_i \equiv_2 \mathbf{q}_i\mathbf{e} \ , \qquad \text{if } \ i \geq 2 \ .$$

Next, let $2 \leq k < i$. Using what has just been shown for the third assertion one obtains:

$$\begin{aligned} \mathbf{e}_{0,k}\mathbf{p}_i \ &\equiv' \ \mathbf{p}_{(1,k)}\mathbf{e}\mathbf{p}_{(1,k)}\mathbf{q}_i \equiv' \mathbf{p}_{(1,k)}\mathbf{e}\mathbf{q}_i\mathbf{p}_{(1,k)} \equiv_2 \mathbf{p}_{(1,k)}\mathbf{q}_i\mathbf{e}\mathbf{p}_{(1,k)} \\ &\equiv' \ \mathbf{q}_i\mathbf{p}_{(1,k)}\mathbf{e}\mathbf{p}_{(1,k)} \equiv' \mathbf{q}_i\mathbf{e}_{0,k} \ . \end{aligned}$$

From this and from QPE_2 there follows that

$$\mathbf{e}_{0,k}\mathbf{q}_i \equiv_2 \mathbf{q}_i\mathbf{e}_{0,k} \ , \quad \text{if } \ 1 \leq k < i \ .$$

Let $1 \leq j < k < i$. Using what has just been shown and an argument similar to the one just used one can show:

$$\mathbf{e}_{j,k}\mathbf{q}_i \equiv_2 \mathbf{q}_i\mathbf{e}_{j,k} \ .$$

From the last two paragraphs there follows part (a).

(a)$'$ Since $\mathbf{q}_i^{-1}\mathbf{e}_{j,k} \equiv' (\mathbf{e}_{j,k}\mathbf{q}_i)^{-1}$ and $\mathbf{e}_{j,k}\mathbf{q}_i^{-1} \equiv' (\mathbf{q}_i\mathbf{e}_{j,k})^{-1}$, part (a)$'$ is equivalent to part (a).

(b) Using PQE_2 for the fifth assertion one obtains:

$$\begin{aligned} \mathbf{e}_{0,1}\mathbf{q}_1 \quad &= \quad \mathbf{eqp}_0 \equiv' \mathbf{ep}_0^2\mathbf{qp}_0 \equiv' \mathbf{eqp}_1^2\mathbf{p}_0 \equiv' \mathbf{eqp}_0\mathbf{p}_1\mathbf{p}_1 \\ &\equiv_2 \quad \mathbf{qp}_0\mathbf{p}_1\mathbf{ep}_1 = \mathbf{q}_1\mathbf{e}_{0,2} \ . \end{aligned}$$

For $k \geq 1$ assume as inductive hypothesis that $\mathbf{e}_{0,k}\mathbf{q}_1 \equiv_2 \mathbf{q}_1\mathbf{e}_{0,k+1}$. Then, using this hypothesis for the third assertion one obtains:

$$\begin{aligned} \mathbf{e}_{0,k+1}\mathbf{q}_1 \quad &\equiv' \quad \mathbf{p}_k\mathbf{e}_{0,k}\mathbf{p}_k\mathbf{q}_1 \equiv' \mathbf{p}_k\mathbf{e}_{0,k}\mathbf{q}_1\mathbf{p}_{k+1} \\ &\equiv_2 \quad \mathbf{p}_k\mathbf{q}_1\mathbf{e}_{0,k+1}\mathbf{p}_{k+1} \equiv' \mathbf{q}_1\mathbf{p}_{k+1}\mathbf{e}_{0,k+1}\mathbf{p}_{k+1} \equiv' \mathbf{q}_1\mathbf{e}_{0,k+2} \ . \end{aligned}$$

Thus, by induction

$$\mathbf{e}_{0,k}\mathbf{q}_1 \equiv_2 \mathbf{q}_1\mathbf{e}_{0,k+1} \ , \quad \text{if} \ \ k \geq 1 \ .$$

Next, for $1 \leq i < k$ assume as inductive hypothesis that $\mathbf{e}_{0,k}\mathbf{q}_i \equiv_2 \mathbf{q}_i\mathbf{e}_{0,k+1}$. Then, using this hypothesis for the second assertion one obtains:

$$\mathbf{e}_{0,k}\mathbf{q}_{i+1} \equiv' \mathbf{e}_{0,k}\mathbf{q}_i\mathbf{p}_i \equiv_2 \mathbf{q}_i\mathbf{e}_{0,k+1}\mathbf{p}_i \equiv' \mathbf{q}_i\mathbf{p}_i\mathbf{e}_{0,k+1} = \mathbf{q}_{i+1}\mathbf{e}_{0,k+1} \ .$$

By induction there follows:

$$\mathbf{e}_{0,k}\mathbf{q}_i \equiv_2 \mathbf{q}_i\mathbf{e}_{0,k+1} \ , \quad \text{if} \ \ 1 \leq i \leq k \ .$$

Finally, assume as inductive hypothesis that if $0 \leq j < i \leq k$ then $\mathbf{e}_{j,k}\mathbf{q}_i \equiv_2 \mathbf{q}_i\mathbf{e}_{j,k+1}$. Suppose that $j+1 < i \leq k$. Then, using the inductive hypothesis and an argument that, by now, should be familiar one can show that $\mathbf{e}_{j,k}\mathbf{q}_i \equiv_2 \mathbf{q}_i\mathbf{e}_{j+1,k+1}$. By induction, there now follows part (b).

(b)$'$ This is equivalent to part (b).

(c) Since $\mathbf{eqp}_0 \equiv_2 \mathbf{qp}_0\mathbf{p}_1\mathbf{ep}_1$, as we saw in the beginning of the proof of part (b), therefore

$$\mathbf{e}_{0,1}\mathbf{q}_0 = \mathbf{eq} \equiv' \mathbf{eqp}_0\mathbf{p}_0 \equiv_2 \mathbf{qp}_0\mathbf{p}_1\mathbf{ep}_1\mathbf{p}_0 = \mathbf{q}_0\mathbf{e}_{1,2} \ .$$

Using this result and an argument by induction similar to the ones used earlier, one next shows that $\mathbf{e}_{0,k}\mathbf{q}_0 \equiv_2 \mathbf{q}_0\mathbf{e}_{1,k+1}$, if $k \geq 1$. Using this result, in turn, and an argument by induction similar to ones used earlier, one then shows that $\mathbf{e}_{j,k}\mathbf{q}_0 \equiv_2 \mathbf{q}_0\mathbf{e}_{j+1,k+1}$, if $0 \leq j < k$. Using this result, and again an argument by induction similar to earlier ones, one finally shows:

$$\mathbf{e}_{j,k}\mathbf{q}_i \equiv \mathbf{q}_i\mathbf{e}_{j+1,k+1} \ , \quad \text{if} \ \ 0 \leq i < j < k \ .$$

(c)′ This is equivalent to part (c).

(d),(d)′,(e),(e)′,(f),(f)′. Since $\equiv_2$ is a symmetric relation, these are equivalent to (a),(a)′, (b),(b)′,(c),(c)′ respectively. □

A consequence of Lemmas 4(a)′, 4(b)′, 4(c)′ is the following lemma. According to parts (d) and (d)′, if $j < k$ then, by suitably changing subscripts, one can move $\mathbf{q}_i^{-1}$ to the left of $\mathbf{q}_j\mathbf{e}_{j,k}$ or $\mathbf{q}_k\mathbf{e}_{j,k}$.

LEMMA 5. *Let $\equiv_2$ be the congruence on $\mathbf{F}(\Gamma(\tilde{\mathbf{p}}, \mathbf{q}, \mathbf{e}))$ that is generated by $\equiv' \cup PQE_2$. Let $j < k$.*

(a) $\mathbf{q}_j\mathbf{e}_{j,k}\mathbf{q}_i^{-1} \equiv_2 \mathbf{q}_{i-1}^{-1}\mathbf{q}_j\mathbf{e}_{j,k}$ *and*

(a)′ $\mathbf{q}_k\mathbf{e}_{j,k}\mathbf{q}_i^{-1} \equiv_2 \mathbf{q}_{i-1}^{-1}\mathbf{q}_k\mathbf{e}_{j,k}$, *if* $k < i$.

(b) $\mathbf{q}_j\mathbf{e}_{j,k}\mathbf{q}_i^{-1} \equiv_2 \mathbf{q}_{i-1}^{-1}\mathbf{q}_j\mathbf{e}_{j,k-1}$ *and*

(b)′ $\mathbf{q}_k\mathbf{e}_{j,k}\mathbf{q}_i^{-1} \equiv_2 \mathbf{q}_i^{-1}\mathbf{q}_{k-1}\mathbf{e}_{j,k-1}$, *if* $j < i < k$.

(c) $\mathbf{q}_j\mathbf{e}_{j,k}\mathbf{q}_i^{-1} \equiv_2 \mathbf{q}_i^{-1}\mathbf{q}_{j-1}\mathbf{e}_{j-1,k-1}$ *and*

(c)′ $\mathbf{q}_k\mathbf{e}_{j,k}\mathbf{q}_i^{-1} \equiv_2 \mathbf{q}_i^{-1}\mathbf{q}_{k-1}\mathbf{e}_{j-1,k-1}$, *if* $i < j$.

(d) $\mathbf{q}_j\mathbf{e}_{j,k}\mathbf{q}_i^{-1} \equiv_2 \mathbf{q}_{i'}^{-1}\mathbf{q}_{j'}\mathbf{e}_{j',k'}$ *for some* i', j', k' *such that* $j' < k'$.

(d)′ $\mathbf{q}_k\mathbf{e}_{j,k}\mathbf{q}_i^{-1} \equiv_2 \mathbf{q}_{i'}^{-1}\mathbf{q}_{k'}\mathbf{e}_{j',k'}$, *for some* i', j', k' *such that* $j' < k'$.

Proof. (a) and (a)′. In the two cases below the first assertion follows from Lemma 4(a)′ and the second from $j < i$ or from $k < i$, respectively:

$$\mathbf{q}_j\mathbf{e}_{j,k}\mathbf{q}_i^{-1} \equiv_2 \mathbf{q}_j\mathbf{q}_i^{-1}\mathbf{e}_{j,k} \equiv' \mathbf{q}_{i-1}^{-1}\mathbf{q}_j\mathbf{e}_{j,k} \ .$$
$$\mathbf{q}_k\mathbf{e}_{j,k}\mathbf{q}_i^{-1} \equiv_2 \mathbf{q}_k\mathbf{q}_i^{-1}\mathbf{e}_{j,k} \equiv' \mathbf{q}_{i-1}^{-1}\mathbf{q}_k\mathbf{e}_{j,k} \ .$$

(b) and (b)′. In the two cases below, the first assertion follows from Lemma 4(b)′ and the second from $j < i$ or from $k > i$, respectively:

$$\mathbf{q}_j\mathbf{e}_{j,k}\mathbf{q}_i^{-1} \equiv_2 \mathbf{q}_j\mathbf{q}_i^{-1}\mathbf{e}_{j,k-1} \equiv' \mathbf{q}_{i-1}^{-1}\mathbf{q}_j\mathbf{e}_{j,k-1} \ .$$
$$\mathbf{q}_k\mathbf{e}_{j,k}\mathbf{q}_i^{-1} \equiv_2 \mathbf{q}_k\mathbf{q}_i^{-1}\mathbf{e}_{j,k-1} \equiv' \mathbf{q}_{i-1}^{-1}\mathbf{q}_{k-1}\mathbf{e}_{j,k-1} \ .$$

(c) and (c)′. In the two cases below, the first assertion follows from Lemma 4(c)′ and the second from $j > i$ or from $k > i$, respectively:

$$\mathbf{q}_j\mathbf{e}_{j,k}\mathbf{q}_i^{-1} \equiv_2 \mathbf{q}_j\mathbf{q}_i^{-1}\mathbf{e}_{j-1,k-1} \equiv' \mathbf{q}_i^{-1}\mathbf{q}_{j-1}\mathbf{e}_{j-1,k-1} \ .$$
$$\mathbf{q}_k\mathbf{e}_{j,k}\mathbf{q}_i^{-1} \equiv_2 \mathbf{q}_k\mathbf{q}_i^{-1}\mathbf{e}_{j-1,k-1} \equiv' \mathbf{q}_i^{-1}\mathbf{q}_{k-1}\mathbf{e}_{j-1,k-1} \ .$$

(d) This follows from parts (a), (b), and (c).

(d)$'$ This follows from parts (a)$'$, (b)$'$, and (c)$'$. □

Each of the recent lemmas ***gives explicitly*** a certain relation on $F(\Gamma(\tilde{\mathbf{p}}, \mathbf{q}, \mathbf{e}))$. For example, the relation thus given by Lemma 5(a) is the following:

$$\{\langle \mathbf{q}_j \mathbf{e}_{j,k} \mathbf{q}_i^{-1}, \mathbf{q}_{i-1}^{-1} \mathbf{q}_j \mathbf{e}_{j,k} \rangle : j < k < i < \omega\}$$

The closure under monotonicity of this relation is the following:

$$\big\{\langle v\mathbf{q}_j \mathbf{e}_{j,k} \mathbf{q}_i^{-1} w, v\mathbf{q}_{i-1}^{-1} \mathbf{q}_j \mathbf{e}_{j,k} w \rangle : j<k<i<\omega,\ v \in \{\emptyset\} \cup F(\Gamma(\tilde{\mathbf{p}}, \mathbf{q}, \mathbf{e})),$$
$$w \in \{\emptyset\} \cup F(\Gamma(\tilde{\mathbf{p}}, \mathbf{q}, \mathbf{e}))\,\big\}\ .$$

It properly includes the relation of which it is the closure and, since it is neither symmetric nor transitive it is properly included in the congruence on $\mathbf{F}(\Gamma(\tilde{\mathbf{p}}, \mathbf{q}, \mathbf{e}))$ that is generated by the relation that is explicitly given by Lemma 5(a).

Given any i, j, j', k, k' such that $j' < k' < j < i < k$, consider the sequence of four assertions that is given below. The first, second, third, and fourth of these holds by Lemma 4(e)$'$, Lemma 4(c), Lemma VI.23, or Lemma 4(d), respectively.

$$\begin{aligned} \mathbf{e}_{j,k} \mathbf{q}_i^{-1} \mathbf{q}_j \mathbf{e}_{j',k'} &\equiv \mathbf{q}_i^{-1} \mathbf{e}_{j,k-1} \mathbf{q}_j \mathbf{e}_{j',k'} \equiv \mathbf{q}_i^{-1} \mathbf{q}_j \mathbf{e}_{j+1,k} \mathbf{e}_{j',k'} \\ &\equiv \mathbf{q}_i^{-1} \mathbf{q}_j \mathbf{e}_{j',k'} \mathbf{e}_{j+1,k} \equiv \mathbf{q}_i^{-1} \mathbf{e}_{j',k'} \mathbf{q}_j \mathbf{e}_{j+1,k}\ . \end{aligned}$$

Let $u_0, \ldots, u_4$ be such that, for $0 \le i < 5$, the ith assertion above is the assertion $u_i \equiv u_{i+1}$. Then the pair $\langle u_0, u_1 \rangle, \langle u_1, u_2 \rangle, \langle u_2, u_3 \rangle, \langle u_3, u_4 \rangle$, belongs to the closure under monotonicity of the relation that is given explicitly by Lemma 4(e)$'$, Lemma 4(c), Lemma VI.23, and Lemma 4(d), respectively. We shall say of the sequence $\langle u_0, \ldots, u_4 \rangle$, and also of the sequence of four assertions displayed above, that it is ***based on*** Lemmas 4(e)$'$, 4(c), VI.23, and 4(d). For other lemmas and theorems, we shall use the notion of a sequence based on them in a similar manner.

It will be convenient to admit sequences $\langle u_0 \rangle$ consisting of only one word in $F(\Gamma(\tilde{\mathbf{p}}, \mathbf{q}, \mathbf{e}))$ and to have them based on any set of lemmas or theorems, including the empty set. A sequence $\langle u_0, \ldots, u_{n-1} \rangle$ such that $n \ge 1$ shall be ***from*** u_0 ***to*** u_{n-1}. There follows that for any set of lemmas or theorems the following two conditions are equivalent:

(i) There is a sequence from v to w that is based on members of the set.

(ii) $\langle v, w \rangle$ belongs to the least quasi-ordering of $F_{SI}(\Gamma(\tilde{\mathbf{p}}, \mathbf{q}, \mathbf{e}))$ that includes each relation given explicitly by one of the lemmas or theorems in the set and that is compatible with $\mathbf{F}_{SI}(\Gamma(\tilde{\mathbf{p}}, \mathbf{q}, \mathbf{e}))$.

LEMMA 6. *Consider any* $v = v_0 \ldots v_{n-1}$, $n \geq 1$ *such that each* $v_{n'}$ *is either some* $\mathbf{q}_i$ *or some* $\mathbf{e}_{j,k}$, $j \neq k$, *and such that among the* $v_{n'}$ *there is at least one* $\mathbf{q}_i$ *and at least one* $\mathbf{e}_{j,k}$. *Let* $\mathbf{q}_{i_0} \ldots \mathbf{q}_{i_{m-1}}$, $m < n$, *be the word that results from* v *when all the* $\mathbf{e}_{j,k}$ *are deleted (and all gaps are closed). Then for some word* $w = w_0 \ldots w_{n-m-1}$ *such that each* $w_{m'}$ *is some* $\mathbf{e}_{j',k'}$ *satisfying* $j' \neq k'$, *there is a sequence based on Lemmas 4(a), 4(b), and 4(c), from* v *to* $\mathbf{q}_{i_0} \ldots \mathbf{q}_{i_{m-1}} w_0 \ldots w_{n-m-1}$. *Also, if* $v^{\#}$ *results from* v^{-1} *by replacing each* $\mathbf{e}_{j,k}^{-1}$ *by* $\mathbf{e}_{j,k}$, *then there is a sequence based on Lemmas 4(a)′, 4(b)′, and 4(c)′, from* $v^{\#}$ *to* $w_{n-m-1} \ldots w_0 \mathbf{q}_{i_{m-1}}^{-1} \ldots \mathbf{q}_{i_0}^{-1}$. *There follows that, if* $\equiv_2$ *is the congruence on* $F(\Gamma(\tilde{\mathbf{p}}, \mathbf{q}, \mathbf{e}))$ *that is generated by* $\equiv' \cup PQE_2$, *then*

$$\text{(i)} \quad v \equiv_2 \mathbf{q}_{i_0} \ldots \mathbf{q}_{i_{m-1}} w_0 \ldots w_{n-m-1} \quad \textit{and}$$
$$\text{(ii)} \quad v^{-1} \equiv' v^{\#} \equiv_2 w_{n-m-1} \ldots w_0 \mathbf{q}_{i_{m-1}}^{-1} \ldots \mathbf{q}_{i_0}^{-1}.$$

Proof. From the inductive hypothesis that the lemma holds for $v = v_0 \ldots v_{n-1}$ where $n \geq 1$ there follows that it holds for any v' such that, for some i or some $j \neq k$, $v' = v\mathbf{q}_i$ or $v' = v\mathbf{e}_{j,k}$. □

Given any $\mathbf{e}_{j_0,k_0} \mathbf{q}_{i_0}^{-1} \ldots \mathbf{q}_{i_{n-1}}^{-1}$ such that $j_0 < k_0$ and $n \geq 1$, we shall say that $\mathbf{e}_{j_0,k_0} \mathbf{q}_{i_0}^{-1} \ldots \mathbf{q}_{i_{n-1}}^{-1}$ ***lends itself to full use of Lemmas*** 4(d)′, 4(e)′, and 4(f)′ if and only if the following holds:

> For every m, $1 \leq m \leq n$, there is a sequence $\langle u_0, \ldots, u_m \rangle$ based on Lemmas 4(d)′, 4(e)′ and 4(f)′ such that, for some j_m and k_m satisfying $j_m < k_m$, $\langle u_0, \ldots, u_m \rangle$ is from
> $\mathbf{e}_{j_0,k_0} \mathbf{q}_{i_0}^{-1} \ldots \mathbf{q}_{i_{m-1}}^{-1}$ to $\mathbf{q}_{i_0}^{-1} \ldots \mathbf{q}_{i_{m-1}}^{-1} \mathbf{e}_{j_m,k_m}$.

Evidently, if $\mathbf{e}_{j,k} \mathbf{q}_{i_0}^{-1} \ldots \mathbf{q}_{i_{n-1}}^{-1}$ lends itself to such use then, for every $m \leq n$, j_m and k_m are unique, $k_m \geq 1$, and by these lemmas
$\mathbf{e}_{j_0,k_0} \mathbf{q}_{i_0}^{-1} \ldots \mathbf{q}_{i_{m-1}}^{-1} \equiv_2 \mathbf{q}_{i_0}^{-1} \ldots \mathbf{q}_{i_{m-1}}^{-1} \mathbf{e}_{j_m,k_m}$.

LEMMA 7. *For* $j_0 < k_0$ *and* $n \geq 1$ *consider any* $\mathbf{e}_{j_0,k_0} \mathbf{q}_{i_0}^{-1} \ldots \mathbf{q}_{i_{n-1}}^{-1}$ *that does not lend itself to full use of Lemmas 4(d)′, 4(e)′ and 4(f)′, so that there is a least* m, $1 \leq m \leq n$, *such that* $\mathbf{e}_{j_0,k_0} \mathbf{q}_{i_0}^{-1} \ldots \mathbf{q}_{i_{m-1}}^{-1}$ *does not lend itself to such use and hence so that if* $w = \mathbf{q}_{i_0}^{-1} \ldots \mathbf{q}_{i_{m-2}}^{-1}$ *(and hence* $w = \emptyset$ *if* $m = 1$*) then there hold both* $\mathbf{e}_{j_0,k_0} w \equiv_2 w \mathbf{e}_{j_{m-1},k_{m-1}}$ *and either (1)* $i_{m-1} = j_{m-1}$ *or (2)* $i_{m-1} = k_{m-1}$. *Then, in case (1), where* $\mathbf{q}_{i_{m-1}}^{-1} = \mathbf{q}_{j_{m-1}}^{-1}$ *there hold (a), (b), and (c) below and, in case (2) where* $\mathbf{q}_{i_{m-1}}^{-1} = \mathbf{q}_{k_{m-1}}^{-1}$ *there hold (d), (e), and (f) below,* $\equiv_{1,2}$ *being the congruence on* $\mathbf{F}(\Gamma(\tilde{\mathbf{p}}, \mathbf{q}, \mathbf{e}))$ *that is generated by* $\equiv' \cup PQE_1 \cup PQE_2$.

(a) $\mathbf{q}_{j_0} \mathbf{e}_{j_0,k_0} w \mathbf{q}_{j_{m-1}}^{-1} \equiv_{1,2} w \mathbf{q}_{k_{m-1}} \mathbf{q}_{k_{m-1}}^{-1}$.

(b) $\mathbf{q}_{k_0}\mathbf{e}_{j_0,k_0}w\mathbf{q}^{-1}_{j_{m-1}} \equiv_{1,2} w\mathbf{q}_{k_{m-1}}\mathbf{q}^{-1}_{k_{m-1}}$, *if* $k_{m-1} - j_{m-1} = 1$.

(c) $\mathbf{q}_{k_0}\mathbf{e}_{j_0,k_0}w\mathbf{q}^{-1}_{j_{m-1}} \equiv_{1,2} w\mathbf{p}_{j_{m-1}}\cdots\mathbf{p}^{-1}_{k_{m-1}-2}$, *if* $k_{m-1} - j_{m-1} \geq 2$.

(d) $\mathbf{q}_{k_0}\mathbf{e}_{j_0,k_0}w\mathbf{q}^{-1}_{k_{m-1}} \equiv_{1,2} w\mathbf{q}_{k_{m-1}}\mathbf{q}^{-1}_{k_{m-1}}$.

(e) $\mathbf{q}_{j_0}\mathbf{e}_{j_0,k_0}w\mathbf{q}^{-1}_{k_{m-1}} \equiv_{1,2} w\mathbf{q}_{k_{m-1}}\mathbf{q}^{-1}_{k_{m-1}}$, *if* $k_{m-1} - j_{m-1} = 1$.

(f) $\mathbf{q}_{j_0}\mathbf{e}_{j_0,k_0}w\mathbf{q}^{-1}_{k_{m-1}} \equiv_{1,2} w\mathbf{p}_{k_{m-1}-2}\cdots\mathbf{p}_{j_{m-1}}$, *if* $k_{m-1} - j_{m-1} \geq 2$.

Proof. We again let $\equiv_1$ and $\equiv_2$ be the congruence on $\mathbf{F}(\Gamma(\tilde{\mathbf{p}},\mathbf{q},\mathbf{e}))$ that is generated by $\equiv' \cup PQE_1$ or by $\equiv' \cup PQE_2$ respectively.

(a) The first assertion below holds by Lemma 5 (and induction) and the second by Lemma 3(a).

$$\mathbf{q}_{j_0}\mathbf{e}_{j_0,k_0}w\mathbf{q}^{-1}_{j_{m-1}} \equiv_2 w\mathbf{q}_{j_{m-1}},\mathbf{e}_{j_{m-1},k_{m-1}}\mathbf{q}^{-1}_{j_{m-1}} \equiv_1 w\mathbf{q}_{k_{m-1}}\mathbf{q}^{-1}_{k_{m-1}}$$

(b) Assume that $k_{m-1} - j_{m-1} = 1$. Then the second assertion below holds by Lemma 3(b), while the first holds by Lemma 5.

$$\mathbf{q}_{k_0}\mathbf{e}_{j_0,k_0}w\mathbf{q}^{-1}_{j_{m-1}} \equiv_2 w\mathbf{q}_{k_{m-1}}\mathbf{e}_{j_{m-1},k_{m-1}}\mathbf{q}^{-1}_{j_{m-1}} \equiv_1 w\mathbf{q}_{k_{m-1}}\mathbf{q}^{-1}_{k_{m-1}}$$

(c) Assume that $k_{m-1} - j_{m-1} \geq 2$. Then the second assertion below holds by Lemma 3(d), while the first again holds by Lemma 5.

$$\mathbf{q}_{k_0}\mathbf{e}_{j_0,k_0}w\mathbf{q}^{-1}_{j_{m-1}} \equiv_2 w\mathbf{q}_{k_{m-1}}\mathbf{e}_{j_{m-1},k_{m-1}}\mathbf{q}^{-1}_{j_{m-1}} \equiv_1 w\mathbf{p}_{j_{m-1}}\cdots\mathbf{p}_{k_{m-1}-2}$$

(d) Similar to the proof of (a).

(e) Similar to the proof of (b).

(f) Similar to the proof of (c) except that one uses Lemma 3(e). □

An $\langle\mathbf{e},\mathbf{q}^{-1},\mathbf{p},\mathbf{q},\mathbf{e}\rangle$ ***word*** shall be any $v^\frown w^\frown x^\frown y^\frown z = vwxyz$ such that the following five conditions are satisfied:

(1) Either $v = \emptyset$ or $v = \mathbf{e}_{j_0,k_0}\cdots\mathbf{e}_{j_{m-1},k_{m-1}}$ for some $n \geq 1$ and some $\mathbf{e}_{j_0,k_0},\ldots,\mathbf{e}_{j_{n-1},k_{n-1}}$.

(2) Either $w = \emptyset$ or $w = \mathbf{q}^{-1}_{i_0}\cdots\mathbf{q}^{-1}_{i_{n-1}}$ for some $n \geq 1$ and some $\mathbf{q}^{-1}_{i_0},\ldots,\mathbf{q}^{-1}_{i_{n-1}}$.

(3) Either $x = \emptyset$ or $x = \mathbf{p}_{i_0}\cdots\mathbf{p}_{i_{n-1}}$ for some $n \geq 1$ and some $\mathbf{p}_{i_0},\ldots,\mathbf{p}_{i_{n-1}}$.

(4) Either $y = \emptyset$ or $y = \mathbf{q}_{i_0}\cdots\mathbf{q}_{i_{n-1}}$ for some $n \geq 1$ and some $\mathbf{q}_{i_0},\ldots,\mathbf{q}_{i_{n-1}}$.

(5) Either $z = \emptyset$ or $z = \mathbf{e}_{j_0,k_0}\cdots\mathbf{e}_{j_{n-1},k_{n-1}}$ for some $n \geq 1$ and some $\mathbf{e}_{j_0,k_0},\ldots,\mathbf{e}_{j_{n-1},k_{n-1}}$.

Also, for example, an $\langle\mathbf{e}\rangle$ ***word*** shall be any v for which (1) holds, and a $\langle\mathbf{q}^{-1},\mathbf{p},\mathbf{q}\rangle$ ***word*** shall be any wxy for which (2), (3), and (4) hold. Thus, a ***nonempty*** $\langle\mathbf{e},\mathbf{q}^{-1},\mathbf{p},\mathbf{q},\mathbf{e}\rangle$ word or, for example, a ***nonempty*** $\langle\mathbf{q}^{-1},\mathbf{q}\rangle$ word is an element of $F(\Gamma(\tilde{\mathbf{p}},\mathbf{q},\mathbf{e}))$ or of $F(\Gamma(\tilde{\mathbf{q}}))$, respectively, that is of a certain special form.

Henceforth, in using $vwxyz$ to refer to an $\langle \mathbf{e}, \mathbf{q}^{-1}, \mathbf{p}, \mathbf{q}, \mathbf{e}\rangle$ word we will assume that v, w, x, y, z satisfy the condition (1),(2),(3),(4),(5), respectively. Similar conventions will apply when we refer, for example, to a $\langle \mathbf{q}^{-1}, \mathbf{q}\rangle$ word.

LEMMA 8. *Let* $\equiv_{1,2}$ *be the congruence on* $\mathbf{F}(\Gamma(\tilde{\mathbf{p}}, \mathbf{q}, \mathbf{e}))$ *that is generated by* $\equiv' \cup PQE_1 \cup PQE_2$. *Then, for any* u *in* $F(\Gamma(\tilde{\mathbf{p}}, \mathbf{q}, \mathbf{e}))$ *either* $u \equiv_{1,2} \mathbf{q}\mathbf{q}^{-1} = \mathbf{q}_0\mathbf{q}_0^{-1}$ *or there is a nonempty* $\langle \mathbf{e}, \mathbf{q}^{-1}, \mathbf{p}, \mathbf{q}, \mathbf{e}\rangle$ *word* $vwxyz$ *such that* $u \equiv_{1,2} vwxyz$.

Proof. If $u = \mathbf{e}_{j,k}$ then $u = vwxyz$ where $v = \mathbf{e}_{j,k}$ and $w = x = y = z = \emptyset$. If $u = \mathbf{e}_{j,k}^{-1}$ then $\mathbf{e}_{j,k}^{-1} \equiv' \mathbf{e}_{j,k} = vwxyz$ for the same v, w, x, y, z. Similarly, if $u = \mathbf{p}_i$ or $u = \mathbf{p}_i^{-1}$, then $u \equiv' vwxyz$ where $x = \mathbf{p}_i$ and $v = w = y = z = \emptyset$. If $u = \mathbf{q}_i^{-1}$, then $u = v\mathbf{q}_i^{-1}xyz$ where $v = x = y = z = \emptyset$, and if $u = \mathbf{q}_i$ then $u = vwx\mathbf{q}_iz$ where $v = w = x = z = \emptyset$. Furthermore, if $u \equiv_{1,2} \mathbf{q}\mathbf{q}^{-1}$ then u again satisfies the conclusion of the lemma.

For a proof by induction, we now consider any u such that either $u = \mathbf{q}\mathbf{q}^{-1}$ or $u = vwxyz$ where $vwxyz$ is a nonempty $\langle \mathbf{e}, \mathbf{q}^{-1}, \mathbf{p}, \mathbf{q}, \mathbf{e}\rangle$ word. Since there hold $\mathbf{e}_{j,k}\mathbf{q}\mathbf{q}^{-1} \equiv' \mathbf{e}_{j,k}, \ldots, \mathbf{q}_i\mathbf{q}\mathbf{q}^{-1} \equiv' \mathbf{q}_i$ and since $\mathbf{e}_{j,k}^{-1} \equiv' \mathbf{e}_{j,k}$ and $\mathbf{p}_i^{-1} \equiv' \mathbf{p}_i$, it suffices to show the following:

> If u' is a word $\mathbf{e}_{j,k}vwxyz$, $\mathbf{q}_i^{-1}vwxyz$, $\mathbf{p}_ivwxyz$, or $\mathbf{q}_ivwxyz$ other than $\mathbf{q}\mathbf{q}^{-1}$ then there is a nonempty $\langle \mathbf{e}, \mathbf{q}^{-1}, \mathbf{p}, \mathbf{q}, \mathbf{e}\rangle$ word $v'w'x'y'z'$ such that $u' \equiv_{1,2} v'w'x'y'z'$.

Let $u' = \mathbf{e}_{j,k}vwxyz$. Since $v' = \mathbf{e}_{j,k}v$ is an $\langle \mathbf{e}\rangle$ word, therefore u' is the nonempty $\langle \mathbf{e}, \mathbf{q}^{-1}, \mathbf{p}, \mathbf{q}, \mathbf{e}\rangle$ word $v'wxyz$.

Let $u' = \mathbf{q}_i^{-1}vwxyz$. By Lemmas 4(a)′, 4(b)′, and 4(c)′, there is some $\langle \mathbf{e}\rangle$ word v' such that $\mathbf{q}_i^{-1}v \equiv_2 v'\mathbf{q}_i^{-1}$. Let $w' = \mathbf{q}_i^{-1}w$. Then $v'w'xyz$ is a nonempty $\langle \mathbf{e}, \mathbf{q}^{-1}, \mathbf{p}, \mathbf{q}, \mathbf{e}\rangle$ word and $u' \equiv_2 v'w'xyz$.

Let $u' = \mathbf{p}_ivwxyz$. By Theorem VI.20, there is some $\langle \mathbf{e}\rangle$ word v' such that $\mathbf{p}_iv \equiv' v'\mathbf{p}_i$. By Lemmas VI.6(a)′, VI.6(e), and VI.6(d)′ and by induction on the length of w there follows that there is a nonempty $\langle \mathbf{p}\rangle$ word x' such that $\mathbf{p}_iw \equiv wx'$. There follows that if $x'' = x'x$, then $v'wx''yz$ is a nonempty $\langle \mathbf{e}, \mathbf{q}^{-1}, \mathbf{p}, \mathbf{q}, \mathbf{e}\rangle$ word and $u' \equiv' v'wx''yz$.

There remains the case where $u' = \mathbf{q}_ivwxyz$ and $u' \neq \mathbf{q}\mathbf{q}^{-1}$. First assume that either $v = \emptyset$ or, for some $n \geq 1$, $v = \mathbf{e}_{j_0,k_0} \ldots \mathbf{e}_{j_{n-1},k_{n-1}}$ and, for each $m < n$, either $j_m = k_m$ or $i \notin \{j_m, k_m\}$. Then by Lemmas 4(d), 4(e),and 4(f), and since $\mathbf{q}_i\mathbf{e}_{j,j} \equiv' \mathbf{e}_{j,j}\mathbf{q}_i$ for each j, there is an $\langle \mathbf{e}\rangle$ word v' such that $\mathbf{q}_iv \equiv_2 v'\mathbf{q}_i$.

By Lemma VI.7, $\mathbf{q}_i wxy \equiv' \mathbf{q}_s \mathbf{q}_s^{-1} w'x'y'$ for some $\langle \mathbf{q}^{-1}, \mathbf{p}, \mathbf{q} \rangle$ word $w'x'y' \neq \emptyset$ and some $s \geq i$. If $s = 0$, then $\mathbf{q}_s \mathbf{q}_s^{-1} w'x'y' \neq \emptyset$ and $u' \equiv' v'w'x'y'z$. If $s \geq 1$ then there is an $\langle \mathbf{e} \rangle$ word $v'' \neq \emptyset$ such that $v'\mathbf{q}_s \mathbf{q}_s^{-1} \equiv' v''$ and $u' \equiv' v''w'x'y'z$.

For the rest of the proof assume that $v = \mathbf{e}_{j_0,k_0} \ldots \mathbf{e}_{j_{n-1},k_{n-1}}$ and that for at least one $m < n$, $j_m \neq k_m$ and $i \in \{j_m, k_m\}$. By Lemma VI.23 we may assume without loss of generality that $\mathbf{e}_{j_{n-1},k_{n-1}} = \mathbf{e}_{i,k_{n-1}}$ and $i \neq k_{n-1}$. If $n = 1$, let $v'' = \emptyset$. If $n > 1$ then let v'' be the $\langle \mathbf{e} \rangle$ word that results from $\mathbf{e}_{j_0,k_0} \ldots \mathbf{e}_{j_{n-2},k_{n-2}}$ when each $\mathbf{e}_{j_m,k_m}$ such that $j_m \neq k_m$ and $i \in \{j_m, k_m\}$ is replaced by $\mathbf{e}_{k_{n-1},k_m}$ if $i = j_m$ and by $\mathbf{e}_{j_m,k_{n-1}}$ if $i = k_m$. From Lemma VI.26 there follows that $v \equiv' v''\mathbf{e}_{i,k_{n-1}}$. Also, by Lemmas 4(d), 4(e) and 4(f), and by $\mathbf{q}_i \mathbf{e}_{j,j} \equiv' \mathbf{e}_{j,j} \mathbf{q}_i$, there is an $\langle \mathbf{e} \rangle$ word v' such that $\mathbf{q}_i v'' \equiv' v'\mathbf{q}_i$. Hence $\mathbf{q}_i v \equiv' v'\mathbf{q}_i \mathbf{e}_{i,k_{n-1}}$.

First, consider the case where either $w = \emptyset$ or $\mathbf{e}_{i,k_{n-1}} w$ lends itself to full use of Lemmas 4(d)′, 4(e)′, and 4(f)′. Then for some j and k such that $j \neq k$ there holds $\mathbf{e}_{i,k_{n-1}} w \equiv_2 w\mathbf{e}_{j,k}$ and hence also

$$u' \;=\; \mathbf{q}_i vwxyz \;\equiv'\; v'\mathbf{q}_i \mathbf{e}_{i,k_{n-1}} wxyz \;\equiv_2\; v'\mathbf{q}_i w\mathbf{e}_{j,k} xyz \;.$$

For some $\langle \mathbf{p} \rangle$ word x'' and some j'' and k'' such that $j'' \neq k''$ there hold $\mathbf{e}_{j,k} x \equiv' x''\mathbf{e}_{j',k'}$. Also, by Lemma 4, for some j' and k' such that $j' \neq k'$ there holds $\mathbf{e}_{j'',k''} y \equiv_2 y\mathbf{e}_{j',k'}$. Hence

$$u' \;\equiv_2\; v'\mathbf{q}_i wx''y\,\mathbf{e}_{j',k'} z \;.$$

Now either $\mathbf{q}_i w \equiv' w'\mathbf{q}_{i'}$ for some $\langle \mathbf{q}^{-1} \rangle$ word w' and some i' or else there are $\langle \mathbf{q}^{-1} \rangle$ words w'' and w''' and there is some i' such that $\mathbf{q}_i w \equiv' w''\mathbf{q}_{i'} \mathbf{q}_{i'}^{-1} w'''$, so that, for some i'', there holds $\mathbf{q}_{i'} \mathbf{q}_{i'}^{-1} w''' \equiv' w'''\mathbf{q}_{i''} \mathbf{q}_{i''}^{-1}$ and hence, for the $\langle \mathbf{q}^{-1} \rangle$ word $w' = w''w'''$ there holds $\mathbf{q}_i w \equiv' w'\mathbf{q}_{i''} \mathbf{q}_{i''}^{-1}$. First suppose that $\mathbf{q}_i w \equiv' w'\mathbf{q}_{i'}$. There is a $\langle \mathbf{p} \rangle$ word x' such that $\mathbf{q}_{i'} x'' \equiv' x'\mathbf{q}_{i'}$. Let y' be the $\langle \mathbf{q} \rangle$ word $\mathbf{q}_{i'} y$ and let z' be the $\langle \mathbf{e} \rangle$ word $\mathbf{e}_{j',k'} z$. Then $v'w'x'y'z'$ is a nonempty $\langle \mathbf{e}, \mathbf{q}^{-1}, \mathbf{p}, \mathbf{q}, \mathbf{e} \rangle$ word and there holds

$$\begin{aligned} u' \;&\equiv_2\; v'\mathbf{q}_i wx''y\mathbf{e}_{j',k'} z \equiv' v'w'\mathbf{q}_{i'} x''y\mathbf{e}_{j',k'} z \\ &\equiv'\; v'w'x'\mathbf{q}_{i'} y\mathbf{e}_{j',k'} z \equiv' v'w'x'y'z' \;. \end{aligned}$$

Now suppose that $\mathbf{q}_i w \equiv' w'\mathbf{q}_{i''} \mathbf{q}_{i''}^{-1}$ and hence $\mathbf{q}_i wx'' \equiv' w'x''\mathbf{q}_{i''} \mathbf{q}_{i''}^{-1}$. If $i'' = 0$, then $\mathbf{q}_{i''} \mathbf{q}_{i''}^{-1} y\mathbf{e}_{j',k'} \equiv' y\mathbf{e}_{j',k'}$. If $i'' \neq 0$ and hence $\mathbf{q}_{i''} \mathbf{q}_{i''}^{-1} = \mathbf{e}_{i''-1,i''-1}$, then, for some j'', $\mathbf{q}_{i''} \mathbf{q}_{i''}^{-1} y\mathbf{e}_{j',k'} \equiv' y\mathbf{e}_{j'',j''} \mathbf{e}_{j',k'}$. Thus, in either case, there is some nonempty $\langle \mathbf{e} \rangle$ word z' such that $\mathbf{q}_{i''} \mathbf{q}_{i''}^{-1} y\mathbf{e}_{j',k'} z \equiv' yz'$, $v'w'x''yz'$ is a nonempty

$\langle \mathbf{e}, \mathbf{q}^{-1}, \mathbf{p}, \mathbf{q}, \mathbf{e}\rangle$ word, and

$$\begin{aligned} u' &\equiv_2 \; v'\mathbf{q}_i w x'' y \mathbf{e}_{j',k'} z \equiv' v'w'\mathbf{q}_{i''}\mathbf{q}_{i''}^{-1} x'' y \mathbf{e}_{j',k'} z \\ &\equiv \; v'w'x''\mathbf{q}_{i''}\mathbf{q}_{i''}^{-1} y \mathbf{e}_{j',k'} z \equiv' v'w'x''yz' \,. \end{aligned}$$

Now consider the case where $\mathbf{e}_{i,k_{n-1}}w$ does not lend itself to full use of Lemmas 4(d)$'$, 4(e)$'$, and 4(f)$'$. Let $w = \mathbf{q}_{i_0}^{-1} \ldots \mathbf{q}_{i_{s-1}}^{-1}$. Then there is a least m, $1 \leq m \leq s$, such that $\mathbf{e}_{i,k_{n-1}}\mathbf{q}_{i_0}^{-1} \ldots \mathbf{q}_{i_{m-1}}^{-1}$ does not lend itself to such use . Let $w' = \mathbf{q}_{i_0}^{-1} \ldots \mathbf{q}_{i_{m-2}}^{-1}$ and $w'' = \mathbf{q}_{i_{m-1}}^{-1} \ldots \mathbf{q}_{i_{s-1}}^{-1}$, so that $w'' \neq \emptyset$ and $w = w'w''$. Then for some j_{m-1} and k_{m-1} such that $j_{m-1} < k_{m-1}$ there hold both $\mathbf{e}_{i,k_{n-1}}w' \equiv_2 w'\mathbf{e}_{j_{m-1},k_{m-1}}$ and either (1) $i_{m-1} = j_{m-1}$ or (2) $i_{m-1} = k_{m-1}$. From Lemma 7 there follows that one of the following holds:

(i) $\mathbf{q}_i\mathbf{e}_{i,k_{n-1}}w' \equiv_{1,2} w'\mathbf{q}_{k_{m-1}}\mathbf{q}_{k_{m-1}}^{-1}$,
(ii) $\mathbf{q}_i\mathbf{e}_{i,k_{n-1}}w' \equiv_{1,2} w'\mathbf{p}_{j_{m-1}} \cdots \mathbf{p}_{k_{m-1}-2}$,
(iii) $\mathbf{q}_i\mathbf{e}_{i,k_{n-1}}w' \equiv_{1,2} w'\mathbf{p}_{k_{m-1}-2} \cdots \mathbf{p}_{j_{m-1}}$.

In case (i), an argument similar to one used earlier in connection with $\mathbf{q}_{i''}\mathbf{q}_{i''}^{-1}$ shows that there is an $\langle \mathbf{e}\rangle$ word z' such that $\mathbf{q}_{k_{n-1}}\mathbf{q}'_{k_{n-1}}w''xyz \equiv' w''xyz'$. Then $v'w'w''xyz'$ is a nonempty $\langle \mathbf{e}, \mathbf{q}^{-1}, \mathbf{p}, \mathbf{q}, \mathbf{e}\rangle$ word, and

$$\begin{aligned} u' = vwxyz &\equiv' \; v'\mathbf{q}_i\mathbf{e}_{i,k_{n-1}}wxyz \\ &\equiv_{1,2} \; v'w'\mathbf{q}_{k_{m-1}}\mathbf{q}'_{k_{m-1}}w''xyz \equiv' v'w'w''xyz' \,. \end{aligned}$$

In case (ii), there is a $\langle \mathbf{p}\rangle$ word x'' such that $\mathbf{p}_{j_{m-1}} \cdots \mathbf{p}_{k_{m-1}}w'' \equiv' w''x''$. Then, for $x' = x''x$, $v'w'w''x'yz$ is a nonempty $\langle \mathbf{e}, \mathbf{q}^{-1}, \mathbf{p}, \mathbf{q}, \mathbf{e}\rangle$ word, and

$$\begin{aligned} u' = vwxyz &\equiv' \; v'\mathbf{q}_i\mathbf{e}_{i,k_{n-1}}wxyz \\ &\equiv_{1,2} \; v'w'\mathbf{p}_{j_{m-1}} \cdots \mathbf{p}_{k_{m-1}-2}w''xyz \equiv' v'w'w''x'yz \,. \end{aligned}$$

In case (iii), a similar argument shows that there is a nonempty $\langle \mathbf{e}, \mathbf{q}^{-1}, \mathbf{p}, \mathbf{q}, \mathbf{e}\rangle$ word $v'w'w''x'yz$ such that $u' \equiv_{1,2} v'w'w''x'yz$. □

Let $vwxyz$ be any $\langle \mathbf{e}, \mathbf{q}^{-1}, \mathbf{p}, \mathbf{q}, \mathbf{e}\rangle$ word. There is a unique special triple ${}_{s_1}\langle M_1\rangle_{t_1}$ such that, if $w \neq \emptyset$, then $h(w) = {}_{s_1}[\![M_1]\!]_{t_1}$ and, if $w = \emptyset$ then ${}_{s_1}\langle M_1\rangle_{t_1} = {}_0\langle\emptyset\rangle$ and hence ${}_{s_1}[\![M_1]\!]_{t_1} = {}_{[0,\omega)}\overset{\circ}{1}$. Also, there is a unique special triple ${}_{s_2}\langle M_2\rangle$ such that, if $x \neq \emptyset$, then $h(x) = {}_{s_2}[\![M_2]\!]$ and, if $x = \emptyset$, then ${}_{s_2}\langle M_2\rangle = {}_0\langle\emptyset\rangle$. Further, there is a unique special triple ${}_{s_3}\langle M_3\rangle_{t_3}$ such that, if $y \neq \emptyset$, then $h(y) = {}_{s_3}[\![M_3]\!]_{t_3}$ and, if $y = \emptyset$, then ${}_{s_3}\langle M_3\rangle_{t_3} = {}_0\langle\emptyset\rangle$. While $gd\, v = \langle s_0, s_0\rangle$ and $s_0 \geq 2$ together imply that there are many special triples ${}_{s_0}\langle M_0\rangle$ such that $h(v) = {}_{s_0}[\![M_0]\!]$ there is always a unique special triple ${}_{s_0}\langle M_0\rangle$ such that, if $v \neq \emptyset$, then $M_0 = eq_{s_0}M_0$

and $h(v) = {}_{s_0}[\![M_0]\!]$ and, if $v = \emptyset$, then ${}_{s_0}\langle M_0\rangle = {}_0\langle\emptyset\rangle$. Likewise, there is a unique special triple ${}_{s_4}\langle M_4\rangle$ such that, if $z \neq \emptyset$, then $M_4 = eq_{s_4} M_4$ and $h(z) = {}_{s_4}[\![M_4]\!]$ and, if $z = \emptyset$, then ${}_{s_4}\langle M_4\rangle = {}_0\langle\emptyset\rangle$. We call ${}_{s_0}\langle M_0\rangle$, ${}_{s_1}\langle M_1\rangle_{t_1}$, ${}_{s_2}\langle M_2\rangle$, ${}_{s_3}\langle M_3\rangle_{t_3}$, and ${}_{s_4}\langle M_4\rangle$ ***the special triple associated with*** v, w, x, y, z respectively.

Let $\langle s', t'\rangle = \langle s_1, t_1\rangle \odot \langle s_2, t_2\rangle \odot \langle s_3, t_3\rangle$, where $\langle \omega\times\omega, \odot\rangle$ is the bicyclic semigroup (discussed before Lemma II.5). Let ${}_{s_1'}\langle M_1'\rangle_{t_1'}$, ${}_{s_2'}\langle M_2'\rangle_{t_2'}$, ${}_{s_3'}\langle M_3'\rangle_{t_3'}$, and ${}_{s'}\langle N'\rangle_{t'}$ be the following special triple respectively:

$$\begin{aligned}
{}_{s_1'}\langle M_1'\rangle_{t_1'} &= {}_{s'}\langle M_1 \cup \{\langle s_1+n, t_1+n\rangle : n < s'-s_1\}\rangle_{t_1+s'-s_1}\\
{}_{s_2'}\langle M_2'\rangle_{t_2'} &= {}_{t_1+s'-s_1}\langle M_2 \cup \{\langle s_2+n, t_2+n\rangle : n < t_1+s'-s_1-s_2\}\rangle_{s_3+t'-t_3}\\
{}_{s_3'}\langle M_3'\rangle_{t_3'} &= {}_{s_3+t'-t_3}\langle M_3 \cup \{\langle s_3+n, t_3+n\rangle : n < t'-t_3\}\rangle_{t'}\\
{}_{s'}\langle N'\rangle_{t'} &= {}_{s'}\langle M_1' \circ M_2' \circ M_3'\rangle_{t'}
\end{aligned}$$

When $wxy \neq \emptyset$, then using Theorem II.7(g) and Lemma III.9(a) for the second assertion below and Lemma III.8(b) for the third one obtains:

$$\begin{aligned}
h(wxy) &= {}_{s_1}[\![M_1]\!]_{t_1} \circ {}_{s_2}[\![M_2]\!]_{t_2} \circ {}_{s_3}[\![M_3]\!]_{t_3}\\
&= {}_{s_1'}[\![M_1']\!]_{t_1'} \circ {}_{s_2'}[\![M_2']\!]_{t_2'} \circ {}_{s_3'}[\![M_3']\!]_{t_3'} = {}_{s'}[\![N']\!]_{t'} \ .
\end{aligned}$$

The special triple associated with wxy shall be ${}_{s'}[\![N']\!]_{t'}$ if $wxy \neq \emptyset$ and ${}_0\langle\emptyset\rangle_0 = {}_0\langle\emptyset\rangle$ if $wxy = \emptyset$.

Let $vwxyz$ be any $\langle \mathbf{e}, \mathbf{q}^{-1}, \mathbf{p}, \mathbf{q}, \mathbf{e}\rangle$ word and let ${}_{s_0}\langle M_0\rangle$, ${}_{s'}\langle N'\rangle_{t'}$, and ${}_{s_4}\langle M_4\rangle$ be the special triple associated with v, wxy, or z, respectively. Then $vwxyz$ shall be ***frameable*** if and only if

$$vwxyz \neq \emptyset \quad \text{and} \quad s_0 - s' = s_4 - t' \geq 0 \ .$$

Now assume that $vwxyz$ is frameable. Let

$$N = N' \cup \{\langle s'+n, t'+n\rangle : n < s_0 - s'\} \ .$$

Then $\langle M_0, N, M_4\rangle$ is a framing triple. It shall be ***the framing triple associated with*** $vwxyz$.

Assume that $vwxyz$ is frameable and $\langle M_0, N, M_4\rangle$ is the framing triple associated with $vwxyz$. Then, using Theorem II.7(g) and Lemma III.9(a) for the second assertion below and Lemma III.6 for the third one obtains:

$$\begin{aligned}
h(vwxyz) &= {}_{s_0}[\![M_0]\!] \circ {}_{s'}[\![N']\!]_{t'} \circ {}_{s_4}[\![M_4]\!]\\
&= {}_{s_0}[\![M_0]\!] \circ {}_{s_0}[\![N]\!]_{s_4} \circ {}_{s_4}[\![M_4]\!] = [\![M_0, N, M_4]\!] \ .
\end{aligned}$$

LEMMA 9. *For any $\langle \mathbf{e}, \mathbf{q}^{-1}, \mathbf{p}, \mathbf{q}, \mathbf{e}\rangle$ word $vwxyz$ that is nonempty there is some v' and some z' such that $v'wxyz'$ is frameable and such that $vwxyz \equiv' v'wxyz'$.*

Proof. Let ${}_{s_0}\langle M_0\rangle$, ${}_{s'}\langle N'\rangle_{t'}$, ${}_{s_4}\langle M_4\rangle$ be the special triple associated with v, wxy, or z, respectively. If $s_0 - s' \geq 0$ let $v'' = v$ and if $s_0 - s' < 0$ and hence $s' > 0$, let $v'' = v\mathbf{e}_{s'-1,s'-1}$. Since $wxy \equiv' \mathbf{q}_{s'}\mathbf{q}_{s'}^{-1}wxy \equiv' \mathbf{e}_{s'-1,s'-1}wxy$ if $s' > 0$, therefore $vwxy \equiv' v''wxy$. Similarly, let $z'' = z$ if $s_4 - t' \geq 0$ and let $z'' = \mathbf{e}_{t'-1,t'-1}z$ if $s_4 - t' < 0$. Then $wxyz \equiv' wxyz''$. Let $gd\, v'' = \langle s_0'', s_0''\rangle$ and $gd\, z'' = \langle s_4'', s_4''\rangle$. Then $s_0'' - s' \geq 0$ and $s_4'' - t' \geq 0$.

If $s_0'' - s' = s_4'' - t'$, then $v''wxyz''$ is frameable. Now suppose that $s_0'' - s' < s_4'' - t'$. Let $s_0' = s_4'' - t' + s'$ and let $v' = v''\mathbf{e}_{s_0'-1,s_0'-1}$. Since $z'' \equiv' \mathbf{q}_{s_4''}\mathbf{q}_{s_4''}^{-1}z''$, therefore

$$v''wxyz'' \equiv' v''\mathbf{q}_{s_0'}\mathbf{q}_{s_0'}^{-1}wxyz'' \equiv' v''\mathbf{e}_{s_0'-1,s_0'-1}wxyz'' = v'wxyz'' \ .$$

Since $s_0' - s' = s_4'' - t' \geq 0$, therefore $v'wxyz''$ is frameable. If $s_4'' - t' < s_0'' - t'$ then, similarly, there is some z' such that $v''wxyz'' \equiv' v''wxyz'$ and such that $v''wxyz'$ is frameable. □

LEMMA 10. *For any frameable $v'w'x'y'z'$ there is a frameable $vwxyz$ that satisfies $vwxyz \equiv_2 v'w'x'y'z'$ and the following conditions:*

(1) *Either $w = \emptyset$ or w is strictly descending.*
(2) *Either $y = \emptyset$ or y is strictly ascending.*
(3) *If $v \equiv' v\mathbf{e}_{j,k}$ and $\mathbf{e}_{j,k}wxy \equiv_2 wxy\mathbf{e}_{j',k'}$, then $z \equiv' \mathbf{e}_{j',k'}z$.*
(4) *If $z \equiv' \mathbf{e}_{j',k'}z$ and $wxy\mathbf{e}_{j',k'} \equiv_2 \mathbf{e}_{j,k}wxy$, then $v \equiv' v\mathbf{e}_{j,k}$.*

Proof. By Lemma V.4(a), or by Theorem V.10, there is some w such that either w' and w are both empty or $w' \neq \emptyset$, $w \equiv' w'$, and w is strictly descending. Similarly, there is some y such that either y' and y are both empty or $y' \neq \emptyset$, $y \equiv' y'$, and y is strictly ascending. In any case, $v'w'x'y'z' \equiv' v'wxyz'$.

To obtain from $v'wxyz'$ some $vwxyz$ that satisfies (3) and (4), in addition to (1) and (2), one adjoins to v' and to z' sufficiently often an appropriate $\mathbf{e}_{j,k}$ or $\mathbf{e}_{j',k'}$, respectively. Suppose that there hold $v' \equiv' v'\mathbf{e}_{j_1,k_1}, \ldots, v' \equiv' \mathbf{e}_{j_m,k_m}$ and also $\mathbf{e}_{j_1,k_1}wxy \equiv_2 wxy\mathbf{e}_{j_1',k_1'}, \ldots, \mathbf{e}_{j_m,k_m}wxy \equiv_2 wxy\mathbf{e}_{j_m',k_m'}$, but that there hold none of $z' \equiv' \mathbf{e}_{j_1',k_1'}z', \ldots, z' \equiv' \mathbf{e}_{j_m',k_m'}z'$. Then a suitable step is to let $z'' = \mathbf{e}_{j_1',k_1'} \ldots \mathbf{e}_{j_m',k_m'}z'$, so that $v'wxyz''$ is frameable and $v'wxyz' \equiv_2 v'wxyz''$. Now suppose that there hold $z'' \equiv' \mathbf{e}_{j_{m+1}',k_{m+1}'}z'', \ldots, z'' \equiv' \mathbf{e}_{j_n',k_n'}z''$ and also $xyz\mathbf{e}_{j_{m+1}',k_{m+1}'} \equiv_2 \mathbf{e}_{j_{m+1},k_{m+1}}xyz, \ldots, xyz\mathbf{e}_{j_n',k_n'} \equiv_2 \mathbf{e}_{j_n,k_n}xyz$, but that there hold none of $v' \equiv' v'\mathbf{e}_{j_{m+1},k_{m+1}}, \ldots, v' \equiv' v'\mathbf{e}_{j_n,k_n}$. Then a suitable next step is to let $v'' = v'\mathbf{e}_{j_{m+1},k_{m+1}} \ldots \mathbf{e}_{j_n,k_n}$. Alternating in this manner as often as needed one eventually obtains some v and z such that (3) and (4) hold, $vwxyz$ is frameable, and $vwxyz \equiv_2 v'wxyz'$, and hence $vwxyz \equiv_2 v'w'x'y'z'$. □

LEMMA 11. *Assume that $vwxyz$ is frameable and satisfies conditions* $(1), \ldots, (4)$ *of Lemma* 10. *Let $\langle M_0, N, M_4\rangle$ be the framing triple associated with $vwxyz$. Then $\langle M_0, N, M_4\rangle$ is precanonical.*

Proof. Let ${}_{s_1}\langle M_1\rangle_{t_1}$, ${}_{s_2}\langle M_2\rangle$, and ${}_{s_3}\langle M_3\rangle_{t_3}$ be the special triple associated with w, x, or y, respectively. Consider any $\langle j, j'\rangle$ in N. Then $j \in Do\, N$ and $j' \in Rg\, N$. Since, by condition (1), $w \neq \emptyset$ implies that w is strictly descending, therefore if $\mathbf{q}_j^{-1}$ occurred in w then one would have $j < s_1$ and $j \notin Do\, M_1$ and hence also $j \notin Do\, N$. Hence $\mathbf{q}_j^{-1}$ does not occur in w. Similarly, by condition (2), $\mathbf{q}_{j'}$ does not occur in y.

Assume that jNj' and kNk'. Then, as we just saw, $\mathbf{q}_j^{-1}$ and $\mathbf{q}_k^{-1}$ do not occur in w. Suppose that $w \neq \emptyset$. Then, since w is strictly descending and $\mathbf{q}_j^{-1}$ and $\mathbf{q}_k^{-1}$ do not occur in it, therefore $\mathbf{e}_{j,k}w$ lends itself to full use of Lemmas 4(d)$'$, 4(e)$'$, and 4(f)$'$.

Using induction on the length of w one can show that $\mathbf{e}_{j,k}w \equiv_2 w\mathbf{e}_{j'',k''}$, where jM_1j'' and kM_1k''. If $w = \emptyset$ then, for $j'' = j$ and $k'' = k$, there again holds $\mathbf{e}_{j,k}w \equiv_2 w\mathbf{e}_{j'',k''}$. Also, for $j''M_2j'''$ and $k''M_2k'''$ there holds $\mathbf{e}_{j'',k''}x \equiv' x\mathbf{e}_{j''',k'''}$. Further, by Lemmas 4(a), 4(b), and 4(c), if $j'''M_3j'$ and $k'''M_3k'$, and hence jNj' and kNk', there holds $\mathbf{e}_{j''',k'''}y \equiv_2 y\mathbf{e}_{j',k'}$. There now follows that $\mathbf{e}_{j,k}wxy \equiv_2 wxy\mathbf{e}_{j',k'}$.

Assume, in addition, that jM_0k. Then $v \equiv' v\mathbf{e}_{j,k}$. By condition (3), there now follows that $z \equiv' \mathbf{e}_{j',k'}z$ and hence that $j'M_4k'$. Thus, jNj', kNk', and jM_0k together imply that $j'M_4k'$.

A similar argument, using condition (4) instead of (3), shows that jNj', kNk', and $j'M_4k'$ together imply that jM_0k. There now follows that if jNj' and kNk', then jM_0k if and only if $j'M_4k'$. Hence $\langle M_0, N, M_4\rangle$ is precanonical. □

LEMMA 12. *Let $\equiv_1'$ be the congruence on $\mathbf{F}(\Gamma(\tilde{\mathbf{p}}, \mathbf{q}, \mathbf{e}))$ that is generated by $\equiv' \cup\, QE_1$. Assume that $vwxyz$ satisfies conditions* $(1), \ldots, (4)$ *of Lemma 10 and is frameable, that $\langle M_0, N, M_4\rangle$ is the associated framing triple, that $N' \subseteq N$, and that $[\![M_0, N', M_4]\!] = [\![M_0, N, M_4]\!]$. Then there are w' and y' such that $vw'xy'z$ satisfies* $(1), \ldots, (4)$ *and is frameable, such that the framing triple associated with $vw'xy'z$ is $\langle M_0, N', M_4\rangle$, and such that $vw'xy'z \equiv_1' vwxyz$.*

Proof. Since N is finite and there are only finitely many pairs in $N \cap -N'$, therefore it suffices to prove the lemma for the special case where in $N \cap -N'$ there is exactly one pair, since the lemma itself follows from this special case by induction.

Assume that $\langle i, i' \rangle \in N$ and that $N' = N \cap -\{\langle i, i' \rangle\}$. Since $\langle i, i' \rangle \in N$ therefore $h(wxy\mathbf{q}_{i'}) = h(\mathbf{q}_i wxy)$ and hence $wxy\mathbf{q}_{i'} \equiv' \mathbf{q}_i wxy$. Let w' be such that $w' \equiv' \mathbf{q}_i^{-1} w$ and w' is strictly descending and let y' be such that $y' \equiv' y\mathbf{q}_{i'}$ and y' is strictly ascending. Then

$$w'xy' \equiv' \mathbf{q}_i^{-1}\mathbf{q}_i wxy \ .$$

Also $vw'xy'z$ satisfies conditions (1),... ,(4) of Lemma 10 and is frameable, and the associated framing triple is $\langle M_0, N', M_4 \rangle$. There remains to show that $vw'xy'z \equiv_1' vwxyz$.

Since $[\![M_0, N', M_4]\!] = [\![M_0, N, M_4]\!]$ and since $\overline{\overline{U}} \geq 2$, there is a pair $\langle j, j' \rangle$ in N' such that $\langle i, j \rangle \in M_0$ and $\langle i', j' \rangle \in M_4$. Since N is one-one, therefore $i \neq j$ and $i' \neq j'$. Since $\langle i, j \rangle \in M_0$, therefore $v \equiv' v\mathbf{e}_{i,j}$. Since $\langle i', j' \rangle \in M_4$, therefore $z \equiv' \mathbf{e}_{i',j'}z$. To show that $vw'xy'z \equiv_1' vwxyz$, it therefore suffices to show that $\mathbf{e}_{i,j}\mathbf{q}_i^{-1}\mathbf{q}_i wxy\mathbf{e}_{i',j'} \equiv_1' \mathbf{e}_{i,j}wxy\mathbf{e}_{i',j'}$. Thus, since $\mathbf{e}_{i,j} \equiv' \mathbf{p}_{(0,i)}\mathbf{p}_{(i,j)}\mathbf{e}_{0,1}\mathbf{p}_{(i,j)}\mathbf{p}_{(0,i)}$ and $\mathbf{e}_{i',j'} \equiv' \mathbf{p}_{(0,i')}\mathbf{p}_{(i,j')}\mathbf{e}_{0,1}\mathbf{p}_{(0,i')}\mathbf{p}_{(1,j')}$, it suffices to show the following:

$$\text{(1)} \qquad \begin{aligned} &\mathbf{p}_{(0,i)}\mathbf{p}_{(1,j)}\mathbf{e}_{0,1}\mathbf{p}_{(1,j)}\mathbf{p}_{(0,i)}\mathbf{q}_i^{-1}\mathbf{q}_i wxy\mathbf{p}_{(0,i')}\mathbf{p}_{(1,j')}\mathbf{e}_{0,1}\mathbf{p}_{(1,j')}\mathbf{p}_{(0,i')} \\ &\equiv_1' \mathbf{p}_{(0,i)}\mathbf{p}_{(1,j)}\mathbf{e}_{0,1}\mathbf{p}_{(1,j)}\mathbf{p}_{(0,i)}wxy\mathbf{p}_{(0,i')}\mathbf{p}_{(1,j')}\mathbf{e}_{0,1}\mathbf{p}_{(1,j')}\mathbf{p}_{(0,i')} \ . \end{aligned}$$

We now assume that $j \neq 0$. With minor changes, an argument similar to the one to be given can, but will not, be carried out for the case where $j = 0$. Since $j \neq 0$, therefore $\mathbf{p}_{(1,j)}\mathbf{q}_0^{-1}\mathbf{q}_0 \equiv' \mathbf{q}_0^{-1}\mathbf{q}_0\mathbf{p}_{(1,j)}$.
Since also $\mathbf{p}_{(0,i)}\mathbf{q}_i^{-1}\mathbf{q}_i \equiv' \mathbf{q}_0^{-1}\mathbf{q}_0\mathbf{p}_{(0,i)}$, therefore $\mathbf{p}_{(1,j)}\mathbf{p}_{(0,i)}\mathbf{q}_i^{-1}\mathbf{q}_i$
$\equiv' \mathbf{q}_0^{-1}\mathbf{q}_0\mathbf{p}_{(1,j)}\mathbf{p}_{(0,i)}$. Hence assertion (1) follows from the following:

$$\text{(2)} \qquad \begin{aligned} &\mathbf{e}_{0,1}\mathbf{q}_0^{-1}\mathbf{q}_0\mathbf{p}_{(1,j)}\mathbf{p}_{(0,i)}wxy\mathbf{p}_{(0,i')}\mathbf{p}_{(1,j')}\mathbf{e}_{0,1} \\ &\equiv_1' \mathbf{e}_{0,1}\mathbf{p}_{(1,j)}\mathbf{p}_{(0,i)}wxy\mathbf{p}_{(0,i')}\mathbf{p}_{(1,j')}\mathbf{e}_{0,1}n \ . \end{aligned}$$

Let N'' be the relation that is given below. Then for $gd(vwxyz) = \langle s, t \rangle$, there holds the equality that is given underneath.

$$N'' = (1, j) \circ (0, i) \circ N \circ (0, i') \circ (1, j') \ .$$

$$h(\mathbf{p}_{(1,j)}\mathbf{p}_{(0,i)}wxy\mathbf{p}_{(0,i')}\mathbf{p}_{(1,j')}) = {}_s[\![N'']\!]_t \ .$$

Since N is one-one, therefore N'' is one-one. Since $\langle i, i' \rangle$ and $\langle j, j' \rangle$ are in N, therefore $\langle 0, 0 \rangle$ and $\langle 1, 1 \rangle$ are in N'' and $N'' \cap -\{\langle 0, 0 \rangle, \langle 1, 1 \rangle\}$ is included in $\{\langle m, n \rangle : 2 \leq m < \omega,\ 2 \leq n < \omega\}$. There follows:

${}_s[\![N'']\!]_t = h(w''x''y'')$, for some w'', x'', y'' such that
(i) $w'' = \emptyset$ or w'' is a $\langle \mathbf{q}^{-1} \rangle$ word that does not contain $\mathbf{q}_0^{-1}$ or $\mathbf{q}_1^{-1}$,
(ii) $x'' = \emptyset$ or x'' is a $\langle \mathbf{p} \rangle$ word that does not contain $\mathbf{p}_0$ or $\mathbf{p}_1$,
(iii) $y'' = \emptyset$ or y'' is a $\langle \mathbf{q} \rangle$ word that does not contain $\mathbf{q}_0$ or $\mathbf{q}_1$.

The above two equalities involving ${}_s[\![N'']\!]_t$ together yield the equality that is shown below. Since it does not involve $\mathbf{e}$, from it there follows the assertion (3) that is given underneath it.

$$h(\mathbf{p}_{(1,j)}\mathbf{p}_{(0,i)}wxy\mathbf{p}_{(0,i')}\mathbf{p}_{(1,j')}) = h(w''x''y'') \; .$$

$$\mathbf{p}_{(1,j)}\mathbf{p}_{(0,i)}wxy\mathbf{p}_{(0,i')}\mathbf{p}_{(1,j')} \equiv' w''x''y'' \; . \tag{3}$$

In view of (3) the proof of (2) can be reduced to proving the following assertion about $w''x''y''$:

$$\mathbf{e}_{0,1}\mathbf{q}_0^{-1}\mathbf{q}_0 w''x''y''\mathbf{e}_{0,1} \equiv_1' \mathbf{e}_{0,1}w''x''y''\mathbf{e}_{0,1} \; . \tag{4}$$

Since by conditions (i), (ii), (iii),

$$w''x''y''\mathbf{e}_{0,1} \equiv' \mathbf{e}_{0,1}w''x''y'' \; ,$$

to show that (4), it suffices to show that

$$\mathbf{e}_{0,1}\mathbf{q}_0^{-1}\mathbf{q}_0\mathbf{e}_{0,1} \equiv_1' \mathbf{e}_{0,1}\mathbf{e}_{0,1} \; , \tag{5}$$

which holds by QE_1 and $\mathbf{e}_{0,1}\mathbf{e}_{0,1} \equiv' \mathbf{e}_{0,1}$. □

Recall that $\equiv$ is the congruence on $\mathbf{F}(\Gamma(\tilde{\mathbf{p}}, \mathbf{q}, \mathbf{e}))$ that is generated by $\equiv' \cup QE_1 \cup PQE_1 \cup PQE_2$.

LEMMA 13. *For any u in $F(\Gamma(\tilde{\mathbf{p}}, \mathbf{q}, \mathbf{e}))$ either $u \equiv \mathbf{q}\mathbf{q}^{-1}$ or there is a frameable $vwxyz$ whose associated framing triple is both precanonical and trim such that $u \equiv vwxyz$.*

Proof. This follows from Lemmas 8, 9, 10, 11, and 12 and the fact that Lemma 12 holds in particular for those N' such that $\langle M_0, N', M_4 \rangle$ is trim. □

LEMMA 14. *Assume that $vwxyz$ and $v'w'x'y'z'$ are frameable, that their respective associated framing triple is both precanonical and trim, and that $h(vwxyz) = h(v'w'x'y'z')$. Then $vwxyz \equiv' v'w'x'y'z'$.*

Proof. Let $\langle M_0, N, M_4\rangle$ and $\langle M_0', N', M_4'\rangle$ be the framing triple associated with $vwxyz$ or $v'w'x'y'z'$ respectively. Since $h(vwxyz) = h(v'w'x'y'z')$, therefore $[\![M_0, N, M_4]\!] = [\![M_0', N', M_4']\!]$. Also, for some s and t such that either $s \geq 1$ or $t \geq 1$, $gd(vwxyz) = gd(v'w'x'y'z') = \langle s,t\rangle$. Further, since $\overline{\overline{U}} \geq 2$ and since $\langle M_0, N, M_4\rangle$ and $\langle M_0', N', M_4'\rangle$ are precanonical therefore, by Theorem III.3(c), $M_0 = M_0'$ and $M_4 = M_4'$. Moreover, by Lemma III.1(a) and Theorem III.3(b), $M_0 \circ N \circ M_4 = M_0 \circ N' \circ M_4$.

First, assume that $N = N'$. For some M and some r such that $0 \leq r \leq \min\{s,t\}$, the special triple associated with wxy is ${}_{s-r}\langle M\rangle_{t-r}$ and $N = M \cup \{\langle s - r + n,\ t - r + n\rangle : n < r\}$. Also, for some M' and some r' such that $0 \leq r' \leq \min\{s,t\}$, the special triple associated with $w'x'y'$ is ${}_{s-r'}\langle M'\rangle_{t-r'}$ and $N' = M' \cup \{\langle s - r' + n,\ t - r' + n\rangle : n < r'\}$. Since $N = N'$ and hence ${}_s[\![N]\!]_t = {}_s[\![N']\!]_t$ there follows that $h(\mathbf{q}_s\mathbf{q}_s^{-1}wxy) = h(\mathbf{q}_s\mathbf{q}_s^{-1}w'x'y')$. Hence, by Theorem VI.10, $\mathbf{q}_s\mathbf{q}_s^{-1}wxy \equiv' \mathbf{q}_s\mathbf{q}_s^{-1}w'x'y'$. Since both $vwxyz$ and $v'w'x'y'z'$ have grade $\langle s,t\rangle$, therefore $vwxyz \equiv' v\mathbf{q}_s\mathbf{q}_s^{-1}wxyz$ and $v'w'x'y'z' \equiv' v'\mathbf{q}_s\mathbf{q}_s^{-1}w'x'y'z'$. There follows that $vwxyz \equiv' v'wxyz'$. Since ${}_s\langle M_0\rangle$ is associated with both v and v', therefore either $v = v' = \emptyset$ or $v \equiv' v'$. Since ${}_t\langle M_4\rangle$ is associated with both z and z', therefore either $z = z' = \emptyset$ or $z \equiv' z'$. There now follows that $vwxyz \equiv' v'w'x'y'z'$.

Now assume that $N \neq N'$. We want to show that for some w'', x'', and y'', $vw''x''y''z$ is frameable, the associated framing triple is $\langle M_0, N', M_4\rangle$, and $vwxyz \equiv' vw''x''y''z$. By what has just been shown, $vw''x''y''z \equiv' v'w'x'y'z'$. Then there follows that $vwzyx \equiv' v'w'x'y'z'$.

To show that there is some $vw''x''y''z$ that satisfies the above conditions, it suffices to show that this holds when $\overline{\overline{N \cap -N'}} = 1$. For $\overline{\overline{N \cap -N'}} \geq 1$, the same conclusion then follows by induction on $\overline{\overline{N \cap -N'}}$. Assume therefore that $\overline{\overline{N \cap -N'}} = 1$ Let $\langle i,j\rangle \in N \cap -N'$. Since $\langle M_0, N, M_4\rangle$ and $\langle M_0, N', M_4\rangle$ are precanonical and trim and since $M_0 \circ N \circ M_4 = M_0 \circ N' \circ M_4$, therefore for some i' and j' there hold the following:

$$\begin{aligned} &\langle i,i'\rangle \in M_0, \quad \langle j,j'\rangle \in M_4, \quad N' = (N \cap -\{\langle i,j\rangle\}) \cup \{\langle i',j'\rangle\}\ , \\ &\text{and} \quad N = (N' \cap -\{\langle i',j'\rangle\}) \cup \{\langle i,j\rangle\}\ . \end{aligned}$$

Since $h(vwxyz) = [\![M_0, N, M_4]\!]$, there follows that $h(v\mathbf{p}_{(i,i')}wxy\mathbf{p}_{(j,j')}z) = [\![M_0, N', M_4]\!]$. Since $\langle i,i'\rangle \in M_0$, therefore $v \equiv' v\mathbf{e}_{i,i'}$. Since $\mathbf{e}_{i,i'}\mathbf{p}_{(i,i')} \equiv' \mathbf{e}_{i,i'}$, there follows that $v \equiv' v\mathbf{p}_{(i,i')}$. Since $\langle j,j'\rangle \in M_4$ therefore, similarly $z \equiv' \mathbf{p}_{(j,j')}z$. There now follows that

$$vwxyz \equiv' v\mathbf{p}_{(i,i')}wxy\mathbf{p}_{(j,j')}z\ .$$

There are w'', x'', and y'' such that $vw''x''y''z$ is frameable and such that

$$v\mathbf{p}_{(i,i')}wxy\mathbf{p}_{(j,j')}z \equiv' vw''x''y''z \ .$$

The framing triple associated with $vw''x''y''z$ is $\langle M_0, N', M_4\rangle$. Also, $vwxyz \equiv' vw''x''y''z$. □

We can now give a presentation of $\mathbf{S}_{pqe} = \langle S_{pqe}, \circ, ^{\smile}\rangle$.

THEOREM 15. *For any u and u' in $F(\Gamma(\tilde{\mathbf{p}}, \mathbf{q}, \mathbf{e}))$, $h(u) = h(u')$ if and only if $u \equiv u'$. There follows that $\mathbf{S}_{pqe} = \mathbf{F}(\Gamma(\tilde{\mathbf{p}}, \mathbf{q}, \mathbf{e}))/\equiv$.*

Proof. By Lemmas 13 and 14, if $h(u) = h(u')$, then $u \equiv u'$. By Lemma 1(b), if $u \equiv u'$, then $h(u) = h(u')$. □

In order to present $\langle \mathbf{S}_{pqe}, \subseteq\rangle$ we let $\leq\!\cdot$ be the least quasi-ordering of $F(\Gamma(\tilde{\mathbf{p}}, \mathbf{q}, \mathbf{e}))$ that includes $\equiv$, the relations $PQ_5 = \{\langle \mathbf{q}\mathbf{p}_0^2\mathbf{q}^{-1}, \mathbf{q}\mathbf{q}^{-1}\rangle\}$ and $PQ_6 = \{\langle \mathbf{q}\mathbf{p}_0^2\mathbf{q}^{-1}, \mathbf{q}^{-1}\mathbf{q}\rangle\}$ used in presenting $\langle \mathbf{S}_{pq}, \subseteq\rangle$, and the relations $P_4 = \{\langle \mathbf{p}_i^2\mathbf{p}_i, \mathbf{p}_i\rangle : i < \omega\}$ and $PE_5 = \{\langle \mathbf{e}, \mathbf{p}_0^2\rangle\}$ used in presenting $\langle \mathbf{S}_{pe}, \subseteq\rangle$. Thus, $\leq\!\cdot$ is also the least quasi-ordering of $F(\Gamma(\tilde{\mathbf{p}}, \mathbf{q}, \mathbf{e}))$ that includes the least quasi-ordering of $F(\Gamma(\tilde{\mathbf{p}}, \mathbf{q}))$ used in presenting $\langle \mathbf{S}_{pq}, \subseteq\rangle$, the least quasiordering of $F(\mathbf{p}, \mathbf{e})$ used in presenting $\langle \mathbf{S}_{pe}, \subseteq\rangle$, the three relations $QE_1 = \{\langle \mathbf{e}\mathbf{q}^{-1}\mathbf{q}\mathbf{e}, \mathbf{e}\rangle\}$, $PQE_1 = \{\langle \mathbf{q}\mathbf{e}\mathbf{q}^{-1}, \mathbf{q}\mathbf{p}_0^2\mathbf{q}^{-1}\rangle\}$, $PQE_2 = \{\langle \mathbf{e}\mathbf{q}\mathbf{p}_0\mathbf{p}_1, \mathbf{q}\mathbf{p}_0\mathbf{p}_1\mathbf{e}\rangle\}$ used in presenting $\mathbf{S}_{pqe}$, and their converses.

THEOREM 16. *For any u and u' in $F(\Gamma(\mathbf{p}, \mathbf{q}, \mathbf{e}))$, $h(u) \subseteq h(u')$ if and only if $u \leq\!\cdot\, u'$. There follows that $\langle \mathbf{F}(\Gamma(\mathbf{p}, \mathbf{q}, \mathbf{e}), \leq\!\cdot\rangle/\equiv$ is isomorphic to $\langle \mathbf{S}_{pqe}, \subseteq\rangle$.*

Proof. If $\langle u, u'\rangle$ is in $\equiv$, PQ_5, PQ_6, P_4, or PE_5 then, as we saw earlier at the relevant places, $h(u) \subseteq h(u')$. Hence, if $u \leq\!\cdot\, u'$ then $h(u) \subseteq h(u')$.

Assume that $h(u) \subseteq h(u'')$. Let $gd\,u = \langle s, t\rangle$. Then also $h(u) \subseteq {}_{[s,\omega)}\mathring{1} \circ h(u'')$. Let $u' = \mathbf{q}_s\mathbf{q}_s^{-1}u''$. Then $gd\,u' = \langle s, t\rangle$. Also $h(u) \subseteq h(u')$. Further, by Theorem VI.11 and the definition of $\leq\!\cdot$, $\mathbf{q}_s\mathbf{q}_s^{-1}u'' \leq\!\cdot\, u''$ and hence $u' \leq\!\cdot\, u''$. Thus, it suffices to show that $u \leq\!\cdot\, u'$.

First suppose that $u' \equiv \mathbf{q}\mathbf{q}^{-1}$ and hence $h(u') = {}_{[0,\omega)}\mathring{1}$. Since $h(u) \subseteq h(u')$, therefore $h(u) = {}_{[n,\omega)}\mathring{1}$ for some n. By Theorem 15, $u \equiv \mathbf{q}_n\mathbf{q}_n^{-1}$. By Theorem VI.11 and the definition of $\leq\!\cdot$, $\mathbf{q}_n\mathbf{q}_n^{-1} \leq\!\cdot\ \mathbf{q}\mathbf{q}^{-1}$. Hence $u \leq\!\cdot\, u'$.

Now suppose that not $u' \equiv \mathbf{q}\mathbf{q}^{-1}$ and hence also not $u \equiv \mathbf{q}\mathbf{q}^{-1}$. Then, by Lemma 13, for some $vwxyz$ and $v'w'x'y'z'$ that are frameable and whose associated framing triple is both precanonical and trim, there holds $u \equiv vwxyz$ and $u' \equiv v'w'x'y'z'$. Let $\langle M_0, N, M_4\rangle$ and $\langle M_0', N', M_4'\rangle$ be the triple associated with

$vwxyz$ and $v'w'x'y'z'$, respectively. Since $h(vwxyz) \subseteq h(v'w'x'y'z')$ and hence $[\![M_0, N, M_4]\!] \subseteq [\![M'_0, N', M'_4]\!]$, and since $[\![M_0, N, M_4]\!]$ and $[\![M'_0, N, M'_4]\!]$ have the same grade $\langle s, t\rangle$, therefore, by Theorem III.3(c), $M'_0 \subseteq M_0$ and $M'_4 \subseteq M_4$. There follows that ${}_s[\![M_0]\!] \subseteq {}_s[\![M'_0]\!]$, and hence $h(v) \subseteq h(v')$, and that ${}_t[\![M_4]\!] \subseteq {}_4[\![M'_4]\!]$, and hence $h(z) \subseteq h(z')$. Then, by Lemma VI.32 and the definition of $\lessdot$, $v \lessdot v'$ and $z \lessdot z'$. There follows that $vwxyz \lessdot v'xwyz'$. Thus to show that $u \lessdot u'$, it suffices to show that $v'wxyz' \lessdot v'w'x'y'z'$.

Since $vwxyz$ is frameable and the associated triple is precanonical, therefore $v'xwyz'$ is frameable, its associated triple is $\langle M'_0, N, M'_4\rangle$, and $\langle M'_0, N, M'_4\rangle$ is precanonical. Also, from ${}_s[\![M_0]\!] \subseteq {}_s[\![M'_0]\!]$, ${}_t[\![M_4]\!] \subseteq {}_t[\![M'_4]\!]$, and $[\![M_0, N, M_4]\!] \subseteq [\![M'_0, N', M'_4]\!]$ there follows that $[\![M'_0, N, M'_4]\!] \subseteq [\![M'_0, N', M'_4]\!]$. Then there is some N'' such that $\langle M'_0, N'', M'_4\rangle$ is precanonical, $[\![M'_0, N'', M'_4]\!] = [\![M'_0, N, M'_4]\!]$, and $N' \subseteq N''$. Also, there is some $v'w''x''y''z'$ that is frameable whose associated triple is $\langle M'_0, N'', M'_4\rangle$. By Theorem 15, or by Lemma 14, $v'wxyz' \equiv v'w''x''y''z'$. Since $N' \subseteq N''$ and hence ${}_{s'}[\![N'']\!]_{t'} \subseteq {}_{s'}[\![N']\!]_{t'}$, therefore $h(w''x''y'') \subseteq h(w'x'y')$. From Theorem VI.11 and the definition of $\lessdot$ there follows that $w''x''y'' \lessdot w'x'y'$ and hence $v'w''x''y''z' \lessdot v'w'x'y'z'$. There now follows that $v'wxyz' \lessdot v'w'x'y'z'$. □

We now let $\equiv^+$ be the congruence on $\mathbf{F}(\Gamma(\tilde{\mathbf{p}}, \mathbf{q}, \mathbf{e}))$ that is generated by $\equiv \cup P_4$. Thus, $\equiv^+$ is also the least congruence on $\mathbf{F}(\Gamma(\tilde{\mathbf{p}}, \mathbf{q}, \mathbf{e}))$ that includes the congruence used in presenting ${}_{[\omega]}\mathbf{S}_{pq}$, the congruence used in presenting ${}_{[\omega]}\mathbf{S}_{pe}$, QE_1, PQE_1, and PQE_2. Moreover, PQE_1 can be replaced in this characterization by the simpler relation $\{\langle \mathbf{qeq}^{-1}, \mathbf{qq}^{-1}\rangle\}$, since $\{\langle \mathbf{qp}_0^2\mathbf{q}^{-1}, \mathbf{qq}^{-1}\rangle\}$ is included in the congruence used in presenting ${}_{[\omega]}\mathbf{S}_{pq}$.

We now also let $\lessdot^+$ be the least quasiordering of $F(\Gamma(\tilde{\mathbf{p}}, \mathbf{q}, \mathbf{e}))$ that is compatible with $\mathbf{F}(\Gamma(\tilde{\mathbf{p}}, \mathbf{q}, \mathbf{e}))$ and includes $\equiv^+$ and $PQ_6 = \{\langle \mathbf{qp}_0^2\mathbf{q}^{-1}, \mathbf{q}^{-1}\mathbf{q}\rangle\}$ or, alternatively, $\equiv^+$ and the simpler relation $\{\langle \mathbf{qq}^{-1}, \mathbf{q}^{-1}\mathbf{q}\rangle\}$.

A presentation of ${}_{[\omega]}\mathbf{S}_{pqe} = \langle {}_{[\omega]}S_{pqe}, \circ, {}^\smile\rangle$ and one of $\langle {}_{[\omega]}\mathbf{S}_{pqe}, \subseteq\rangle$ can now be given as follows.

THEOREM 17. (a) *For any u and u' in $F(\Gamma(\tilde{\mathbf{p}}, \mathbf{q}, \mathbf{e}))$, $\overset{\omega}{\to}(h(u)) = \overset{\omega}{\to}(h(u'))$ if and only if $u \equiv^+ u'$. There follows that ${}_{[\omega]}\mathbf{S}_{pqe}$ is isomorphic to $\mathbf{F}(\Gamma(\tilde{\mathbf{p}}, \mathbf{q}, \mathbf{e}))/\equiv^+$.*

(b) *For any u and u' in $F(\Gamma(\tilde{\mathbf{p}}, \mathbf{q}, \mathbf{e}))$, $\overset{\omega}{\to}(h(u)) \subseteq \overset{\omega}{\to}(h(u'))$ if and only if $u \lessdot^+ u'$. There follows that $\langle {}_{[\omega]}\mathbf{S}_{pqe}, \subseteq\rangle$ is isomorphic to $\langle \mathbf{F}(\Gamma(\tilde{\mathbf{p}}, \mathbf{q}, \mathbf{e})), \lessdot^+\rangle/\equiv^+$.*

Proof. (a) From what we saw earlier, there follows that $u \equiv^+ u'$ implies that $\overset{\omega}{\to}(h(u)) = \overset{\omega}{\to}(h(u'))$. Now assume that $\overset{\omega}{\to}(h(u)) = \overset{\omega}{\to}(h(u'))$. Let $gd\,u = \langle s, t\rangle$ and $gd\,u' = \langle s', t'\rangle$. Then either $\langle s+n, t+n\rangle = \langle s', t'\rangle$ for some n and $h(\mathbf{q}_{s'}\mathbf{q}_{s'}^{-1}u) = h(u')$

or else $\langle s, t\rangle = \langle s'+n, t'+n\rangle$ for some n and $h(u) = h(\mathbf{q}_s\mathbf{q}_s^{-1}u')$. Since $\mathbf{q}_{s'}\mathbf{q}_{s'}^{-1}u \equiv^+ u$ and $\mathbf{q}_s\mathbf{q}_s^{-1}u' \equiv^+ u'$, therefore, by Theorem 15, in either case $u \equiv^+ u'$.

(b) The proof is similar to that of part (a), except for using Theorem 16 in place of Theorem 15. □

Appendix. Presentation of ${}_\delta\mathbf{S}_q$ and Related Structures

At the end of Chapter V we noted a certain conservational aspect of our presentation of $\mathbf{S}_p$. Our aim in this Appendix is to show that the presentation of $\mathbf{S}_q = \langle S_q, \circ, {}^\smile \rangle$ that was given in Chapter V is in a similar sense conservational. For $2 \le \delta \le \omega$ and ${}_\delta S_q$ the closure under $\circ$ and ${}^\smile$ of $\{q_i : i < \delta\}$, this will yield a presentation of ${}_\delta\mathbf{S}_q = \langle {}_\delta S_q, \circ, {}^\smile \rangle$.

Throughout, δ shall be an ordinal such that $2 \le \delta \le \omega$. We let ${}_\delta\Gamma(\tilde{\mathbf{q}}) = \{\mathbf{q}_i : i < \delta\}$, $F({}_\delta\Gamma(\tilde{\mathbf{q}})) = F_{SI}({}_\delta\Gamma(\tilde{\mathbf{q}}))$, and $\mathbf{F}({}_\delta\Gamma(\tilde{\mathbf{q}})) = \mathbf{F}_{SI}({}_\delta\Gamma(\tilde{\mathbf{q}}))$. Also, for any relation R on $F({}_\delta\Gamma(\tilde{\mathbf{q}}))$, we let ${}_\delta R = R \cap (F({}_\delta\Gamma(\tilde{\mathbf{q}})) \times F({}_\delta\Gamma(\tilde{\mathbf{q}})))$. Thus, for example, if R is the relation Q_2 of Chapter V, then ${}_3Q_2$ consists of the three pairs $\langle \mathbf{q}_0\mathbf{q}_1^{-1}, \mathbf{q}_0^{-1}\mathbf{q}_0 \rangle$, $\langle \mathbf{q}_0\mathbf{q}_2^{-1}, \mathbf{q}_1^{-1}\mathbf{q}_0 \rangle$, $\langle \mathbf{q}_1\mathbf{q}_2^{-1}, \mathbf{q}_1^{-1}\mathbf{q}_1 \rangle$. Throughout, ${}_\delta\!\equiv$ will be the congruence on $F({}_\delta\Gamma(\tilde{\mathbf{q}}))$ that is generated by ${}_\delta Q_1 \cup {}_\delta Q_2 \cup {}_\delta Q_3 \cup {}_\delta Q_4$, where Q_1, Q_2, Q_3, Q_4 are the relations on $F({}_\omega\Gamma(\tilde{\mathbf{q}}))$ described before Lemma 2 of Chapter V. Also, $\equiv$ will throughout be the congruence on $F({}_\omega\Gamma(\tilde{\mathbf{q}}))$ that has been used in our presentation of $\mathbf{S}_q$ and is generated by $Q_1 \cup Q_2 \cup Q_3 \cup Q_4$. Our aim is to show that $\equiv$ is conservational in the following sense:

$$ {}_\delta\!\equiv \text{ coincides with } \equiv \cap (F({}_\delta\Gamma(\tilde{\mathbf{q}})) \times F({}_\delta\Gamma(\tilde{\mathbf{q}}))) \ . $$

We begin by generalizing Lemmas V.3 and V.4 as follows: We replace $\equiv$ by ${}_\delta\!\equiv$, in the hypothesis of V.4 we replace $F(\Gamma(\tilde{\mathbf{q}}))$ by $F({}_\delta\Gamma(\tilde{\mathbf{q}}))$, and to both V.3 and V.4 we add the hypothesis that every $\mathbf{q}_r$ and every $\mathbf{q}_r^{-1}$ involved satisfies $r < \delta$. A proof of these two generalizations is obtained from their original proof by replacing everywhere $Q_1, Q_2, Q_3, Q_4, \equiv$ by ${}_\delta Q_1, {}_\delta Q_2, {}_\delta Q_3, {}_\delta Q_4, {}_\delta\!\equiv$ respectively.

Lemma V.3, thus generalized, is used in proving part (a) of the following lemma.

LEMMA 1. *Consider any word* $x = \mathbf{q}_{i_0}^{-1} \dots \mathbf{q}_{i_{m-1}}^{-1}$ *formed from* $\{\mathbf{q}_i^{-1} : i < \delta\}$. *For any* m', $0 \le m' < m$, *let* $s_{m'} = i_{m'} + m'$ *and* $t_{m'} = \max(0, i_{m'} - (m - m' - 1))$. *Let* $s = \max\{s_m : m' < m\}$ *and* $t = \max\{t_{m'} : m' < m\}$.

(a) $x \;{}_\delta\!\equiv\; x\mathbf{q}_{t_{m'}}\mathbf{q}_{t_{m'}}^{-1}$, *if* $m' < m$. *In particular,* $x \;{}_\delta\!\equiv x\mathbf{q}_t\mathbf{q}_t^{-1}$.

(b) $x \equiv \mathbf{q}_{s_{m'}}\mathbf{q}_{s_{m'}}^{-1}x$, *if* $m' < m$. *In particular,* $x \equiv \mathbf{q}_s\mathbf{q}_s^{-1}x$.

(c) $gd\ x = \langle s, t\rangle$.

Proof. (a) Suppose that $0 < k \leq i < \delta$. Then, by generalized Lemmas V.3(d) and V.3(f), respectively; $\mathbf{q}_k\mathbf{q}_k^{-1}\mathbf{q}_i^{-1}\ {}_\delta{\equiv}\ q_i^{-1}\ {}_\delta{\equiv}\ q_i^{-1}\mathbf{q}_{k-1}\mathbf{q}_{k-q}^{-1}$. Also, if $i < k < \delta$ then, by generalized Lemma V.3(b) there again holds $\mathbf{q}_k\mathbf{q}_k^{-1}\mathbf{q}_i^{-1}\ {}_\delta{\equiv}\ \mathbf{q}_i^{-1}\mathbf{q}_{k-1}\mathbf{q}_{k-1}^{-1}$. Hence, for $i_m - (m - m' - 1) \geq 0$, part (a) follows by induction on the length $m - m' - 1$ of $\mathbf{q}_{i_{m'}}^{-1} \ldots \mathbf{q}_{i_{m-1}}^{-1}$. If $i_{m'} - (m - m' - 1) < 0$, then one uses, in addition, that, for any w in $F({}_\delta\Gamma(\tilde{\mathbf{q}}))$, $\mathbf{q}_0\mathbf{q}_0^{-1}w\ {}_\delta{\equiv}\ w\ {}_\delta{\equiv}\ w\mathbf{q}_0\mathbf{q}_0^{-1}$.

(b) Suppose that $k < i$. Then, by Lemmas V.3(f) and V.3(d), respectively, one has $\mathbf{q}_i^{-1}\mathbf{q}_k\mathbf{q}_k^{-1} \equiv \mathbf{q}_i^{-1} \equiv \mathbf{q}_{k+1}\mathbf{q}_{k+1}^{-1}\mathbf{q}_i^{-1}$. Also, if $i \leq k$, then, by Lemma V.3(b), one again has $\mathbf{q}_i\mathbf{q}_k\mathbf{q}_k^{-1} \equiv \mathbf{q}_{k+1}\mathbf{q}_{k+1}^{-1}\mathbf{q}_i^{-1}$. Hence, part (b) follows by Q_3 and induction on the length $m' + 1$ of $\mathbf{q}_{i_0}^{-1} \ldots \mathbf{q}_{i_{m'}}^{-1}$.

(c) This follows from parts (a) and (b) and Lemma V.2. It also can be verified directly, using the definitions of *gd* and of q_i^{-1} and $q_iq_i^{-1}$, $i < \omega$. □

Note that in part (b) of Lemma 1 one cannot replace $\equiv$ by ${}_\delta{\equiv}$. For example, suppose that $\delta < \omega$, $0 \leq i_0 \leq i_1 = \delta - 1$, and $x = \mathbf{q}_{i_0}^{-1}\mathbf{q}_{i_1}^{-1}$. Then $s_1 = i_1 + 1$. Hence according to part (b), $x \equiv \mathbf{q}_{i_1+1}\mathbf{q}_{i_1+1}^{-1}x$. In contrast, since $\delta = i + 1$ and since $\mathbf{q}_{i_1+1}$ and $\mathbf{q}_{i_1+1}^{-1}$ are not in $F({}_{i+1}\Gamma(\tilde{\mathbf{q}}))$, therefore $\langle x, \mathbf{q}_{i+1}\mathbf{q}_{i+1}^{-1}x\rangle$ is not in ${}_\delta{\equiv}$.

This contrast between Lemmas 1(a) and 1(b) illustrates that certain notions do not lend themselves to dealing with ${}_\delta{\equiv}$ when $\delta < \omega$. Among these are the following: strictly descending, strictly ascending, $\mathbf{q}$-normal. For example, suppose that $\delta < \omega$ and consider the word $\mathbf{q}_{\delta-1}^{-1}\mathbf{q}_{\delta-1}^{-1}$ of $F({}_\delta\Gamma(\tilde{\mathbf{q}}))$, which is not strictly descending. For the congruence $\equiv$ on $F(\Gamma(\tilde{\mathbf{q}}))$, there holds $\mathbf{q}_{\delta-1}^{-1}\mathbf{q}_{\delta-1}^{-1} \equiv \mathbf{q}_\delta^{-1}\mathbf{q}_{\delta-1}^{-1}$, where $\mathbf{q}_\delta^{-1}\mathbf{q}_{\delta-1}^{-1}$ is strictly descending. However, $\mathbf{q}_\delta^{-1}\mathbf{q}_{\delta-1}^{-1}$ does not belong to $F({}_\delta\Gamma(\tilde{\mathbf{q}}))$. In fact, there is no word w in $F({}_\delta\Gamma(\tilde{\mathbf{q}}))$, that is strictly descending such that $\mathbf{q}_{\delta-1}^{-1}\mathbf{q}_{\delta-1}^{-1} \equiv w$. Thus, in place of the three notions mentioned one has to find notions that do not take one beyond $F({}_\delta\Gamma(\tilde{\mathbf{q}}))$.

A word x in $F({}_\delta\Gamma(\tilde{\mathbf{q}}))$ shall be ***nonstrictly ascending*** if and only if it is a word $\mathbf{q}_{i_0}^{-1} \ldots \mathbf{q}_{i_{m-1}}^{-1}$ formed from $\{\mathbf{q}_i^{-1} : i < \delta\}$ such that, whenever $m'' < m' < m$, then $i_{m''} \leq i_{m'}$. A word y in $F({}_\delta\Gamma(\tilde{\mathbf{q}}))$ shall be ***nonstrictly descending*** if and only if it is a word $\mathbf{q}_{j_0} \ldots \mathbf{q}_{j_{n-1}}$ formed from $\{\mathbf{q}_j : j < \delta\}$ such that, whenever $n'' < n' < n$, then $j_{n''} \geq j_{n'}$. A word w in $F({}_\delta\Gamma(\tilde{\mathbf{q}}))$ shall be ***in conservational normal form*** if and only if $w = x\mathbf{q}_r\mathbf{q}_r^{-1}y$ for some x, r, y such that x is either empty or nonstrictly ascending, y is either empty or nonstrictly descending, $r \geq t_x$ if $x \neq \emptyset$ and $gd\ x = \langle s_x, t_x\rangle$, and $r \geq s_y$ if $y \neq \emptyset$ and $gd\ y = \langle s_y, t_y\rangle$.

LEMMA 2. *For any v in $F({}_\delta\Gamma(\tilde{\mathbf{q}}))$ there is some $x\mathbf{q}_r\mathbf{q}_r^{-1}y$ in conservational normal form such that $v \;{}_\delta\!\equiv\; x\mathbf{q}_r\mathbf{q}_r^{-1}y$.*

Proof. Consider any v in $F({}_\delta\Gamma(\tilde{\mathbf{q}}))$. By the proof, generalized to $F({}_\delta\Gamma(\tilde{\mathbf{q}}))$, of Sublemma 1 of Lemma V.5, there are x' and y' in $F({}_\delta\Gamma(\tilde{\mathbf{q}}))$ such that $v \;{}_\delta\!\equiv x'y'$, x' contains no noncoupled occurrence of any $\mathbf{q}_j$, and y' contains no noncoupled occurrence of any $\mathbf{q}_i^{-1}$.

To avoid the need for certain case distinctions later on, a slightly stronger assertion will be useful. Suppose that $x' \neq \emptyset$. Then x' ends with some $\mathbf{q}_i^{-1}$, $i < \delta$. Let $k \leq i$. Then, by generalized Lemma V.3(f),$\mathbf{q}_i^{-1} \;{}_\delta\!\equiv\; \mathbf{q}_i^{-1}\mathbf{q}_k\mathbf{q}_k^{-1} \;{}_\delta\!\equiv\; \mathbf{q}_i^{-1}\mathbf{q}_k\mathbf{q}_k^{-1}\mathbf{q}_k\mathbf{q}_k^{-1}$. There follows that for some $k < \delta$, $x'y' \;{}_\delta\!\equiv\; x\mathbf{q}_k\mathbf{q}_k^{-1}\mathbf{q}_k\mathbf{q}_k^{-1}y'$. Now suppose that $y' \neq \emptyset$. Then y' begins with some $\mathbf{q}_j$, $j < \delta$. Let $k \leq j$. Then, by generalized Lemma V.3(c), $\mathbf{q}_j \;{}_\delta\!\equiv\; \mathbf{q}_{k'}\mathbf{q}_{k'}^{-1}\mathbf{q}_j \;{}_\delta\!\equiv\; \mathbf{q}_k\mathbf{q}_k^{-1}\mathbf{q}_k\mathbf{q}_k^{-1}y'$. Then there again follows that, for some $k < \delta$, $x'y' \;{}_\delta\!\equiv\; x'\mathbf{q}_k\mathbf{q}_k^{-1}\mathbf{q}_k\mathbf{q}_k^{-1}y'$. Thus, we can now make the following slightly stronger assertion: There are x' and y' such that $v \;{}_\delta\!\equiv\; x'y'$ and, in addition to the above condition on noncoupled occurrences, x' satisfies the further condition that it ends with some $\mathbf{q}_k\mathbf{q}_k^{-1}$, $k < \delta$, and y' satisfies the further condition that it begins with some $\mathbf{q}_k\mathbf{q}_k^{-1}$, $k < \delta$.

We let $gd\ x' = \langle s_{x'}, t_{x'}\rangle$ and $gd\ y' = \langle s_{y'}, t_{y'}\rangle$. Also, we let x'' consist of the noncoupled occurrences of $\mathbf{q}_i^{-1}$, $i < \delta$, if any, in the order in which they occur in x'. Further, we let y'' consist of the noncoupled occurrences of $\mathbf{q}_j$, $j < \delta$, if any, in the order in which they occur in y'.

SUBLEMMA 1. (a) *$x' \;{}_\delta\!\equiv\; x''z$, for some word z formed from $\{q_kq_k^{-1} : k < \delta\}$.*
(b) *$y' \;{}_\delta\!\equiv\; z'y''$ for some word z' formed from $\{q_kq_k^{-1} : k < \delta\}$.*

Proof. (a) First suppose that there is no occurrence in x' of any $q_kq_k^{-1}$ to the left of any noncoupled occurrence of any $\mathbf{q}_i^{-1}$, $i < \delta$. Then $x' = x''z$ for some word z formed from $\{q_kq_k^{-1} : k < \delta\}$. Now suppose that there are such occurrences. Then one applies sufficiently often generalized Lemmas V.3(d) and V.3(b). In other words, one replaces as often as needed an occurrence of $q_kq_k^{-1}\mathbf{q}_i^{-1}$ such that $k \leq i+1$ by one of $\mathbf{q}_i^{-1}$ and an occurrence of $q_kq_k^{-1}\mathbf{q}_i^{-1}$ such that $k > i+1$ by one of $\mathbf{q}_i^{-1}\mathbf{q}_{k-1}\mathbf{q}_{k-1}^{-1}$. After sufficiently many replacements of this kind one obtains a word $x''z$ such that $x' \;{}_\delta\!\equiv\; x''z$, and z is a word formed from $\{q_kq_k^{-1} : k < \delta\}$.

(b) First suppose that there is no occurrence in y' of any $\mathbf{q}_k\mathbf{q}_k^{-1}$ to the right of any noncoupled occurrence of any $\mathbf{q}_j$, $j < \delta$. Then $y' = z'y''$ for some word z' formed from $\{\mathbf{q}_k\mathbf{q}_k^{-1} : k < \delta\}$. Now suppose that there are such occurrences. Then one applies sufficiently often generalized Lemmas V.3(c) and V.3(a). In other words, we replace as often as needed an occurrence of $\mathbf{q}_j\mathbf{q}_k\mathbf{q}_k^{-1}$ such that $k \leq j+1$ by one of $\mathbf{q}_j$ and an occurrence of $\mathbf{q}_j\mathbf{q}_k\mathbf{q}_k^{-1}$ such that $k > j+1$ by one of $\mathbf{q}_{k-1}\mathbf{q}_{k-1}^{-1}\mathbf{q}_j$. After sufficiently many replacements of this kind one obtains a word $z'y''$ such that $y' \;{}_\delta{\equiv}\; z'y''$ and such that z' is a word formed from $\{q_k q_k^{-1} : k < \delta\}$. □

Recall that $gd\ x' = \langle s_{x'}, t_{x'}\rangle$ and $gd\ y' = \langle s_{y'}, t_{y'}\rangle$. Let x'' and y'' be as in Sublemma 1.

SUBLEMMA 2. (a) $x' \;{}_\delta{\equiv}\; x''\mathbf{q}_{t_{x'}}\mathbf{q}_{t_{x'}}^{-1}$.

(b) $y' \;{}_\delta{\equiv}\; \mathbf{q}_{s_{y'}}\mathbf{q}_{s_{y'}}^{-1}y''$.

(c) $x'y' \;{}_\delta{\equiv}\; x''\mathbf{q}_r\mathbf{q}_r^{-1}y''$, *where* $r = \max(t_{x'}, s_{y'})$, *so that* $r \geq t_{x''}$, *if* $x'' \neq \emptyset$ *and* $gd\ x'' = \langle s_{x''}, t_{x''}\rangle$, *and* $r \geq s_{y''}$ *if* $y'' \neq \emptyset$ *and* $gd\ y'' = \langle s_{y''}, t_{y''}\rangle$.

Proof. (a) Since z is a word formed from $\{\mathbf{q}_k\mathbf{q}_k^{-1} : k < \delta\}$, therefore $gd\ z = \langle s_z, s_z\rangle$ for some $s_z < \delta$. Hence, by generalized Lemma V.3(g) and induction on the length of z, $z \;{}_\delta{\equiv}\; \mathbf{q}_{s_z}\mathbf{q}_{s_z}^{-1}$. If $x'' = \emptyset$ then $x' = z$, $\langle s_z, s_z\rangle = \langle s_{x'}, t_{x'}\rangle$, and $x' = x''z = z \;{}_\delta{\equiv}\; \mathbf{q}_{s_z}\mathbf{q}_{s_z}^{-1} = x''\mathbf{q}_{t_{x'}}\mathbf{q}_{t_{x'}}^{-1}$. Now assume that $x'' \neq \emptyset$. Let $gd\ x'' = \langle s_{x''}, t_{x''}\rangle$. Then, by parts (a) and (c) of Lemma 1, $x'' \;{}_\delta{\equiv}\; x''\mathbf{q}_{t''_x}\mathbf{q}_{t''_x}^{-1}$. Also, since $\langle s_{x'}, t_{x'}\rangle = gd\ x' = gd\ x'' \odot gd\ z = \langle s_{x''}, t_{x''}\rangle \circ \langle s_z, t_z\rangle$ and hence $t_{x'} = \max(t_{x''}, s_z)$, therefore $\mathbf{q}_{t_{x''}}\mathbf{q}_{t_{x''}}^{-1}\mathbf{q}_{s_z}\mathbf{q}_{s_z}^{-1} \;{}_\delta{\equiv}\; \mathbf{q}_{t_{x'}}\mathbf{q}_{t_{x''}}^{-1}$. Hence, again, $x' \;{}_\delta{\equiv}\; x''z \;{}_\delta{\equiv}\; x''\mathbf{q}_{t_{x''}}\mathbf{q}_{t_{x''}}^{-1}\mathbf{q}_{s_z}\mathbf{q}_{s_z}^{-1} \;{}_\delta{\equiv}\; x''\mathbf{q}_{t_{x'}}\mathbf{q}_{t_{x'}}^{-1}$.

(b) For words $\mathbf{q}_{j_0}\ldots\mathbf{q}_{j_{n-1}}$ formed from $\{\mathbf{q}_j : j < \delta\}$ there holds a dual of Lemma 1. Applying this dual to y', y'', z' one obtains a proof of part (b) by an argument similar to the one just used to prove part (a).

(c) By parts (a) and (b) and generalized Lemma V.3(g). Suppose that $x'' \neq \emptyset$ and $gd\ x'' = \langle s_{x''}, t_{x''}\rangle$. Then $\langle s_{x'}, t_{x'}\rangle = gd\ x' = gd\ x'' \odot gd\ z = \langle s_{x''}, t_{x''}\rangle \odot \langle s_z, s_z\rangle \geq \langle s_{x''}, t_{x''}\rangle$. Hence $t_{x'} \geq t_{x''}$. Since $r = \max(t_{x'}, s_{y'})$, therefore $r \geq t_{x''}$. By a similar argument, if $y'' \neq \emptyset$ and $gd\ y'' = \langle s_{y''}, t_{y''}\rangle$, then $s_{y'} \geq s_{y''}$ and hence $r \geq s_{y''}$. □

SUBLEMMA 3. *Let* ${}_\delta{\equiv}_1$ *be the congruence on* $\mathbf{F}({}_\delta\Gamma(\tilde{\mathbf{q}}))$ *that is generated by* ${}_\delta Q_1$.

(a) *For any word* x'' *formed from* $\{\mathbf{q}_i^{-1} : i < \delta\}$ *there is a word* x *that is nonstrictly ascending and of the same length as* x'' *such that* $x'' \;{}_\delta{\equiv}_1\; x$.

(b) *For any word* y'' *formed from* $\{\mathbf{q}_j : j < \delta\}$ *there is a word* y *that is nonstrictly descending and of the same length as* y'' *such that* $y'' \;{}_\delta{\equiv}_1\; y$.

Proof. (a) Whenever $i < j < \delta$ then, by ${}_\delta Q_1$, one has $\mathbf{q}_j^{-1}\mathbf{q}_i^{-1} \;{}_\delta{\equiv_1}\; (\mathbf{q}_i\mathbf{q}_j)^{-1} \;{}_\delta{\equiv_1}\; (\mathbf{q}_{j-1}\mathbf{q}_i)^{-1} \;{}_\delta{\equiv_1}\; \mathbf{q}_i^{-1}\mathbf{q}_{j-1}^{-1}$, where $\mathbf{q}_i^{-1}\mathbf{q}_{j-1}^{-1}$ is nonstrictly ascending. Hence, by replacing sufficiently often an occurrence of $\mathbf{q}_j^{-1}\mathbf{q}_i^{-1}$ such that $i < j < \delta$ by an occurrence of $\mathbf{q}_i^{-1}\mathbf{q}_{j-1}^{-1}$ one obtains from the given x'' a word x that is nonstrictly ascending and of the same length as x'' such that $x'' \;{}_\delta{\equiv_1}\; x$.

(b) Whenever $i < j < \delta$, then $\mathbf{q}_1\mathbf{q}_j \;{}_\delta{\equiv_1}\; \mathbf{q}_{j-1}\mathbf{q}_i$, where $\mathbf{q}_{j-1}\mathbf{q}_i$ is nonstrictly descending. Hence, by replacing sufficiently often an occurrence of $\mathbf{q}_i\mathbf{q}_j$ such that $i < j < \delta$ by an occurrence of $\mathbf{q}_{j-1}\mathbf{q}_i$ one obtains from the given y'' a word y that is nonstrictly descending and of the same length as y'' such that $y'' \;{}_\delta{\equiv_1}\; y$. □

We have seen in Sublemma 2(c) that $v \;{}_\delta{\equiv}\; x'y' \;{}_\delta{\equiv}\; x''zz'y'' \;{}_\delta{\equiv}\; x''\mathbf{q}_r\mathbf{q}_r^{-1}y''$. If $x'' = \emptyset$ let $x = \emptyset$ and if $x'' \neq \emptyset$ let x be related to x'' as in Sublemma 3(a). If $y'' = \emptyset$ let $y = \emptyset$ and if $y'' \neq \emptyset$ let y be related to y'' as in Sublemma 3(b). There follows that, if $x \neq \emptyset$ and $gd\ x = \langle s_x, t_x\rangle$, then $t_x = t_{x''} \leq r$, and if $y \neq \emptyset$ and $gd\ y = \langle s_y, t_y\rangle$, then $s_y = s_{y''} \leq r$. Thus, $x\mathbf{q}_r\mathbf{q}_r^{-1}y$ is in conservational normal form. Since $v \;{}_\delta{\equiv}\; x''\mathbf{q}_r\mathbf{q}_r^{-1}y''$ and $x''\mathbf{q}_r\mathbf{q}_r^{-1}y'' \;{}_\delta{\equiv}\; x\mathbf{q}_r\mathbf{q}_r^{-1}y$ there now follows that $x\mathbf{q}_r\mathbf{q}_r^{-1}y$ satisfies all the desired conditions. □

There is a certain one-one correspondence between the words that are strictly descending and the words that are nonstrictly ascending that is useful for the study of $\mathbf{S}_q$ and may have uses elsewhere. Given any m, $1 \leq m < \omega$, and any word $x = \mathbf{q}_{i_0}^{-1} \ldots \mathbf{q}_{i_{m-1}}^{-1}$ of length m that is strictly descending, we assign to x the word $\mathbf{q}_{j_0}^{-1} \ldots \mathbf{q}_{j_{m-1}}^{-1}$ of length m formed from $\{\mathbf{q}_j^{-1} : j < \omega\}$ such that

$$j_{m'} = i_{m-1-m'} - m' \,, \quad \text{for any } \; m' < m \,.$$

Also, given any m, $1 \leq m < \omega$, and any word $x' = \mathbf{q}_{j_0}^{-1} \ldots \mathbf{q}_{j_{m-1}}^{-1}$ of length m that is nonstrictly ascending, we assign to x' the word $\mathbf{q}_{i_0}^{-1} \ldots \mathbf{q}_{i_{m-1}}^{-1}$ of length m formed from $\{\mathbf{q}_i^{-1} : i < \omega\}$ such that

$$i_{m'} = j_{m-1-m'} + (m-1-m') \,, \quad \text{for any } \; m' < m \,.$$

Then, for any x that is strictly descending, the word x' assigned to x is nonstrictly ascending and the word assigned to x' is x. Also, for any x' that is nonstrictly ascending, the word x assigned to x' is strictly descending and the word assigned to x is x'. We shall say of x and x' that they ***correspond.***

A certain one-one correspondence between the words that are strictly ascending and the words that are nonstrictly descending will play a related role. Given any m, $1 \leq m < \omega$, and any word $y = \mathbf{q}_{i_0} \ldots \mathbf{q}_{i_{m-1}}$ of length m that is strictly ascending, we assign to y the word $\mathbf{q}_{j_0}\ldots\mathbf{q}_{j_{m-1}}$ of length m formed from $\{\mathbf{q}_j : j<\omega\}$ such that

$$j_{m'} = i_{m-1-m'} - (m-1-m')\ , \quad \text{for any} \quad m'<m\ .$$

Also, given any m, $1\le m<\omega$, and any word $y'=\mathbf{q}_{j_0}\cdots\mathbf{q}_{j_{m-1}}$ of length m that is nonstrictly descending, we assign to y' the word $\mathbf{q}_{i_0}\cdots\mathbf{q}_{i_{m-1}}$ of length m formed from $\{\mathbf{q}_i : i<\omega\}$ such that

$$i_{m'} = j_{m-1-m'} + m'\ , \quad \text{for any} \quad m'<m\ .$$

Then, for any y that is strictly ascending, the word y' assigned to y is nonstrictly descending and the word assigned to y' is y. Also, for any y' that is nonstrictly descending, the word y assigned to y' is strictly ascending and the word assigned to y is y'. We shall say of y and y' that they ***correspond.***

Between the words $\mathbf{q}_{s'}\mathbf{q}_{s'}^{-1}xy$ in $F(\Gamma(\tilde{\mathbf{q}}))$ that are $\mathbf{q}$-normal and the words $x'\mathbf{q}_r\mathbf{q}_r^{-1}y'$ in $F(\Gamma(\tilde{\mathbf{q}}))$ that are in conservational normal form, one can now define a one-one correspondence as follows. Consider any $v=\mathbf{q}_{s'}\mathbf{q}_{s'}^{-1}xy$ that is $\mathbf{q}$-normal. Let m, $0\le m<\omega$, be the length of x and let $r=s'-m$. Let x' and y' correspond to x or y respectively. Then to v there shall ***correspond*** the word $x'\mathbf{q}_r\mathbf{q}_r^{-1}y'$. Now consider any $v'=x'\mathbf{q}_r\mathbf{q}_r^{-1}y'$ that is in conservational normal form. Let m, $0\le m<\omega$, be the length of x' and let $s'=r+m$. Let x and y correspond to x' or y' respectively. Then to v' there shall ***correspond*** the word $\mathbf{q}_{s'}\mathbf{q}_{s'}^{-1}xy$.

LEMMA 3. *In parts* (a) *and* (b), ${}_\delta{\equiv}_1$ *shall be the congruence on* $\mathbf{F}({}_\delta\Gamma(\tilde{\mathbf{q}}))$ *that is generated by* ${}_\delta Q_1$.

(a) *For any x and x' in $F({}_\delta\Gamma(\tilde{\mathbf{q}}))$ such that x is strictly descending and x' is nonstrictly ascending, if x and x' correspond then $x\ {}_\delta{\equiv}_1\ x'$.*

(b) *For any y and y' in $F({}_\delta\Gamma(\tilde{\mathbf{q}}))$ such that y is strictly ascending and y' is nonstrictly descending, if y and y' correspond then $y\ {}_\delta{\equiv}_1\ y'$.*

(c) *For any v and v' in $F({}_\delta\Gamma(\tilde{\mathbf{q}}))$ such that v is $\mathbf{q}$-normal and v' is in conservational normal form, if v and v' correspond then $v\ {}_\delta{\equiv}\ v'$.*

Proof. (a) Let x and x' be of length m. To derive $x\ {}_\delta{\equiv}_1\ x'$ one uses $m-1$ assertions. Since the proof is essentially similar when $m<6$, we assume that $m\ge 6$ and let $x=\mathbf{q}_{i_0}^{-1}\mathbf{q}_{i_1}^{-1}\mathbf{q}_{i_2}^{-1}\cdots\mathbf{q}_{i_{m-3}}^{-1}\mathbf{q}_{i_{m-2}}^{-1}\mathbf{q}_{i_{m-1}}^{-1}$. Then the $m-1$ assertions that one uses are the following, for which the number of times one uses ${}_\delta Q_1$ is

$m-1, m-2, \ldots, 2, 1$, respectively.

$$\begin{array}{rcl} \mathbf{q}_{i_0}^{-1}\mathbf{q}_{i_1}^{-1}\mathbf{q}_{i_2}^{-1}\cdots\mathbf{q}_{i_{m-3}}^{-1}\mathbf{q}_{i_{m-2}}^{-1}\mathbf{q}_{i_{m-1}}^{-1} & {}_\delta{\equiv}_1 & \mathbf{q}_{i_1}^{-1}\mathbf{q}_{i_2}^{-1}\cdots\mathbf{q}_{i_{m-3}}^{-1}\mathbf{q}_{i_{m-2}}^{-1}\mathbf{q}_{i_{m-1}}^{-1}\mathbf{q}_{i_0-(m-1)}^{-1} \\ \mathbf{q}_{i_1}^{-1}\mathbf{q}_{i_2}^{-1}\cdots\mathbf{q}_{i_{m-3}}^{-1}\mathbf{q}_{i_{m-2}}^{-1}\mathbf{q}_{i_{m-1}}^{-1} & {}_\delta{\equiv}_1 & \mathbf{q}_{i_2}^{-1}\cdots\mathbf{q}_{i_{m-3}}^{-1}\mathbf{q}_{i_{m-2}}^{-1}\mathbf{q}_{i_{m-1}}^{-1}\mathbf{q}_{i_1-(m-2)}^{-1} \\ & \vdots & \\ \mathbf{q}_{i_{m-3}}^{-1}\mathbf{q}_{i_{m-2}}^{-1}\mathbf{q}_{i_{m-1}}^{-1} & {}_\delta{\equiv}_1 & \mathbf{q}_{i_{m-2}}^{-1}\mathbf{q}_{i_{m-1}}^{-1}\mathbf{q}_{i_{m-3}-2}^{-1} \\ \mathbf{q}_{i_{m-2}}^{-1}\mathbf{q}_{i_{m-1}}^{-1} & {}_\delta{\equiv}_1 & \mathbf{q}_{i_{m-1}}^{-1}\mathbf{q}_{i_{m-2}-1}^{-1} \end{array}$$

There follows that $x\ {}_\delta{\equiv}_1\ x'$, where $x' = \mathbf{q}_{j_0}^{-1}\cdots\mathbf{q}_{j_{m-1}}^{-1}$ and $j_{m-1-m''} = i_{m''} - (m - 1 - m'')$ for any $m'' < m$, and hence also $j_{m'} = i_{m-1-m'} - m'$ for any $m' < m$.

(b) Instead of moving, for any $m'' < m$, $\mathbf{q}_{i_m}^{-1}$ altogether $m-1-m''$ places to the right, changing it to $\mathbf{q}_{i_{m''}-(m-1-m'')}^{-1}$, one moves, for any $m'' < m$, $\mathbf{q}_{i''_m}$ altogether m'' places to the left, changing it to $\mathbf{q}_{i_{m''}-m''}$. Hence, if $y' = \mathbf{q}_{j_0}\cdots\mathbf{q}_{j_{m-1}}$ is the word thus resulting from $y = \mathbf{q}_{i_0}\cdots\mathbf{q}_{i_{m-1}}$, then, for any $m' < m$, if $m'' = m-1-m'$, then $j_{m'} = i_{m''} - m'' = i_{m-1-m'} - (m-1-m')$.

(c) Let $v = q_{s'}q_{s'}^{-1}xy$, let x be of length m, and let $gd(xy) = \langle s, t\rangle$. For proof, one uses parts (a) and (b) and, if $s' > s$, generalized Lemma V.3(b) altogether m times while, if $s' = s$, one uses ${}_\delta Q_3$. □

LEMMA 4. (a) *If x and x' are nonstrictly ascending and $x \equiv x'$, then $x = x'$.*

(b) *If y and y' are nonstrictly descending and $y \equiv y'$, then $y = y'$.*

(c) *If v and v' are in conservational normal form and $v \equiv v'$, then $v = v'$.*

Proof. (a) Assume that x and x' are nonstrictly ascending and $x \equiv x'$. Let $w = \mathbf{q}_{i_0}^{-1}\cdots\mathbf{q}_{i_{m-1}}^{-1}$ and $w' = \mathbf{q}_{i'_0}^{-1}\cdots\mathbf{q}_{i'_{m'-1}}^{-1}$ be the word that is strictly descending and corresponds to x or to x' respectively. By Lemma 3(a), $w = \mathbf{q}_{i_0}^{-1}\cdots\mathbf{q}_{i_{m-1}}^{-1} \equiv \mathbf{q}_{i'_0}^{-1}\cdots\mathbf{q}_{i'_{m'-1}}^{-1} = w'$. Hence, by Lemma V.2, $h(w) = h(w')$. Since w and w' are strictly descending, $h(w) = q_{\{i_0,\ldots i_{m-1}\}}^{-1}$ and $h(w') = q_{\{i'_0\ldots i'_{m'-1}\}}^{-1}$. Since $q_{\{i_0,\ldots i_{m-1}\}}^{-1} = q_{\{i'_0\ldots i'_{m'-1}\}}^{-1}$ therefore $\{i_0, \ldots, i_{m-1}\} = \{i'_0, \ldots i'_{m'-1}\}$. Hence $w = w'$. Since x corresponds to w, x' corresponds to w', and the correspondence is one-one, therefore $x = x'$.

(b) One uses Lemma 3(b) for a similar proof.

(c) Assume that v and v' are in conservational normal form and that $v \equiv v'$. Let w and w' be the $\mathbf{q}$-normal word that corresponds to v or to v' respectively. By Lemma 3(c), $w \equiv w'$. Hence, by Lemma V.6, $w = w'$. Since v corresponds to w, v' corresponds to w', and the correspondence is one-one, therefore $v = v'$. □

Letting h be the mapping of $F(\Gamma(\tilde{\mathbf{q}}))$ onto S_q used in Chapter V for our presentation of $\mathbf{S}_q$, we can now give a presentation of ${}_\delta\mathbf{S}_q = \langle {}_\delta S_q, \circ, ^{\smile}\rangle$, $2 \leq \delta \leq \omega$.

THEOREM 5. *For any δ such that $2 \leq \delta \leq \omega$, ${}_\delta\!\equiv$ coincides with $(h \circ h^{\smile}) \cap (F({}_\delta\Gamma(\tilde{\mathbf{q}})) \times F({}_\delta\Gamma(\tilde{\mathbf{q}}))$. Hence $\mathbf{F}({}_\delta\Gamma(\tilde{\mathbf{q}}))/{}_\delta\!\equiv$ is isomorphic to ${}_\delta\mathbf{S}_q$.*

Proof. One of the two inclusions holds by Lemma V.2. The other by Lemmas 2 and 4(c). □

The analogue of Theorem 5 for $\delta = 1$ does not hold. Since ${}_1Q_1$, ${}_1Q_2$, and ${}_1Q_4$ are empty, therefore ${}_1\!\equiv$ is the congruence on $\mathbf{F}({}_1\Gamma(\tilde{\mathbf{q}})) = \mathbf{F}\{\mathbf{q}_0\}$ that is generated by ${}_1Q_3 = \{\langle \mathbf{q}_0\mathbf{q}_0^{-1}\mathbf{q}_0, \mathbf{q}_0\rangle\}$. Since there are relations R that satisfy $RR^{\smile}R = R \neq RRR^{\smile}$, there follows that the pair $\langle \mathbf{q}_0\mathbf{q}_0\mathbf{q}_0^{-1}, \mathbf{q}_0\rangle\}$ is not in ${}_1\!\equiv$. In contrast, this pair is in $h \circ h^{\smile}$ and in $F({}_1\Gamma(\tilde{\mathbf{q}})) \times F({}_1\Gamma(\tilde{\mathbf{q}}))$. Note that by ${}_2Q_1, {}_2Q_2, {}_1Q_3$ respectively there holds

$$\mathbf{q}_0\mathbf{q}_0\mathbf{q}_0^{-1} \;{}_2\!\equiv\; \mathbf{q}_0\mathbf{q}_1\mathbf{q}_0^{-1} \;{}_2\!\equiv\; \mathbf{q}_0\mathbf{q}_0^{-1}\mathbf{q}_0 \;{}_2\!\equiv\; \mathbf{q}_0 \;.$$

As we saw in Chapter V, $\mathbf{S}_r = \langle S_r, \circ, ^{\smile}\rangle$ is isomorphic to $\mathbf{S}_p$. Hence ${}_\delta\mathbf{S}_r = \langle {}_\delta S_r, \circ, ^{\smile}\rangle$, where ${}_\delta S_r$ is the closure under $\circ$ and $^{\smile}$ of $\{r_i : i < \delta\}$ is isomorphic to ${}_\delta\mathbf{S}_q$. Hence Theorem 5 also yields a presentation of ${}_\delta\mathbf{S}_r$.

Let ${}_\delta T_q$ and ${}_\delta T_r$ be the closure under $\circ$ of $\{q_i : i < \delta\}$ or $\{r_i : i < \delta\}$, respectively. Let ${}_\delta\mathbf{T}_q = \langle {}_\delta T_q, \circ\rangle$ and ${}_\delta\mathbf{T}_r = \langle {}_\delta T_r, \circ\rangle$. From the relevant parts of the proof of Theorem 5 one can extract the following presentation of ${}_\delta\mathbf{T}_q$ and a similar presentation of ${}_\delta\mathbf{T}_r$, where $F_T({}_\delta\Gamma(\tilde{\mathbf{q}}))$ is the set of words formed from $\{q_i : i < \delta\}$ and $\mathbf{F}_T({}_\delta\Gamma(\tilde{\mathbf{q}})) = \langle F_T({}_\delta\Gamma(\tilde{\mathbf{q}})), ^\frown\rangle$ is the free semigroup with ${}_\delta\Gamma(\tilde{\mathbf{q}})$ as free generating set.

THEOREM 6. *Let ${}_\delta\!\equiv_1'$ be the congruence on $\mathbf{F}_T({}_\delta\Gamma(\tilde{\mathbf{q}}))$ that is generated by $Q_1 \cap (F_T({}_\delta\Gamma(\tilde{\mathbf{q}})) \times F_T({}_\delta\Gamma(\tilde{\mathbf{q}}))$. Then $\mathbf{F}_T({}_\delta\Gamma(\tilde{\mathbf{q}}))/{}_\delta\!\equiv_1'$ is isomorphic to ${}_\delta\mathbf{T}_q$.* □

For our presentation of $\langle \mathbf{S}_q, \subseteq\rangle = \langle \mathbf{S}_q, \circ, ^{\smile}, \subseteq\rangle$ in Chapter V we used the least quasiordering $\preceq$ of $F(\Gamma(\tilde{\mathbf{q}}))$ compatible with $\mathbf{F}(\Gamma(\tilde{\mathbf{q}}))$ that includes $Q_1 \cup \cdots \cup Q_4 \cup Q_1^{\smile} \cup \cdots \cup Q_4^{\smile} \cup Q_5 \cup Q_6$. Also, for our presentation of $\langle \mathbf{S}_r, \subseteq\rangle$ we used the least quasiordering $\preceq'$ of $F(\Gamma(\tilde{\mathbf{r}}))$ compatible with $\mathbf{F}(\Gamma(\tilde{\mathbf{r}}))$ that includes $R_1 \cup \cdots \cup R_4 \cup R_1^{\smile} \cup \cdots \cup R_4^{\smile} \cup R_5 \cup R_6^{\smile}$. For ${}_\delta\!\preceq$ and ${}_\delta\!\preceq'$ defined in the obvious way, there again arises the question of whether ${}_\delta\!\preceq$ coincides with $\preceq \cap (F({}_\delta\Gamma(\tilde{\mathbf{q}})) \times F({}_\delta\Gamma(\tilde{\mathbf{q}}))$ and whether ${}_\delta\!\preceq'$ coincides with $\preceq' \cap (F({}_\delta\Gamma(\tilde{\mathbf{r}})) \times F({}_\delta\Gamma(\tilde{\mathbf{r}}))$. Although we believe that in both cases the answer is affirmative, the construction of Lemma V.10(a) cannot be used for a proof, since it makes use of words that are not in $F({}_\delta\Gamma(\tilde{\mathbf{q}}))$. Perhaps a construction more closely related to conservational normal forms will work.

Whether there are conservational results regarding our presentation of $\mathbf{S}_{pq}$, $\mathbf{S}_{pe}$, $\mathbf{S}_{pqe}$ or related structures, we have not investigated.

Bibliography

[Br 91] D. A. Bredikhin, On relation algebras with general superpositions, in *Algebraic Logic*, edited by H.Andréka, J.D.Monk, and I.Németi, (North-Holland, 1991), 111–124.

[Br 94] D. A. Bredikhin, Representation of inverse semigroups by difunctional multipermutations, in *Proceeedings of Essex Conf. on Transformation Semigroups*, edited by Peter M. Higgins (Essex Math Dept, 1994), 1–10.

[BS] Stanley Burris and H. P. Sankappanavar, *A Course in Universal Algebra*, (Springer-Verlag, 1981), xvi + 276 pp.

[CM] H.S.M. Coxeter and W.O.J. Moser, *Generators and Relations for Discrete Groups*, 3rd edition. Ergebnisse der Mathematik und ihrer Grenzgebiete, vol. 14 (Springer-Verlag), ix + 161 pp.

[Cr 74a] William Craig, *Logic in Algebraic Form*, (North-Holland, Amsterdam, 1974), viii + 204 pp.

[Cr 74b] William Craig, Diagonal relations, in *Proceedings of the Tarski Symposium*, (American Mathematical Society, Providence, RI, 1974), 92–104.

[DGP] J. Michael Dunn, Mai Gehrke, and Alessandra Palmigliano, Canonical extensions and relational completeness of some structural logics, *Journal of Symbolic Logic*, vol. **70** (2005),713–740.

[GJ] Mai Gehrke and Bjarni Jónsson, Monotone bounded distributive lattice expansions, *Mathematica Japonica* **52** (2000), 197–213.

[Go] R. I. Goldblatt, Varieties of complex algebras, *Annals of Pure and Applied Logic*, vol. **38** (1989), 173–241.

[Ha] P. R. Halmos, *Algebraic Logic*, (Chelsea, New York, 1962), 271 pp.

[HMT] L. Henkin, D. Monk, and A. Tarski, *Cylindric Algebras*, (North-Holland, Amsterdam). Part I (1971), vi + 508 pp. Part II (1985), vii + 302 pp.

[HT] L. Henkin and A. Tarski, Cylindric algebras, in *Lattice Theory*, (American Mathematical Society, Providence, RI, 1961), 83–113.

[Ho] John M. Howie, *Fundamentals of Semigroup Theory*, (Clarendon Press, Oxford, 1995), x + 351 pp.

[Jo] D.L.Johnson, *Topics in the Theory of Group Presentations*, (Cambridge University Press, 1980), London Math. Soc. Lecture Notes 42, vii + 311 pp.

[JT] B. Jónsson and A. Tarski, Boolean algebras with operators, Part I, *Am. J. Math.* vol. **73** (1951), 891–939; part II, *Am. J. Math.* vol. **74** (1952), 127–162.

[Lo] M. Lothaire, *Applied Combinatorics on Words*, (Cambridge University Press, 2005), xv + 610 pp.

[MKS] W. Magnus, A. Karass and D. Solitar, *Combinatorial Group Theory: Presentation of Groups in Terms of Generators and Relations*, (John Wiley & Sons, 1966), xii + 444 pp.

[Ma] A. I. Mal'cev, *Algebraic Systems* (Gundlehren der Mathematischen Wissenschaften), vol. 192 (Springer-Verlag, 1973), xii + 317 pp.

[MMT] R.McKenzie, G.McNulty and W.Taylor, *Algebras, Lattices, Varieties*, vol. I (Wadsworth & Brooks/Cole, 1987), xii + 361 pp.

[Mo] Donald J. Monk, On an algebra of sets of finite sequences, *Journal of Symbolic Logic* **35** (1970), 19–28.

[Moo] E. H. Moore, Concerning the abstract group of order $k!$ and $\frac{1}{2}k!$, *Proc. London Math. Society*, vol. **28** (1897), 357–366.

[NS] T. E. Nordahl and H. E. Scheiblich, Regular $*$-semigroups, *Semigroup Forum* **16** (1978), 369–377.

[Pe] Mario Petrich, *Inverse Semigroups*, (John Wiley & Sons, 1984), x + 674 pp.

[Ri] Jacques Riguet, Quelques propriétes des relations difonctionelles, *C.R. Acad. Sci. Paris* **230** (1950), 1999–2000.

[Ro] Joseph J. Rotman, *An Introduction to the Theory of Groups*, Fourth Edition, (Springer-Verlag, 1995), xv + 513 pp.

[Sch 87] Boris M. Schein, Multigroups, *Journal of Algebra* **111** (1987), 114–132.

[Sch 88] Boris M. Schein, On certain classes of semigroups of binary relations, *AMS Transl.* (2) **139**, 117–137.

[Ta] A. Tarski, *Logic, Semantics, Metamathematics*, (Oxford U. Press, 1956), xiv + 471 pp. VI. On Definable Sets of Real Numbers. VIII. The Concept of Truth in Formalized Languages.

Index of Symbols

Symbols with different meanings on different pages are listed accordingly. For symbols that denote a relation involved in one of the major presentations there is a separate list. Some symbols that are used only fairly briefly are not listed. Symbols that are used in the Overview but also in the main text are listed according to where they appear in the main text.

xiv	$\dagger$	conjugate
xiv	$\subseteq$	inclusion-associated ordering of operations
xxii	Acs	the set of almost constant shifts
xxiii	$\mathring{1}_A$	identity relation on A
xxiv	$\mathcal{K}_{p\dot{q}e}$	the class of unary representations of $\langle \mathbf{S}_{pqe}, \subseteq \rangle$
xxiv	$\mathbf{C_A}$	the complex algebra of $\mathbf{A}$
1	$R^*(X)$	the direct image of X under R
1	$\emptyset$	null set, sequence of length 0
1	$\{\emptyset\}$	the set whose only element is $\emptyset$
2	$R^{\smile}$	the converse of the relation R
2	$R \circ R'$, RR'	the relative product of R and R'
2	$Do\, R$	the domain of R
3	$Rg\, R$	the range of R
3	R^n	the nth power of R
3	R_x, $R(x)$	the value y, if $\langle x, y \rangle \in R$ and R is single-valued
3	$[\beta, \delta]$	closed interval of ordinals
3	$[\beta, \delta)$	half-open interval of ordinals
3	ω	first infinite ordinal
3	$\beta, \gamma, \ldots$	ordinals (usually $\leq \omega$)
3	$i, j, \ldots$	finite ordinals

3	U	universe, base set
3	$\overset{=}{X}$	the cardinality of X
3	${}^{\gamma}U$	the set of those sequences formed from elements of U, that are of length γ
3	$v, z, \ldots$	elements of $\bigcup\{{}^{\gamma}U : 0 \leq \gamma \leq \omega\}$
3	$\langle x_0, \ldots, x_{n-1}\rangle$	sequence x of length n
3	$\langle x_0, \ldots, x_n, \ldots\rangle$	sequence x of length ω
3	$\frown$	concatenation
3	$x^\frown y,\ xy$	concatenation of the sequences x and y
3	$\lvert x\rvert$	length of the sequence x
4	${}^{[m,\gamma)}U$	the set of sequences of length at least m and less than γ
4	${}^{[m,\gamma)}U$	the set of sequences of length at least m and at most γ
4	${}_{[m,\gamma)}\overset{\circ}{1}$	identity relation on ${}^{[m,\gamma)}U$
4	${}_{[m,\gamma]}\overset{\circ}{1}$	identity relation on ${}^{[m,\gamma]}U$
4	${}_{\gamma}\overset{\circ}{1}$	identity relation on ${}^{\gamma}U$
4	$p_{(i,j)}$	(i,j)-interchange (applied to finite sequences)
4	$e_{(i,j)}, e_{i,j}, e_{ij}$	identity relation on the (i,j) diagonal set
4	$\dot{q}_i, q_i^{\smile}$	excision at place i, i-excision
4	$q_i, \dot{q}_i^{\smile}$	multivalued i-insertion
4	d_{ij}	(i,j)-diagonal set (of finite sequences)
5	$c_i, q_i^{\smile} \circ q_i$	an equivalence relation on ${}^{[i+1,\omega)}U$
5	$r_i, q_i \circ e_{i,i+1}$	replication at place i
5	$\dot{r}_i, e_{i,i+1} \circ \dot{q}_i$	i-excision restricted to $d_{i(i+1)}$
6	e^*_{ij}	intersection with d_{ij}
6	$p^*_{(i,j)}$	(i,j)-interchange (applied to sets of finite sequences)
6	q^*_i	i-insertion (applied to sets of finite sequences)
6	$\dot{q}^*_i$	i-excision (applied to sets of finite sequences)
7	${}_{[i,\omega)}\overset{\circ}{1}{}^*$	intersection with ${}^{[i,\omega)}U$
7	c^*_i	partial i-cylindrification (applied to sets of finite sequences)
7	r^*_i	i-replication (applied to sets of finite sequences)

7	$\dot{r}_i^*$	restricted i-excision (applied to sets of finite sequences)
8	S_q	the closure under $\circ$ and $\smile$ of $\{q_i : i < \omega\}$
8	p_i	abbreviation of $p_{(i,i+1)}$
8	S_p	the closure under $\circ$ of $\{q_i : i < \omega\}$, closed also under $\smile$
8	S_e	the closure under $\circ$ of $\{e_{ij} : \{i, j\} \neq \{\emptyset\}\}$, closed also under $\smile$
9	$S_{pq}, S_{pe}, S_{pqe}, \ldots$	the closure under $\circ$ of $S_p \cup S_q, S_p \cup S_e, S_p \cup S_q \cup S_e, \ldots$
9	S_r	the closure under $\circ$ and $\smile$ of $\{r_i : i < \omega\}$
9	S_c	the closure under $\circ$ of $\{c_i : i < \omega\}$, closed also under $\smile$
9	T_q	the closure under $\circ$ of $\{q_i : i < \omega\}$
9	T_r	the closure under $\circ$ of $\{r_i : i < \omega\}$
9	T_{pq}	the closure under $\circ$ of $\{p_i : i < \omega\} \cup \{q_i : i < \omega\}$
9	O_{pqe}	the set of operations R^* such that R is in S_{pqe}
9	$O_p, O_q, O_e, O_{pq}, \ldots$	similarly related to $S_p, S_q, S_e, S_{pq}, \ldots$
9	$q, \dot{q}, r, \dot{r}, e$	abbreviation of $q_0, \dot{q}_0, r, \dot{r}_0, e_{0,1}$ respectively
10	$X \underset{2}{-} Y$	the complement of Y relative to X
10	$-X, \underset{1}{-}X$	the complement of X (relative to a set that is understood)
11	$\langle U, X_\delta \rangle_{\delta<\eta}$	a relational structure
11	$\rho(\delta)$	the rank of X_δ
11	$\approx$	symbol for equality
11	$\mathbf{P}_\delta$	predicate symbol
11	$\mathbf{v}_i$	individual variable
11	$\wedge$	conjunction symbol
11	$\neg$	negation symbol
11	$\exists$	existential quantifier
19	κ	the cardinality $\overline{\overline{U}}$ of U
19	u_λ	the λth member of U
19	ι_λ	$\{\langle x, x \rangle : 1 \leq \lvert x \rvert < \omega,\ x_0 = u_\lambda\}$
19	$\iota_{(j/\lambda)}$	$\{\langle x, x \rangle : j < \lvert x \rvert < \omega,\ x_j = u_\lambda\}$
19	$\iota^*_{(j/\lambda)}(X)$	$X \cap \{x : j < \lvert x \rvert < \omega,\ x_j = u_\lambda\}$

19	O_i	the closure under $\circ$ of $\{\iota^*_{(j/\lambda)} : \lambda < \kappa,\ j < \omega\}$
19	O_{pqei}	the closure under $\circ$ of $O_{pqe} \cup O_i$
19	S_i	the closure under $\circ$ of $\{\iota_{(j/\lambda)} : \lambda < \kappa,\ j < \omega\}$
19	S_{pqei}	the closure under $\circ$ of $S_{pqe} \cup S_i$
19	$s_{\lambda,j}$	insertion of u_λ at place j
20	$s^{\smile}_{\lambda,j}$, $\dot{s}_{\lambda,j}$	excision of u_λ at place j
20	$\langle U, X_\delta, u_\lambda\rangle_{\delta<\eta,\lambda<\kappa}$	a relational structure with distinguished elements
20	$\mathbf{c}_\lambda$	individual constant
22	b_i, ${}_{[0,i)}\overset{\circ}{1} \cup \{q_i^{\smile} \circ q_i\}$	an equivalence relation on ${}^{[0,\omega)}U$
22	b_i^*	i-cylindrification (applied to sets of finite sequences)
26	$\underset{(1)}{\overset{\gamma}{\rightarrow}} X$	the γ-prolongation of the subset X of ${}^{[0,\omega)}U$
27	$\overset{\gamma}{\rightarrow} X$, $\underset{(2)}{\overset{\gamma}{\rightarrow}} X$	the γ-prolongation of the relation X on ${}^{[0,\omega)}U$
27	$\overset{[0,\omega)}{\longrightarrow} X$	the union $\bigcup\{\overset{m}{\rightarrow} X : m < \omega\}$ of the m-prolongations of X
28	Rk	the set of relations on ${}^{[0,\omega)}U$ that are ranked
29	${}_{[0,\omega)}Un$	the set of relations that are uniform on ${}^{[0,\omega)}U$
29	${}_{[\omega]}Un$	the set of relations that are uniform on ${}^{\omega}U$
30	$\dot{0}$	the grade of the empty relation $\emptyset$
30	$gd\,X$	the grade of X, if X is in ${}_{[0,\omega)}Un$
31	$\overleftrightarrow{t}$	inversion
31	$t^{\nwarrow}$	counterclockwise rotation
31	$\overset{\frown}{e}$	left-right identity
31	$q^{\#}$	right-end insertion
31	$\dot{q}^{\#}$	right-end excision
31	$S^{\#}$	the closure under $\circ$ of $\{q, \dot{q}, q^{\#}, \dot{q}^{\#}, \overset{\frown}{e}\}$
31	$O^{\#}$	the closure under $\circ$ of $\{q^*, \dot{q}^*, q^{\#*}, \dot{q}^{\#*}, \overset{\frown}{e}^*\}$
35	sc	the successor function on ω
35	$\precsim$, $\precsim_g$	the ordering generated by (near-) striate g
35	$\approx, \approx_g$	the equivalence generated by (near-) striate g
36	$x/\approx$	the equivalence class $\{y : x \approx y\}$

38	$\odot$	binary operation on $(\omega \times \omega) \cup \{\dot{0}\}$
38	$\langle \omega \times \omega, \odot \rangle$	the bicyclic semigroup
38	$\langle (\omega \times \omega) \cup \{\dot{0}\}, \odot \rangle$	the bicyclic semigroup with zero
38	$\overset{\bullet}{\vee}$	a unary operation on $(\omega \times \omega) \cup \{\dot{0}\}$
38	$\overset{\bullet}{\to}$	a unary operation on $(\omega \times \omega) \cup \{\dot{0}\}$
38–59	$\geq$	an ordering of $(\omega \times \omega) \cup \{\dot{0}\}$
38–59	$\leq$	the converse of $\geq$
47	$lgd\ X$	the least grade of X, if X is in ${}_{[\omega]}Un$ and $\bar{\bar{U}} \geq 2$
48	$\overset{\omega}{\to} p_i,\ \overset{\omega}{\to} q_i, \dots$	the ω-prolongation of $p_i, q_i, \dots$
56	${}_{[\omega]}S_p$	the closure under $\circ$ of $\{\overset{\omega}{\to} p_i : i < \omega\}$, closed also under $\smile$
56	${}_{[\omega]}S_q$	the closure under $\circ$ and $\smile$ of $\{\to q_i : i < \omega\}$
56	${}_{[\omega]}S_e$	the closure under $\circ$ of $\{\overset{\omega}{\to} e_{i,j} : i, j < \omega\}$, closed also under $\smile$
56	${}_{[\omega]}S_{pq}, {}_{[\omega]}S_{pe}, {}_{[\omega]}S_{pqe}, \dots$	the closure under $\circ$ of ${}_{[\omega]}S_p \cup {}_{[\omega]}S_q, {}_{[\omega]}S_p \cup {}_{[\omega]}S_e, {}_{[\omega]}S_p \cup {}_{[\omega]}S_q \cup {}_{[\omega]}S_{pe}, \dots$
61	$M, N, \dots$	(binary) relations on ω
61	$\overset{\circ}{1}_\gamma$	the identity relation on $\gamma = \{k : k < \gamma\}$
61	$eq_\gamma M$	the equivalence relation generated by $\overset{\circ}{1}_\gamma \cup (M \cap (\gamma \times \gamma))$
61	$\langle M_0, M_1, M_2 \rangle$	a framing triple
62	$[M_0, M_1, M_2]$, $[\langle M_0, M_1, M_2 \rangle]$	the (diagonal) relation determined by $\langle M_0, M_1, M_2 \rangle$
62	$[\![M_0, M_1, M_2]\!]$, $[\![\langle M_0, M_1, M_2 \rangle]\!]$	the union $\bigcup\{\overset{m}{\to} [M_0, M_1, M_2] : m < \omega\}$, if $M_0 \cup M_1 \cup M_2$ is finite
63	${}_{[0,\omega)}D$	the set of diagonal relations in ${}_{[0,\omega)}Un$
63	${}_{\langle \gamma, \delta \rangle}D$	the set of diagonal relations in ${}^\gamma U \times {}^\delta U$
63	${}_{[\gamma]}D$	abbreviation of ${}_{\langle \gamma, \gamma \rangle}D$
64	$\langle M_0, M_1, M_2 \rangle^\smile$	abbreviation of $\langle M_2, M_1^\smile, M_0 \rangle$
71	${}_\gamma\langle M \rangle_\delta$	abbreviation of $\langle 1_\gamma, M, \overset{\circ}{1}_\delta \rangle$

71	${}_{\gamma}[M]_{\delta}$	abbreviation of $[\mathring{1}_{\gamma}, M, \mathring{1}_{\delta}]$
71	${}_{s}[\![M]\!]_{t}$	abbreviation of $[\![\mathring{1}_{s}, M, \mathring{1}_{t}]\!]$
71	${}_{\gamma}[M]$	abbreviation of ${}_{\gamma}[M]_{\gamma}$
71	${}_{s}[\![M]\!]$	abbreviation of ${}_{s}[\![M]\!]_{s}$
71	${}_{\gamma}[\ ]_{\delta}$	mapping of $\{M : M \subseteq \gamma \times \delta\}$ into ${}_{\langle\gamma,\delta\rangle}D$
71	${}_{s}[\![\ ]\!]_{t}$	mapping of $\{M : M \subseteq s \times t\}$ into ${}_{[0,\omega)}D$
72	$dfc\ R$	the difunctional closure of R
77	(i, j)	the permutation of ω that transposes i and j
78	${}_{\langle\gamma,\delta\rangle}DF$	the set of relations in ${}_{\langle\gamma,\delta\rangle}D$ that are full
78	${}_{[\gamma]}DF$	abbreviation of ${}_{\langle\gamma,\gamma\rangle}DF$
78	${}_{[0,\omega)}DF$	the set of relations in ${}_{[0,\omega)}D$ that are upward full
78	${}_{[\gamma]}DI$	the set of subidentities in ${}_{[\gamma]}D$
78	${}_{[0,\omega)}DI$	the set of subidentities in ${}_{[0,\omega)}D$
78	${}_{[\gamma]}DP$	the set of permutations in ${}_{[\gamma]}D$
78	${}_{[0,\omega)}DP$	the set of upward full permutations in ${}_{[0,\omega)}D$
78	$Q \bullet R$	the difunctional product $dfc(Q \circ R)$ of Q and R
93	$R(W)$	$\{\langle x, x\rangle : \langle x, y\rangle \in W \text{ for some } y\}$
95	$f, g, \ldots$	relations on ω that are single-valued
98	${}_{[0,\omega)}DE$	the set of excisions in ${}_{[0,\omega)}D$
98	${}_{\langle\gamma,\delta\rangle}DE$	the set of excisions in ${}_{\langle\gamma,\delta\rangle}D$
98	${}_{[\gamma]}DE$	the set of excisions in ${}_{[\gamma]}D$
105	$\widehat{+}$	one-one correspondence between two kinds of relations on ω
111	${}_{[\omega]}UD$	the set ${}_{[\omega]}Un \cap {}_{[\omega]}D$
111	${}_{[\omega]}UDF$	the set ${}_{[\omega]}Un \cap {}_{[\omega]}DF$
111	${}_{[\omega]}UDP$	the set ${}_{[\omega]}Un \cap {}_{[\omega]}DP$
111	${}_{[\omega]}UDE$	the set ${}_{[\omega]}Un \cap {}_{[\omega]}DE$
111	${}_{[\omega]}UDI$	the set ${}_{[\omega]}Un \cap {}_{[\omega]}DI$
111	Bsl	the set of relations on ω that are bisectable
112	Bst	the set of bisections

112	$\langle N, \langle s, t\rangle\rangle$	a bisection
117	$I, J, K, \ldots$	subsets of ω
117	f_J	the strictly increasing total function on ω with range $\omega \mathbin{\bar{_2}} J$
117	$f_{\{i\}}$	the successor function that delays until i
120	${}_{[\omega]}S_r$	the closure under $\circ$ and $\smile$ of $\{\overset{\omega}{\to} r_i : i < \omega\}$
120	${}_{[\omega]}S_{prc}$	the closure under $\circ$ and $\smile$ of $\{\{\overset{\omega}{\to} p_i, \overset{\omega}{\to} r_i, \overset{\omega}{\to} c_i\} : i < \omega\}$
121	$\mathbf{B}$	the expanded bicyclic semigroup
123	$\mathbf{Bsl}$	the structure $\langle Bsl, \circ, \smile, id, \supseteq\rangle$
123	$\mathbf{Bst}$	the structure $\langle Bst, \odot, \overset{\bullet}{\vee}, \overset{\bullet}{\to}, \geq\rangle$
126	$[\![N, \langle s, t\rangle]\!]$, $[\![\langle N, \langle s, t\rangle\rangle]\!]$	the relation mediately determined by $\langle N, \langle s, t\rangle\rangle$
135	$\mathbf{S}_{pqe}$	the semigroup with involution $\langle S_{pqe}, \circ, \smile\rangle = \langle {}_{[0,\omega)}D, \circ, \smile\rangle$
135	$\langle \mathbf{S}_{pqe}, \subseteq\rangle$	the structure $\langle S_{pqe}, \circ, \smile, \subseteq\rangle$
135	$\mathbf{S}_p, \mathbf{S}_q, \mathbf{S}_{pq}, \mathbf{S}_{pe}$	the semigroup with involution whose universe is S_p, S_q, S_{pq}, S_{pe}
135	${}_{[\omega]}\mathbf{S}_{pqe}$	the semigroup with involution $\langle {}_{[\omega]}S_{pqe}, \circ, \smile\rangle$
135	${}_{[\omega]}\mathbf{S}_p, {}_{[\omega]}\mathbf{S}_q, {}_{[\omega]}\mathbf{S}_{pq}, {}_{[\omega]}\mathbf{S}_{pe}$	the semigroup with involution whose universe is ${}_{[\omega]}S_p, {}_{[\omega]}S_q, {}_{[\omega]}S_{pq}, {}_{[\omega]}S_{pe}$
135	Γ	a free generating set
135	SI	the class of semigroups with involution
135	$\mathbf{F}_{SI}(\Gamma), \mathbf{F}(\Gamma)$	the free semigroup with involution freely generated by Γ
135	$F_{SI}(\Gamma)$	the universe of $\mathbf{F}_{SI}(\Gamma)$
136	$\mathbf{F}_{SI}(\Gamma)/\equiv$	the quotient algebra of $\mathbf{F}_{SI}(\Gamma)$ modulo the congruence $\equiv$
137	$\langle x_0, \ldots, x_{n-1}\rangle^{-1}$	the element $\langle x_{n-1}^{-1}, \ldots, x_0^{-1}\rangle$ of $F_{SI}(\Gamma)$
138	$(\widehat{f^{+}_{\{i\}}})^{\smile}$	the predecessor function that delays until i
138	$\mathbf{T}_q, {}_{[\omega]}\mathbf{T}_q, \mathbf{T}_r, {}_{[\omega]}\mathbf{T}_r$	the semigroup whose universe is $T_q, {}_{[\omega]}T_q, T_r, {}_{[\omega]}T_r$

138	$\mathbf{F}_T(\Gamma)$	the free semigroup freely generated by Γ
138	$\Gamma(\tilde{\mathbf{q}})$	the set $\{\mathbf{q}_i : i<\omega\}$, each $\mathbf{q}_i$ used for naming
138	$\Gamma(\tilde{\mathbf{r}})$	the set $\{\mathbf{r}_i : i<\omega\}$, each $\mathbf{r}_i$ used for naming
140	$\mathbf{F}_{SI}(\Gamma(\tilde{\mathbf{q}})), \mathbf{F}(\Gamma(\tilde{\mathbf{q}}))$	the free semigroup with involution freely generated by $\Gamma(\tilde{\mathbf{q}})$
140	$F_{SI}(\Gamma(\tilde{\mathbf{q}})), F(\Gamma(\tilde{\mathbf{q}}))$	the universe of $\mathbf{F}_{SI}(\Gamma(\tilde{\mathbf{q}}))$
140	$\mathbf{F}_{SI}(\Gamma(\tilde{\mathbf{r}})), \mathbf{F}(\Gamma(\tilde{\mathbf{r}}))$	the free semigroup with involution freely generated by $\Gamma(\tilde{\mathbf{r}})$
140	$F_{SI}(\Gamma(\tilde{\mathbf{r}})), F(\Gamma(\tilde{\mathbf{r}}))$	the universe of $\mathbf{F}_{SI}(\Gamma(\tilde{\mathbf{r}}))$
141–157	$\equiv$	the congruence on $\mathbf{F}_{SI}(\Gamma(\tilde{\mathbf{q}}))$ generated by $Q_1 \cup Q_2 \cup Q_3 \cup Q_4$
141–157	$\equiv'$	the congruence on $\mathbf{F}_{SI}(\Gamma(\tilde{\mathbf{r}}))$ generated by $R_1 \cup R_2 \cup R_3 \cup R_4$
141–157	h	the homomorphism of $\mathbf{F}_{SI}(\Gamma(\tilde{\mathbf{q}}))$ onto $\mathbf{S}_q$ such that $h(\mathbf{q}_i) = q_i$
141–157	h'	the homomorphism of $\mathbf{F}_{SI}(\Gamma(\tilde{\mathbf{r}}))$ onto $\mathbf{S}_r$ such that $h(\mathbf{r}_i) = r_i$
144	$gd\,w$	the grade associated with the word w
151	$\langle \mathbf{S}_q, \subseteq \rangle$	the structure $\langle S_q, \circ, ^{\smile}, \subseteq \rangle$
151	$\langle \mathbf{S}_r, \subseteq \rangle$	the structure $\langle S_r, \circ, ^{\smile}, \subseteq \rangle$
151–157	$\lessdot$	the least quasiordering of $F(\Gamma(\tilde{\mathbf{q}}))$ that is compatible with $\mathbf{F}(\Gamma(\tilde{\mathbf{q}}))$ and includes $\equiv \cup Q_5 \cup Q_6$
151–157	$\lessdot'$	the least quasiordering of $F(\Gamma(\tilde{\mathbf{r}}))$ that is compatible with $\mathbf{F}(\Gamma(\tilde{\mathbf{r}}))$ and includes $\equiv' \cup R_5 \cup R_6^{\smile}$
156	$\langle {}_{[\omega]}\mathbf{S}_q, \subseteq \rangle$	the structure $\langle {}_{[\omega]}S_q, \circ, ^{\smile}, \subseteq \rangle$
156–157	$\equiv^+$	the congruence on $\mathbf{F}_{SI}(\Gamma(\tilde{\mathbf{q}}))$ generated by $Q_1 \cup Q_2 \cup Q_3 \cup Q_5$
156–157	$\lessdot^+$	the least quasiordering of $F_{SI}(\Gamma(\tilde{\mathbf{q}}))$ that is compatible with $\mathbf{F}_{SI}(\Gamma(\tilde{\mathbf{q}}))$ and includes $\equiv^+ \cup Q_6$

157	$S_{p,\delta}, S_\delta$	the closure under $\circ$ of $\{(i, i+1) \cap (\delta \times \delta) : i+1 < \delta\}$, also closed under $\smile$
157	$\mathbf{S}_{p,\delta}, \mathbf{S}_\delta$	the group whose universe is $S_{p,\delta}$
158	${}_\delta\Gamma(\tilde{\mathbf{p}}), {}_\delta\Gamma$	the set $\{\mathbf{p}_i : i+1 < \delta\}$, each $\mathbf{p}_i$ used for naming
158	$\Gamma(\tilde{\mathbf{p}}))$	abbreviation of ${}_\omega\Gamma(\tilde{\mathbf{p}})$
158	$\mathbf{F}_{SI}({}_\delta\Gamma(\tilde{\mathbf{p}})), \mathbf{F}({}_\delta\Gamma(\tilde{\mathbf{p}}))$	the free semigroup with involution freely generated by ${}_\delta\Gamma(\tilde{\mathbf{p}})$
159	$\mathbf{F}_{Gp}({}_\delta\Gamma)$	the free group freely generated by ${}_\delta\Gamma(\tilde{\mathbf{p}})$
165	$\mathbf{q}$	abbreviation of $\mathbf{q}_0$
165	$\Gamma(\tilde{\mathbf{p}}, \mathbf{q})$	the set $\{\mathbf{p}_i : i < \omega\} \cup \{\mathbf{q}\}$
165–179	$\equiv_P$	the congruence on $\mathbf{F}_{SI}(\Gamma(\tilde{\mathbf{p}}, \mathbf{q}))$ generated by $P_1 \cup P_2 \cup P_3 \cup P_4^{\#} \cup P_5 \cup P_6$
165–179	$\equiv$	the congruence on $\mathbf{F}_{SI}(\Gamma(\tilde{\mathbf{p}}, \mathbf{q}))$ generated by $\equiv_P \cup {}_1Q_3 \cup PQ_1 \cup PQ_2 \cup PQ_3 \cup PQ_4$
165–179	h	the homomorphism of $\mathbf{F}_{SI}(\Gamma(\tilde{\mathbf{p}}, \mathbf{q}))$ onto $\mathbf{S}_{pq}$ such that $h(\mathbf{q}) = q_0$ and $h(\mathbf{p}_i) = p_i = p_{(i,i+1)}$ for each i
166	$\mathbf{q}_{i+1}$	abbreviation of $\mathbf{q}_i\mathbf{p}_i$
167	$\mathbf{F}(\{\mathbf{q}_i : i < \omega\})$	the subalgebra of $\mathbf{F}_{SI}(\Gamma(\tilde{\mathbf{p}}, \mathbf{q}))$ generated by $\{\mathbf{q}_i : i < \omega\})$
177–178	$\lessdot$	the least quasiordering of $F_{SI}(\Gamma(\tilde{\mathbf{p}}, \mathbf{q}))$ that is compatible with $\mathbf{F}_{SI}(\Gamma(\tilde{\mathbf{p}}, \mathbf{q}))$ and includes $\equiv \cup P_4 \cup PQ_5 \cup PQ_6$
179–180	$\equiv^+$	the congruence on $\mathbf{F}_{SI}(\Gamma(\tilde{\mathbf{p}}, \mathbf{q}))$ that is generated by $\equiv \cup P_4 \cup PQ_5$
179–180	$\lessdot^+$	the least quasiordering of $F_{SI}(\Gamma(\tilde{\mathbf{p}}, \mathbf{q}))$ that is compatible with $\mathbf{F}_{SI}(\Gamma(\tilde{\mathbf{p}}, \mathbf{q}))$ and includes $\equiv^+ \cup PQ_6$

179–180 h' the homomorphism of $\mathbf{F}_{SI}(\Gamma(\tilde{\mathbf{p}}, \mathbf{q}))$ onto ${}_{[\omega]}\mathbf{S}_{pq}$ such that $h'(\mathbf{q}) = \stackrel{\omega}{\rightarrow} q_0$ and $h'(\mathbf{p}_i) = \stackrel{\omega}{\rightarrow} p_i$ for each $i < \omega$

179–180 h'' the homomorphism of $\mathbf{F}_{SI}(\Gamma(\tilde{\mathbf{p}}, \mathbf{q}))$ onto $\langle\{N \in Bsl : N \text{ one-one}\}, \circ, ^{\smile}\rangle$ such that $h''(\mathbf{q}) = f_{\{0\}}$ and $h''(\mathbf{p}_i) = (i, i+1)$ for each $i < \omega$

180 $\mathbf{e}$ abbreviation of $\mathbf{e}_{0,1}$

180 $\Gamma(\tilde{\mathbf{p}}, \mathbf{e})$ the set $\{\mathbf{p}_i : i < \omega\} \cup \{\mathbf{e}\}$

180 $\mathbf{F}_{SI}(\Gamma(\tilde{\mathbf{p}}, \mathbf{e}))$, $\mathbf{F}(\Gamma(\tilde{\mathbf{p}}, \mathbf{e}))$ the free semigroup with involution freely generated by $\Gamma(\tilde{\mathbf{p}}, \mathbf{e}))$

180–207 $\equiv_P$ the congruence on $\mathbf{F}_{SI}(\Gamma(\tilde{\mathbf{p}}, \mathbf{e}))$ generated by $P_1 \cup P_2 \cup P_3 \cup P_4^{\#} \cup P_5 \cup P_6$

180–207 $\equiv$ the congruence on $\mathbf{F}_{SI}(\Gamma(\tilde{\mathbf{p}}, \mathbf{e}))$ generated by $\equiv_P \cup E_1 \cup E_2 \cup PE_1 \cup PE_2 \cup PE_3 \cup PE_4$

180–207 h the homomorphism of $\mathbf{F}_{SI}(\Gamma(\tilde{\mathbf{p}}, \mathbf{e}))$ onto S_{pe} such that $h(\mathbf{e}) = e = e_{0,1}$ and $h(\mathbf{p}_i) = p_i = p_{(i,i+1)}$ for each $i < \omega$

182 $\mathbf{e}_{j,k}$ an abbreviation

182 $\mathbf{p}_{(j,k)}$ an abbreviation

190–207 $\equiv'$ the congruence on $\mathbf{F}_{SI}(\Gamma(\tilde{\mathbf{p}}, \mathbf{e}))$ that is generated by $\equiv_P \cup E_1 \cup PE_1 \cup PE_2$

195–199 $\equiv''$ the congruence on $\mathbf{F}_{SI}(\Gamma(\tilde{\mathbf{p}}, \mathbf{e}))$ that is generated by $\equiv' \cup PE_3 \cup PE_4$

198–204 $\mathbf{F}(\Gamma(\tilde{\mathbf{e}}))$ the subalgebra of $\mathbf{F}(\Gamma(\tilde{\mathbf{p}}, \mathbf{e}))$ that is generated by $\{\mathbf{e}_{j,k} : \{j,k\} \neq 0\}$

202 $\mathbf{p}_{(i_0,\ldots,i_{n-1})}$ an abbreviation

205–206 $\lessdot$ the least quasiordering of $F(\Gamma(\tilde{\mathbf{p}}, \mathbf{e}))$ that is compatible with $\mathbf{F}(\Gamma(\tilde{\mathbf{p}}, \mathbf{e}))$ and includes $\equiv \cup P_4 \cup PE_5$

206–207 $\equiv^+$ the congruence on $\mathbf{F}(\Gamma(\tilde{\mathbf{p}}, \mathbf{e}))$ that is generated by $\equiv \cup P_4$

206–207 $\lessdot^+$ the least quasiordering of $F(\Gamma(\tilde{\mathbf{p}}, \mathbf{e}))$ that is compatible with $\mathbf{F}(\Gamma(\tilde{\mathbf{p}}, \mathbf{e}))$ and includes $\lessdot \cup P_4^{\smile}$

209	$\Gamma,\ \Gamma(\tilde{\mathbf{p}},\mathbf{q},\mathbf{e}))$	the union $\Gamma(\tilde{\mathbf{p}},\mathbf{q}) \cup \Gamma(\tilde{\mathbf{p}},\mathbf{e})$
209	$\mathbf{F}(\Gamma), \mathbf{F}_{SI}(\Gamma(\tilde{\mathbf{p}},\mathbf{q},\mathbf{e})$	the free semigroup with involution that is freely generated by $\Gamma(\tilde{\mathbf{p}},\mathbf{q},\mathbf{e}))$
209–232	$\equiv'$	the congruence on $F(\Gamma(\tilde{\mathbf{p}},\mathbf{q},\mathbf{e}))$ that is generated by the union of the two congruences used for presenting $\mathbf{S}_{pq}$ or $\mathbf{S}_{pe}$ respectively
209–232	$\equiv$	the congruence on $F(\Gamma(\tilde{\mathbf{p}},\mathbf{q},\mathbf{e}))$ that is generated by $\equiv' \cup QE_1 \cup PQE_1 \cup PQE_2$
209–232	h	the homomorphism of $\mathbf{F}(\Gamma)$ onto $\mathbf{S}_{pqe}$ such that $h(\mathbf{q}) = q = q_0$, $h(\mathbf{e}) = e = e_{0,1}$, and $h(\mathbf{p}_i) = p_i = p_{(i,i+1)}$ for each $i < \omega$
229–230	$\lessdot$	the least quasiordering of $F(\Gamma(\tilde{\mathbf{p}},\mathbf{q},\mathbf{e}))$ that includes $\equiv \cup PQ_5 \cup PQ_6 \cup P_4 \cup PE_5$
230–231	$\equiv^+$	the congruence on $\mathbf{F}(\Gamma(\tilde{\mathbf{p}},\mathbf{q},\mathbf{e}))$ that is generated by $\equiv \cup P_4$
230–231	$\lessdot^+$	the least quasiordering of $F(\Gamma(\tilde{\mathbf{p}},\mathbf{q},\mathbf{e}))$ that is compatible with $\mathbf{F}(\Gamma(\tilde{\mathbf{p}},\mathbf{q},\mathbf{e}))$ and includes $\equiv^+ \cup PQ_6$
233	${}_\delta S_q$	the closure under $\circ$ and $\smile$ of $\{q_i : i < \delta\}$
233	${}_\delta \mathbf{S}_q$	the semigroup with involution $\langle {}_\delta S_q, \circ, \smile \rangle$
233	${}_\delta\Gamma(\tilde{\mathbf{q}})$	the set $\{\mathbf{q}_i : i < \delta\}$
233	$\mathbf{F}({}_\delta\Gamma(\tilde{\mathbf{q}})), \mathbf{F}_{SI}({}_\delta\Gamma(\tilde{\mathbf{q}}))$	the free semigroup with involution freely generated by ${}_\delta\Gamma(\tilde{\mathbf{q}})$
233	$F({}_\delta T(\tilde{\mathbf{q}})), F_{SI}({}_\delta\Gamma(\tilde{\mathbf{q}}))$	the universe of $\mathbf{F}({}_\delta\Gamma(\tilde{\mathbf{q}}))$
233	${}_\delta R$	the restriction $\{\langle v, w\rangle \in R : v, w, \in {}_\delta\Gamma(\tilde{\mathbf{q}})\}$ of R to ${}_\delta\Gamma(\tilde{\mathbf{q}})\}$
233–242	${}_\delta\equiv$	the congruence on $\mathbf{F}({}_\delta\Gamma(\tilde{\mathbf{q}}))$ generated by ${}_\delta Q_1 \cup {}_\delta Q_2 \cup {}_\delta Q_3 \cup {}_\delta Q_4$
233–242	$\equiv$	the congruence on $\mathbf{F}({}_\omega\Gamma(\tilde{\mathbf{q}}))$ generated by $Q_1 \cup Q_2 \cup Q_3 \cup Q_4$
240	${}_\delta T_q$	the closure of $\{q_i : i < \delta\}$ under $\circ$
240	${}_\delta \mathbf{T}_q$	the semigroup $\langle {}_\delta T_q, \circ \rangle$
240	$\mathbf{F}_T({}_\delta\Gamma(\tilde{\mathbf{q}}))$	the free semigroup $\langle F_T({}_\delta\Gamma(\tilde{\mathbf{q}})), \frown \rangle$ freely generated by ${}_\delta\Gamma(\tilde{\mathbf{q}})$
240	${}_\delta\equiv_1'$	the congruence on $\mathbf{F}_T({}_\delta\Gamma(\tilde{\mathbf{q}}))$ generated by $Q_1 \cap (F_T({}_\delta\Gamma(\tilde{\mathbf{q}})) \times F_T({}_\delta\Gamma(\tilde{\mathbf{q}}))$

Index of Phrases and Subjects

adequate for first-order logic, xvi, **11**, **20**
adequately linked, **82**, 84
algebraic first-order logics, xvi, xxii, 23
almost constant shift, xxii
attuned, **41**, 42

based on, **212**
base set/universe, xiii, **xv**, **3**, 135
bicyclic semigroup, **38**
  expanded, **121**
  with zero, **38**
biordered, **104**
bisectable relation, **xvi**, **111**, 122
bisection, **112**, 122, 123
bisector, **112**, 122
  least, **112**, 125
Boolean operations, xxii, 2, 3, **10**, 14, 16, 18, 22, 23

canonical triple, **64**, 71
cardinal, **3**
cardinality of U, xv, **3**, **19**, 46, 50, 56, 57, 74, 85, 135, 230
clone of operations, 19, 21–23, 31
closed under/with respect to:
  conjugate operation, xii, 9
  intersection, xxi, 54, 107–109
  relative product, xix, 41, 50, 73, 85, 94, 99, 100
  transitivity, **200**
complementation:
  1-ary, xxii, **10**, 17
  relative, xxii, **10**, 15, 17
complex algebra of a relational structure, **xxiv**
composition of operations, xiv, 9
congruence on $\mathbf{F}(\Gamma)$ generated by a relation on $F(\Gamma)$, 135
concatenation of sequences, **3**
conjugate, **xiv**, 12
conservative extension, 163, 233
conservational, 207, 231
conservational normal form, **234**
converse of a relation, **2**
correlated, **104**
corresponds, **238**
counterclockwise rotation, **31**
coupled occurrence, **148**
covers, **202**
cylindric algebra, xiii, xxiii, 23
cylindrification, xiv, 22
  partial, **7**

denotes, names, 136
determines, **xx**, 61, **62**
diagonal closure, **107**
diagonal relation, **xx**, **62**, 85–93
  determined by, **xx**, **62**
  mediately determined by, **xx**, **62**, 126
  on ${}^{\gamma}U$/included in ${}^{\gamma}U \times {}^{\gamma}U$, **63**, 64–109
  on ${}^{[0,\omega)}U$, xx, **63**, 64–93, 108, 111, 121–133
  on ${}^{\omega}U$ and uniform, **xx**, **111**, 112–121

equivalence, 75
excision, **4**, 95–98, 112–119, 127, 130
excision-insertion product, **99**, 115–116, 117–118
full/upward full, **74**, **78**, 85, 94 111–116, 119, 127, 171
insertion, **5**, **7**, 97
one-one, 75, 94
permutation/upward full permutation, 74,**78**, 111–114,119,127,131
special, xvi, xix, **73**, 74–82, 94, 127–130
subidentity, 75, **78**, 79, 82, 85, 94, 112, 113, 115–116, 119, 132-133
subidentity-permutation product, 100, 115–116
diagonal set, xiv, **4**, **61**
diagram:
atomic, **136**
positive atomic, **136**
difunctional:
closure, **72**
product, **78**
relation, **72**
domain of relation, **2**
downward full, **77**
dual conditions, xxi, 74

earlier/earliest triple, **70**
eligible extension, **17**, 18
equicardinality condition, **100**
equivalence relation, **74**
generated by a relation, 35–36
obtained from $\langle \gamma, N \rangle$, 61
partial, 5
also listed under:
diagonal relation,
special
excision:
applied to sets of sequences, **7**
of u_λ, **20**
restricted to $Do\, e_{i,i+1}$, **6**
right-end, **31**
see also: diagonal relation, special
first-order definable:
with individual parameters, **20**
without individual parameters, 11–15
first-order language, 11, 20
fixed point(s), xxv, **3**, **35**
frameable word, **223**
the associated triple, **223**
framing triple, xix **61–62**
finite, **62**
special, **71**
free:
generating set, 135, 137, 159
group, 159
semigroup, 138
semigroup with involution, 135, 136
full relation, xxi, **74**, 75, **78**, 85
functional element, **xxiii**

generators, xiv, xvii, xxii, 20, 112, 117, 119, 121, 131
gives explicitly, **217**
grade of a relation:
on ${}^{[0,\omega)}U$, **xix**, **30**, 31–35, 38, 42, 48, 50, 57
on ${}^{\omega}U$, **46**, 51–55
least, **47**, 51
associated with a word, **144**, **160**

homomorphisms (some), xvi, 38, 56, 113

idempotent, **93**
identity element, xviii, xxiii, 5, 8
identity relation:
left-right, **31**
on a subset/subidentity, 4, 5, 19, **74**, 78
on d_{ij}, xiv, **4**
on ${}^{[n,\omega)}U$, **xxiii**, **4**, 5, 31, 36, 41
on ${}^{\omega}U$, **xxiii**, **4**, 50
individual constant, **20**
individual variable, **11**
induces/induced, **xiii**, **1**, **3**, 6, 9,
initial section, **xix**, **30**

insertion:
- applied to sets of sequences, **7**
- of elements u_λ, **19**
- right-end, **31**
- see also: diagonal relation, special

intersection with d_{ij}, xiv, **7**
intersection with ${}^{[0,\omega)}U$, **7**
invariant/logical, **6**, 9, 19, 63
inversion, **31**
involution, xiv, 85, 93
isomorphisms (some), xiv, xvii, xxi, 59, 87, 94, 99, 100, 105, 115–116, 121 127, 138–139, 150, 155–156

length-bounded set of sequences, xxii, **15**
length of a sequence, **3**

match, **200**
mediately determines, **xx**, **62**, **126**
monoid with involution, xviii

near-striate, **35**, 42
neat embedding, xxiii
normal:
- form, xxi, 25
- **q**-normal, **147**
- **r**-normal, **147**
- $\{\tilde{\mathbf{p}}, \mathbf{q}\}$-normal, **174**

null sequence/its singleton, 2, 12, 15

operation on sets of sequences, **xiii**, xvi, 1
ordered algebra, xv, **41**, 42, 50, 85
ordered partition into intervals,**105**
ordering of operations on sets, **xiv**
ordinal finite/infinite, **3**

partly independent of U, **7**
periodic, r-, from n on/eventually, **51**
permutation, **74**, 75, **78**
precanonical triple, **64**, 65–70, 225
predecessor function that delays, **138**
predicate symbol, **11**
presentation, xvii, xviii
- of a group, 159
- of a semigroup, 138–140
- of a semigroup with involution, **135–136**
- of an ordered semigroup with involution, **135–136**

prolongation of a set of sequences, **26**,
prolongation of a relation between sequences:
- γ-prolongation, xv, **xviii**, **27**, 80–81
- ω-prolongation, xv, 48–50, 56–58, 113
- 1-prolongation, **27**, 36–37, 38, 41–42, 45–46, 85
- equivalence generated, xv–xvi, **36**, 56, 57–58
- one-many, **27**
- ordering generated, **35**, 40

prototype, **36**

quasiordering, 136

range of a relation, **3**
rank:
- associated with a formula, **11**
- of a relation between sequences, **28**, 48, 50
- of a set of sequences, 1, 8, 15

ranked relation, **28**
ranked set, **2**, 7, 11
relation, xiii–xiv
- difunctional, **72**
- full/upward full, **74**
- inducing an operation, xiii, **1**, 9
- invariant/logical, **6**
- subidentity, **74**

relation on ω:
- almost constant shift, **xxii**
- bisectable, xvi, **111**, 122
- difunctional, 104
- difunctional and full, 75, 94
- difunctional, full, biordered, 104–107
- equivalence, 74, 79, 94, 112
- one-one, 75, 78, 94, 99
- one-one and bisectable, 114, 127
- one-one and strictly increasing,

98, 104–107
permutation, 75, 99, 112
single-valued and strictly increasing, 95–99
relational structure, **11**
non/vacuous, **11**, 12, 15, 17
with distinguished elements, **20**
relative complement, xxi, **10**, 11, 16
relative product, xiii, **xiv**, **2**
replication, **5**, 17, 120

semigroup, 9, 137–138
inverse semigroup, xviii, **93**, 94, 131
with inverse-forming involution, **93**
with involution, xiv, 8–9, 41, 85, 112
free, 136–137
generated by functional elements, xxiii
ordered, xiv, xxii, 41, 85, 136
sequence, 1, **3**
of length γ, xv–xvi, **3**
range of sequence, **3**
striate function, **35**, 36
strictly ascending, **139**, **174**
strictly descending, **147**, **174**
strictly increasing, 95, 98
subidentity: listed under diagonal relation, special
subsequence, 96
support, 26, 121

transitive closure, **72**
transmits computability, **7**
transposition, **77**
triple:
associated with a word, **175**, **223**
canonical, 64–71
$\langle\gamma,\delta\rangle$ triple, **61**, 82
finite, **62**, 113
framing, **61**
precanonical, 64–70
special **71**, 113
trim, **64**, 227

unary representation, xxiii–xxiv
uniform, **29**
evenly, **29**
on ${}^{[0,\omega)}U$, xix, **29**–46, 56–81
on ${}^{\omega}U$, **xix**, **29**, 46–81
universe/base set, xiii, **xv**, **3**, 135
upward full, see: diagonal relation

word:
associated grade, **144**, **160**
closed with respect to transitivity, **200**
frameable, **223**
in conservational normal form, **234**
in $F_{SI}(\Gamma)$, **137**
nonstrictly ascending, **234**
nonstrictly descending, **234**
normal:
$\mathbf{q}$-normal, **147**
$\mathbf{r}$-normal, **147**
$\{\tilde{\mathbf{p}},\mathbf{q}\}$-normal, **174**
$\langle\mathbf{q}^{-1}\rangle$-, $\langle\mathbf{p}\rangle$-, $\langle\mathbf{q}^{-1},\mathbf{q}\rangle$-, $\langle\mathbf{q}^{-1},\mathbf{p},\mathbf{q}\rangle$-word, **173–174**
$\langle\mathbf{q}^{-1},\mathbf{p},\mathbf{q}\,\rangle$-, $\langle\mathbf{e},\mathbf{q}^{-1},\mathbf{p},\mathbf{q},\mathbf{e}\rangle$-word, **219**
strictly acscending, 139, 147
strictly descending, **147**

List of Relations Involved in Presentations

141 $Q_1 = \{\langle \mathbf{q}_i\mathbf{q}_j, \mathbf{q}_{j-1}\mathbf{q}_i\rangle : i < j < \omega\}$

$Q_2 = \{\langle \mathbf{q}_i\mathbf{q}_j^{-1}, \mathbf{q}_{j-1}^{-1}\mathbf{q}_i\rangle : i < j < \omega\}$

$Q_3 = \{\langle \mathbf{q}_i\mathbf{q}_i^{-1}\mathbf{q}_i, \mathbf{q}_i\rangle : i < \omega\}$

$Q_4 = \{\mathbf{q}\mathbf{q}_i^{-1}\mathbf{q}_{i+1}, \mathbf{q}_{i+1}\rangle : i < \omega\}$

151 $Q_5 = \{\langle \mathbf{q}_{i+1}\mathbf{q}_{i+1}^{-1}, \mathbf{q}_i\mathbf{q}_i^{-1}\rangle : i < \omega\}$

$Q_6 = \{\langle \mathbf{q}_{i+1}\mathbf{q}_{i+1}^{-1}, \mathbf{q}_i^{-1}\mathbf{q}_i\rangle : i < \omega\}$

233 ${}_\delta Q_1, \ldots, {}_\delta Q_6$: obtained from $Q_1, \ldots, Q_6$
by restriction to $\mathbf{q}_k$ and $\mathbf{q}_k^{-1}$, $k < \delta$

141, 151 $R_1, \ldots, R_6$: obtained from $Q_1, \ldots, Q_6$
by replacing each $\mathbf{q}_k$ by $\mathbf{r}_k$.

151, 155 $S_{\lambda,1}, \ldots, S_{\lambda,6}$: obtained from $Q_1, \ldots, Q_6$
by replacing each $\mathbf{q}_k$ by $\mathbf{s}_{\lambda,k}$.

158 $P_1 = \{\langle \mathbf{p}_i^{-1}, \mathbf{p}_i\rangle : i < \omega\}$

$P_2 = \{\langle \mathbf{p}_i^2\mathbf{p}_i, \mathbf{p}_i\rangle : i < \omega\}$

$P_3 = \{\langle \mathbf{p}_i^2\mathbf{p}_{i+1}, \mathbf{p}_{i+1}\rangle : i < \omega\}$

$P_4 = \{\langle \mathbf{p}_{i+1}^2\mathbf{p}_i, \mathbf{p}_i\rangle : i < \omega\}$

$P_5 = \{\langle \mathbf{p}_i\mathbf{p}_{i+1}\mathbf{p}_i, \mathbf{p}_{i+1}\mathbf{p}_i\mathbf{p}_{i+1}\rangle : i < \omega\}$

$P_6 = \{\langle \mathbf{p}_i\mathbf{p}_j, \mathbf{p}_j\mathbf{p}_i\rangle : i \le j - 2 < \omega\}$

161 $P_4^{\#} = \{\langle \mathbf{p}_{i+1}^2\mathbf{p}_i, \mathbf{p}_i\mathbf{p}_{i+1}^2\rangle : i < \omega\}$

158 ${}_\delta P_1, \ldots, {}_\delta P_6$: obtained from $P_1, \ldots, P_6$
by restriction to $\mathbf{p}_k$, $k < \delta$.

165 ${}_1Q_3 = \{\mathbf{qq}^{-1}\mathbf{q}, \mathbf{q}\rangle\}$
$PQ_1 = \{\langle \mathbf{qq}^{-1}\mathbf{p}_0, \mathbf{p}_0\rangle\}$
$PQ_2 = \{\langle \mathbf{qp}_0\mathbf{q}^{-1}, \mathbf{q}^{-1}\mathbf{q}\rangle\}$
$PQ_3 = \{\langle \mathbf{q}^2\mathbf{p}_0, \mathbf{q}^2\rangle\}$
$PQ_4 = \{\langle \mathbf{p}_i\mathbf{q}, \mathbf{qp}_{i+1}\rangle : i < \omega\}$
177 $PQ_5 = \{\langle \mathbf{qp}_0^2\mathbf{q}^{-1}, \mathbf{qq}^{-1}\rangle\}$
$PQ_6 = \{\langle \mathbf{qp}_0^2\mathbf{q}^{-1}, \mathbf{q}^{-1}\mathbf{q}\rangle\}$
180 $E_1 = \{\langle \mathbf{e}^{-1}, \mathbf{e}\rangle\}$
$E_2 = \{\langle \mathbf{e}^2, \mathbf{e}\rangle\}$
$PE_1 = \{\langle \mathbf{p}_0\mathbf{e}, \mathbf{e}\rangle\}$
$PE_2 = \{\langle \mathbf{p}_i\mathbf{e}, \mathbf{ep}_i\rangle : 2 \leq i < \omega\}$
$PE_3 = \{\langle \mathbf{ep}_1\mathbf{ep}_1, \mathbf{p}_1\mathbf{ep}_1\mathbf{e}\rangle\}$
$PE_4 = \{\langle \mathbf{ep}_1\mathbf{p}_0\mathbf{p}_2\mathbf{p}_1\mathbf{ep}_1\mathbf{p}_2\mathbf{p}_0\mathbf{p}_1, \mathbf{p}_1\mathbf{p}_0\mathbf{p}_2\mathbf{p}_1\mathbf{ep}_1\mathbf{p}_2\mathbf{p}_0\mathbf{p}_1\mathbf{e}\rangle\}$
205 $PE_5 = \{\langle \mathbf{e}, \mathbf{p}_0^2\rangle\}$

209 $QE_1 = \{\langle \mathbf{eq}^{-1}\mathbf{qe}, \mathbf{e}\rangle\}$
$PQE_1 = \{\langle \mathbf{qeq}^{-1}, \mathbf{qp}_0^2\mathbf{q}^{-1}\rangle\}$
$PQE_2 = \{\langle \mathbf{eqp}_0\mathbf{p}_1, \mathbf{qp}_0\mathbf{p}_1\mathbf{e}\rangle\}$

Synopsis of Presentations

For ordered structures, the relations cited generate a quasi-ordering. In all other cases they generate a congruence relation. Every ${}_{[\omega]}\mathbf{T}$, ${}_{[\omega]}\mathbf{S}$, and $\langle {}_{[\omega]}\mathbf{S}, \subseteq \rangle$, except when ${}_{[\omega]}\mathbf{S}$ is ${}_{[\omega]}\mathbf{S}_{pqe}$, has an isomorph under $({}_{\omega}[\ \ ])^{\smile}$. Presentations of ${}_{\delta}T_q$, ${}_{\delta}S_q$, and $\langle {}_{\delta}\mathbf{S}_q, \subseteq \rangle$ are obtained by restriction to $\mathbf{q}_k$ and $\mathbf{q}_k^{-1}$, $k < \omega$.

Theorem; structure presented; relations involved.

V.1; $\mathbf{T}_q$ and ${}_{[\omega]}\mathbf{T}_q$; Q_1.

V.1; $\mathbf{T}_r$ and ${}_{[\omega]}\mathbf{T}_r$; R_1.

V.7(a); $\mathbf{S}_q$; Q_1, Q_2, Q_3, Q_4.

V.7(b); $\mathbf{S}_r$; R_1, R_2, R_3, R_4.

V.7(d); $\mathbf{S}_{s,\lambda}$; $S_{\lambda,1}, S_{\lambda,2}, S_{\lambda,3}, S_{\lambda,4}$.

V.12(a); $\langle \mathbf{S}_q, \subseteq \rangle$; $\equiv$ of V.7(a), Q_5, Q_6.

V.12(b); $\langle \mathbf{S}_r, \subseteq \rangle$; $\equiv$ of V.7(b), $R_5, R_6^{\smile}$.

V.12(d); $\langle \mathbf{S}_{s,\lambda}, \subseteq \rangle$; $\equiv$ of V.7(d), $S_{\lambda,5}, S_{\lambda,6}^{\smile}$.

V.13(a); ${}_{[\omega]}\mathbf{S}_q$; Q_1, Q_2, Q_3, Q_5.

V.13(b); $\langle {}_{[\omega]}\mathbf{S}_q, \subseteq \rangle$; $\equiv^+$ of V.13(a), Q_6.

V.15(c); ${}_{[\omega]}\mathbf{S}_p$; $P_1, P_2, P_3, P_4, P_5, P_6$.

V.18(a); $\mathbf{S}_p$; $P_1, P_2, P_3, P_4^{\#}, P_5, P_6$.

V.18(b); $\langle \mathbf{S}_p, \subseteq \rangle$; $\equiv$ of V.18(a), P_4.

VI.10; $\mathbf{S}_{pq}$; $\equiv$ of V.18(a), ${}_1Q_3, PQ_1, PQ_2, PQ_3, PQ_4$.

VI.11; $\langle \mathbf{S}_{pq}, \subseteq \rangle$; $\equiv$ of VI.10, P_4, PQ_5, PQ_6.

VI.12(a); ${}_{[\omega]}\mathbf{S}_{pq}$; $\equiv$ of VI.10, P_4, PQ_5.

VI.12(b); $\langle {}_{[\omega]}\mathbf{S}_{pq}, \subseteq \rangle$; $\equiv^+$ of VI.12(a), PQ_6.

VI.31; $\mathbf{S}_{pe}$; $\equiv$ of V.18(a), $E_1, E_2, PE_1, PE_2, PE_3, PE_4$.

VI.33; $\langle \mathbf{S}_{pe}, \subseteq \rangle$; $\equiv$ of VI.31, P_4, PE_5.

VI.34(a); ${}_{[\omega]}\mathbf{S}_{pe}$; $\equiv$ of VI.31, P_4.

VI.34(b); $\langle {}_{[\omega]}\mathbf{S}_{pe}, \subseteq \rangle$; $\equiv^+$ of VI.34(a), PE_5.

VII.15; $\mathbf{S}_{pqe}$; $\equiv$ of VI.10, $\equiv$ of VI.31, QE_1, PQE_1, PQE_2.

VII.16; $\langle \mathbf{S}_{pqe}, \subseteq \rangle$; $\equiv$ of VII.15, PQ_5, PQ_6, P_4, PE_5.

VII.17(a); ${}_{[\omega]}\mathbf{S}_{pqe}$; $\equiv$ of VII.15, P_4.

VII.17(b); $\langle {}_{[\omega]}\mathbf{S}_{pqe}, \subseteq \rangle$; $\equiv^+$ of VII.17(a), PQ_6.

Editorial Information

To be published in the *Memoirs*, a paper must be correct, new, nontrivial, and significant. Further, it must be well written and of interest to a substantial number of mathematicians. Piecemeal results, such as an inconclusive step toward an unproved major theorem or a minor variation on a known result, are in general not acceptable for publication. Papers appearing in *Memoirs* are generally at least 80 and not more than 200 published pages in length. Papers less than 80 or more than 200 published pages require the approval of the Managing Editor of the Transactions/Memoirs Editorial Board.

As of July 31, 2006, the backlog for this journal was approximately 11 volumes. This estimate is the result of dividing the number of manuscripts for this journal in the Providence office that have not yet gone to the printer on the above date by the average number of monographs per volume over the previous twelve months, reduced by the number of volumes published in four months (the time necessary for preparing a volume for the printer). (There are 6 volumes per year, each containing at least 4 numbers.)

A Consent to Publish and Copyright Agreement is required before a paper will be published in the *Memoirs*. After a paper is accepted for publication, the Providence office will send a Consent to Publish and Copyright Agreement to all authors of the paper. By submitting a paper to the *Memoirs*, authors certify that the results have not been submitted to nor are they under consideration for publication by another journal, conference proceedings, or similar publication.

Information for Authors

Memoirs are printed from camera copy fully prepared by the author. This means that the finished book will look exactly like the copy submitted.

The paper must contain a *descriptive title* and an *abstract* that summarizes the article in language suitable for workers in the general field (algebra, analysis, etc.). The *descriptive title* should be short, but informative; useless or vague phrases such as "some remarks about" or "concerning" should be avoided. The *abstract* should be at least one complete sentence, and at most 300 words. Included with the footnotes to the paper should be the 2000 *Mathematics Subject Classification* representing the primary and secondary subjects of the article. The classifications are accessible from `www.ams.org/msc/`. The list of classifications is also available in print starting with the 1999 annual index of *Mathematical Reviews*. The Mathematics Subject Classification footnote may be followed by a list of *key words and phrases* describing the subject matter of the article and taken from it. Journal abbreviations used in bibliographies are listed in the latest *Mathematical Reviews* annual index. The series abbreviations are also accessible from `www.ams.org/publications/`. To help in preparing and verifying references, the AMS offers MR Lookup, a Reference Tool for Linking, at `www.ams.org/mrlookup/`. When the manuscript is submitted, authors should supply the editor with electronic addresses if available. These will be printed after the postal address at the end of the article.

Electronically prepared manuscripts. The AMS encourages electronically prepared manuscripts, with a strong preference for $\mathcal{A}\mathcal{M}\mathcal{S}$-LaTeX. To this end, the Society has prepared $\mathcal{A}\mathcal{M}\mathcal{S}$-LaTeX author packages for each AMS publication. Author packages include instructions for preparing electronic manuscripts, the *AMS Author Handbook*, samples, and a style file that generates the particular design specifications of that publication series. Though $\mathcal{A}\mathcal{M}\mathcal{S}$-LaTeX is the highly preferred format of TeX, author packages are also available in $\mathcal{A}\mathcal{M}\mathcal{S}$-TeX.

Authors may retrieve an author package from e-MATH starting from `www.ams.org/tex/` or via FTP to `ftp.ams.org` (login as `anonymous`, enter username as password, and type `cd pub/author-info`). The *AMS Author Handbook* and the *Instruction Manual* are available in PDF format following the author packages link from `www.ams.org/tex/`. The author package can also be obtained free of charge by sending email to tech-support@ams.org (Internet) or from the Publication Division, American Mathematical Society, 201 Charles St., Providence, RI 02904-2294, USA. When requesting an author package, please specify $\mathcal{A}_{\mathcal{M}}\mathcal{S}$-LaTeX or $\mathcal{A}_{\mathcal{M}}\mathcal{S}$-TeX and the publication in which your paper will appear. Please be sure to include your complete mailing address.

Sending electronic files. After acceptance, the source file(s) should be sent to the Providence office (this includes any TeX source file, any graphics files, and the DVI or PostScript file).

Before sending the source file, be sure you have proofread your paper carefully. The files you send must be the EXACT files used to generate the proof copy that was accepted for publication. For all publications, authors are required to send a printed copy of their paper, which exactly matches the copy approved for publication, along with any graphics that will appear in the paper.

TeX files may be submitted by email, FTP, or on diskette. The DVI file(s) and PostScript files should be submitted only by FTP or on diskette unless they are encoded properly to submit through email. (DVI files are binary and PostScript files tend to be very large.)

Electronically prepared manuscripts can be sent via email to pub-submit@ams.org (Internet). The subject line of the message should include the publication code to identify it as a Memoir. TeX source files, DVI files, and PostScript files can be transferred over the Internet by FTP to the Internet node e-math.ams.org (130.44.1.100).

Electronic graphics. Comprehensive instructions on preparing graphics are available at www.ams.org/jourhtml/graphics.html. A few of the major requirements are given here.

Submit files for graphics as EPS (Encapsulated PostScript) files. This includes graphics originated via a graphics application as well as scanned photographs or other computer-generated images. If this is not possible, TIFF files are acceptable as long as they can be opened in Adobe Photoshop or Illustrator. No matter what method was used to produce the graphic, it is necessary to provide a paper copy to the AMS.

Authors using graphics packages for the creation of electronic art should also avoid the use of any lines thinner than 0.5 points in width. Many graphics packages allow the user to specify a "hairline" for a very thin line. Hairlines often look acceptable when proofed on a typical laser printer. However, when produced on a high-resolution laser imagesetter, hairlines become nearly invisible and will be lost entirely in the final printing process.

Screens should be set to values between 15% and 85%. Screens which fall outside of this range are too light or too dark to print correctly. Variations of screens within a graphic should be no less than 10%.

Inquiries. Any inquiries concerning a paper that has been accepted for publication should be sent directly to the Electronic Prepress Department, American Mathematical Society, 201 Charles St., Providence, RI 02904, USA.

Editors

This journal is designed particularly for long research papers, normally at least 80 pages in length, and groups of cognate papers in pure and applied mathematics. Papers intended for publication in the *Memoirs* should be addressed to one of the following editors. In principle the Memoirs welcomes electronic submissions, and some of the editors, those whose names appear below with an asterisk (*), have indicated that they prefer them. However, editors reserve the right to request hard copies after papers have been submitted electronically. Authors are advised to make preliminary email inquiries to editors about whether they are likely to be able to handle submissions in a particular electronic form.

***Algebra** to ALEXANDER KLESHCHEV, Department of Mathematics, University of Oregon, Eugene, OR 97403-1222; email: `ams@noether.uoregon.edu`

Algebra and its application to MINA TEICHER, Emmy Noether Research Institute for Mathematics, Bar-Ilan University, Ramat-Gan 52900, Israel; email: `teicher@macs.biu.ac.il`

Algebraic geometry to DAN ABRAMOVICH, Department of Mathematics, Brown University, Box 1917, Providence, RI 02912; email: `amsedit@math.brown.edu`

***Algebraic number theory** to V. KUMAR MURTY, Department of Mathematics, University of Toronto, 100 St. George Street, Toronto, ON M5S 1A1, Canada; email: `murty@math.toronto.edu`

***Algebraic topology** to ALEJANDRO ADEM, Department of Mathematics, University of British Columbia, Room 121, 1984 Mathematics Road, Vancouver, British Columbia, Canada V6T 1Z2; email: `adem@math.ubc.ca`

***Combinatorics** to JOHN R. STEMBRIDGE, Department of Mathematics, University of Michigan, Ann Arbor, Michigan 48109-1109; email: `FRS@umich.edu`

Complex analysis and harmonic analysis to ALEXANDER NAGEL, Department of Mathematics, University of Wisconsin, 480 Lincoln Drive, Madison, WI 53706-1313; email: `nagel@math.wisc.edu`

***Differential geometry and global analysis** to LISA C. JEFFREY, Department of Mathematics, University of Toronto, 100 St. George St., Toronto, ON Canada M5S 3G3; email: `jeffrey@math.toronto.edu`

Dynamical systems and ergodic theory to AMIE WILKINSON, Department of Mathematics, Northwestern University, 2033 Sheridan Road, Evanston, IL 60208-2730; email: `transactions@math.northwestern.edu`

***Functional analysis and operator algebras** to MARIUS DADARLAT, Department of Mathematics, Purdue University, 150 N. University St., West Lafayette, IN 47907-2067; email: `mdd@math.purdue.edu`

***Geometric analysis** to TOBIAS COLDING, Courant Institute, New York University, 251 Mercer St., New York, NY 10012; email: `traneditor@cims.nyu.edu`

***Geometric analysis** to MLADEN BESTVINA, Department of Mathematics, University of Utah, 155 South 1400 East, JWB 233, Salt Lake City, Utah 84112-0090; email: `bestvina@math.utah.edu`

Harmonic analysis, representation theory, and Lie theory to ROBERT J. STANTON, Department of Mathematics, The Ohio State University, 231 West 18th Avenue, Columbus, OH 43210-1174; email: `stanton@math.ohio-state.edu`

***Logic** to STEFFEN LEMPP, Department of Mathematics, University of Wisconsin, 480 Lincoln Drive, Madison, Wisconsin 53706-1388; email: `lempp@math.wisc.edu`

***Ordinary differential equations, and applied mathematics** to PETER W. BATES, Department of Mathematics, Michigan State University, East Lansing, MI 48824-1027; email: `bates@math.msu.edu`

***Partial differential equations** to GUSTAVO PONCE, Department of Mathematics, South Hall, Room 6607, University of California, Santa Barbara, CA 93106; email: `ponce@math.ucsb.edu`

***Probability and statistics** to KRZYSZTOF BURDZY, Department of Mathematics, University of Washington, Box 354350, Seattle, Washington 98195-4350; email: `burdzy@math.washington.edu`

***Real analysis and partial differential equations** to DANIEL TATARU, Department of Mathematics, University of California, Berkeley, Berkeley, CA 94720; email: `tataru@math.berkeley.edu`

All other communications to the editors should be addressed to the Managing Editor, ROBERT GURALNICK, Department of Mathematics, University of Southern California, Los Angeles, CA 90089-1113; email: `guralnic@math.usc.edu`.

Titles in This Series

868 **Gelu Popescu,** Entropy and multivariable interpolation, 2006

867 **Vilmos Totik,** Metric properties of harmonic measures, 2006

866 **William Craig,** Semigroups underlying first-order logic, 2006

865 **Nathanial P. Brown,** Invariant means and finite representation theory of C^*-algebras, 2006

864 **John M. Lee,** Fredholm operators and Einstein metrics on conformally compact manifolds, 2006

863 **M. Lübke and A. Teleman,** The Universal Kobayashi-Hitchin correspondence on Hermitian manifolds, 2006

862 **Alberto Canonaco,** The Beilinson complex and canonical rings of irregular surfaces, 2006

861 **Leon A. Takhtajan and Lee-Peng Teo,** Weil-Petersson metric on the universal Teichmüller space, 2006

860 **Thomas M. Fiore,** Pseudo limits, biadjoints and pseudo algebras: Categorical foundations of conformal field theory, 2006

859 **N. Arcozzi, R. Rochberg, and E. Sawyer,** Carleson measures and interpolating sequences for Besov spaces on complex balls, 2006

858 **Enrico Valdinoci, Berardino Sciunzi, and Vasile Ovidiu Savin,** Flat level set regularity of p-Laplace phase transitions, 2006

857 **Donatella Danielli, Nocola Garofalo, and Duy-Minh Nhieu,** Non-doubling Ahlfors measures, perimeter measures, and the characterization of the trace spaces of Sobolev functions in Carnot-Carathéodory spaces, 2006

856 **Vladimir Bolotnikov and Harry Dym,** On boundary interpolation for matrix valued Schur functions, 2006

855 **Yevgenia Kashina, Yorck Sommerhäuser, and Yongchang Zhu,** On higher Frobenius-Schur indicators, 2006

854 **Noam Greenberg,** The role of true finiteness in the admissible recursively enumerable degrees, 2006

853 **Joachim Krieger,** Stability of spherically symmetric wave maps, 2006

852 **Viorel Barbu, Irena Lasiecka, and Roberto Triggiani,** Tangential boundary stabilization of Navier-Stokes equations, 2006

851 **Jie Wu,** On maps from loop suspensions to loop spaces and the shuffle relations on the Cohen groups, 2006

850 **Siegfried Echterhoff, S. Kaliszewski, John Quigg, and Iain Raeburn,** A categorical approach to imprimitivity theorems for C^*-dynamical systems, 2006

849 **Katsuhiko Kuribayashi, Mamoru Mimura, and Tetsu Nishimoto,** Twisted tensor products related to the cohomology of the classifying spaces of loop groups, 2006

848 **Bob Oliver,** Equivalences of classifying spaces completed at the prime two, 2006

847 **Eric T. Sawyer and Richard L. Wheeden,** Hölder continuity of weak solutions to subelliptic equations with rough coefficients, 2006

846 **Victor Beresnevich, Detta Dickinson, and Sanju Velani,** Measure theoretic laws for lim–sup sets, 2006

845 **Ehud Friedgut, Vojtech Rödl, Andrzej Ruciński, and Prasad V. Tetali,** A Sharp threshold for random graphs with a monochromatic triangle in every edge coloring, 2006

844 **Amadeu Delshams, Rafael de la Llave, and Tere M. Seara,** A geometric mechanism for diffusion in Hamiltonian systems overcoming the large gap problem: Heuristics and rigorous verification on a model, 2006

843 **Denis V. Osin,** Relatively hyperbolic groups: Intrinsic geometry, algebraic properties, and algorithmic problems, 2006

842 **David P. Blecher and Vrej Zarikian,** The calculus of one-sided M-ideals and multipliers in operator spaces, 2006

841 **Enrique Artal Bartolo, Pierrette Cassou-Noguès, Ignacio Luengo, and Alejandro Melle Hernández,** Quasi-ordinary power series and their zeta functions, 2005
840 **Sławomir Kołodziej,** The complex Monge-Ampère equation and pluripotential theory, 2005
839 **Mihai Ciucu,** A random tiling model for two dimensional electrostatics, 2005
838 **V. Jurdjevic,** Integrable Hamiltonian systems on complex Lie groups, 2005
837 **Joseph A. Ball and Victor Vinnikov,** Lax-Phillips scattering and conservative linear systems: A Cuntz-algebra multidimensional setting, 2005
836 **H. G. Dales and A. T.-M. Lau,** The second duals of Beurling algbras, 2005
835 **Kiyoshi Igusa,** Higher complex torsion and the framing principle, 2005
834 **Kenı́chi Ohshika,** Kleinian groups which are limits of geometrically finite groups, 2005
833 **Greg Hjorth and Alexander S. Kechris,** Rigidity theorems for actions of product groups and countable Borel equivalence relations, 2005
832 **Lee Klingler and Lawrence S. Levy,** Representation type of commutative Noetherian rings III: Global wildness and tameness, 2005
831 **K. R. Goodearl and F. Wehrung,** The complete dimension theory of partially ordered systems with equivalence and orthogonality, 2005
830 **Jason Fulman, Peter M. Neumann, and Cheryl E. Praeger,** A generating function approach to the enumeration of matrices in classical groups over finite fields, 2005
829 **S. G. Bobkov and B. Zegarlinski,** Entropy bounds and isoperimetry, 2005
828 **Joel Berman and Paweł M. Idziak,** Generative complexity in algebra, 2005
827 **Trevor A. Welsh,** Fermionic expressions for minimal model Virasoro characters, 2005
826 **Guy Métivier and Kevin Zumbrun,** Large viscous boundary layers for noncharacteristic nonlinear hyperbolic problems, 2005
825 **Yaozhong Hu,** Integral transformations and anticipative calculus for fractional Brownian motions, 2005
824 **Luen-Chau Li and Serge Parmentier,** On dynamical Poisson groupoids I, 2005
823 **Claus Mokler,** An analogue of a reductive algebraic monoid whose unit group is a Kac-Moody group, 2005
822 **Stefano Pigola, Marco Rigoli, and Alberto G. Setti,** Maximum principles on Riemannian manifolds and applications, 2005
821 **Nicole Bopp and Hubert Rubenthaler,** Local zeta functions attached to the minimal spherical series for a class of symmetric spaces, 2005
820 **Vadim A. Kaimanovich and Mikhail Lyubich,** Conformal and harmonic measures on laminations associated with rational maps, 2005
819 **F. Andreatta and E. Z. Goren,** Hilbert modular forms: Mod p and p-adic aspects, 2005
818 **Tom De Medts,** An algebraic structure for Moufang quadrangles, 2005
817 **Javier Fernández de Bobadilla,** Moduli spaces of polynomials in two variables, 2005
816 **Francis Clarke,** Necessary conditions in dynamic optimization, 2005
815 **Martin Bendersky and Donald M. Davis,** V_1-periodic homotopy groups of $SO(n)$, 2004
814 **Johannes Huebschmann,** Kähler spaces, nilpotent orbits, and singular reduction, 2004
813 **Jeff Groah and Blake Temple,** Shock-wave solutions of the Einstein equations with perfect fluid sources: Existence and consistency by a locally inertial Glimm scheme, 2004